AF349269

Progress in Mathematics

Volume 144

Series Editors

H. Bass
J. Oesterlé
A. Weinstein

Sub-Riemannian Geometry

André Bellaïche
Jean-Jacques Risler

Editors

Birkhäuser Verlag
Basel · Boston · Berlin

Editors:

André Bellaïche
Departement de Mathématiques
Université Paris 7 – Denis Diderot
2, place Jussieu
F-75251 Paris 5e

Jean-Jacques Risler
Université Paris VI – Pierre et Marie Curie
F-75252 Paris 5e

1991 Mathematics Subject Classification 53C99, 58E25, 93B29, 49L99

A CIP catalogue record for this book is available from the Library of Congress, Washington D.C., USA

Deutsche Bibliothek Cataloging-in-Publication Data

Sub-Riemannian geometry / André Bellaïche ; Jean-Jacques
Risler ed. – Basel ; Boston ; Berlin : Birkhäuser, 1996
  (Progress in mathematics ; Vol. 144)
    ISBN 3-7643-5476-3 (Basel ...)
    ISBN 0-8176-5476-3 (Boston)
NE: Bellaïche, André [Hrsg.]; GT

© 1996 Birkhäuser Verlag, P.O. Box 133, CH-4010 Basel, Switzerland
Printed on acid-free paper produced of chlorine-free pulp. TCF ∞
Printed in Germany
ISBN 3-7643-5476-3
ISBN 0-8176-5476-3

9 8 7 6 5 4 3 2 1

Preface

Following a suggestion by Héctor J. Sussmann we organized, in the summer of 1992 in Paris, a satellite meeting of the first European Congress of Mathematics. The topic of the meeting was "Nonholonomy", and officially titled:

JOURNÉES NONHOLONOMES
Géométrie sous-riemannienne, théorie du contrôle, robotique

It was held at Université Paris VI–Pierre et Marie Curie (Jussieu), on June 30th and July 1st, 1992.

Sub-Riemannian Geometry (also known as Carnot Geometry in France, and Nonholonomic Riemannian Geometry in Russia) has been a fully-fledged research domain for fifteen years, with motivations and ramifications in several parts of pure and applied mathematics, namely:
- Control Theory;
- Classical Mechanics;
- Riemannian Geometry (of which Sub-Riemannian Geometry constitutes a natural generalization, and where sub-Riemannian metrics may appear as limit cases);
- Gauge theories;
- Diffusions on manifolds;
- Analysis of hypoelliptic operators; and
- Cauchy-Riemann (or CR) Geometry.

Although links between these domains had been foreseen by many authors in the past, it is only in recent years that Sub-Riemannian Geometry has been recognized as a possible common framework for all these topics (e.g., the conference paper by Agrachev at the 1994 International Mathematical Congress in Zurich).

To illustrate this fact, it should be noted that the first editor of this volume was interested in nonholonomy, following encouragement by Robert Azencott to provide a geometric frame for the study of non-elliptic diffusions. The second editor, a specialist in real algebraic geometry, came to the same subject after a collaboration with roboticians, when it became clear that nonholonomy was one of the main problems in robotics.

The first article, by André Bellaïche, is a local study of sub-Riemannian structures. After some general definitions, he examines the notion of tangent space at a point of a sub-Riemannian manifold: this space has a natural structure of nilpotent Lie group with dilations at regular points, and of a quotient of such a group otherwise, showing some similarity with the Riemannian case where the tangent space is a linear space, i.e., a commutative Lie group with dilations.

The next and rather extensive article by M. Gromov is impossible to summarize in few words. It builds on basic facts, given in the preceding paper, and then adresses an impressive number of questions and conjectures, in general inspired by Riemannian geometry. The point of view of Gromov is roughly the following: if one "lives" inside a metric space, whose metric comes from a sub-Riemannian structure on a manifold, what are the properties of the distribution (differentiable structure, dimension of the distribution, dimension of its derived distributions, etc.) that one can recover?

The paper by R. Montgomery describes the phenomenon of abnormal extremals (or abnormal geodesics) by surveying their properties, and by describing the first example (due to the author) of such a (minimizing) geodesic.

This subject is also taken up by H. Sussmann, who here gives yet further exhaustive examples in dimension 4, showing that the phenomenon of abnormal extremals is "generic" in sub-Riemannian geometry.

Finally, J.-M. Coron's paper takes a different approach, since he deals with stabilization and feedback laws, closer to control theory than sub-Riemannian geometry. Nevertheless, the reader will find this paper to be remarkably consistent with the previous ones.

The interest and coherence of the conference papers induced us to bring these texts together in the present volume. Publication comes late—for which we apologize—but we hope the reader will find the waiting worthwhile.

We thank the five authors for the confidence they have shown in this project.

We thank Héctor Sussmann for initiating the "Journées" and Jean-Paul Laumond, from the LAAS in Toulouse, who awakened or, better, renewed our interest in these questions, and who also was one of the initiators of this meeting. We thank also Birkhäuser for publishing these texts.

André Bellaïche, Jean-Jacques Risler

Contents

The tangent space
in sub-Riemannian geometry

André Bellaïche

Carnot-Carathéodory spaces
seen from within*

Mikhael Gromov

Survey of singular geodesics

Richard Montgomery

---

*A detailed table of contents of this contribution appears on pages 79–84.

CONTENTS

A cornucopia of four-dimensional abnormal sub-Riemannian minimizers

HÉCTOR J. SUSSMANN

Stabilization of controllable systems

JEAN-MICHEL CORON

Progress in Mathematics, Vol. 144, © 1996 Birkhäuser Verlag Basel/Switzerland

The tangent space
in sub-Riemannian geometry

ANDRÉ BELLAÏCHE[*]

Tangent spaces of a sub-Riemannian manifold are themselves sub-Riemannian manifolds. They can be defined as metric spaces, using Gromov's definition of tangent spaces to a metric space, and they turn out to be sub-Riemannian manifolds. Moreover, they come with an algebraic structure: nilpotent Lie groups with dilations. In the classical, Riemannian, case, they are indeed vector spaces, that is, abelian groups with dilations. Actually, the above is true only for regular points. At singular points, instead of nilpotent Lie groups one gets quotient spaces G/H of such groups G.

The proof of these facts is based on the definition and the construction of the homogeneous nilpotent approximation $(\widehat{X}_1, \ldots, \widehat{X}_m)$ of a system of vector fields $(X_1, \ldots, X_m)$ at a given point. It uses a precise comparison of the sub-Riemannian distances d and $\widehat{d}$ attached to these systems of vector fields. Both the notion of approximation we introduce in §4 and the distance estimates of §7 may be of some interest in control theory. They may also bring some simplifications in the study of partial differential operators of the form $X_1^2 + \cdots + X_m^2$.

Before becoming a problem in geometry, questions of the kind studied in this paper were first studied by Guy Métivier, Elias Stein and other workers in the field of hypoelliptic differential equations (see [12,17,22,28]). The main notions (dilations, weights) were introduced in the paper [28] of Rothschild and Stein, with perhaps slightly different definitions. However, the use of arbitrary privileged coordinates, introduced in §4, allows us to get stronger results, as well as a real clarification in proofs. An analysis of the nilpotent Lie group structure for the tangent space at regular points has been previously published by J. Mitchell in [23].

◇ ◇ ◇

The article is organized as follows. We begin by discussing, in §1, the possible definitions of sub-Riemannian metrics (also called Carnot-Carathéodory metrics) on a smooth manifold. Our definition is general

---

[*]Département de Mathématiques. Université Paris 7-Denis Diderot. 2, place Jussieu, 75251 Paris 5e, France.

enough to meet the needs of Control Theory and the theory of subelliptic second-order differential operators. However, it excludes—for lack of smoothness—those metrics which are naturally associated to operators which are not "sums of squares", that is, which cannot be written locally as $L = X_1^2 + \cdots + X_m^2$ plus lower-order terms.

In §2, we review known results of Accessibility Theory, namely Sussmann and Stefan's Theorem, Chow's Theorem and related results. We give more modern and simpler versions of the original demonstrations. Recall that Chow's theorem asserts that if M is connected, and the vector fields $X_1, \ldots, X_m$, together with their iterated brackets, $[X_i, X_j], \ldots, [[X_i, X_j], X_k], \ldots$ span the tangent space T_pM at every point p, then any two points of M can be connected by (a concatenation of) integral curves of the X_i. If d is the sub-Riemannian distance associated to the X_i, Chow's theorem asserts simply that $d(p, q)$ is finite for any p and any q.

In §3, we describe two examples of sub-Riemannian manifolds: the Grušin plane, and the Heisenberg group H_3. The latter stands for a paradigm for the theory. The Grušin plane has too few dimensions to exhibit real sub-Riemannian features, but as it is, it is the simplest example showing singular points. As we shall see, it is strongly connected with H_3. We introduce by these examples our main objects of interest: non-isotropic dilations, and distance estimates in terms of coordinates.

In §4, we define the notion of privileged coordinates around some point p, which is the key technical notion in our work.

From privileged coordinates, we next define, in §5, dilations centered at p, and a quasi-homogeneous structure on the tangent space T_pM. Actually, we obtain thus the first elements of a calculus adapted to the sub-Riemannian structure, namely a notion of local order at p for functions and, more generally, for differential operators. Replacing vector fields X_i by their principal parts at point p, we define for each p in M a graded nilpotent Lie algebra $\mathfrak{g}_p$, called the tangent Lie algebra, which acts infinitesimally on the tangent space at p. This gives rise to an actual action of G_p, the simply connected Lie group corresponding to $\mathfrak{g}_p$, on T_pM. If Chow's condition holds, this action will be transitive. It is a free action at regular points, which form an open dense set in M. So, at regular points, T_pM is naturally isomorphic to G_p. At singular points, the dimension of G_p is usually greater than that of M, and T_pM is only a homogeneous space of G_p. In both cases, T_pM has a natural sub-Riemannian structure.

A striking observation is that (at regular points for the sake of simplicity) the structure of T_pM is similar to the structure of a real vector space: we have a group structure and an action of $\mathbb{R}$, both interrelated. Only, the group is not abelian, but nilpotent.

More striking, the algebraic structure of the tangent space stems from the distance structure on M. We give in §6 the definition, due to Gromov, of the tangent space at some point to a metric space. The main object of §7 is to identify the metric tangent space, as defined by Gromov, with the sub-Riemannian manifold T_pM previously defined. Finally, in §8, we try to understand in general how a group structure may be extracted from purely metric data.

Of course, the tangent space to a metric space need not exist without very special properties of the distance. The identification of T_pM as a Gromov tangent space is based on a precise comparison of distances in M (near p) and in T_pM. The main difficulty in establishing this comparison is to majorize $d(p,q)$ in terms of privileged coordinates around q.

◇ ◇ ◇

Let us conclude this introduction by returning to Chow's theorem. We will not ignore its concrete significance: by using alternately X, Y, $-X$, $-Y$, you can move into the $[X, Y]$ direction. Everybody parking his own car uses this fact.

But we are interested in a more abstract point of view. Recalling that the conclusion of Chow's theorem may be simply stated as: $d(p,q) < \infty$ for all p and q, we may interpretate the proof of our local estimate for d as an effective proof of Chow's theorem. Generally, one proves Chow's theorem by showing that, if "Chow's condition" is verified, then the end-point map—which for each p is defined on some Banach space, namely the space of finite length paths starting at p, or some control function space—is open at 0, *although* its derivative at 0 is not surjective. But, if we used the notion of derivative that is natural in our context, defined from non-isotropic dilations in M, or in T_pM, then Chow's condition would simply be the assumption that the derivative of the end-point map is onto. Thus Chow's theorem appears as a plain generalization of the submersion theorem. As for the proof of the needed estimates for d, it appears to be similar to an iterative proof of the submersion theorem, using only a Newton-type method instead of a simple iterative method.

Roger Brockett notes in [6] that "it seems that the intuitive content of Riemannian geometry is sufficiently robust so as to withstand modifications [such as generalization into sub-Riemannian geometry] and still provide a reasonably 'geometric' picture". It seems today that this observation has been largely confirmed, and that we may now extend it to some formal aspects of differential calculus: Taylor limited and series expansions, tangent spaces and tangent maps.

1. Sub-Riemannian manifolds

1.1. Definition of sub-Riemannian distances

A sub-Riemannian manifold is often defined as a manifold M of dimension n together with a distribution D of m-planes ($m \leq n$) and a Riemannian metric on D. From this structure one derives a distance on M: the length of an absolutely continuous path tangent to D is defined via the Riemannian metric on D, and the distance $d(p, q)$ of two points of M is in turn defined as the infimum of the lengths of absolutely continuous paths which are tangent to D and join p to q. If no such path exists, one sets $d(p, q) = +\infty$.

However, in particular for the needs of applications, this definition must be enlarged: it is necessary to relax the hypothesis of constant rank for D.

In Control Theory, indeed, one is interested in systems of differential equations of the form

$$\dot{x} = \sum_{i=1}^{m} u_i(t) X_i(x), \tag{1}$$

where $X_1, \ldots, X_m$ are given vector fields on M, and the $u_1, \ldots, u_m$ are variable L^1 functions on some bounded interval. These functions are called *control functions* or *controls*. Any path obtained by integrating (1) is called a controlled path. One often refers to x as to the state of some system, and, if $x(a) = p$, $x(b) = q$, one says that the controls $u_1, \ldots, u_m$ steer the system from state p to state q.

When the rank of the system of vector fields $X_1, \ldots, X_m$ is constant, controlled paths coincide with the absolutely continuous paths tangent to the distribution

$$D = \langle X_1, \ldots, X_m \rangle$$

generated by $X_1, \ldots, X_m$. Conversely, any rank m distribution D can, locally, be written as $D = \langle X_1, \ldots, X_m \rangle$. Recall why the adverb *locally* is needed: in general, for global topological reasons, one cannot find smooth vector fields which may serve as a basis of D on all M, even for $D = TM$. For example, on $M = S^2$, any continuous vector field must vanish at some point.

In writing D as $D = \langle X_1, \ldots, X_m \rangle$ on some open set U, one may even assume that $X_1, \ldots, X_m$ form an orthonormal basis of D, for the Riemannian metric of D, at each point. The length of a controlled path $x(t)$ ($a \le t \le b$) situated in U is then given by the formula

$$\mathrm{length}(x) = \int_a^b \left(\left(u_1(t) \right)^2 + \cdots + \left(u_m(t) \right)^2 \right)^{1/2} dt.$$

Now, for many systems of interest in Control Theory, the rank of $X_1, \ldots, X_m$ is not constant, but the above can be easily generalized. We are led to the following definitions.

Definition 1.1. Let $X_1, \ldots, X_m$ be smooth, i.e., C^∞, vector fields on a manifold M. For $x \in M$ and $v \in T_x M$, we set

$$g(x, v) = g_x(v) = \inf\{ u_1^2 + \cdots + u_m^2 \mid u_1 X_1(x) + \cdots + u_m X_m(x) = v \}.$$

Then, g_x is a positive definite quadratic form on the subspace $F_x = \langle X_1(x), \ldots, X_m(x) \rangle$ of $T_x M$. If v is outside F_x, we have $g_x(v) = +\infty$.

(Proof: Consider the mapping $\sigma_x : \mathbb{R}^m \to T_x M$ which maps $(u_1, \ldots, u_m)$ to $u_1 X_1(x) + \cdots + u_m X_m(x)$. Then, the restriction of σ_x to $(\ker \sigma_x)^\perp$ is a linear isomorphism onto F_x. Let $\rho_x : F_x \to (\ker \sigma_x)^\perp$ be the inverse mapping. We have $g_x(v) = \|\rho_x(v)\|^2$ if $v \in F_x$, $g_x(v) = +\infty$ otherwise.)

We will say that g is the *sub-Riemannian metric associated to the system* $X_1, \ldots, X_m$.

Given g, we set

$$\|v\|_x = g(x, v)^{1/2}$$

and we define the *length* of a absolutely continuous path $c(t)$ ($a \le t \le b$) in M as

$$\mathrm{length}(c) = \int_a^b \|\dot{c}(t)\|_{c(t)} \, dt.$$

Finally, the *distance d associated to the system of vector fields* $X_1, \ldots,$ X_m is defined by

$$d(p, q) = \inf \operatorname{length}(c), \tag{2}$$

where the infimum is taken on all the absolutely continuous paths joining p to q.

A necessary condition for a path c to have finite length is that $\|\dot{c}(t)\|$ be finite for almost any t. It is the same to say that $c(t)$ satisfies the differential equation

$$\dot{x} = \sum_{i=1}^{m} u_i(t) X_i(x) \qquad \text{a.e. on } [a, b], \tag{3}$$

with measurable control functions $u_1, \ldots, u_m$. Such a path will be said *admissible*, or *controlled*. The phrase *horizontal path* is also used, in reference to the important case of the distribution of horizontal planes in a fiber bundle with connection.

When the rank of $X_1, \ldots, X_m$ is constant and equal to m, a controlled path has finite length if, and only if, the control functions—which are unique—are in L^1. Otherwise, when $X_1, \ldots, X_m$ are not supposed to be independent at each point, a given path may be defined by different systems of control functions. But it has finite length if, and only if, one of these systems, at least, consists of functions in L^1: for each t, indeed, one can choose the $u_i(t)$ so as to ensure $\sum (u_i(t))^2 = \|\dot{c}(t)\|^2$. Thus, we have

$$\operatorname{length}(x) = \inf \int_a^b \left((u_1(t))^2 + \cdots + (u_m(t))^2 \right)^{1/2} dt, \tag{4}$$

the infimum being taken on all m-tuples $(u_1, \ldots, u_m)$ of L^1 functions for which (3) holds.

We come to the general definition of a sub-Riemannian distance.

Definition 1.2. A smooth *sub-Riemannian metric on M* is a function $g : TM \to [0, +\infty]$ which, locally, may be defined as the metric associated to some system of smooth vector fields. A *sub-Riemannian distance on M* is a distance which can be defined, via the length of paths, from such a metric.

Observe that a sub-Riemannian distance on M can always be defined by a single locally finite (non necessarily finite) system of vector fields. One could prefer an equivalent, more formal, definition:

Definition 1.3. A smooth *sub-Riemannian metric* on a manifold M is a function $g : TM \to [0, +\infty]$ obtained by the following construction: Let E be a vector bundle over M endowed with a Euclidean metric and let

$$\sigma : E \to TM$$

be a morphism of vector bundles. For each x in M and $v \in T_x M$, set

$$g(x, v) = g_x(v) = \inf\{\|u\|^2 \mid u \in E_x,\ \sigma(u) = v\}.$$

Starting from g, the notions of length of a path, and of distance (*sub-Riemannian distance*) are defined, as above, in the same way as in Riemannian Geometry.

When $E = M \times \mathbb{R}^m$, one can write $\sigma(x, u)$ as $\sigma(x, u) = u_1 X_1(x) + \cdots + u_m X_m(x)$, and one recovers the definition of sub-Riemannian distance attached to a system of vector fields $X_1, \ldots, X_m$.

For the sake of simplicity, we shall always suppose in the sequel that sub-Riemannian distances are defined by a given system of vector fields $X_1, \ldots, X_m$.

Of course, this assumption is perfectly legitimate in all purely local questions.

1.2. The dual form: smoothness and the "sum of squares" hypothesis

Sub-Riemannian metrics as Legendre transforms of semi-elliptic symbols. Sub-Riemannian metrics arise in PDE theory as *Legendre transforms* of smooth semi-positive quadratic forms $a(x, \xi)$ on the cotangent bundle T^*M, namely principal symbols of semi-elliptic second order operators:

$$\tfrac{1}{2} g(x, v) = \sup_{\xi \in T_x^* M} \left(\langle v, \xi \rangle - \tfrac{1}{2} a(x, \xi) \right) \tag{5}$$

In this case, we set $g = a^*$ (we have also $a = g^*$ since Legendre transform is involutive). In order to describe the restriction, denoted by g_x, of g to $T_x M$, introduce the linear map

$$\alpha_x : T_x^* M \to T_x M$$

associated with the quadratic form a_x. Then g_x is a positive definite quadratic form on $F_x = \operatorname{Im} \alpha_x = (\operatorname{Ker} \alpha_x)^\circ$ and is equal to $+\infty$ outside

F_x, as simple computations show: Choose coordinates such that $\langle v, \xi \rangle = v_1 \xi_1 + \cdots + v_n \xi_n$, $a_x(\xi) = \xi_1^2 + \cdots + \xi_r^2$. Then

$$\tfrac{1}{2} g(x, v) = \sup_{\xi_1, \ldots, \xi_r} \left((v_1 \xi_1 + \cdots + v_r \xi_r) - \tfrac{1}{2}(\xi_1^2 + \cdots + \xi_r^2) \right) +$$

$$\sup_{\xi_{r+1}, \ldots, \xi_n} (v_{r+1} \xi_{r+1} + \cdots + v_n \xi_n),$$

and hence

$$g(x, v) = \begin{cases} v_1^2 + \cdots + v_r^2 & \text{if } v_{r+1} = \cdots = v_n = 0; \\ +\infty & \text{otherwise.} \end{cases}$$

We thus get a metric on TM which can be used as above to define a notion of length and a distance on M.

The sub-Riemannian metric attached to a system of smooth vector fields can be obtained in this way, from a quadratic form on T^*M: Starting from $X_1, \ldots, X_m$, we define the quadratic form

$$a(x, \xi) = \sum_{i=1}^{m} \langle X_i(x), \xi \rangle^2, \tag{6}$$

which, incidentally, is the principal symbol of the operator $A = \sum_{i=1}^{m} X_i^2$. The generalized quadratic form $g(x, v)$ obtained from $a(x, \xi)$ by the Legendre transformation is precisely the sub-Riemannian metric attached to $X_1, \ldots, X_m$.

"Sums of squares." If all smooth semi-positive quadratic forms $a(x, \xi)$ could be locally decomposed into a sum of squares of linear forms such as (6), everything would fit well together: we would have a perfect correspondence between smooth semi-positive quadratic forms on T^*M and smooth sub-Riemannian metrics (as we have defined them in Definition 1.3).

In coordinate and matrix terms, denoting by $a(x) = \big(a_{ij}(x)\big)$ the matrix of the quadratic form $a(x, \xi)$, the existence of smooth $X_1, \ldots, X_m$ such that (6) holds is equivalent to the existence of a $n \times m$ matrix $\sigma(x) = \big(\sigma_{ij}(x)\big)$, depending smoothly of x, such that

$$a(x) = {}^t\sigma(x)\, \sigma(x). \tag{7}$$

(Take as columns of $\sigma(x)$ the coordinates of vectors $X_1(x), \ldots, X_m(x)$.) It turns out that such a smooth decomposition does not exist for all a. However, counter-examples are not so easy to build. See [24].

To mention positive results, note that a decomposition of type (7) does always exist, locally, when $\operatorname{rank} a(x)$ is constant: in some coordinate chart, apply the Gauss method to the quadratic form $a(x, \xi) = \sum a_{ij}(x)\xi_i\xi_j$.) It does exist also in the neighbourhood of singular points in many cases, for example for

$$a(x_1, x_2) = \begin{pmatrix} 1 & 0 \\ 0 & x_1^2 \end{pmatrix}$$

In the general case, if constant rank is not assumed, one can only guarantee the existence of a *Lipschitzian* $\sigma(x)$: to get a decomposition with Lipschitzian $\sigma(x)$ or X_i's, take for $(\sigma_{ij}(x))$ the symmetric square root of $(\sigma_{ij}(x))$, and take for X_i's the columns of (σ_{ij}).

To sum up, retain that the existence of a decomposition with smooth X_i's must be assumed as an additional hypothesis. This is the "sum of squares" hypothesis.

For the operator $A = \sum a_{ij}(x)\partial_{x_i}\partial_{x_j}$ with principal symbol a, this hypothesis asserts that A can be written as

$$A = \sum_{i=1}^{m} X_i^2 + Y,$$

with smooth $X_1, \ldots, X_m, Y$.

These "sums of squares" are precisely the operators which correspond via the Legendre transform (of the principal symbol) to our smooth sub-Riemannian metrics. The smoothness of the X_i's allows to compute the Lie brackets of these vector fields, and to bring into the picture (in the analytic side as well as in the geometric side) all the machinery of Lie algebras and Lie groups. A good reason for sticking to our Definition 1.2 (or) and not trying to enlarge it. It is well-known that the study of semi-elliptic operators which are not sums of squares is much harder (see [9,24]). From the geometric side, the (very singular) corresponding metrics are still *terra incognita*.

1.3. Terminology

Sub-Riemannian metrics appear in the literature under a variety of names: singular Riemannian metrics [18], Carnot-Carathéodory metrics [3,13,14,25], sub-Riemannian metrics [33], nonholonomic Riemannian metrics [39]. In the realm of hypoelliptic PDE, they are also used in work by Stein, Fefferman and co-workers, but they are not given a name.

2. Accessibility

2.1. The theorems of Chow and Sussmann

We shall deduce the classical theorem of Chow from a more precise
result by Sussmann. As a first step, we will prove Sussmann's theorem
using L^1 controls. Next, we will show that the results obtained are, to a
great extent, independent of the class of control used.

Consider a symmetric[1] control system on M, as described above,

$$\dot{x} = \sum_{i=1}^{m} u_i X_i(x). \tag{8}$$

For $p \in M$ and $T > 0$, let $\Omega_{p,T}$ be the space of controlled paths with
origin p, parametrized by $[0, T]$. Every such path may be obtained by
integrating the differential system

$$\begin{cases} \dot{x} = \displaystyle\sum_{i=1}^{m} u_i(t) X_i(x), & 0 \le t \le T \\[2mm] x(0) = p \end{cases} \tag{9}$$

for some control function $u \in L^1([0, T], \mathbb{R}^m)$. Actually, given p, the differential system (9) has a well defined solution $x_u(t)$ for $u \in U_{p,T}$, where
$U_{p,T}$ is an open set containing the origin in $L^1([0, T], \mathbb{R}^m)$.

Definition 2.4. We will denote by

$$E_{p,T} : U_{p,T} \to M$$

the mapping which maps u to $x_u(T)$. We will call $E_{p,T}$, or E for short,
the *end-point map*.

Now, the *accessible set* A_p (the set of points accessible in finite time
from p, regardless of time) is exactly the image of $E_{p,T}$ for a chosen T.
Indeed, every controlled path $x : [0, T'] \to M$, defined by the control $u :
[0, T'] \to M$ may be reparametrized by $[0, T]$, at the price of multiplying
the controls by T'/T.

---

[1] Systems such as (8), with no restrictions on the u_i, are called *symmetric*, or
reversible, as every trajectory followed backwards is also a trajectory.

Definition 2.5. For a given u, integrating (9) with p as initial point gives rise to a diffeomorphism $p \mapsto x_u(T)$, which we will denote by Φ_u.

We will call the map $u \mapsto \Phi_u$ the *flow of the controlled vector field* $\sum_{i=1}^{m} u_i X_i$. Of course, Φ_u need not be defined on all of M; its domain may even be empty.

For $u \in L^1([0,T], \mathbb{R}^m)$ and $v \in L^1([0,T'], \mathbb{R}^m)$, we denote by $u * v$ the *concatenation* of u and v, i.e., the control function defined by

$$(u * v)(t) = \begin{cases} u(t) & \text{if } 0 \leq t < T; \\ v(t-T) & \text{if } T \leq t \leq T + T'. \end{cases}$$

We denote by $\check{u}$ the *return* or *inverse* control:

$$\check{u}(t) = -u(T-t) \qquad 0 \leq t \leq T.$$

If u steers p to q, then $\check{u}$ steers back q to p. Thus, we have

$$\Phi_{u*v} = \Phi_v \circ \Phi_u,$$
$$\Phi_{\check{u}} = \left(\Phi_u\right)^{-1}.$$

Sussmann's theorem asserts that the set of points accessible from a given point is an immersed submanifold. We shall prove this theorem using arguments from differential calculus in Banach spaces, taking advantage of the fact that the end-point map is a differentiable mapping (from an open set in L^1) into M, a finite dimensional manifold.

Definition 2.6. Let $\rho(p)$ the maximal rank of the end-point map $E_{p,T} : U_{p,T} \to M$. We say that a control function $u \in U_{p,T}$ is *normal* if the rank of $E_{p,T}$ at u is equal to $\rho(p)$. We shall say that the path x_u defined by u is a normal path. Otherwise, we speak of *abnormal* control, and abnormal path. A point which can be joined to p by a normal path is said to be *normally accessible from p*.

Lemma 2.1. *If u is a normal control, so is $u * v$.*

Proof. We use the language of Calculus of Variations. Let δu be an infinitesimal variation of u. We consider variations of $u * v$ of the form $(u + \delta u) * v$. Let $q = \Phi_u(p)$ and $r = \Phi_{u*v}(p) = \Phi_v(q)$.

When δu describes L^1, δq describes a subspace of $T_q M$ of dimension $\rho(p)$, since u is normal. Therefore, since Φ_v is a diffeomorphism, $\delta r =$

$D\Phi_v(\delta q)$ describes a subspace of $T_r M$ of the same dimension $\rho(p)$. Now, the rank of E_p at $u * v$ is the dimension of the space of the variations of r corresponding to all variations of $u * v$. Since we get already the maximal dimension by using only special variations, this rank is equal to $\rho(p)$. ∎

Lemma 2.2. *Every point accessible from p is normally accessible from p.*

Proof. Suppose q is attained from p by means of a control u. Choose a normal control v steering p to some point q'. The control $v * \check{v} * u$ steers p to q, and it is normal. ∎

The trick of using $v * \check{v} * u$, which appears in [35], has been called the *return method* by Coron (this volume).

So, the accessible sets are images of constant rank maps. It would be tempting to use a global form of the rank theorem to deduce from this fact that they are submanifolds of M. Such a theorem can be found in Bourbaki [5]: Let $F : N \to M$ be a differentiable map having constant rank. One can put a natural manifold structure on the quotient set $N/\mathcal{R}$, where $x\mathcal{R}x'$ is the following equivalence relation "$F(x) = F(x')$ and x and x' belong to the same connected component of $F^{-1}(F(x))$", so that the induced map $\overline{F} : N/\mathcal{R} \to M$ is an immersion.

Therefore, the accessible sets are images of immersions. One needs a little more, if only to show that these immersions are injective, and to get a well-defined manifold structure on their images.

Definition 2.7. An *immersed submanifold* of a manifold M is a subset A of M, endowed with a manifold structure, such that

(a) The inclusion map $i : A \to M$ is an immersion;

(b) Any continuous map $f : P \to M$, where P is a manifold, taking its values in A, is already continuous when considered as a map $f : P \to A$, where A is endowed with its manifold topology.

Roughly speaking, the role of Condition (b) in this definition is to prevent the existence of curves in A, or branches of A, asymptotic in M to some point $a \in A$, while staying far from a from the point of view of the own topology of A. The leaves of a foliation, the Lie subgroups of a Lie group are immersed submanifolds, while they need not be submanifolds in the classical sense, i.e., locally closed submanifolds. One shows (see [29,30]) that Condition (b) yields:

(b') Any C^k map $f : P \to M$, where P is a manifold, taking its values in A, is already C^k when considered as a map $f : P \to A$, where A is endowed with its own manifold structure.

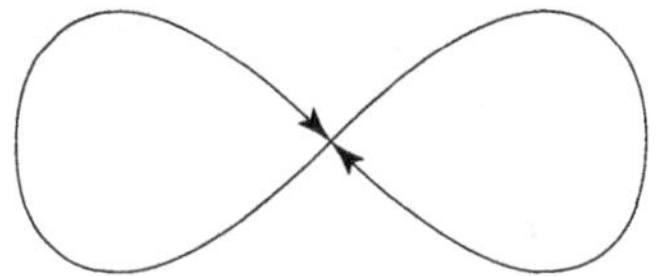

Fig. 1: An immersion of the line violating Condition (b)

As a consequence, the smooth manifold structure on A is unique. It may not be unique when A is only supposed to be the image of some immersion, and Condition (b) fails. See figure 1 (this figure is borrowed from Spivak [29], p. 63). That's why we propose to keep the phrase *immersed submanifold* for the case where Condition (b) holds.

Theorem 2.3. (Sussmann[35], Stefan[30]) *The set A_p of points accessible from a given point p in M is an immersed submanifold.*

Proof. Fix p in M, and set $\rho = \rho(p)$. The normal controls form an open subset, say $N_{p,T}$ of the set $U_{p,T}$ for which the control equation can be integrated with initial value p. By Lemma 2.2, the accessible set A_p is the image of $N_{p,T}$ by a constant rank map.

Since $E_{p,T} : N_{p,T} \to M$ has constant rank ρ, there exists, for each $x \in A_p$ and each $u \in N_{p,T}$ such that $E_{p,T}(u) = x$, a triple (V, U, ψ) such that:

(i) V is a (locally closed) submanifold of dimension ρ of M, and $x \in V$;

(ii) U is a neighbourhood of u;

(iii) ψ is a diffeomorphism from U onto $V \times B$, where B is an open set in some Banach space, such that $E_{p,T}\big(\psi^{-1}(y, b)\big) = y$ for all $y \in V$ and $b \in B$.

We define a topology $\mathcal{T}$ on A_p by saying that the inclusion maps $V \subset A_p$, where V comes from such a triple (V, U, ψ), are open. One can also describe $\mathcal{T}$ as the coarsest topology on A_p for which the end-point map

$E_{p,T}$ is open. Naturally, the point p can be replaced by another point p of A_p.

We also define a smooth manifold structure on A_p by saying that the inclusion maps $V \subset A_p$ are open immersions. Some compatibility must be checked. It suffices to prove that $V'' = V \cap V'$ is an open submanifold of both V and V' if V and V' are given as above. So, let $x = E_{p,T}(u) = E_{p,T}(u')$. Since $E_{p,3T}$ has rank ρ at $u'' = u * \breve{u} * \breve{u}'$, the image of a small neighbourhood of u'' is a small submanifold of M, passing through x, and whose germ at x contains the germ of V and the germ of V'. Since all have the same dimension, this shows that these germs are equal, and $V \cap V'$ is a neighbourhood of x for $\mathcal{T}$.

When A_p is endowed of the manifold structure just described, the inclusion map is an immersion, since any point has a neighbourhood of the form V.

We prove now that A_p satisfies Condition(b) . Let q_0 a point of A_p, and let $u_0 \in N_{p,T}$ be a normal control steering p to q. Since $E_{p,T}$ has constant rank ρ near u_0, there exists a submanifold W_0 of L^1, passing through 0, and a submanifold V_0 of M such that $E_{p,T}$ maps $u_0 + W_0$ diffeomorphically onto V_0. We can also say that $E_{q_0,2T}$ maps $\breve{u}_0 * (u_0 + W_0)$ diffeomorphically onto V_0.

Actually, for any q near q_0, the map

$$E_{q,2T} : \breve{u}_0 * (u_0 + W_0) \to M$$

is a diffeomorphism onto a submanifold $V(q)$ of M. Moreover, if q is in A_p, then $V(q)$ is entirely contained in A_p, and it is an open set in A_p.

Choose a small transversal Q to A_p at q_0. The mapping

$$\psi : Q \times W_0 \to M$$

defined by

$$\psi(q, w) = E_{q,2T}(\breve{u}_0 * (u_0 + w))$$

is a diffeomorphism onto an open neighbourhood of q_0. We have $\psi(\{q\} \times W_0) = V(q)$, and, for given q, either $\psi(\{q\} \times W_0) \subset A_p$, or $\psi(\{q\} \times W_0)$ does not meet A_p.

Now, the topology of A_p has a countable basis. Say, because L^1 is separable. Therefore, since the sets $\psi(\{q\} \times W_0)$, $q \in Q$ are mutually disjoints, only countably many of them can be contained in A_p. Consider a continuous path $x(t)$ contained in A_p, such that $x(t_0) = q_0$. One must have

$x(t) = q_0$ for t near t_0, otherwise $x(t)$ would meet uncountably many slices $\psi(\{q\} \times W_0)$. Then $x(t)$ is contained in the submanifold V_0, and, in a neighbourhood of t_0, it is continuous when considered as a path in A_p, assuming A_p is endowed with its manifold topology. Since continuity is a local property, this proves that $x(t)$ is continuous for all t, if one considers it as a path in A_p. The same argument works for any continuous map $f : P \to M$ taking its values in A_p.   ∎

Since the relation $q \in A_p$ is clearly an equivalence relation, we can speak of *accessibility components*. Since

$$q \in A_p \iff d(p, q) < \infty,$$

the set A_p is the union of open balls $B(p, R)$ (for d), so it is itself an open set. Whence it results that the accessibility components are also the connected components of M for the topology defined by d.

The same process by which we have defined a manifold structure on the accessible set may be used to define a new manifold structure on M, whose connected components are the A_p. Observe that the dimension of this manifold may vary, and that it is in general $< n$, in which case it has uncountably many connected components. Beware that the underlying manifold topology may be different of the distance topology defined by d, although it has the same connected components. See the example on page 18.

Theorem 2.4. (Chow's theorem, Rashevsky[26], Chow[7]) *Suppose M is connected and the following condition holds:*
(C) *The vector fields $X_1, \ldots, X_m$ and their iterated brackets $[X_i, X_j]$, $[[X_i, X_j], X_k]$, etc. span the tangent space $T_x M$ at every point of M.*

Then every two points of M are accessible from one another.

Condition (C) is called Chow's condition.

Proof. It is clear that the accessibility components of M are stable under the flow e^{tX_i} of vector fields X_i $(i = 1, \ldots, m)$. Therefore, since the accessibility component through x is an immersed submanifold, these vector fields are tangent to it at every point and can be considered as vector fields on the manifold A_x (here we use the fact that Condition (b') holds for A_x). It follows that $[X_i, X_j]$, $[[X_i, X_j], X_k]$, etc. are also tangent to A_x.

If Chow's condition holds, then we have

$$T_y A_x = T_y M$$

at every point y of A_x. It follows that the accessibility components have dimension n, so they are open. Since M is connected, there can be only one accessibility component. This proves Chow's theorem. ∎

When $M = A_p$ for some p, one says that system (8) is *controllable*. The reciprocal of Chow's theorem—i.e., if (8) is controllable, the X_i's and their iterated brackets span the tangent space at every point of M— is true if M and the vector fields are analytic [35], and false in the C^∞ case (see the example below in §2.2).

In the analytic case, M is the union of its accessibility components, each one being an immersed submanifold on which Chow's condition holds. Without doubt, the same property is generically true in the C^∞ case. Since for all questions concerning the sub-Riemannian distance we may work on one accessibility component, we may as well suppose that Chow's condition holds on M.

Chow's Condition is also known under the name of Lie Algebra Rank Condition (LARC) since it states that the "rank" at every point x of the Lie algebra generated by the X_i's is full. In the context of PDE, it is known under the name of Hörmander's Condition: when it holds, the differential operator $X_1^2 + \cdots + X_m^2$ is hypoelliptic (Hörmander's Theorem [20]). Conversely, when M and the vector fields $X_1, \ldots, X_m$ are analytic, the hypoellipticity of $X_1^2 + \cdots + X_m^2$ implies Chow's condition. As a matter of fact, the formulation of Chow's condition in terms of Lie brackets is due to Hörmander.

2.2. Openness of the end-point mapping and continuity of the sub-Riemannian distance

Theorem 2.5. *Suppose Chow's condition holds. Then the end-point mapping is open.*

Proof. Let $\rho(p, \varepsilon)$ denote the maximal rank of the end-point mapping E_p on the open ball $B(0, \varepsilon)$ in $L^1([0, T], \mathbb{R}^m)$. We have $\rho(p, \varepsilon) > 0$ for all $\varepsilon > 0$. Indeed, to suppose $\rho(p, \varepsilon) = 0$ implies that E_p is constant on $B(0, \varepsilon)$, whence it follows that $X_1(p) = \cdots = X_m(p) = 0$, and that all the brackets of the X_i's vanish at p, in contradiction with Chow's condition. Since $\rho(p, \varepsilon)$ decreases when ε decreases, there exist ε_0 and ρ_0, a positive integer, such that

$$\rho(p, \varepsilon) = \rho_0 \quad \text{for} \quad 0 < \varepsilon \le \varepsilon_0.$$

In other words, every neighbourhood of 0 contains a control u, on which the rank of E_p is a positive integer ρ_0. By considering $u * \breve{u}$ (and increasing the velocity by a factor 2, so that we are dealing in fact with end-point mapping having the same end-time), we see that every neighbourhood of 0 in L^1 contains a control yielding a *closed* path, on which the rank of E_p is ρ_0.

Let W denote the set of controls in $B(0, \varepsilon_0)$ on which the rank of E_p is equal to ρ_0. Then W is an open set, and its image under E_p is an immersed sub-manifold N, containing p. Now, the flows of vector fields $X_1, \ldots, X_m$ preserve that sub-manifold: the rank of the end-point map at some given path (control) is preserved or increased if one concatenates this path with some piece of integral curve of X_i. Then $X_1, \ldots, X_m$ are tangent to N and so are their brackets. Thus, by Chow's condition, the dimension of $N = E_p(W)$ is equal to n, that is, N is open.

This proves that E_p is open at 0. To show that E_p is open at some point u in $L^1([0, T], \mathbb{R}^m)$, choose first some $T_0 > 0$ and $\varepsilon_0 > 0$. The image of the ball $B(0, \varepsilon_0)$ in $L^1([0, T_0], \mathbb{R}^m)$ is a neighbourhood of 0. Then $B(0, \varepsilon_0) * u = \{v * u \mid v \in B(0, \varepsilon_0)\}$ is a set of controls, defined on $[0, T_0 + T]$, which steer 0 onto a neighbourhood of $E_p(u)$. Now, by choosing T_0 and ε_0 small enough one can ensure that $B(0, \varepsilon_0) * u$, after reparametrization by $[0, T]$, is contained in any given ball $B(u, \varepsilon)$ in $L^1([0, T], \mathbb{R}^m)$. So, the image under E_p of any ball $B(u, \varepsilon)$ is a neighbourhood of $E_p(u)$. $\blacksquare$

Remark. An important observation is that, even when Chow's condition holds and the end-point map is consequently surjective on M and open, it need not be a submersion. Actually, the image of the differential of $E_{p,T}$ at $u = 0$ is the space spanned by $X_1(p), \ldots, X_m(p)$ in T_pM. This follows from the integration of (9), which gives

$$x(T) = X_1(p) \int_0^T u_1(t)\, dt + \cdots + X_m(p) \int_0^T u_m(t)\, dt + O\left(\|u\|^2\right).$$

Example. Consider on $\mathbb{R}^2$ the system

$$X_1 = \begin{pmatrix} 1 \\ 0 \end{pmatrix}, \quad X_2 = \begin{pmatrix} 0 \\ f(x) \end{pmatrix},$$

where f is a C^∞ function such that $f(x) = 0$ for negative x and $f(x) > 0$ for positive x. Observe that this system has rank one in the $x < 0$ half-plane, and rank two in the $x > 0$ half-plane. Admissible paths must be horizontal, in the $x < 0$ half-plane, while any path is admissible in the $x > 0$ half-plane. We illustrate through this example several points discussed in this Section. Notice first that any two points are mutually accessible, although Chow's condition does not hold.

a) *How the return method works.* Fix a point $p_0 = (x_0, y_0)$ in the left-hand half-plane. Any other point of the plane is accessible from p_0. However, to reach from p_0 a point with $x < 0$, $y \neq y_0$, it is needed to pass through the $x > 0$ half-plane. Actually, normal paths starting at p_0 are those which go through the $x > 0$ half-plane. It is clear how the return method works: if one concatenates some path γ starting at p_0 with another path δ starting at p_0, passing through the $x > 0$ half-plane, and returning to p_0, then the end of the path $\delta * \gamma$ so obtained may be moved freely in both x and y directions.

b) *The end-point map is not open.* Keep the same point p_0. Although one can find for any p controls u steering p_0 to p such that the end-point map E_{p_0} is open at u, the end-point map is not open. The image of a small ball in the control space consists of a line segment only.

c) *The distance topology is not the usual topology of* $\mathbb{R}^2$. The distance topology gives rather to $\mathbb{R}^2$ the appearance of a comb.

Corollary 2.6. *Suppose Chow's condition holds. The topology defined by the sub-Riemannian distance d is the original topology of M.*

Proof. Any ball $B^d(p, \varepsilon)$ is a neighbourhood of p. Indeed, $B^d(p, \varepsilon)$ is clearly the image under E_p of the ball $B(0, \varepsilon)$ in L^1. Therefore, it is an open set in M.

Conversely, any neighbourhood U of p contains a ball $B^d(p, \varepsilon)$: Since E_p is continuous at 0, there exists $\varepsilon > 0$ such that E_p maps $B^{L^1}(0, \varepsilon)$ into U.   ∎

2.3. Geodesics and the Hopf-Rinow theorem

Suppose Chow's condition holds, so d defines the topology of M. One can prove that the length of an admissible path $c(t)$ $(a \leq t \leq b)$ is given by

$$\text{length}(c) = \sup \sum_{i=1}^{N} d\big(c(t_{i-1}), c(t_i)\big), \tag{10}$$

where the supremum is taken over all N and over all choices of (t_i) such that $a = t_0 < t_1 < \cdots < t_N = b$. We may extend, using (10), the notion of length to all continuous paths. It turns out that the only paths having finite length are the paths with are differentiable a.e., and for which the integral (4) is finite, and the paths obtained by continuous reparametrization of such paths.

With such a definition, the length is a lower semicontinuous functional on the space of continuous paths $C([a, b], M)$.

Definition 2.8. A path $c : [a, b] \to M$, with constant velocity, such that

$$\text{length}(c) = d\left(c(a), c(b)\right)$$

is called a *minimizing geodesic*.

Theorem 2.7. (Hopf-Rinow theorem for sub-Riemannian manifolds.) *Suppose Chow's condition holds. Then*

(i) *Sufficiently near points can be joined by a minimizing geodesic;*

(ii) *If M is a complete metric space for d, any two points can be joined by a minimizing geodesic.*

Proof. We prove (i), and we leave the proof of (ii) to the reader. Consider a point p, and choose $\varepsilon > 0$ such that the closed ball with center p and radius 3ε is compact. Fix points q and q' in $B(p, \varepsilon)$, and set $T = d(q, q')$.

Let us say that a continuous path c has velocity $\leq k$ if $d(c(s), c(t)) \leq k|s - t|$. This is equivalent to ask c to be a controlled path having $\|\dot{c}\| \leq k$. Denote by P the set of paths $c : [0, T] \to M$ with velocity ≤ 2 and joining q to q'. We make the following observations:

(a) P is non-empty. In fact, it contains, up to reparametrization, all paths with length $\leq 2T$ joining q to q';

(b) All paths in P are contained in $B(0, 3\varepsilon)$;

(c) P is equicontinuous in $C([0, T], M)$;

(d) P is closed in $C([0, T], M)$.

So, by Ascoli-Arzelà theorem, P is a compact subset of $C([0, T], M)$. It follows that the length functional attains its infimum on P. In other words, there exists a path with length $T = d(q, q')$ joining q to q'. ∎

Remarks. 1. We do not assert, either that for q, q' belonging to $B(p, \varepsilon)$ the geodesic joining q and q' is unique, or that it is contained in $B(p, \varepsilon)$. See the examples in §3.

2. Assuming Chow's condition, the conclusion of (ii) holds, in several important cases: when M is compact, when $M = \mathbb{R}^n$ and the X_i are bounded, and when M is a Lie group and the X_i are left-invariant vector fields. Indeed, M is complete in these three cases.

2.4. Direct, effective proofs of Chow's theorem

At this point, it is impossible not to mention the existence of proofs of Chow's Theorem, more effective than the one we have given.

Here and in the sequel, it will be convenient to note on the right the action of diffeomorphisms: the action of e^{tX} on point p will result in $p\, e^{tX}$. This notation is consistent with our notation for the concatenation of paths, and complies with the fact that all diffeomorphisms we shall use on Lie groups in the sequel come from flows of left-invariant vector fields, and so are defined by right multiplications.

For the case $n = 3$, $m = 2$, where one assumes that X_1, X_2 and $[X_1, X_2]$ span the tangent space at p, one proves that the mapping

$$(t_1, t_2, t_3) \mapsto p\, e^{t_1 X_1} e^{t_2 X_2} e^{|t_3|^{1/2} X_1} e^{t_3^{1/2} X_2} e^{-|t_3|^{1/2} X_1} e^{-t_3^{1/2} X_2}, \qquad (11)$$

where we write $t^{1/2}$ for $\mathrm{sgn}(t)|t|^{1/2}$, is tangent to the mapping

$$(t_1, t_2, t_3) \mapsto p\, e^{t_1 X_1} e^{t_2 X_2} e^{t_3 [X_1, X_2]}$$

at $t = 0$. This shows that the end-point mapping E_p is open at the origin. At the same time one gets local estimates for the sub-Riemannian distance: given any Riemannian metric δ, there exist a neighbourhood U of p and constants C, C' such that

$$C\delta(q, q') \le d(q, q') \le C'\delta(q, q')^{1/2}$$

for any q, q' in U.

The effective proof of Chow's theorem in case $n = 3$, $m = 2$ is based on Campbell-Hausdorff formula (see Lobry [21]). It can be generalized to all cases where Chow's condition holds (see Gromov, this volume), and one gets similar estimates with $1/2$ replaced by $1/r$, where r is the smallest integer for which the tangent space is spanned by brackets of length $\leq r$ of $X_1, \ldots, X_m$ (the degree of nonholonomy, see §4.1). We shall prove more precise estimates in Section 7.

Observe that mappings such as (11) give a mean to construct a non-differentiable section, actually Hölderian of order $1/2$, of the end-point map.

2.5. Accessibility does not depend on the class of controls used

The set of points accessible from a given point in M by means of control functions belonging to any reasonable class of control functions, ranging from piecewise constant to L^1, is independent of the class of controls used. More precisely, we have the following theorem.

Theorem 2.8. *Let $\mathcal{C}$ be a class of control functions such that $\mathcal{C}([0, T], \mathbb{R}^m)$ is a dense subspace in $L^1([0, T], \mathbb{R}^m)$. Then any point accessible from p is accessible from p by means of controls of class $\mathcal{C}$.*

Proof. Let q be a point accessible from p. Using Lemma 2.2, choose a *normal* control $u \in L^1([0, T], \mathbb{R}^m)$ steering p to q. For simplicity, we denote $E_{p,T}$ by π. Since u is normal, the linear mapping

$$d\pi_u : L^1 \to T_q A_p$$

is surjective. Fix a sequence of finite-dimensional spaces

$$H_1 \subset H_2 \subset \cdots \subset H_k \subset \cdots$$

of $\mathcal{C}([0, T], \mathbb{R}^m)$, with strictly increasing dimension, such that $\bigcup H_k$ is dense in L^1. (The existence of such a sequence of subspaces stems from the separability of L^1.) For some integer k_0, we have

$$d\pi_u(H_{k_0}) = T_q A_p.$$

Choose a linear subspace V of H_{k_0}, of the same dimension as A_p, such that one has still

$$d\pi_u(V) = T_q A_p.$$

Then there exists $\varepsilon > 0$ such that

(i) π is defined on $B^V(0, 2\varepsilon)$;

(ii) The mapping $\phi : h \mapsto \pi(u + h)$ is a diffeomorphism of $\overline{B} = \overline{B}^V(0, \varepsilon)$, the closed ball with center 0 and radius ε, onto $\phi(\overline{B}) = \pi(u + \overline{B})$.

Now, let $u_k \in H_k$ $(k = 1, 2, \dots)$ a sequence of control functions converging to u. For k large enough, the mapping

$$\phi_k : h \mapsto \pi(u_k + h)$$

is defined on all of $\overline{B}$, and it converges uniformly, in the C^1 sense, to the mapping ϕ. Using Lemma 2.9 below, applied to the sequence of mappings $f_k = \phi^{-1} \circ \phi_k$ $(k = 1, 2, \dots)$, one shows that q is in the image of ϕ_k. Since $\phi_k(\overline{B}) = \pi(u_k + \overline{B})$ consists of images by π of elements in $\bigcup H_k$, it results that q is accessible by means of a control function in the class $\mathcal{C}$. ∎

Lemma 2.9. *Let $\overline{B}$ be the closed ball in $\mathbb{R}^n$ of center q and radius ε, and let $f_k : \overline{B} \to \mathbb{R}^n$ $(n = 1, 2, \dots)$ be a sequence of differentiable mappings converging uniformly, in the C^1 sense, to the identity map of $\overline{B}$. Then, for k large enough, the image $f_k(\overline{B})$ contains q.*

Proof. It suffices to prove that, for any differentiable mapping $g : \overline{B} \to \mathbb{R}^n$ verifying

$$\|g(x) - x\| \le \frac{\varepsilon}{2}, \quad \|dg_x - 1\| \le \frac{1}{2}$$

for all x in $\overline{B}$, then the image of g contains q.

For that purpose, consider the sequence in $\overline{B}$ defined by

$$x_0 = q, \quad x_{i+1} = q + x_i - g(x_i) \quad (i = 1, 2, \dots).$$

It is well defined, that is, one can prove inductively that $x_i \in \overline{B}$. Indeed, we have

$$\|x_i - q\| \le \|x_1 - q\| + \|x_2 - x_1\| + \cdots + \|x_i - x_{i-1}\| \le \frac{\varepsilon}{2} + \frac{\varepsilon}{4} + \cdots + \frac{\varepsilon}{2^i} < \varepsilon.$$

The same computation proves that the series

$$q + (x_1 - q) + (x_2 - x_1) + \cdots + (x_i - x_{i-1}) + \cdots$$

is convergent. In other words, x_i converges to some x_∞. Clearly, we have $\|x_\infty - q\| \le \varepsilon$ and $g(x_\infty) = q$. ∎

The same result is still true, with a proof à la Brouwer, if one supposes only that the f_k are continuous, and C^0 convergence holds. (The use of such a fixed-point argument in the proof of theorem 2.8 has been suggested to me by Héctor Sussmann.)

Remark. The minimal notion of accessibility is obtained by using only concatenations of integral curves of the vector fields $X_1, \ldots,$ X_m, which amounts to use concatenations of controls of the form $(0, \ldots, 0, \pm 1, 0, \ldots, 0)$, the so-called bang-bang controls (in fact, one should use controls $(0, \ldots, 0, \pm t_i, 0, \ldots, 0)$ if one wants to keep a fixed time interval). Bang-bang controls do not form a vector subspace of L^1 but the same conclusion as in Theorem 2.8 holds, with a slightly modified proof.

3. Two examples

Examples of sub-Riemannian manifold include Riemannian manifold—the case $D = TM$ and Riemannian foliations—the case where D is integrable.

More genuine examples are the Grušin plane below—which is almost Riemannian, and really sub-Riemannian along some singular line only, and, most important, the Heisenberg group, where the role of nonholonomy appears clearly, and which serves as a paradigm for the theory.

3.1. The Grušin plane G_2

We take as underlying manifold of G_2 the $\mathbb{R}^2$ plane (with coordinates x, y) and consider the sub-Riemannian metric defined by the vector fields

$$X_1 = \begin{pmatrix} 1 \\ 0 \end{pmatrix}, \quad X_2 = \begin{pmatrix} 0 \\ x \end{pmatrix}.$$

These vector fields span the tangent space everywhere, except along the line $x = 0$, where adding

$$[X_1, X_2] = \begin{pmatrix} 0 \\ 1 \end{pmatrix}$$

is needed. So Chow's condition holds. Outside the line $x = 0$, the sub-Riemannian metric is in fact Riemannian, and is equal to

$$ds^2 = dx^2 + \frac{1}{x^2} dy^2.$$

Any path has finite length, provided its tangent is parallel to the x-axis when crossing the y-axis.

This example of sub-Riemannian manifold is named after Grušin, who was the first to study the analytic properties of the operator $L = X_1^2 + X_2^2 = \partial_x^2 + x^2 \partial_y^2$, and of its multidimensional generalizations [15,16].

Dilations and distance estimates. A very important feature is the existence of a one-parameter group of dilations for G_2 : if we set

$$\delta_\lambda(x,y) = (\lambda x, \lambda^2 y),$$

then we have

$$(\delta_\lambda)^* X_1 = \lambda^{-1} X_1, \quad (\delta_\lambda)^* X_2 = \lambda^{-1} X_2$$

for $\lambda \neq 0$. Therefore, the length of a controlled path is multiplied by $|\lambda|$ under the action of δ_λ. It follows that

$$d(\delta_\lambda p, \delta_\lambda q) = |\lambda| d(p,q)$$

for all $p, q \in G_2$ and $\lambda \in \mathbb{R}$.

It is easy to bound $d\big((0,0),(x,y)\big)$ on the boundary of the square $|x| \leq 1$, $|y| \leq 1$: it is ≥ 1 and ≤ 3. Using homogeneity under the action of δ_λ we get the estimates

$$\sup\big(|x|, |y|^{1/2}\big) \leq d\big((0,0),(x,y)\big) \leq 3\sup\big(|x|, |y|^{1/2}\big). \qquad (12)$$

Instead of (12), one may prefer to use

$$\tfrac{1}{2}\big(|x| + |y|^{1/2}\big) \leq d\big((0,0),(x,y)\big) \leq 3\big(|x| + |y|^{1/2}\big). \qquad (13)$$

In geometric terms, (12) means that balls $B(0,\varepsilon)$ are roughly of the shape

$$[-\varepsilon, \varepsilon] \times [-\varepsilon^2, \varepsilon^2].$$

More precisely, we have

$$\tfrac{1}{9}\,[-\varepsilon, \varepsilon] \times [-\varepsilon^2, \varepsilon^2] \subset B(0,\varepsilon) \subset [-\varepsilon, \varepsilon] \times [-\varepsilon^2, \varepsilon^2].$$

Similar estimates hold around (x_0, y_0) when $x_0 = 0$, but not around regular points, when $x_0 \neq 0$. In this case, there are no dilations centered at (x_0, y_0), and one has only local estimates: balls centered at regular points $p = (x,y)$, that is $x \neq 0$, have the overall form

$$[-\varepsilon, \varepsilon] \times [-\varepsilon, \varepsilon]$$

for small ε, since the metric is Riemannian near those points.

3.2. The Heisenberg group

Consider now $\mathbb{R}^3$, with the sub-Riemannian metric defined by

$$X_1 = \begin{pmatrix} 1 \\ 0 \\ 0 \end{pmatrix}, \quad X_2 = \begin{pmatrix} 0 \\ 1 \\ x \end{pmatrix}.$$

The vector fields X_1, X_2 and

$$[X_1, X_2] = \begin{pmatrix} 0 \\ 0 \\ 1 \end{pmatrix}$$

span $\mathbb{R}^3$ everywhere. Here, $\mathbb{R}^3$ can be identified with the Heisenberg group H_3, so every point can be reached from any other point. Setting $X_3 = [X_1, X_2]$, we see that

$$[X_1, X_3] = [X_2, X_3] = 0.$$

Therefore, the Lie algebra generated by X_1 and X_2 is isomorphic to the Heisenberg Lie algebra $\mathfrak{h}_3$. We can actually identify $\mathbb{R}^3$ with the Heisenberg group H_3 by letting (x, y, z) map to $e^{zX_3}e^{yX_2}e^{xX_1}$. The product operation on $\mathbb{R}^3$ is given by

$$(x, y, z)(x', y', z') = (x + x', y + y', z + z' + xy'). \tag{14}$$

In this picture, the distance defined by X_1 and X_2 becomes left-invariant, i.e., we have $d(gg', gg'') = d(g', g'')$. These statements may be easily proved by using the identity $e^a e^b = e^b e^a e^{[a,b]}$, holding in H_3.

Dilations and distance estimates. Here also there exists a one-parameter group of dilations

$$\delta_\lambda : (x, y, z) \mapsto (\lambda x, \lambda y, \lambda^2 z),$$

so, as in the Grušin case, one can prove estimates of the form

$$C\big(|x| + |y| + |z|^{1/2}\big) \leq d\big(0, (x, y, z)\big) \leq C'\big(|x| + |y| + |z|^{1/2}\big). \tag{15}$$

On the set $|x| + |y| + |z|^{1/2} = 1$ the function $d\big(0, (x, y, z)\big)$ is positive and finite. Since the set is compact and d is continuous on $\mathbb{R}^3$, there exists positive, finite constants C, C' such that $C \leq d\big(0, (x, y, z)\big) \leq C'$ for $|x| + |y| + |z|^{1/2} = 1$. Using dilations, we get (15).

It follows that balls $B(0, \varepsilon)$ look roughly like

$$[-\varepsilon, \varepsilon] \times [-\varepsilon, \varepsilon] \times [-\varepsilon^2, \varepsilon^2].$$

Exact distance estimates. We can give precise bounds for d. First, from the formula

$$(x, y, z) = (0,0,0)e^{-z^{1/2}X_2}e^{-z^{1/2}X_1}e^{z^{1/2}X_2}e^{z^{1/2}X_1}e^{yX_2}e^{xX_1}$$

for $z \geq 0$, and from a similar formula for $z \leq 0$, one constructs a concatenation of integral curves of X_1 and X_2 of total length $|x| + |y| + 4|z|^{1/2}$, leading from the origin to (x, y, z). This gives an upper bound for d. To get a lower bound, we observe that finite length paths starting at the origin are obtained by integrating the system $\dot{p} = u_1(t)X_1 + u_2(t)X_2$, $p(0) = 0$, that is,

$$\begin{cases} \dot{x} = u_1(t) \\ \dot{y} = u_2(t) \\ \dot{z} = xu_2(t) \end{cases} \qquad \begin{cases} x(0) = 0 \\ y(0) = 0 \\ z(0) = 0. \end{cases}$$

Integrating gives

$$x(T) = \int_0^T u_1(t)\, dt, \quad y(T) = \int_0^T u_2(t)\, dt, \quad z(T) = \int_0^T \left(\int_0^t u_1(\tau)\, d\tau\right) u_2(t)\, dt.$$

If we choose controls such that $u_1(t)^2 + u_2(t)^2 = 1$, we obtain the estimates

$$|x(T)| \leq T, \quad |y(T)| \leq T, \quad |z(T)| \leq T^2$$

Since $d(0, p)$ is the infimum of T such that there exists a path with velocity 1, parameterized by $[0, T]$, and joining 0 to p, it follows that

$$|x| \leq d\big(0, (x, y, z)\big), \quad |y| \leq d\big(0, (x, y, z)\big), \quad |z| \leq d\big(0, (x, y, z)\big)^2.$$

Summing up, we obtain

$$\tfrac{1}{3}\big(|x| + |y| + |z|^{1/2}\big) \leq d\big(0, (x, y, z)\big) \leq 4\big(|x| + |y| + |z|^{1/2}\big). \qquad (16)$$

Observe that, because of group invariance, all points of H_3 play the same role. So, every point of H_3 is the center of a 1-parameter group of dilations. Estimates similar to (16) hold for $d\big((x, y, z), (x', y', z')\big)$. See §7, Eq. (52).

3.3. The Heisenberg group using exponential coordinates

There is an alternative presentation of H_3: One may use the coordinates given by the exponential mapping

$$(x, y, z) \mapsto \exp(xX_1 + yX_2 + z[X_1, X_2])$$

(canonical coordinates of the first kind). In these coordinates, vector fields X_1 and X_2 read as

$$X_1 = \begin{pmatrix} 1 \\ 0 \\ -\frac{x}{2} \end{pmatrix}, \quad X_2 = \begin{pmatrix} 0 \\ 1 \\ \frac{x}{2} \end{pmatrix}.$$

The group law can be computed using Campbell-Hausdorff formula, which in the case of H_3 is simply $e^a e^b = e^{a+b+\frac{1}{2}[a,b]}$. We get

$$(x, y, z)(x'y'z') = \left(x + x', y + y', z + z' + \tfrac{1}{2}(xy' - yx')\right).$$

The dilations keep the same form

$$\delta_\lambda : (x, y, z) \mapsto (\lambda x, \lambda y, \lambda^2 z)$$

than in the former system of coordinates. So, estimates of the form

$$C\left(|x| + |y| + |z|^{1/2}\right) \leq d\left(0, (x, y, z)\right) \leq C'\left(|x| + |y| + |z|^{1/2}\right). \qquad (17)$$

still holds.

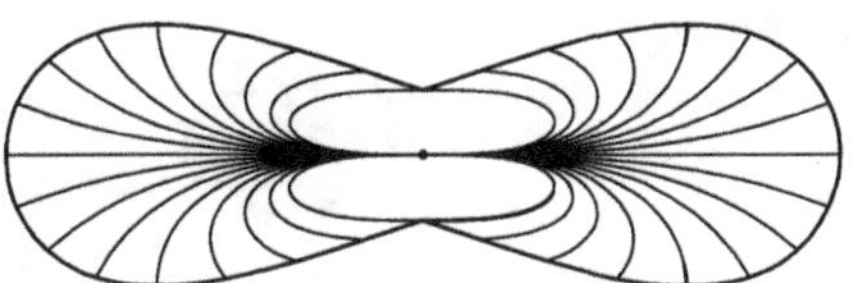

Fig. 2: The Grušin ball and some geodesics

3.4. Geodesics of the Grušin plane and of the Heisenberg group

Minimizing geodesics can be computed in both examples. For the Grušin plane, it's only Riemannian geometry. In the Heisenberg case, the problem is equivalent to the classical isoperimetric problem in the plane, as observed, in substance, by Gaveau [10,11].

Actually, the Pontrjagin Maximum Principle allows to show that, in some class of sub-Riemannian manifolds, to which our examples belong, the geodesics—constant velocity *locally* minimizing curves—are the projections of bicharacteristic curves in T^*M (see the papers of Montgomery and Sussmann in this volume).

Namely, geodesics originating at x_0 are the projections of solutions of Hamilton's equations

$$\dot{x} = \frac{1}{2}\frac{\partial a}{\partial \xi}, \quad \dot{\xi} = -\frac{1}{2}\frac{\partial a}{\partial x}, \quad x(0) = x_0, \quad \xi(0) = \xi_0. \qquad (18)$$

where $a = a(x, \xi) = \sum_{i=1}^{m} \langle X_i(x), \xi \rangle^2$. Moreover $a(x, \xi) = \|\dot{x}\|^2$ is constant along solutions of (18). Thus, the set of geodesics issued from a given point x_0 is parametrized by $\xi \in T^*_{x_0} M$, and the set of geodesics with velocity 1 is parametrized by the "cylinder" $a(x_0, \xi) = 1$ in $T^*_{x_0} M$.

So, in the Grušin case, geodesics starting at 0 depend on a vector $(a, b) \in \mathbb{R}^2$. They are given by

$$x(t) = \frac{a}{b}\sin bt, \quad y(t) = \frac{a}{b}\left(\frac{t}{2} - \frac{\sin 2bt}{4b}\right).$$

The velocity is equal to $|a|$.

The sphere of radius 1 and center 0 is the set of end-points of geodesics starting at 0, defined on $[0, 1]$, having velocity 1 (which imposes $a = \pm 1$) and which are minimizing between $t = 0$ and $t = 1$. The last condition imposes the restriction $b \leq \pi$.

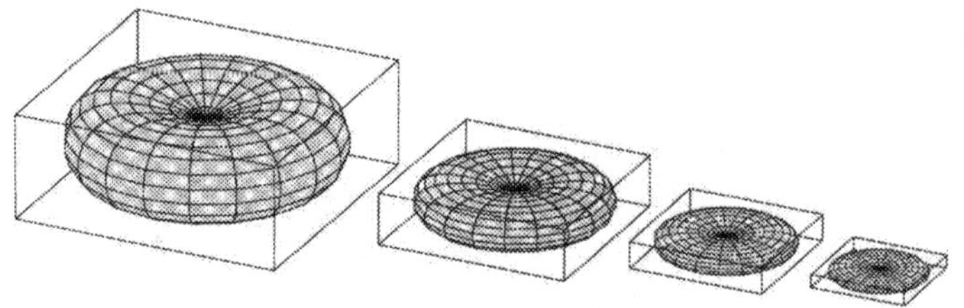

Fig. 3: Balls of different sizes in the Heisenberg group

To deal with the Heisenberg group, we use exponential coordinates. Geodesics starting at 0 depend on $(a, b, c) \in \mathbb{R}^3$ and are given by

$$x = \frac{a\sin ct - b(1 - \cos ct)}{c}, \quad y = \frac{b\sin ct + a(1 - \cos ct)}{c}, \quad z = \frac{(a^2 + b^2)(ct - \sin ct)}{2c^2}.$$

They have velocity 1 if $a^2 + b^2 = 1$, and they are minimizing over $[0, 1]$ if $|b| \leq 2\pi$. The sphere of radius 1 and center 0 is thus the image of a cylinder.

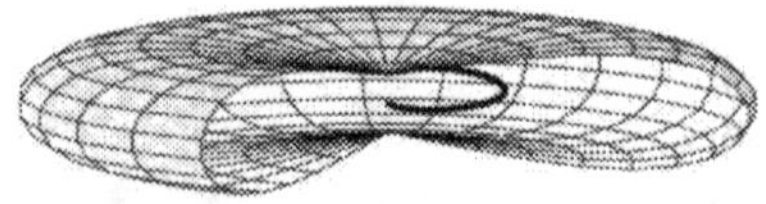

Fig. 4: The ball of radius 1 and a geodesic

3.5. Coverings by ε-balls and Hausdorff dimension

Suppose $M = H_3$. From the form of dilations δ_λ and the group invariance, we deduce

$$\operatorname{Vol} B(p, \varepsilon) = \varepsilon^4 \operatorname{Vol} B(0, 1), \tag{19}$$

for any p in H_3, where Vol denotes the usual volume (Lebesgue measure) in $\mathbb{R}^3$. Observe that Vol is a Haar measure for H_3, i.e., Vol is invariant by left translations (and, as it turns out, by right translations also). Denote by $N(K, \varepsilon)$ the minimal number of balls of radius ε needed to cover a given compact subset K of M, with non-empty interior. As a consequence of (19), we have

$$N(K, \varepsilon) \asymp \varepsilon^{-4}.$$

The proof goes along the following lines: It is easy to give a lower bound for $N(K, \varepsilon)$, just by volume computations. To get the upper bound, consider a maximal family B_i, $i = 1, \ldots, \nu$ of disjoint balls of radius $\varepsilon/2$ contained in K. We have

$$\nu \operatorname{Vol} B(0, 1)(\varepsilon/2)^4 \le \operatorname{Vol} K.$$

Now the balls with same centers as the B_i's and radius ε make up a covering of K, otherwise (B_i) would not be maximal. So we have

$$N(K, \varepsilon) \le \nu \le \frac{16 \operatorname{Vol} K}{\operatorname{Vol} B(0, 1)} \varepsilon^{-4}.$$

For the Grušin plane G_2, things are not so simple. The area (Lebesgue measure) of balls $B(p, \varepsilon)$ may take any real value $\ge C\varepsilon^3$, where $C = \operatorname{Vol} B(0, 1)$, and area arguments cannot help in counting elements in coverings. One could think of using the Riemannian volume $dx\, dy/|x|$, which is actually the Hausdorff measure of G_2, but it assigns infinite measure to open sets meeting the singular axis.

Consider yet a compact set K in G_2, with non-empty interior. When K does not meet the singular line, we have of course

$$N(K, \varepsilon) \asymp \varepsilon^{-2}. \tag{20}$$

But when K is a compact set whose interior meets the singular line $x = 0$, then $N(K, \varepsilon)$ satisfies the estimate

$$N(K, \varepsilon) \asymp \varepsilon^{-2} \log \varepsilon^{-1}. \tag{21}$$

To show this, it suffices to deal with $K = [0, 1] \times [0, 1]$. Fix $\varepsilon = 1/N$ and, for $i = 0, 1, \ldots, 2N - 1$, consider the vertical bands $B_i = [i\varepsilon/2, (i +$

1)$\varepsilon/2]\times[0,1]$. For $i \geq 1$, each band has width $\varepsilon/2$ and height of the order of $2/(i\varepsilon)$ (for the sub-Riemannian distance d). So one needs roughly $4/(i\varepsilon^2) \asymp N^2/i$ balls of radius ε to get a covering of B_i. By putting together these coverings, adding perhaps N^2 balls to cover B_0, one gets a covering of K using

$$N^2 + N^2 \sum_{i=1}^{2N-1} \frac{1}{i} \asymp N^2 \log N$$

balls of radius ε. This covering is roughly minimal, whence (21).

This shows that the Hausdorff dimension of G_2 is 2, but asymptotic bounds such as (21) have a much more precise content. For example (21) is needed to estimate the growth of eigenvalues of the operator $L = X_1^2 + X_2^2$.

4. Privileged coordinates

In all the sequel, we will fix a manifold M, of dimension n, a system of vector fields $X_1, \ldots, X_m$ on M, and a point p of M. We will suppose that $X_1, \ldots, X_m$ verify Chow's condition. We will denote by d the distance defined on M by means of vector fields $X_1, \ldots, X_m$.

4.1. Regular and singular points

Let $\mathcal{L}^1 = \mathcal{L}^1(X_1, \ldots, X_k)$ be the set of linear combinations, with smooth coefficients, of the vector fields $X_1, \ldots, X_k$. We define recursively

$$\mathcal{L}^s = \mathcal{L}^{s-1} + [\mathcal{L}^1, \mathcal{L}^{s-1}], \tag{22}$$

so that $\mathcal{L}^s$ is generated by the vector fields

$$X_\alpha = [X_{\alpha_1}, [X_{\alpha_2}, \ldots, [X_{\alpha_{\ell-1}}, X_{\alpha_\ell}] \ldots]]$$

with $1 \leq \ell \leq s$. Observe that, due to Jacobi identity, we have $[\mathcal{L}^i, \mathcal{L}^j] \subset \mathcal{L}^{i+j}$.

We will denote by $L^s(p)$ the subspace of T_pM which consists of the values $Y(p)$ taken, at the point p, by the vector fields Y belonging to $\mathcal{L}^s$. In other words, $L^s(p)$ is the subspace of T_pM spanned by values at p of the brackets of length $\leq s$ of vector fields $X_1, \ldots, X_m$.

Chow's condition, which we supposed to hold, states that for each point $p \in M$, there is a smallest integer $r = r(p)$ such that $L^{r(p)}(p) = T_pM$.

This integer is called the *degree of nonholonomy* at p. Notice that r is an upper continuous function, that is $r(q) \leq r(p)$ for q near p. For each point $p \in M$, there is in fact an increasing sequence of vector subspaces, or flag:

$$\{0\} = L^0(p) \subset L^1(p) \subset \cdots \subset L^s(p) \subset \cdots \subset L^{r(p)}(p) = T_pM.$$

Definition 4.9. We say that p is a *regular point* if the integers $n_s(q) = \dim L^s(q)$ ($s = 1, 2, \ldots$) remain constant for q in some neighbourhood of p. Otherwise we say that p is a *singular point*.

For example in the Grušin plane, the points on the line $x = 0$ are singular, while the other points in the plane are regular. One can also give examples of systems of vector fields for which $\dim L^1(q)$ remains constant, but having yet singular points. This is the case for the system defined on $\mathbb{R}^3$ by

$$X_1 = \begin{pmatrix} 1 \\ 0 \\ 0 \end{pmatrix}, \quad X_2 = \begin{pmatrix} 0 \\ 1 \\ x^2 \end{pmatrix}.$$

Since

$$[X_1, X_2] = \begin{pmatrix} 0 \\ 0 \\ 2x \end{pmatrix},$$

we have $\dim L^1(x, y, z) = 2$ everywhere, while $\dim L^2(x, y, z)$ is equal to 3 if $x \neq 0$, and to 2 if $x = 0$. So, all points with $x = 0$ are singular.

In the Heisenberg group, all points are regular, owing to group invariance.

Some further observations are in order. At a regular point p, we have

$$0 < n_1(p) < n_2(p) < \cdots < n_{r(p)}(p) = n. \tag{23}$$

In other words, the sequence $\dim L^s(p)$, $s = 0, 1, 2, \ldots, r(p)$ is strictly increasing. The proof goes by noticing first that, if $L^s(p) = L^{s+1}(p)$, we also have $L^s(q) = L^{s+1}(q)$ for q near p. Consider then vector fields $Y_1, \ldots, Y_{n_s}$, obtained as brackets of order $\leq s$ of the X_i, and forming a basis of the (constant rank) distribution L^s. We have $[X_i, Y_j] \in L^{s+1}(q) = L^s(q)$ at all neighbour points, so one may write $[X_i, Y_j](q) = \sum f_{ijk}(q) Y_k(q)$, with smooth coefficients $f_{ijk}(q)$. It follows that $L^s(q) = L^{s+1}(q) = L^{s+2}(q) = \cdots$

Besides, regular points form an open dense set in M, and, in the analytic case, if M is connected, the sequence (23) is the same for all regular points.

4.2. Distance estimates and privileged coordinates

Now, fix a point p in M, regular or singular. We set $n_s = n_s(p) = \dim L^s(p)$, $s = 0, 1, \ldots, r$.

Consider a system of coordinates centered at p, such that the differentials $dy_1, \ldots, dy_n$ form a basis of $T_p^* M$ adapted to the flag

$$\{0\} = L^0(p) \subset L^1(p) \subset \cdots \subset L^s(p) \subset \cdots \subset L^r(p) = T_p M. \qquad (24)$$

Such a coordinate system will be said *linearly adapted* at p.

The estimates we have proved for the sub-Riemannian distance in the Grušin and Heisenberg examples can be generalized, as local estimates, to all cases where $r = 2$. Using linearly adapted coordinates, and setting $n_1 = \dim L^1(p)$, one can prove without much difficulty that

$$d\left(0, (y_1, \ldots, y_n)\right) \asymp |y_1| + \cdots + |y_{n_1}| + |y_{n_1+1}|^{1/2} + \cdots + |y_n|^{1/2}, \qquad (25)$$

for y near $p = (0, \ldots, 0)$. Coordinates $y_1, \ldots, y_{n_1}$ are said to be of weight 1, and coordinates $y_{n_1+1}, \ldots, y_n$ are said to be of weight 2. We shall not give a proof of (25) now, and we will content ourselves with the examples in §3, since (25) will be superseded by more general statements. (See Theorem 7.34, taking in account the remark following Theorem 4.15.)

To define the notion of weight in the general case, observe that the structure of a flag such that

$$\{0\} = V_0 \subset V_1 \subset \cdots \subset V_r = V$$

may be described by two non-decreasing sequences of integers. The first one is the sequence

$$0 = n_0 \leq n_1 \leq n_2 \leq \cdots \leq n_r = n$$

of dimensions of subspaces which form the flag. The second one is the sequence

$$w_1 \leq w_2 \leq \cdots \leq w_n$$

which is best understood by using a basis $v_1, \ldots, v_n$ adapted to the flag. One sets $w_j = s$ if v_j belongs to V_s and do not belong to V_{s-1}.

For the flag (24) we define $w_1 \leq w_2 \leq \cdots \leq w_n$ by the same recipe, just replacing V_s by $L^s(p)$. Notice we always have $w_1 = 1$. Otherwise all the

X_i would vanish at p, which would imply that all their brackets vanish at p, in contradiction with Chow's condition. Moreover $w_n = r$, the degree of nonholonomy at p.

Thus the j-th vector of an adapted basis of $T_p M$ can be written as a linear combination of (values at p of) brackets of order w_j of $X_1, \ldots, X_m$, but it cannot be obtained from lesser order brackets. If $y_1, \ldots, y_n$ form a system of linearly adapted coordinates at p, their differentials $dy_1, \ldots, dy_n$ form the dual basis of such an adapted basis of $T_p M$. The integer w_j can be characterized by the fact that dy_j vanish on $L^{w_j-1}(p)$ and does not vanish identically on $L^{w_j}(p)$.

Definition 4.10. We shall say that w_j is the *weight* of coordinate y_j.

With this definition, the proper generalization of (25) would be

$$d\left(0, (y_1, \ldots, y_n)\right) \asymp |y_1|^{1/w_1} + \cdots + |y_n|^{1/w_n}. \tag{26}$$

It turns out that this estimate is generically false as soon as $r \geq 3$ for linearly adapted coordinates. This is the motivation for introducing privileged coordinates.

A simple counter-example is given by the system

$$X_1 = \begin{pmatrix} 1 \\ 0 \\ 0 \end{pmatrix}, \quad X_2 = \begin{pmatrix} 0 \\ 1 \\ x^2 + y \end{pmatrix} \tag{27}$$

on $\mathbb{R}^3$. We have

$$L^1(0) = L^2(0) = \mathbb{R}^2 \times \{0\}, \quad L^3(0) = \mathbb{R}^3,$$

so that $y_1 = x$, $y_2 = y$, $y_3 = z$ are linearly adapted coordinates at 0 and have weight 1, 1 and 3. In this case, estimate (26) cannot be true. Indeed, this would imply

$$d(0, (x, y, z)) \geq \mathrm{const}\left(|x| + |y| + |z|^{1/3}\right),$$

whence

$$\left|z\left(e^{tX_2}(0)\right)\right| \leq \mathrm{const}\, |t|^3$$

(since $d(0, e^{tX_2}(0)) \leq |t|$), but this is impossible since

$$\left. \frac{d^2}{dt^2} z\left(e^{tX_2}(0)\right)\right|_{t=0} = (X_2^2 z)(0) = (X_2(x^2 + y))(0) = 1.$$

However a slight nonlinear change of coordinates allows for (26) to hold. It is sufficient to replace y_1, y_2, y_3 by $z_1 = x$, $z_2 = y$, $z_3 = z - y^2/2$.

In the above example, the point under consideration is singular, but one can give similar examples with regular p in dimension ≥ 4.

To formulate conditions on coordinate systems under which estimates like (26) may hold, we introduce some definitions.

Definition 4.11. Call $X_1 f, \ldots, X_m f$ the *nonholonomic partial derivatives of order* 1 *of* f (with respect to the system $(X_1, \ldots, X_m)$).

If the manifold under study was $M = \mathbb{R}^n$ with its Euclidean metric, one would have $m = n$, and one could take $X_1 = \partial_{x_1}, \ldots, X_n = \partial_{x_n}$. The nonholonomic derivations will thus play a role analogous to that of $\partial_{x_1}, \ldots, \partial_{x_n}$ on $\mathbb{R}^n$

Call further $X_i X_j f$, $X_i X_j X_k f$, $\ldots$, the nonholonomic derivatives of order 2, 3, $\ldots$ of f.

Proposition 4.10. *Let s be a non-negative integer. For a smooth function f defined near p, the following conditions are equivalent:*
(i) One has $f(q) = O\left(d(p,q)^s\right)$ for q near p;

(ii) The nonholonomic derivatives of order $\leq s - 1$ of f all vanish at p

Proof. (i) $\Rightarrow$ (ii) . We have

$$\left(X_{i_1} \ldots X_{i_k} f\right)(p) = \left.\frac{\partial^k}{\partial t_1 \ldots \partial t_k} f(p\, e^{t_1 X_{i_1}} \ldots e^{t_k X_{i_k}})\right|_{t=0}.$$

Since

$$d(p, p\, e^{t_1 X_{i_1}} \ldots e^{t_k X_{i_k}}) \leq |t_1| + \cdots + |t_k|,$$

we have

$$f(p\, e^{t_1 X_{i_1}} \ldots e^{t_k X_{i_k}}) = O\left((|t_1| + \cdots + |t_k|)^s\right).$$

Therefore

$$\left(X_{i_1} \ldots X_{i_k} f\right)(p) = 0$$

if $k \leq s - 1$.

(ii) $\Rightarrow$ (i) . We argue by induction on s. For $s = 0$, there is nothing to prove. So let $s > 0$, and assume that

$$\left(X_{i_1} \ldots X_{i_k} f\right)(p) = 0 \tag{28}$$

whenever $k \leq s - 1$. We have also

$$\left(X_{i_1} \ldots X_{i_{k-1}}\right)\left(X_i f\right)(p) = 0$$

for any choices of i, and of $i_1, \ldots, i_{k-1}$. So, applying the induction hypothesis to $X_i f$ $(1 \leq i \leq m)$, we see that there exist $\varepsilon > 0$ and $C > 0$ such that

$$\left(X_i f\right)(q) \leq C \, d(p, q)^{s-1}$$

if $d(p, q) \leq \varepsilon$. Choose any $q \in B(p, \varepsilon)$. Let $T = d(p, q)$ and let $\gamma : [0, T] \to M$ a minimizing geodesic joining p to q, with velocity 1. Denote by u_i $(1 \leq i \leq m)$ the corresponding control functions. We have

$$\frac{d}{dt} f\left(\gamma(t)\right) = \sum_{i=0}^{m} \left(X_i f\right)\left(\gamma(t)\right) u_i(t) \quad \text{a.e.},$$

with $\sum_{i=1}^{m} u_i(t)^2 = 1$ a.e. It follows that

$$\left|\frac{d}{dt} f\left(\gamma(t)\right)\right| \leq mC \, d\left(p, \gamma(t)\right)^{s-1} = mCt^{s-1}.$$

By (28), applied for $k = 0$, we have $f(0) = 0$, so by integrating we obtain

$$\left|f\left(\gamma(t)\right)\right| \leq \frac{mC}{s} t^s$$

and

$$|f(q)| \leq \frac{mC}{s} \, d(p, q)^s,$$

proving thus the proposition. ∎

Definition 4.12. If Condition (i), or (ii), holds, we say that f is *of order* $\geq s$ at p. We say that f is *of order s at p* if it is of order $\geq s$, and not of order $\geq s + 1$.

Definition 4.13. We call *system of privileged coordinates* a system of local coordinates $z_1, \ldots, z_n$ centered at p such that:

(i) $z_1, \ldots, z_n$ are linearly adapted at p;

(ii) The order of z_j at p is exactly w_j.

If we suppose only that $z_1, \ldots, z_n$ are linearly adapted, then the order of z_j is always $\leq w_j$: Fix j, and set $s = w_j$. For some choice of of the indices $i_1, i_2, \ldots, i_s$, we have

$$\langle [X_{i_1}, [X_{i_2}, \ldots [X_{i_{s-1}}, X_{i_s}] \ldots]], dz_j \rangle \neq 0.$$

Now,

$$\langle [X_{i_1}, [X_{i_2}, \ldots [X_{i_{s-1}}, X_{i_s}] \ldots]], dz_j \rangle = \left([X_{i_1}, [X_{i_2}, \ldots [X_{i_{s-1}}, X_{i_s}] \ldots]] \cdot z_j \right)(p)$$

is a linear combination of nonholonomic derivatives of order s of z_j. One of this derivatives must be non zero. So z_j cannot be of order $\geq s + 1$, and must be of order $\leq s = w_j$. But it may well happen that the order of z_j be $< w_j$: for the system (27), the order of coordinate $y_3 = z$ at 0 is 2, while $w_3 = 3$.

Our goal in the sequel is to show that the estimate

$$d\left(0, (y_1, \ldots, y_n)\right) \asymp |y_1|^{1/w_1} + \cdots + |y_n|^{1/w_n}.$$

holds near p if and only if $y_1, \ldots, y_n$ form a system of privileged coordinates at p (see Theorem 7.34).

4.3. Construction of privileged coordinates

To prove in an effective way the existence of privileged coordinates, we first choose vector fields $Y_1, \ldots, Y_n$ whose values at p form a basis of T_pM:

First, choose among $X_1, \ldots, X_m$ a number n_1 of vector fields such that their values form a basis of $L^1(p)$. Call them $Y_1, \ldots, Y_{n_1}$. Then for each s ($s = 2, \ldots, r$) choose vector fields of the form $[X_{i_1}, [X_{i_2}, \ldots [X_{i_{s-1}}, X_{i_s}] \ldots]]$ which form a basis of $L^s(p) \bmod L^{s-1}(p)$, and call them $Y_{n_{s-1}+1}, \ldots, Y_{n_s}$.

We obtain in this way a sequence of vector fields $Y_1, \ldots, Y_n$ whose values at p—and at points near p—form a basis of the tangent space. At p—but not at neighbour points, if p is singular—this basis is adapted to the flag (24).

Lemma 4.11. *Any vector field $Y \in \mathcal{L}^s(X_1, \ldots, X_m)$ can be written near p as*

$$Y = \sum_{j=1}^{n} c_j Y_j, \tag{29}$$

where each c_j is a smooth function, of order $\geq w_j - s$ at point p. In particular, $c_j(p) = 0$ if $w_j > s$.

If p is a regular point, we have

$$Y = \sum_{j=1}^{n_s} c_j Y_j, \tag{30}$$

that is, (29) holds with c_j identically zero for $w_j > s$.

Note that having order $\geq w_j - s$ at p is a restrictive condition only when $w_j > s$.

Proof. Let $Y \in \mathcal{L}^s(X_1, \ldots, X_m)$. For all q in a neighbourhood of p, tangent vectors $Y_1(q), \ldots, Y_n(q)$ form a basis of $T_q M$, depending smoothly on q. Whence (29), with smooth c_j.

In the case of a regular p, tangent vectors $Y_1(q), \ldots, Y_{n_s}(q)$ are independent for q near p, so they form a basis of $L^s(q)$ depending smoothly on q. On the other side, if a vector field Y belongs to $\mathcal{L}^s$, we have $Y(q) \in L^s(q)$ for all q near p. Whence (30). It can be noted that, conversely, a vector field Y such that $Y(q) \in L^s(q)$ for all q near p is in $\mathcal{L}^s$ (regular case only).

What we have left to prove for singular p amounts to the following: For all $k \geq 0$, the function c_j is of order $\geq k$ at p whenever $w_j \geq s + k$.

We use induction on k. Our assertion is trivial for $k = 0$. Assume that it holds for $0, 1, \ldots, k$. We have

$$[X_{i_1}, [\cdots, [X_{i_k}, Y] \cdots]](p) = \sum_{j=1}^{n} (X_{i_1} \ldots X_{i_k} c_j)(p) Y_j(p) + \tag{31}$$
$$\sum_{j=1}^{n} \sum_{\alpha, \beta} (X_{\alpha_1} \ldots X_{\alpha_\ell} c_j)(p) [X_{\beta_1}, [\cdots, [X_{\beta_{k-\ell}}, Y_j] \cdots]](p),$$

where the last sum is taken on all partitions of the sequence $(i_1, \ldots, i_k)$ into subsequences $\alpha = (\alpha_1, \ldots, \alpha_\ell)$ and $\beta = (\beta_1, \ldots, \beta_{k-\ell})$ such that $0 \leq \ell \leq k - 1$.

In (31), the right hand side is a tangent vector belonging to $L^{s+k}(p)$. Let us write $T_{j\alpha\beta}$ as an abbreviation for

$$\left(X_{\alpha_1} \ldots X_{\alpha_\ell} c_j \right)(p) [X_{\beta_1}, [\cdots, [X_{\beta_{k-l}}, Y_j] \cdots]](p) .$$

If $l \geq w_j - s$, then $T_{j\alpha\beta} \in L^{s+k}(p)$. If, on the contrary, $l < w_j - s$, we have $w_j \geq s + l + 1$. By the induction hypothesis, the function c_j is of order $\geq l + 1$ at p, and the coefficient $\left(X_{\alpha_1} \ldots X_{\alpha_\ell} c_j \right)(p)$ vanishes. Therefore, in any case, we have $T_{j\alpha\beta} \in L^{s+k}(p)$. We conclude that the tangent vector

$$\sum_{j=1}^{n} \left(X_{i_1} \ldots X_{i_k} c_j \right)(p) Y_j(p)$$

belongs to $L^{s+k}(p)$. Now, tangent vectors $Y_j(p)$ with $w_j \geq s + k + 1$ are independent mod L^{s+k}. We get immediately that

$$\left(X_{i_1} \ldots X_{i_k} c_j \right)(p) = 0 \tag{32}$$

whenever $w_j \geq s + k + 1$.

Let us take j such that $w_j \geq s + k + 1$. We know from the induction hypothesis that c_j is of order $\geq k$ at p. Since (32) holds for all choices of indices $i_1, \ldots, i_k$, the function is of order $\geq k+1$, as desired.   ∎

It will be convenient, in the following lemma and in the sequel, to introduce the notation $w(\alpha) = w_1\alpha_1 + \cdots + w_n\alpha_n$,

Lemma 4.12. (i) *Any product $X_{i_1} X_{i_2} \ldots X_{i_s}$, where $i_1, \ldots, i_s$ are integers, can be written as a linear combination of ordered monomials*

$$\sum c_{\alpha_1 \ldots \alpha_n} Y_1^{\alpha_1} \ldots Y_n^{\alpha_n},$$

where the $c_{\alpha_1 \ldots \alpha_n}$ are smooth functions, and $c_{\alpha_1 \ldots \alpha_n}$ is of order $\geq w(\alpha) - s$. In particular, $c_{\alpha_1 \ldots \alpha_n}(p) = 0$ if $w(\alpha) > s$.

If p is a regular point, one may take $c_{\alpha_1 \ldots \alpha_n}(p)$ identically zero for $w(\alpha) > s$.

(ii) *A function f is of order $> s$ at p if and only if*

$$\left(Y_1^{\alpha_1} \ldots Y_n^{\alpha_n} f \right)(p) = 0$$

for all $\alpha = (\alpha_1, \ldots, \alpha_n)$ such that $w(\alpha) \leqslant s$.

Proof. First, note that by Lemma 4.11 with $s = 1$, we have $X_i = \sum_{j=1}^{n} \lambda_{ij} Y_j$, where the λ_{ij} are smooth functions of order $\geq w_j - 1$ at p. One deduces easily that

$$X_{i_1} X_{i_2} \ldots X_{i_s} = \sum \mu_{j_1 j_2 \ldots j_q} Y_{j_1} Y_{j_2} \ldots Y_{j_q},$$

where the sum is taken on sequences $(j_1, \ldots, j_q)$ such that $q \leq s$ and the $\mu_{j_1 j_2 \ldots j_q}$ are smooth functions of order $\geq w_{j_1} + w_{j_2} + \cdots + w_{j_q} - s$ at p. Therefore, it is enough to prove that any product $Y_{j_1} \ldots Y_{j_q}$, with $w_{j_1} + w_{j_2} + \cdots + w_{j_q} = \rho$ can be written as a linear combination

$$\sum c_\alpha Y_1^{\alpha_1} \ldots Y_n^{\alpha_n},$$

where each c_α is smooth and of order $\geq w(\alpha) - \rho$ at p.

We will argue by double induction, first on q, next on the number ι of inversions in the sequence $(j_1, \ldots, j_q)$. Let us take two indices such that $j_k > j_{k+1}$ (if they don't exist, there is nothing to prove). Using Lemma

4.11, we obtain

$$Y_{j_k} Y_{j_{k+1}} = Y_{j_{k+1}} Y_{j_k} + [Y_{j_k}, Y_{j_{k+1}}]$$

$$= Y_{j_{k+1}} Y_{j_k} + \sum_{j=1}^{n} \nu_j Y_j,$$

where the order of ν_j at p is $\geq w_j - w_{j_k} - w_{j_{k+1}}$. Now, we replace $Y_{j_k} Y_{j_{k+1}}$ in the product $Y_{j_1} \ldots Y_{j_q}$ by the right hand side in the last equality. This yields

$$Y_{j_1} \ldots Y_{j_q} = Y_{j_1} \ldots Y_{j_{k+1}} Y_{j_k} \ldots Y_{j_q} + \sum_{j=1}^{n} Y_{j_1} \ldots Y_{j_{k-1}} (\nu_j Y_j) Y_{j_{k+2}} \ldots Y_{j_q},$$

$$= Y_{j_1} \ldots Y_{j_{k+1}} Y_{j_k} \ldots Y_{j_q} +$$

$$\sum_{j=1}^{n} \sum_{\lambda,\mu} (Y_{\lambda_1} \ldots Y_{\lambda_\ell} \nu_j) Y_{\mu_1} \ldots Y_{\mu_{k-1}}{}_\ell Y_{j_{k+2}} \ldots Y_{j_q},$$

where in the last sum (λ, μ) runs on the set of all partitions of the sequence $(j_1, \ldots, j_{k-1})$ into subsequences $\lambda = (\lambda_1, \ldots, \lambda_\ell)$ and $\mu = (\mu_1, \ldots, \mu_{k-1-\ell})$.

To compute the order of $Y_{\lambda_1} \ldots Y_{\lambda_\ell} \nu_j$ at point p, observe that $X_i \varphi$ is of order $\geq s - 1$ at p if φ is of order $\geq s$, so $Y_j \varphi$ is of order $\geq s - w_j$ at p, and a factor like $Y_{\lambda_1} \ldots Y_{\lambda_\ell} \nu_j$ is of order $\geq w_j - w_{j_k} - w_{j_{k+1}} - w_{\lambda_1} - \cdots - w_{\lambda_\ell}$.

Applying the induction hypothesis on ι for the term $Y_{j_1} \ldots Y_{j_{k+1}} Y_{j_k} \cdots Y_{j_q}$, and the induction hypothesis on q for the terms in the last sum, we see that $Y_{j_1} \ldots Y_{j_q}$ is a linear combination with smooth coefficients of $Y_1^{\alpha_1} \ldots Y_n^{\alpha_n}$, as was to be shown, the coefficients c_α having the desired orders.

The case of a regular point is treated along the same lines, noticing only that one may write in this case

$$[Y_{j_k}, Y_{j_{k+1}}] = \sum_{\{j \,|\, w_j \leq w_{j_k} + w_{j_{k+1}}\}} \nu_j Y_j.$$

The second part of the lemma is an immediate consequence of (i), which proves the sufficiency, and of the fact that products $Y_1^{\alpha_1} \ldots Y_n^{\alpha_n}$ are themselves noncommutative polynomials of the $X_i's$ of degree $\leq w_1 \alpha_1 + \cdots + w_n \alpha_n$, which proves the necessity. $\blacksquare$

I owe the idea of the proof of Lemma 4.12 to J.-J. Risler [4].

Choose now any system of coordinates $y_1, \ldots, y_n$ such that

$$\langle y_j, Y_k \rangle = \delta_{jk} \qquad \text{at } p.$$

These coordinates are linearly adapted at p.

Lemma 4.13. *Let $P(y)$ be a homogeneous polynomial of degree q. Then, we have*

$$\left(Y_1^{\alpha_1} \ldots Y_n^{\alpha_n} P \right)(0) = \left(\partial_{y_1}^{\alpha_1} \ldots \partial_{y_n}^{\alpha_n} P \right)(0). \tag{33}$$

if $q = \alpha_1 + \cdots + \alpha_n$. If $q > \alpha_1 + \cdots + \alpha_n$, both sides are 0.

Proof. The lemma will be proved if we show that

$$Y_1^{\alpha_1} \ldots Y_n^{\alpha_n} = \partial_{y_1}^{\alpha_1} \ldots \partial_{y_n}^{\alpha_n} + \sum_{\{\beta \,|\, \beta_1 \leq \alpha_1, \ldots, \beta_n \leq \alpha_n, \beta \neq \alpha\}} a_{\alpha\beta}(y) \partial_{y_1}^{\beta_1} \ldots \partial_{y_n}^{\beta_n} + Q, \tag{34}$$

where $a_{\alpha\beta}(0) = 0$ for all β occurring in the sum, and Q is a differential operator of (usual) order $< \alpha_1 + \cdots + \alpha_n$, without constant term. This is proved, first, by noticing that each Y_i can be written as

$$Y_i = \partial_{y_i} + \sum_{j=1}^{n} a_{ij}(y) \partial_{y_j}$$

where $a_{ij}(0) = 0$, $i = 1, \ldots, n$, then, by applying repeatedly the formula

$$\partial_{y_l} Y_i = \partial_{y_l y_i}^2 + \sum_{j=1}^{n} a_{ij}(y) \partial_{y_l y_j}^2 + \sum_{j=1}^{n} (\partial_{y_l} a_{ij})(y) \, \partial_{y_j} \qquad \blacksquare$$

Lemma 4.14. *Let f be a linear form in the variables of weight $> s$, that is*

$$f = a_{n_s+1} y_{n_s+1} + \cdots + a_n y_n.$$

Then there exists a polynomial h in the variables $y_1, \ldots, y_{n_s}$, having only terms of order ≥ 2, such that the function

$$g(y) = h(y_1, \ldots, y_{n_s}) + a_{n_s+1} y_{n_s+1} + \cdots + a_n y_n$$

has local order $\geq s+1$ at point p. Moreover, the polynomial can be chosen of the form

$$h(y_1, \ldots, y_{n_s}) = \sum_{w_1 \alpha_1 + \cdots + w_{n_s} \alpha_{n_s} \leq s} a_{\alpha_1 \ldots \alpha_{n_s}} y_1^{\alpha_1} \ldots y_n^{\alpha_n}$$

and can be obtained by an effective procedure.

Proof. By Lemma 4.12, the function g will have local order $\geq s+1$ at p if

$$\left(Y_1^{\alpha_1} \ldots Y_n^{\alpha_n} h\right)(p) = 0 \tag{35}$$

for all $\alpha = (\alpha_1, \ldots, \alpha_n)$ such that $w_1\alpha_1 + \cdots + w_n\alpha_n \leq s$. Since $w_i > s$ when $i > n_s$, we may content ourselves to ask for (35) with n replaced by n_s.

We shall construct h as a sum $h = h_1 + h_2 + \cdots + h_s$ where h_q is a homogeneous polynomial of degree q in the variables $y_1, \ldots, y_{n_s}$, such that for $q = 1, \ldots, s$, the relation

$$\left(Y_1^{\alpha_1} \ldots Y_{n_s}^{\alpha_{n_s}} (f + h_1 + h_2 + \cdots + h_q)\right)(p) = 0 \tag{36}$$

holds for all $\alpha = (\alpha_1, \ldots, \alpha_{n_s}, 0, \ldots, 0)$ such that $w_1\alpha_1 + \cdots + w_{n_s}\alpha_{n_s} \leqslant s$ and $\alpha_1 + \cdots + \alpha_{n_s} \leq q$. We need precisely that this condition be satisfied for $q = s$.

We proceed by induction on q. We take $h_1 = 0$. The induction hypothesis is then satisfied for $q = 1$, since $(Y_j f)(p) = 0$ for any j with $j \leq n_s$. Suppose $h_2, \ldots, h_{q-1}$ have been found. We have to find h_q such that (36) holds. We know by the induction hypothesis that

$$\left(Y_1^{\alpha_1} \ldots Y_{n_s}^{\alpha_{n_s}} (f + h_2 + \cdots + h_{q-1})\right)(p) = 0$$

and, by Lemma 4.13, that

$$\left(Y_1^{\alpha_1} \ldots Y_{n_s}^{\alpha_{n_s}} (h_q)\right)(p) = 0$$

if $w_1\alpha_1 + \cdots + w_{n_s}\alpha_n \leq s$ and $\alpha_1 + \cdots + \alpha_{n_s} \leq q - 1$.

So, we have only to find h_q such that

$$\left(Y_1^{\alpha_1} \ldots Y_{n_s}^{\alpha_{n_s}} (f + h_2 + \cdots + h_q)\right)(p) = 0,$$

or

$$\left(Y_1^{\alpha_1} \ldots Y_{n_s}^{\alpha_{n_s}} h_q\right)(p) = -\left(Y_1^{\alpha_1} \ldots Y_{n_s}^{\alpha_{n_s}} (f + h_2 + \cdots + h_{q-1})\right)(p)$$

for all α with $w_1\alpha_1 + \cdots + w_{n_s}\alpha_n \leq s$ and $\alpha_1 + \cdots + \alpha_{n_s} = q$ exactly. The problem boils down to the construction of a homogeneous polynomial of degree q having some partial derivatives of degree q specified at the origin. This construction is immediate, using Taylor's formula. ∎

Theorem 4.15. *One can in an effective way, compute for each j ($j = 1, \ldots, n$) a polynomial H_j in the variables $y_1, \ldots, y_{\nu(j)}$, without linear term, nor constant term, such that the functions $z_j = y_j + H_j(y_1, \ldots, y_{\nu(j)})$ form a system of privileged coordinates at p.*

Proof. Apply Lemma 4.14 to $f = y_j$, $s = w_j - 1$, yielding $g = z_j$. The functions z_j obtained in this way vanish at p, and have the same linear parts as the y_j, so they form a system of coordinates around z_j. They have order $\geq w_j$ by Lemma 4.14, and, since $Y_j z_j = Y_j y_j = 1$, they have order $\leq w_j$ in virtue of Lemma 4.12. The theorem is therefore proved. $\blacksquare$

Remark. The coordinates y_j having weight 1 need not be changed. In case $r = 2$, no change at all is needed, the coordinates y_j (with weight 1 and 2) form already a system of privileged coordinates, as it follows immediately from the definition.

Notice that the coordinates $z_1, \ldots, z_n$ supplied by the construction of Theorem 4.15 are given from original coordinates by expressions of the form

$$z_1 = y_1$$
$$z_2 = y_2 + \mathrm{pol}(y_1)$$
$$\cdots$$
$$z_n = y_n + \mathrm{pol}(y_1, \ldots, y_{n-1})$$

where pol denotes a polynomial, without constant or linear term. It is easy to see that the reciprocal change of coordinates has exactly the same form.

Other ways of getting privileged coordinates are to use the mappings

$$(z_1, \ldots, z_n) \mapsto p \exp(z_1 Y_1 + \cdots + z_n Y_n) \quad \text{(compare [12,28])},$$
$$(z_1, \ldots, z_n) \mapsto p \exp(z_n Y_n) \cdots \exp(z_1 Y_1) \quad \text{(compare [19])}.$$

Following the usage in Lie group theory, these coordinates are called *canonical coordinates of the first (resp. second) kind*. We shall not prove here that canonical coordinates of the first or second kind are privileged coordinates, as we will not use them in the sequel. One of the points of this paper is indeed to show that otherwise unspecified privileged coordinates, or privileged coordinates obtained from a simple polynomial change of coordinates, are better suited than canonical coordinates in many kinds of computations.

5. The tangent nilpotent Lie algebra and the algebraic structure of the tangent space

5.1. Computation of the local order of a function

Using privileged coordinates $z_1, \ldots, z_n$, one can compute the order at p of a smooth function f in a purely algebraic way.

Let $w_1, \ldots, w_n$ be the *weights* assigned to $z_1, \ldots, z_n$. We will say that $w(\alpha) = w_1 \alpha_1 + \cdots + w_n \alpha_n$ is the the *weighted degree*, or simply the degree, of the monomial

$$z^\alpha = z_1^{\alpha_1} \ldots z_n^{\alpha_n}.$$

Then the order of z^α at p is $w(\alpha)$, as it appears from Leibnitz's rule. If f has Taylor expansion

$$f(z) \sim \sum_\alpha a_\alpha z^\alpha,$$

then the order of f is the order of the least degree monomial appearing effectively in the expansion.

We define the 1-parameter group of *dilations*

$$\delta_\lambda : (z_1, \ldots, z_n) \mapsto (\lambda^{w_1} z_1, \ldots, \lambda^{w_n} z_n).$$

So, relative to the chosen system of privileged coordinates, we have a notion of homogeneity: a function f is *homogeneous of degree s* if

$$f(\delta_\lambda z) = \lambda^s f(z).$$

For polynomials it is the same as being a sum of monomials of weighted degree s. Observe that a *smooth* function, homogeneous of degree s, is necessarily a polynomial.

We also can also compute the order of f at p as the least integer s such that

$$f(z) = O(\|z\|^s)$$

when z tends to 0. Here, we set

$$\|z\| = |z_1|^{1/w_1} + \cdots + |z_n|^{1/w_n}.$$

The function $\|z\|$ is homogeneous of degree 1. It is called the *pseudo-norm*, and it will be of great use in the sequel. As a matter of fact, it was used by specialists in PDE, long before they discovered the role of sub-Riemannian metrics (see [31]).

5.2. The local order of a differential operator

Definition 5.14. A differential operator P is said to have order $\geq \sigma$ at point p if Pf has order $\geq \sigma + s$ at p whenever f has order $\geq s$. It has order σ at p if it has order $\geq \sigma$ but not $\geq \sigma - 1$.

In privileged coordinates, computing the order of a vector field X at p is just as simple as for a function: use the Taylor expansion

$$X(z) \sim \sum_{\alpha,j} a_{\alpha,j} z^\alpha \partial_{z_j},$$

and do the same as above for f, only assigning to ∂_{z_j} the weight $-w_j$. This yields that all non-zero vector fields have order $\geq -w_n \ (= -r)$ at p.

Relative to the chosen system of privileged coordinates, we also have a notion of homogeneous differential operator: The differential operator P is weighted homogeneous of degree s if

$$\delta_\lambda^* P = \lambda^s P,$$

where the action of δ_λ on differential operators is given by

$$(\delta_\lambda^* P)(\delta_\lambda^* f) = \delta_\lambda^*(Pf),$$

and for a function f, $\delta_\lambda^* f = f \circ \delta_\lambda$. For a vector field X, e.g., this means that X is a finite sum

$$X(z) = \sum_{\alpha,j} a_{\alpha,j} z^\alpha \partial_{z_j},$$

where all the terms, counting ∂_{z_j} with degree $-w_j$, have degree s .

Proposition 5.16. *Suppose the vector fields X and Y have degree k and l, respectively, at p. Then $[X, Y]$ has order $\geq k + l$ at p.*

If X and Y are homogeneous of degree k and l respectively (in the chosen system of privileged coordinates), then $[X, Y]$ is homogeneous of degree $k + l$, or is zero.

Proof. Clear.   ∎

5.3. The nilpotent approximation

Now, the defining vector fields X_i have order ≥ -1 at p. So they can be expanded in a series of homogeneous vector fields of the form

$$X_i = X_i^{(-1)} + X_i^{(0)} + X_i^{(1)} + X_i^{(2)} + \cdots \tag{37}$$

where $X_i^{(s)}$ has degree s. We set

$$\widehat{X}_i = X_i^{(-1)}, \qquad i = 1, \dots, m.$$

Definition 5.15. We shall call the system of vector fields $(\widehat{X}_1, \dots, \widehat{X}_m)$ the canonical *nilpotent homogeneous approximation* of the system $(X_1, \dots, X_m)$ at p.

Various nilpotent approximations have been used in the study of hypoelliptic partial differential equations, and in nonlinear Control Theory, since the works of Rothschild and Stein [28] and Goodman [12] around 1976. Some of them are very close [27], or equivalent [19] to the approximation presented here. However, in these references, it is not very clear in which sense the $\widehat{X}_i$'s do approximate the X_i's.

We consider now on $\mathbb{R}^n$ the sub-Riemannian distance $\widehat{d}$ defined from the system of vector fields $\widehat{X}_1, \dots, \widehat{X}_m$.

Since the vector fields $\widehat{X}_i$ are homogeneous of degree -1, which can be written

$$(\delta_\lambda)^* \widehat{X}_i = \lambda^{-1} \widehat{X}_i,$$

the length of a curve is multiplied by λ under the action of δ_λ; it follows that

$$\widehat{d}(\delta_\lambda x, \delta_\lambda y) = \lambda \widehat{d}(x, y).$$

Proposition 5.17. *The vector fields* $\widehat{X}_i$, $i = 1, \dots, m$, *generate a nilpotent Lie algebra* $\mathrm{Lie}(\widehat{X}_1, \dots, \widehat{X}_m)$, *of step* $r = w_n$. *They satisfy Chow's condition at every point* $x \in \mathbb{R}^n$, *and the distance* $\widehat{d}(x, y)$ *is finite for every* $x, y \in \mathbb{R}^n$.

Proof. To prove that $\mathrm{Lie}(\widehat{X}_1, \dots, \widehat{X}_m)$ is nilpotent, it is enough to say that a bracket of length s of vector fields $\widehat{X}_i$ is homogeneous of degree $-s$, so it must be zero if $s > r$.

Consider now the vector fields $\widehat{Y}_j$ $(i = 1, \dots, n)$ defined from the $\widehat{X}_i$'s by the same formulas which define Y_j $(i = 1, \dots, n)$ from the X_i's. For each j, the vector field $\widehat{Y}_j$ is the homogeneous component of degree $-w_j$ of Y_j, and we have $\widehat{Y}_j(p) = \partial_{z_j}$. Thus Chow's condition is satisfied at p, and, by continuity, near p. Let us observe now that if some point q is accessible from zero, then $\delta_\lambda(q)$ is also accessible from zero. Indeed, suppose that q is the end-point $q = x(T)$ of some controlled path, solution of the differential equation

$$\dot{x} = u_1 X_1(x) + \cdots + u_m X_m(x), \quad x(0) = 0, \quad 0 \le t \le T.$$

Then $\delta_\lambda q$ is the end-point of the solution of

$$\dot{x} = \lambda u_1 X_1(x) + \cdots + \lambda u_m X_m(x), \quad x(0) = 0, \quad 0 \le t \le T$$

and is thus accessible from zero. So, the set of points accessible from zero is invariant under δ_λ. Since it contains a neighbourhood of zero, it consists of all of $\mathbb{R}^n$. This means that $\widehat{d}$ is finite. ∎

The following proposition will be of great importance in the sequel.

Proposition 5.18. *In privileged coordinates, the system*

$$\dot{x} = \sum_{i=1}^{m} u_i \widehat{X}_i(x)$$

takes the following form

$$\dot{z}_j = \sum_{i=1}^{m} u_i f_{ij}(z_1, \ldots, z_{n_{w_j-1}}), \qquad j = 1, \ldots, n, \tag{38}$$

where the functions f_{ij} are weighted homogeneous polynomials of degree $w_j - 1$.

Of course, $n_{w_j-1} \le j - 1$, since n_{w_j-1} is the maximum index for a variable having weight $< w_j$.

Proof. Since

$$\widehat{X}_i = \sum_{j=1}^{n} f_{ij}(z_1, \ldots, z_n) \partial_{z_j}$$

is homogeneous of degree -1, and ∂_{z_j} is homogeneous of degree $-w_j$, the functions f_{ij} must be homogeneous of degree $w_j - 1$. In particular, they must be polynomials, and they cannot involve variables of weight $\ge w_j$. So, all variables z_k with $w_k \ge w_j$ are excluded. ∎

We say that the control system in (38) is in *triangular chained form*, in fact a block triangular form. In the equation for $\dot{z}_j$ only variables having a weight $< w_j$ appear in the right hand side. So, it is possible to compute the z_j one after the other, only by computing primitives, once given the control functions $u_1(t), \ldots, u_m(t)$.

We state in a Rothschild and Stein-like manner the approximation result we have obtained:

Theorem 5.19. *We have*

$$X_i = \widehat{X}_i + R_i, \qquad i = 1, \ldots, m, \tag{39}$$

where $\widehat{X}_i$ is homogeneous of order -1 and R_i is of order ≥ 0 at p.

In privileged coordinates, the system

$$\dot{x} = \sum_{i=1}^{m} u_i X_i(x)$$

takes the following form

$$\dot{z}_j = \sum_{i=1}^{m} u_i \left[f_{ij}(z_1, \ldots, z_{n_{w_j-1}}) + O(\|z\|^{w_j}) \right] \qquad j = 1, \ldots, n, \tag{40}$$

where the functions f_{ij} are weighted homogeneous polynomials of degree $w_j - 1$.

Proof. Equation (39) is only a rewriting of the series expansion (37). In coordinates, we have

$$R_i = \sum_{j=1}^{n} r_{ij}(z_1, \ldots, z_n) \partial_{z_j},$$

but since R_i has order ≥ 0 at 0, the order of each of its components $r_{ij}(z_1, \ldots, z_n)\partial_{z_j}$ must be ≥ 0, so $r_{ij}(z_1, \ldots, z_n) = O(\|z\|^{w_j})$. Using Proposition 5.18, we get (40).   ■

It is time now to say that the vector fields $\widehat{X}_1, \ldots, \widehat{X}_m$ are independent of the choice of a particular system of privileged coordinates.

Proposition 5.20. *Let $z_1, \ldots, z_n$ and $z'_1, \ldots, z'_n$ be two systems of privileged coordinates around p. Assume that the change of coordinates formulas are*

$$z'_j = \phi_j(z_1, \ldots, z_n), \qquad j = 1, \ldots, n. \tag{41}$$

Denote by $\widehat{X}_1, \ldots, \widehat{X}_m$ and $\widehat{X}'_1, \ldots, \widehat{X}'_m$ respectively the nilpotent approximations of the system $X_1, \ldots, X_m$ defined by means of these coordinates. Then vector fields $\widehat{X}'_1, \ldots, \widehat{X}'_m$ may be obtained from $\widehat{X}_1, \ldots, \widehat{X}_m$ through the change of coordinates

$$z'_j = \widehat{\phi}_j(z_1, \ldots, z_n), \qquad j = 1, \ldots, n. \tag{42}$$

where $\widehat{\phi}_j(z_1, \ldots, z_n)$ is the sum of monomials of weight w_j in the Taylor expansion of $\phi_j(z_1, \ldots, z_n)$.

In particular, (42) gives rise to an isomorphism between Lie algebras $\mathrm{Lie}(\widehat{X}_1, \ldots, \widehat{X}_m)$ and $\mathrm{Lie}(\widehat{X}'_1, \ldots, \widehat{X}'_m)$.

Proof. Clear, considering that if the Taylor expansion of $\phi_j(z_1, \ldots, z_n)$ is written as a sum of homogeneous terms, the first term has degree w_j. ∎

Definition 5.16. We will call $\mathrm{Lie}(\widehat{X}_1, \ldots, \widehat{X}_m)$ the *tangent Lie algebra* of $\mathrm{Lie}(X_1, \ldots, X_m)$ at point p.

Of course, $\mathrm{Lie}(\widehat{X}_1, \ldots, \widehat{X}_m)$ does not depend only of $\mathrm{Lie}(X_1, \ldots, X_m)$ as a Lie algebra of vector fields. It depends essentially of a supplementary datum, namely the filtration defined on $\mathrm{Lie}(X_1, \ldots, X_m)$ by the order of brackets.

Actually, a more natural presentation is by considering the Lie algebra $\mathcal{L}(X_1, \ldots, X_m)$ generated over the ring of smooth functions by $X_1, \ldots, X_m$. It is naturally a filtered algebra, and it appears that $\mathrm{Lie}(\widehat{X}_1, \ldots, \widehat{X}_m)$ depends only of the submodule $\mathcal{L}^1(X_1, \ldots, X_m)$, that is, the module generated by $X_1, \ldots, X_m$ over the smooth functions.

Recall that, when $\mathrm{rank}(X_1, \ldots, X_m)$ is constant, $\mathcal{L}^1(X_1, \ldots, X_m)$ is the module of smooth sections of the distribution generated by $X_1, \ldots, X_m$. In the opposite case, the geometric datum consisting of subspaces $L^1(X_1, \ldots, X_m)(x) \subset T_x M$ do not account faithfully for the properties of the given system of vector fields, and, as it is well known, the role of the distribution must be taken up by the module $\mathcal{L}^1(X_1, \ldots, X_m)$. One may call this sub-module a distribution and say in either case that $\mathrm{Lie}(\widehat{X}_1, \ldots, \widehat{X}_m)$ depends only of the distribution generated by $X_1, \ldots, X_m$.

Definition 5.17. We will call the space $\mathbb{R}^n$ endowed with the sub-Riemannian structure defined by the vector fields $\widehat{X}_1, \ldots, \widehat{X}_m$ the *tangent space of M at p.*

5.4. The tangent space as a homogeneous space

Denote by G_p the group generated by the diffeomorphisms $\exp t\widehat{X}_i$ acting on $\mathbb{R}^n$. Since the Lie algebra $\mathrm{Lie}(\widehat{X}_1, \ldots, \widehat{X}_m)$ is nilpotent, G_p is a simply connected Lie group having $\mathfrak{g}_p = \mathrm{Lie}(\widehat{X}_1, \ldots, \widehat{X}_m)$ as its Lie algebra. We insist that $\mathfrak{g}_p$ is not an "abstract" Lie algebra, but a Lie algebra of vector fields on $\mathbb{R}^n$. It splits into homogeneous components

$$\mathfrak{g}_p = \mathfrak{g}^1 \oplus \cdots \oplus \mathfrak{g}^r$$

where $\mathfrak{g}^s$ consists of vector fields homogeneous of degree $-s$ under the action of δ_λ. Note that $\widehat{X}_1, \ldots, \widehat{X}_m$ span $\mathfrak{g}^1$ and generate $\mathfrak{g}_p$ as a Lie algebra. The action of δ_λ is by automorphisms of $\mathfrak{g}_p$, and extends through the exponential mapping to a 1-parameter group

$$\delta_\lambda : G_p \to G_p$$

of automorphisms of G_p.

The action of G_p on $\mathbb{R}^n$ is transitive, as this is the same as saying that any point is accessible from the origin by using piecewise constant controls. Assigning pg to g gives rise to a map

$$\Psi_p : G_p \to \mathbb{R}^n$$

mapping the identity of G_p to 0. Recall the action of G_p is a right action, denoted as such. Denoting by H_p the isotropy subgroup of p in G_p—recall that p is identified to zero—we thus get a bijection

$$\phi_p : G_p/H_p \to \mathbb{R}^n.$$

Observe now that
$$(\delta_\lambda x)(\delta_\lambda g) = \delta_\lambda xg.$$

This implies that H_p is invariant under dilations. Hence, it is connected and simply connected, and we have

$$H_p = \exp(\mathfrak{h}_p)$$

where $\mathfrak{h}_p$ consists of all vector fields $Z \in \mathfrak{g}_p$ such that $Z(p) = 0$.

The subalgebra $\mathfrak{h}_p$ being invariant under dilations, splits into homogeneous components:

$$\mathfrak{h}_p = \mathfrak{h}^1 \oplus \cdots \oplus \mathfrak{h}^r.$$

We may describe as follows the structure of $\mathfrak{g}_p$ as a Lie algebra of vector fields: $\widehat{Y}_1, \ldots, \widehat{Y}_n$ span a complement of $\mathfrak{h}_p$ in $\mathfrak{g}_p$. A basis of $\mathfrak{g}_p$ may be obtained by adding to the homogeneous vector fields $\widehat{Y}_1, \ldots, \widehat{Y}_n$ a series of homogeneous vector fields $\widehat{Z}_{n+1}, \ldots, \widehat{Z}_{\tilde{n}}$, where $\tilde{n} = \dim \mathfrak{g}_p$. Each $\widehat{Z}_{n+k}$ may be written as

$$\widehat{Z}_{n+k}(z) = \sum_{j=1}^{n} \zeta_{jk}(z) \widehat{Y}_j(z),$$

where the functions $\zeta_{jk}(z)$ vanish at $z = 0$.

Example. For a simple example, recall the Grušin system

$$X_1 = \begin{pmatrix} 1 \\ 0 \end{pmatrix}, \quad X_2 = \begin{pmatrix} 0 \\ x \end{pmatrix}.$$

At $p = (0,0)$, coordinates $z_1 = x$ and $z_2 = y$ may be taken as privileged coordinates. The vector fields X_1 and X_2 are homogeneous of degree -1, so we have $\widehat{X}_1 = X_1$, $\widehat{X}_2 = X_2$. A basis of the tangent space at the origin is given by the values of

$$\widehat{Y}_1 = X_1, \quad \widehat{Y}_2 = [X_1, X_2] = \begin{pmatrix} 0 \\ 1 \end{pmatrix}.$$

To get all of $\mathfrak{g}_p$, we must add

$$\widehat{Z}_3 = X_2 = x\widehat{Y}_2.$$

Thus $\mathfrak{g}_p$ is generated by $\widehat{Y}_1 = X_1$, $\widehat{Y}_2 = [X_1, X_2]$, $\widehat{Z}_3 = X_2$, and is isomorphic to the Heisenberg Lie algebra, while $\mathfrak{h}_p = \mathbb{R}X_2$.

Returning to the general situation, denote by $\widehat{\xi}_1, \ldots, \widehat{\xi}_m$ the elements $\widehat{X}_1, \ldots, \widehat{X}_m$ of $\mathfrak{g}_p$, when they are viewed as left invariant vector fields on G_p, that is, when they act infinitesimally on the right on G_p. These vector fields also act on $G_p/H_p = \{H_p g \mid g \in G_p\}$ under the denomination of $\tilde{\xi}_1, \ldots, \tilde{\xi}_m$. It is now a matter of routine verification that $\phi_p : G_p/H_p \to \mathbb{R}^n$ is a diffeomorphism and makes $\tilde{\xi}_1, \ldots, \tilde{\xi}_m$ correspond to $\widehat{X}_1, \ldots, \widehat{X}_m$.

Recalling results from the preceding Section, we get

Theorem 5.21. *There exists a well-defined graded Lie algebra $\mathfrak{g}_p$, generated by its component of degree 1, say $\mathfrak{g}^1$, and a graded subalgebra $\mathfrak{h}_p$ of $\mathfrak{g}_p$, such that T_pM is isometric to G_p/H_p, where $G_p = \exp(\mathfrak{g}_p)$, $H_p = \exp(\mathfrak{h}_p)$, and G_p/H_p is endowed with the sub-Riemannian metric associated to some basis of $\mathfrak{g}^1$, acting on the right on G_p/H_p.*

Example. The Grušin plane G_2 is such a quotient G/H. We have $G_2 = H_3/\exp(\mathbb{R}X_2)$, where H_3 is the Heisenberg group.

At points (x,y) with $x = 0$, the tangent space to G_2 is isometric to G_2 itself. At points with $x \neq 0$, it is isometric to Euclidean $\mathbb{R}^2$.

5.5. At regular points, the tangent space is a group

Proposition 5.22. *If p is regular, then $H_p = \{0\}$, and T_pM is isometric to the group $G_p = \exp(\mathfrak{g}_p)$.*

Proof. Let $\widehat{Z} \in \mathfrak{h}_p$. We have to prove that $\widehat{Z} = 0$. As already observed, all the homogeneous components of $\widehat{Z}$ belong to $\mathfrak{h}_p$, and thus vanish at 0. So we may suppose $\widehat{Z}$ is homogeneous of degree $-s$, that is

$$\widehat{Z} = \sum_\alpha a_\alpha [\widehat{X}_{\alpha_1}, \dots [\widehat{X}_{\alpha_{s-1}}, \widehat{X}_{\alpha_s}] \dots]$$

with $a_\alpha \in \mathbb{R}$. Set

$$Z = \sum_\alpha a_\alpha [X_{\alpha_1}, \dots [X_{\alpha_{s-1}}, X_{\alpha_s}] \dots]$$

so that $\widehat{Z}$ is the homogeneous component of degree $-s$ of Z.

The regularity hypothesis implies $\dim L^s(q) = \text{const}$ on a neighbourhood of 0. Since $Y_1(q), \dots, Y_{n_s}(q)$ are independent, they form a basis of $L^s(q)$ for all q near 0, and we can write

$$Z = \sum_{j=1}^{n_s} f_j(q) Y_j(q) \tag{43}$$

in a unique way and the f_j are smooth. Since $Z(0) = 0$, we have $f_1(0) = \cdots = f_{n_s}(0) = 0$. But, as $Y_1, \dots, Y_{n_s}$ have order $\geq -s$ at 0, this implies that the right-hand side in (43) has order $\geq -s + 1$. Hence $\widehat{Z} = 0$, and the conclusion follows.   ∎

The converse of Proposition 5.22 is false, as the following example shows: Take $M = \mathbb{R}^3, m = 3$ and $X_1 = \partial_x, X_2 = \partial_y - x\partial_z, X_3 = z^{10}\partial_z$. The origin is a singular point, since rank $L^1(x, y, z)$ is 3 for $\widehat{z} \neq 0$ and 2 for $z = 0$. We can take as privileged coordinates x (order 1), y (order 1) and z (order 2), and we get $\widehat{X}_1 = X_1, \widehat{X}_2 = X_2, \widehat{X}_3 = 0$, so the tangent space is isomorphic to the Heisenberg group.

Example. Let us compute the nilpotent approximation for the system given on $\mathbb{R}^2 \times S^1$ by

$$X = \begin{pmatrix} \cos\theta \\ \sin\theta \\ 0 \end{pmatrix}, \quad Y = \begin{pmatrix} 0 \\ 0 \\ 1 \end{pmatrix}.$$

(The control equation $\dot{q} = uX + vY$ is used in robotics to model the kinematics of car-like robots. In fact, this model is accurate only for unicycles.) We have

$$[X, Y] = \begin{pmatrix} -\sin\theta \\ \cos\theta \\ 0 \end{pmatrix},$$

so the system is controllable. At the origin $(x = y = \theta = 0)$, the coordinates θ and x have weight 1, and y has weight 2. By the remark following Theorem 4.15, they are privileged coordinates. Since Y is homogeneous, we have $\widehat{Y} = Y$. The expansion of X into homogeneous components is

$$X = \cos\theta\,\partial_x + \sin\theta\,\partial_y = \underbrace{\partial_x + \theta\partial_y}_{\text{order } -1} - \underbrace{\left(\frac{1}{2}\theta^2\partial_x + \frac{1}{6}\theta^3\partial_y\right)}_{\text{order } 1} + \underbrace{\cdots}_{\text{order } \geq 3}$$

So we get

$$\widehat{X} = \begin{pmatrix} 1 \\ \theta \\ 0 \end{pmatrix}, \quad \widehat{Y} = \begin{pmatrix} 0 \\ 0 \\ 1 \end{pmatrix},$$

where we recognize a presentation of the Heisenberg group.

5.6. Non-abelian vector spaces.
(Carnot groups and homogeneous spaces.)

At a regular point p, the natural structure of the tangent space T_pM thus consists of

(a) A simply connected nilpotent Lie group structure on T_pM—of a particular kind: the Lie algebra $\mathfrak{g}$ of T_pM is graded and generated by its component of degree 1, say $\mathfrak{g}^1$;

(b) A 1-parameter group (δ_λ) of group automorphisms of T_pM—naturally obtained from the grading of $\mathfrak{g}$;

(c) A left-invariant sub-Riemannian metric on T_pM obtained from a basis of $\mathfrak{g}^1$—or, better, from a positive definite quadratic form on $\mathfrak{g}^1$—on which (δ_λ) acts by dilations.

This is what is called a Carnot group by Pansu in [25] and Gromov in [14], a denomination which goes with that of Carnot spaces, used for sub-Riemannian manifolds. The structure of a Carnot group is strikingly similar to that of a vector space ((a) and (b) , just replacing "nilpotent" by "abelian" in (a)), equipped with a Euclidean metric (c) .

At singular points, (a) gets replaced by a structure of homogeneous space G/H, where G is a Carnot group, and H a connected subgroup associated to a graded subalgebra of $\mathrm{Lie}(G)$. Dilations (b) are compatible with the dilations on G, but not with the action of G, and, likewise, the sub-Riemannian metric, defined from the infinitesimal action of the component of degree 1 of $\mathrm{Lie}(G)$, is not G-invariant. In fact, in this case, T_pM is homogeneous under G as long as the metric plays no role. As soon as it shows in, there are in $T_pM = G/H$ regular and singular points, of which p is the most singular, along with images of p by the centralizer of H. Only these points may be the center of a 1-parameter group of dilations. Maybe G/H could be called a Carnot homogeneous space, since it is homogeneous in both sense, under dilations, and under the action of a group. Moreover, any sub-Riemannian manifold having a 1-parameter group of dilations, centered at a point p is such a G/H, since, as it is easily proved, it is isomorphic to its tangent space at p.

Carnot groups and their quotients play in sub-Riemannian geometry the same role as Euclidean spaces do in Riemannian geometry. Since the algebraic structure of Carnot groups is moreover similar to that of Euclidean spaces, it is really tempting to call them *non-abelian vector spaces*, or *nonholonomic vector spaces*, or *nonholonomic Euclidean spaces* if one wants to take the metric into account.

There is nevertheless one major difference between Euclidean spaces and Carnot groups: they are many algebraically non isomorphic Carnot groups having the same dimension n, uncountably many for $n \geq 6$, as there may be modules in their classification. We note that non-isomorphic Carnot group are not isometric either. This is a consequence of the construction, carried in §8, of the group law from the metric.

For an example consider the Carnot groups associated to Lie algebras

$$\mathfrak{g}_F = \mathbb{R}^m \oplus \wedge^2 \mathbb{R}^m / F$$

where F is a subspace of codimension k of $\wedge^2 \mathbb{R}^m$ and the Lie bracket is defined as

$$[X, Y] = \begin{cases} X \wedge Y \bmod F & \text{if } X, Y \in \mathbb{R}^m; \\ 0 & \text{otherwise.} \end{cases}$$

Clearly, Lie algebras $\mathfrak{g}_F$ and $\mathfrak{g}_{F'}$ are isomorphic if there exists a bijective linear map $\phi : \mathbb{R}^m \to \mathbb{R}^m$ such that $(\wedge^2 \phi)(F) = F'$. When $m = 3$ the isomorphism class of $\mathfrak{g}_F$ depends only of the integer k. But take $m = 4$. Now, the grassmannian manifold of subspaces of codimension 2 of $\wedge^2 \mathbb{R}^m$ has dimension 19, while the linear group $GL(m, \mathbb{R})$ has dimension 16 only. So the classification up to isomorphism of Lie algebras of type $\mathfrak{g}_F$ with $m = 4$, $k = 2$ (and of corresponding Carnot groups) depends on 3 modules at least.

In a given sub-Riemannian manifold, the algebraic structure of the tangent space may be different from one point to another, as in the Grušin plane, but it may also, even in regular situations, vary continuously from point to point (see [38]).

6. Gromov's notion of tangent space

6.1. Tangent cones in $\mathbb{R}^n$

Recall first the notion of Hausdorff distance between two subsets of $\mathbb{R}^n$. We have H-dist$(A, B) \leq \rho$ if any point of A is within distance ρ of B, and any point of B is within distance ρ of A. We say that A_n converges to A in the Hausdorff sense, and we write

$$\lim_{n \to \infty} A_n = A,$$

if, for any compact set K in $\mathbb{R}^n$, we have

$$\lim_{n \to \infty} \text{H-dist}(A_n \cap K, A \cap K) = 0.$$

Now, let S be a closed subset of $\mathbb{R}^n$, and $p \in S$. Consider for $\lambda > 0$ the dilation $\delta_{p\lambda}$ of center p and ratio λ. We call

$$T_p S = \lim_{\lambda \to \infty} \delta_{p\lambda} S$$

the *tangent cone*, or *tangent subspace* to S at p, provided the limit exists.

6.2. Tangent spaces to a metric space

Gromov has shown in [13] how to extend this definition to arbitrary metric spaces. One defines dilations in an abstract way: λM is the metric space with the same underlying set as M and all distances multiplied by λ. The Hausdorff distance between two metric spaces X and Y is defined as follows: H-dist(X, Y) is the infimum of real numbers ρ for which there exists isometric embeddings of X and Y in a same metric space Z, say $i : X \to Z$ and $j : Y \to Z$, such that the Hausdorff distance of $i(X)$ and $j(Z)$ as subsets of Z is $\leq \rho$.

Lemma 6.23. *Let X and Y be metric spaces. If there exist (non necessarily continuous) maps $f : X \to Y$ and $g : Y \to X$ such that*

$$\big|d(x, x') - d(f(x), f(x'))\big| \leq 2\rho, \quad \big|d(g(y), g(y')) - d(y, y')\big| \leq 2\rho \quad (44)$$

for any $x, x' \in X$, $y, y' \in Y$, then

$$\text{H-dist}(X, Y) \leq \rho. \tag{45}$$

Conversely, if H-dist$(X, Y) \leq \rho$, then (44) holds for any $x, x' \in X$, $y, y' \in Y$, if X and Y are compact, and (44) holds with ρ replaced by $\rho + \varepsilon$, for any $\varepsilon > 0$, in the general case.

Proof. To prove the direct assertion, we have only to take $Z = X \amalg Y$ (the disjoint union), and to define a distance d_Z on Z by setting $d_Z(x, x') = d_X(x, x')$, $d_Z(y, y') = d_Y(y, y')$, $d_Z(x, f(x)) = d_Z(g(y), y) = \rho$, for $x, x' \in X$, $y, y' \in Y$. Distance d_Z is only defined on a subset S of $Z \times Z$, but the triangle inequality holds, so we can extend it by taking for $d(z, z')$ the infimum of sums $d(z, z_1) + \cdots + d(z_{n-1}, z')$, where $(z, z_1), \ldots, (z_{n-1}, z)$ are in S.

To prove the converse assertion, define $f(x)$ as a point $y \in Y$ such that $d_Z(x, y) = d_Z(x, Y)$, if this is possible, or else $d_Z(x, y) \leq d_Z(x, y) + \varepsilon$. Define $g(y)$ similarly. ■

Corollary 6.24. *If the metric spaces X and Y have the same underlying set and if $\big|d_X(x, x') - d_Y(x, x')\big| \leq \rho$ for all x and x' in X, then we have H-dist$(X, Y) \leq \rho/2$.*

Thanks to Hausdorff distance, one can define the notion of limit of a sequence of metric spaces. For unbounded spaces, one uses the following definition:

A sequence of *pointed* metric spaces (X_n, x_n) is said to converge to (X, x) if

$$\lim_{n \to \infty} \text{H-dist}\big(B^{X_n}(x_n, R), B^X(x, R)\big) = 0$$

for any positive R.

Definition 6.18. We set

$$(T_p M, 0) = \lim_{\lambda \to \infty} (\lambda M, p) \tag{46}$$

the *tangent space to M at p*, provided the limit exists.

Since

$$\lambda'(T_p M, 0) = \lim_{\lambda \to \infty} (\lambda \lambda' M, p) = (T_p M, 0)$$

the tangent space possesses a 1-parameter group of dilations having 0 as a fixed point. In particular, all balls centered at 0 are similar to $B(0, 1)$.

Taking this into account, and replacing λ by ε^{-1}, one can rewrite (46) as

$$\lim_{\varepsilon \to 0} \varepsilon^{-1} B^M(p, R\varepsilon) = B^{T_p M}(0, R). \tag{47}$$

In fact, $R = 1$ suffices. Thus, the existence of $T_p M$ means simply that small balls $B(p, \varepsilon)$ in M (renormalized to radius 1) get more and more alike when $\varepsilon \to 0$.

When M is a C^1 Riemannian manifold, one thus recovers in a purely metric way the tangent space $T_p M$ with its Euclidean metric.

We shall prove in the following section that the tangent space in the sense of Gromov exists at every point p of a sub-Riemannian manifold M, and that it is isometric to the tangent space we have already defined, i.e., $T_p M$ endowed with its natural sub-Riemannian structure. Until then, we shall denote the former space by Gromov-$T_p M$, if needed.

7. Distance estimates and the metric tangent space

Throughout this Section, we fix some system $x_1, \ldots, x_n$ of privileged coordinates near p. We use these coordinates to identify a neighbourhood of p with a neighbourhood of the origin in $\mathbb{R}^n$. We will denote by $\|x\|_p$, or $\|x\|$, the pseudo-norm $|x_1|^{w_1} + \cdots + |x_n|^{w_n}$.

Let $\widehat{X}_1, \ldots, \widehat{X}_m$ be the nilpotent homogeneous system, defined on $\mathbb{R}^n$ approximating $X_1, \ldots, X_m$ at p. We will denote by $\widehat{d}_p$, or simply by $\widehat{d}$, the corresponding sub-Riemannian distance on $T_p M = \mathbb{R}^n$.

7.1. Distance estimates in the tangent space

Proposition 7.25. (Estimates on $\widehat{d}_p(p,q)$) *There exist positive constants, C, C' such that for all q in $\mathbb{R}^n$ we have*

$$C\big(|q_1|^{1/w_1}+\cdots+|q_n|^{1/w_n}\big) \le \widehat{d}_p(p,q) \le C'\big(|q_1|^{1/w_1}+\cdots+|q_n|^{1/w_n}\big). \quad (48)$$

Proof. Recall $p = (0,\ldots,0)$. The pseudo-norm $q \mapsto \|q\| = |q_1|^{1/w_1}+\cdots+|q_n|^{1/w_n}$ and the function $q \mapsto \widehat{d}_p(p,q)$ are both homogeneous, of degree one, under δ_λ. We know from Proposition 5.17 that $\widehat{d}_p(p,q)$ is positive and finite on the set $\|q\| = 1$. Since this set is compact and $\widehat{d}_p(p,q)$ is continuous on $\mathbb{R}^n$, it follows that there exist numbers C, C', positive and finite, such that

$$C \le \widehat{d}_p(p,q) \le C'$$

for $\|q\| = 1$. Using homogeneity, we get (48). $\blacksquare$

As it is simpler to prove the estimates we have in mind in the case of tangent spaces (or Carnot groups and homogeneous spaces), we will consider this case first. The proof of corresponding estimates in the manifold M will not depend on the results obtained in the case of tangent spaces, but it will follow, more or less, the same lines.

Proposition 7.26. (Estimate for $\widehat{d}_p(q,q')$) *There exists a positive constant C such that for all q, q' in $T_pM = \mathbb{R}^n$ one has*

$$\widehat{d}(q,q') \le C \sum_{\{k,j\,|\,w_k \le w_j\}} \|q\|^{1-w_k/w_j}|q'_k - q_k|^{1/w_j}. \quad (49)$$

An equivalent estimate is

$$\widehat{d}(q,q') \le C\Big(\|q' - q\| + \|q\|^{1-1/r}\|q' - q\|^{1/r}\Big). \quad (50)$$

Proof. (A) We first deal with the case where the action of G_p on T_pM is simply transitive, *i.e.*, the map $g \mapsto gp$ is a bijection. As we have said, this map is an isometry when G_p is endowed with the sub-Riemannian metric defined by $\widehat{X}_1,\ldots,\widehat{X}_m$. So we may as well work on the group $G = G_p$. The results obtained will be useful for the general case also.

Choose canonical coordinates on G defined by means of the mapping

$$(x_1,\ldots,x_n) \mapsto \exp(x_1\widehat{Y}_1 + \cdots + x_n\widehat{Y}_n).$$

Then for x, x' in G, we have

$$x^{-1}x' = \big(F_1(x,x'),\ldots,F_n(x,x')\big),$$

where the j-th coordinate has the form

$$F_j(x,x') = x'_j - x_j + \sum_{\{k\,|\,w_k < w_j\}} (x'_k - x_k)P_{kj}(x,x'),$$

where $P_{kj}(x,x')$ is a weighted homogeneous polynomial of degree $w_j - w_k$.

Indeed, due to the nilpotency of G, the coordinates of $x^{-1}x'$ must be polynomials in the coordinates of x and x'. Using dilations δ_λ, one shows that $F_j(x,x')$ must be weighted homogeneous of degree w_j, and, in particular, a function of the coordinates of x and x' having weight $\le w_j$. Since $F_j(x,x') = 0$ when $x' = x$, one must have

$$F_j(x,x') = \sum_{\{k\,|\,w_k \le w_j\}} (x'_k - x_k)P_{kj}(x,x').$$

The form of $P_{kj}(x,x')$ for $w_j = w_k$ is obtained by looking to the expression of $x^{-1}x'$ when x and x' tend to 0, the identity element.

Since $\widehat{d}$ is left invariant, we have $\widehat{d}(x,x') = \widehat{d}(0, x^{-1}x')$, and we get from Proposition 7.25 the estimates

$$C\sum_j |F_j(x,x')|^{\frac{1}{w_j}} \le \widehat{d}(x,x') \le C'\sum_j |F_j(x,x')|^{\frac{1}{w_j}}.$$

Working with the right hand side we obtain

$$\widehat{d}(x,x') \le \text{const} \sum_j \Big(|x'_j - x_j| + \sum_{\{k\,|\,w_k < w_j\}} |x'_k - x_k|(\|x\| + \|x'\|)^{w_j - w_k} \Big)^{\frac{1}{w_j}},$$

whence, using twice the inequality $(A+B)^\alpha \le A^\alpha + B^\alpha$ $(0 \le \alpha \le 1)$,

$$\widehat{d}(x,x') \le \text{const} \sum_j |x'_j - x_j|^{\frac{1}{w_j}} + \text{const} \sum_{\{j,k\,|\,w_k < w_j\}} |x'_k - x_k|^{\frac{1}{w_j}}(\|x\| + \|x'\|)^{1 - \frac{w_k}{w_j}}$$

and

$$\widehat{d}(x,x') \le \text{const} \sum_j |x'_j - x_j|^{\frac{1}{w_j}} + \text{const} \sum_{\{j,k\,|\,w_k < w_j\}} \|x\|^{1 - \frac{w_k}{w_j}} |x'_k - x_k|^{\frac{1}{w_j}}$$

$$+ \text{const} \sum_{\{j,k\,|\,w_k < w_j\}} \|x'\|^{1 - \frac{w_k}{w_j}} |x'_k - x_k|^{\frac{1}{w_j}}$$

Using the inequality $\|x'\|^\alpha \leq n\|x\|^\alpha + n\|x' - x\|^\alpha$ $(0 \leq \alpha \leq 1)$, we may get rid of the last sum in the preceding equation. We have thus proved (49) for T_pM, p regular, or in case M is a group.

(B) Let us turn now to the general case, when the stabilizer H_p of p in G_p is non-trivial. Then, the vector fields $\widehat{Y}_1, \ldots, \widehat{Y}_n$ do not make up any more a basis of the Lie algebra $\mathfrak{g}_p$. They only span a subspace K which is a complement of $\mathfrak{h}_p$. Then the restriction to $\exp(K)$ of the mapping $g \mapsto pg$ is a diffeomorphism. More precisely, the mapping

$$(x_1, \ldots, x_n) \mapsto p \exp(x_1 \widehat{Y}_1 + \cdots + x_n \widehat{Y}_n)$$

defines privileged coordinates on T_pM.

Consider on G_p the sub-Riemannian structure defined by $\widehat{X}_1, \ldots, \widehat{X}_m$, viewed as left invariant vector fields. For any path $x(t)$, $(0 \leq t \leq T)$ in G_p the path $px(t)$, $(0 \leq t \leq T)$ in T_pM has the same length. It follows that the mapping $g \to gp$ is Lipschitzian:

$$\widehat{d}(pg, pg') \leq d_{G_p}(g, g'). \tag{51}$$

On T_pM, in the coordinates just defined, the points $q = (q_1, \ldots, q_n)$ and $q' = (q'_1, \ldots, q'_n)$ read

$$q = \exp(q_1 \widehat{Y}_1 + \cdots + q_n \widehat{Y}_n + 0\widehat{Z}_{n+1} + \cdots + 0\widehat{Z}_{\tilde{n}})p$$

and

$$q' = \exp(q'_1 \widehat{Y}_1 + \cdots + q'_n \widehat{Y}_n + 0\widehat{Z}_{n+1} + \cdots + 0\widehat{Z}_{\tilde{n}})p,$$

where we have introduced a basis $\widehat{Z}_{n+1}, \ldots, \widehat{Z}_{\tilde{n}}$ of $\mathfrak{h}_p$ consisting of homogeneous vector fields. (Things would be still more clear if we had ordered vector fields in these expressions following their weights.) By applying (49) in G_p and (51) we see that (49) holds in T_pM.

(C) To get the coarser estimate (50), it suffices to remind that $|q'_k - q_k| \leq \|q' - q\|^{w_k}$, which yields

$$\widehat{d}(q, q') \leq C \sum_{\{k, j \mid w_k \leq w_j\}} \|q\|^{1 - w_k/w_j} \|q' - q\|^{w_k/w_j}.$$

Then use the inequality $A^{1-\alpha}B^\alpha \leq A^{1-1/r}B^r + B$, which holds for $1/r \leq \alpha \leq 1$, due to the convexity of $A^{1-x}B^x$. ∎

Remark. Points q and q' do not play the same role, as they should, in estimates (49) and (50). But note that these estimates have been proved with $\|q\| + \|q'\|$ instead of $\|q\|$, and that in the case of (50), e.g., it is immediate to pass from the given form to a symmetric one, and vice versa, keeping in mind the inequalities

$$\|q\| \leq \|q\| + \|q'\| \leq \|q\| + \|q' - q\| + \|q\|^{1-1/r}\|q' - q\|^{1/r}.$$

7.2. Examples: distance estimates in the Heisenberg group and in the Grušin plane

We give three examples.

(a) *The Heisenberg group.* Consider in the Heisenberg group H_3 coordinates given by the mapping

$$(x, y, z) \mapsto \exp(z[X_1, X_2]) \exp(yX_2) \exp(xX_1)$$

(canonical coordinates of the second kind). As we know from §3.2, X_1 and X_2 read as

$$X_1 = \begin{pmatrix} 1 \\ 0 \\ 0 \end{pmatrix}, \quad X_2 = \begin{pmatrix} 0 \\ 1 \\ x \end{pmatrix}.$$

From the form (14) of the group law, we get

$$(x, y, z)^{-1}(x', y', z') = (x' - x, y' - y, z' - z - x(y' - y)).$$

Using (16) and the left invariance of distance d, we obtain

$$d\big((x, y, z), (x', y', z')\big) \leq 4\big(|x' - x| + |y' - y| + |z' - z|^{1/2} + |x|^{1/2}|y' - y|^{1/2}\big). \tag{52}$$

(The lower bound in (16) does not yield a pleasant lower bound for $d\big((x, y, z), (x', y', z')\big)$.) Observe that applying of (49) would yield us to majorize $d\big((x, y, z), (x', y', z')\big)$ by

$$C\Big(|x'-x|+|y'-y|+|z'-z|^{1/2}+\big(|x|+|y|+|z|^{1/2}\big)^{1/2}\big(|x'-x|^{1/2}+|y'-y|^{1/2}\big)\Big).$$

This kind of bound holds actually for any system of privileged coordinates. We see that it is an estimate of the same kind as (52), but having more terms.

(b) *The Heisenberg group in exponential coordinates.* We have seen in §3.3 that in the exponential coordinates defined by

$$(x, y, z) \mapsto \exp(xX_1 + yX_2 + z[X_1, X_2])$$

vector fields X_1 and X_2 read as

$$X_1 = \begin{pmatrix} 1 \\ 0 \\ -\frac{x}{2} \end{pmatrix}, \quad X_2 = \begin{pmatrix} 0 \\ 1 \\ \frac{x}{2} \end{pmatrix}.$$

The product operation is given by

$$(x, y, z)(x', y', z') = \big(x + x', y + y', z + z' + \tfrac{1}{2}(xy' - yx')\big).$$

Since clearly $(x, y, z)^{-1} = (-x, -y, -z)$, we have

$$(x, y, z)^{-1}(x', y', z') = \big(x' - x, y' - y, z' - z + \tfrac{1}{2}(yx' - xy')\big).$$

Using estimates (16), which still hold in exponential coordinates, and observing that $yx' - xy' = y(x' - x) - x(y' - y)$, we deduce that

$$d\big((x, y, z), (x', y', z')\big) \le 4\big(|x|+|y|+|z|^{1/2}+|x|^{1/2}|y'-y|^{1/2}+|y|^{1/2}|x'-x|^{1/2}\big).$$
$$\tag{53}$$

(c) *The Grušin plane.* As we have said earlier, the Grušin plane G_2 may be considered as a quotient of the Heisenberg group : $G_2 = H_3/\exp(\mathbb{R}X_2)$. Observations made in the course of the proof of Proposition 7.26 show that the Grušin plane is in correspondence with the plane $y = 0$ in H_3 if we use exponential coordinates. Moreover, the mapping which maps the element $\exp(xX_1 + z[X_1, X_2])$ of H_3 to $(x, z) \in G_2$ is distance decreasing. Therefore, using (53), we obtain the estimate

$$d\big((x, z), (x', z')\big) \le 4\big(|x' - x| + |z' - z|^{1/2}\big), \tag{54}$$

holding for all (x, z), (x', z') in G_2. We leave a direct proof of (54) as an exercise to the interested reader.

7.3. Local distances estimates in M

Our next goal is to show that estimate (49) still holds when $\widehat{d}$ is replaced by d. We will need smoothly varying systems of privileged coordinates, namely systems of coordinates defined around a variable point q, and smoothly varying with q.

It is hard to imagine how such a smooth varying system of privileged coordinates may exist in the neighbourhood of a singular point. But, when p is regular, we easily obtain one: First, we define a system of coordinates $u_1, \ldots, u_n$ around q, writing

$$q' = q + u_1 Y_1(q) + \cdots + u_n Y_n(q).$$

Recall that we identify M with $\mathbb{R}^n$, using privileged coordinates around p. Since p is regular, tangent vector $Y_1(q), \ldots, Y_n(q)$ form a basis of $T_q M$ adapted to the flag

$$\{0\} = L^0(q) \subset L^1(q) \subset \cdots \subset L^s(q) \subset \cdots \subset L^r(q) = T_q M$$

for all q in a neighbourhood of p. Therefore the u_j form a system of linearly adapted coordinates around q. We emphasize the dependence of u_j in q and q' by writing $u_j = u_j(q, q')$. Then, we turn this system into a system of privileged coordinates around q, by employing the effective construction described in the proof of Theorem 4.15. Call $s_j(q, q')$ the privileged coordinates around q we obtain.

We need to estimate $s_j(q, q')$ in terms of q and $q - q'$. We proceed as follows. First, we observe that we have

$$Y_j(q) = \partial_{q_j} + \sum_{\{k \,|\, w_k > w_j\}} \varphi_{jk}(q)\, \partial_{q_k}, \qquad j = 1, \ldots, n$$

where $\varphi_{jk}(q)$ has order $\geq w_k - w_j$ at $q = 0$. So the change of coordinates formula between the $q'_j - q_j$ and the u_j is

$$q'_k - q_k = u_k + \sum_{\{j \,|\, w_j < w_k\}} \varphi_{jk}(q) u_j, \qquad k = 1, \ldots, n.$$

A simple argument shows that the inverse change of coordinates has the same form:

$$u_k = q'_k - q_k + \sum_{\{j \,|\, w_j < w_k\}} \psi_{jk}(q)(q'_j - q_j), \qquad k = 1, \ldots, n,$$

where the functions $\varphi_{jk}(q)$ have order $\geq w_k - w_j$ at $q = 0$. Now, by Theorem 4.15 and Lemma 4.14, we know that

$$s_k(q, q') = u_k + \sum_{\{\alpha \,|\, w(\alpha) < w_k\}} c_\alpha u_1^{\alpha_1} \ldots u_n^{\alpha_n}, \qquad k = 1, \ldots, n.$$

Replacing in this formula the u_k by their expression above, we obtain the following result.

Lemma 7.27. *Assume p is a regular point. The coordinates $s_k(q, q')$ defined in the preceding discussion may be computed in terms of q and $q' - q$ by means of formula*

$$s_k(q, q') = q'_k - q_k + \sum_{\{j \,|\, w_j < w_k\}} (q'_j - q_j) P_{jk}(q, q'), \qquad k = 1, \ldots, n. \quad (55)$$

where $P_{jk}(q, q')$ is a smooth function of (q, q') having order $w_k - w_j$ at point $(0, 0)$ in $M \times M$.

Using these coordinates, we define a nilpotent approximation of the whole structure at q, that is,

(a) vector fields $\widehat{X}_1^q, \ldots, \widehat{X}_m^q, \widehat{Y}_1^q, \ldots, \widehat{Y}_n^q$, on $\mathbb{R}^n$, in the neighbourhood of q, through the diffeomorphism $\psi_q : q' \mapsto (s_1(q, q'), \ldots, s_n(q, q'))$;

(b) a nilpotent Lie group structure on $\mathbb{R}^n$, denoted by $T_q M$ or G_q;

(c) a distance $\widehat{d^q}$ on $T_q M$ (the sub-Riemannian distance defined by vector fields $\widehat{X}_1^q, \ldots, \widehat{X}_m^q$), and

(d) a pseudo-norm on $T_q M$, which reads in M as

$$\|q'\|_q = \sum_{j=1}^{n} |s_j(q, q')|^{1/w_j}.$$

Lemma 7.28. *Assume p is a regular point. There exists $\varepsilon > 0$ and $C > 0$ such that for $\|q\| \le \varepsilon$ and $\|q'\| \le \varepsilon$, we have*

$$\widehat{d}_q(q, q') \le C \|q'\|_q. \tag{56}$$

As a consequence, we have

$$\widehat{d}_q(q, q') \le C \|q' - q\| + C \|q\|^{1 - 1/r} \|q' - q\|^{1/r}. \tag{57}$$

Proof. From Proposition 7.25, we obtain for each q in some neighbourhood of p the estimate $\widehat{d}_q(q, q') \le C_q \|q'\|_q$ holding for $\|q'\|_q \le \varepsilon_q$. Since everything depends smoothly in q, we can remove the dependence of C_q and ε_q on q, by assigning q to a small neighbourhood of p. We obtain thus (56).

To prove the second part of the lemma, we only read off the $s_j(q, q')$ from. We have

$$\|q'\|_q = \sum |s_j(q, q')|^{1/w_j} \le \sum_{\{j, k \, | \, w_k \le w_j\}} |q'_k - q_k|^{1/w_j} \left(\|q\| + \|q'\| \right)^{1 - w_k/w_j}.$$

We end like in the proof of Proposition 7.26. ∎

We turn to the comparison of trajectories defined by the system $X_1, \ldots, X_m$, and its approximation $\widehat{X}_1, \ldots, \widehat{X}_m$. We do not suppose p to be regular in the next proposition.

Proposition 7.29. *Suppose $x(t)$ and $\widehat{x}(t)$ are controlled paths, starting at the same point q, having velocity 1, and defined by the same control functions, respectively for the system $X_1, \ldots, X_m$ and the approximating system $\widehat{X}_1, \ldots, \widehat{X}_m$. There exists $\varepsilon > 0$ such that:*

(i) If $\|q\| \leq \varepsilon$ and $t \leq \varepsilon$, then we have

$$|x_j(t) - \widehat{x}_j(t)| \leq C\,\varepsilon^{w_j}\,t, \quad j = 1, \ldots, n. \tag{58}$$

and

$$\widehat{d}_p(x(t), \widehat{x}(t)) \leq C\,\varepsilon\,t^{1/r}. \tag{59}$$

(ii) If the starting point is p, and if $|t| \leq \varepsilon$, we have

$$|x_j(t) - \widehat{x}_j(t)| \leq C\,t^{w_j+1}, \quad j = 1, \ldots, n. \tag{60}$$

and

$$\widehat{d}_p(x(t), \widehat{x}(t)) \leq C\,t^{1+1/r}. \tag{61}$$

Proof. (i) We use the triangular form of the approximating system. In privileged coordinates $x(t)$ satisfies the differential system

$$\dot{x}_j = \sum_{i=1}^m u_i f_{ij}(x) + \sum_{i=1}^m u_i r_{ij}(x), \quad j = 1, \ldots, n, \tag{62}$$

where $f_{ij}(x)$ are weighted homogeneous polynomials of degree $w_j - 1$ respectively, and $r_{ij}(x)$ are functions of order w_j at 0. This yields the differential inequations

$$|\dot{x}_j| \leq \text{const}\, \|x\|^{w_j - 1}, \quad j = 1, \ldots, n.$$

To integrate these inequations, it may be convenient to replace $\|x\|$ by the equivalent pseudo-norm

$$\|\!|x|\!\| = \Big(\sum_{j=1}^n x_j^{N/w_j}\Big)^{1/N},$$

where the integer N is chosen so that the N/w_j are even integers. This pseudo-norm has the advantage of being differentiable outside the origin. We obtain in this way

$$\|\!|x(t)|\!\| \leq \|\!|x(0)|\!\| + \text{const}\, t.$$

The homogeneous approximating system reads

$$\dot{\widehat{x}}_j = \sum_{i=1}^{m} u_i f_{ij}(\widehat{x}), \quad j = 1, \dots, n. \tag{63}$$

But subtracting (63) from (62) we get the equations

$$\dot{x}_j - \dot{\widehat{x}}_j = \sum_{i=1}^{m} u_i \big(f_{ij}(x) - f_{ij}(\widehat{x})\big) + \sum_{i=1}^{m} u_i r_{ij}(x), \quad j = 1, \dots, n.$$

which can be written as

$$\dot{x}_j - \dot{\widehat{x}}_j = \sum_{i=1}^{m} u_i \sum_{\{k \mid w_k < w_j\}} (x_k - \widehat{x}_k) Q_{ijk}(x, \widehat{x}) + O\big(\|x\|^{w_j}\big), \quad j = 1, \dots, n.$$

where $Q_{ijk}(x, \widehat{x})$ are polynomials of weighted degree $w_j - w_k - 1$ with respect to x and $\widehat{x}$, or are zero. This yields the differential inequations

$$|\dot{x}_j - \dot{\widehat{x}}_j| \leq \text{const } t\varepsilon^{w_j - 1} + \text{const } \varepsilon^{w_j}, \quad j = 1, \dots, n.$$

By integrating these inequations we obtain (58), which yields

$$\|x(t) - \widehat{x}(t)\| \leq \text{const } \varepsilon t^{1/r}.$$

Further, using Proposition 7.26, we may write

$$\widehat{d}_p\big(x(t), \widehat{x}(t)\big) \leq \text{const} \sum_{\{k,j \mid w_k \leq w_j\}} \varepsilon^{1 - w_k/w_j} |\varepsilon^{w_k} t|^{1/w_j} \leq \text{const } \varepsilon t^{1/r},$$

which is the required result.

(ii) If $x(0) = \widehat{x}(0) = p$, we may enforce (58) and (59) into $|x_j - \widehat{x}_j| \leq$ const $t^{w_j + 1}$, $\quad j = 1, \dots, n$ and $\widehat{d}(x(t), \widehat{x}(t)) \leq$ const $t^{1 + 1/r}$ in the same way, after noticing that in this case we have $\|\!|x(t)|\!\| \leq$ const t. $\qquad\blacksquare$

We come now to the main step of our proof.

Lemma 7.30. *Suppose p is a regular point. Then, there exist $\varepsilon > 0$ and $C > 0$ such that for $\|q\| \leq \varepsilon$ and $\|q'\| \leq \varepsilon$ the following holds:*

$$d(q, q') \leq C\|q' - q\| + C\|q\|^{1 - 1/r}\|q' - q\|^{1/r}. \tag{64}$$

Proof. Let $\widehat{\gamma} : [0, T] \to M$ be a geodesic for $\widehat{d}_q$, having constant velocity 1, and joining q to q'. Let $\gamma : [0, T] \to M$ be the path starting at q and defined by the same control functions, only replacing $\widehat{X}_i$ by X_i. Call $q^{(1)}$ the end-point of γ. Clearly

$$\widehat{d}_q(q, q') = T, \quad d(q, q^{(1)}) \le T.$$

Moreover by Proposition 7.29 we have

$$\|q' - q^{(1)}\| \le CT^{1+1/r}.$$

Using Lemma 7.28, we obtain

$$\widehat{d}_{q^{(1)}}(q^{(1)}, q') \le$$

$$\operatorname{const} T^{1+1/r} + \operatorname{const} T^{1-1/r}\big(T^{1+1/r}\big)^{1/r} = \operatorname{const} T^{1+1/r} + \operatorname{const} T^{1+1/r^2}$$

that is,

$$\widehat{d}_{q^{(1)}}(q^{(1)}, q') \le C'T^{1+1/r^2}$$

by taking T less than some constant.

Let now $\widehat{\gamma}_1 : [0, T_1] \to M$ a geodesic for $\widehat{d}_{q^{(1)}}$, having constant velocity 1, and joining $q^{(1)}$ to q', and let $\gamma_1 : [0, T_1] \to M$ be the path starting at $q^{(1)}$ and defined by the same control functions, replacing $\widehat{X}_i$ by X_i. Call $q^{(2)}$ the end-point of γ_1. In the same fashion as above, we have

$$\widehat{d}_{q^{(1)}}(q^{(1)}, q') = T_1, \quad d(q^{(1)}, q') \le T_1$$

and

$$\widehat{d}_{q^{(2)}}(q^{(2)}, q') \le C'T_1^{1+1/r^2}.$$

Continuing in this way, we construct a sequence $q^{(1)}$, $q^{(2)}$, $q^{(3)}$, ... of points such that

$$d(q^{(k)}, q') \le T_k,$$

with

$$T_k \le C'(T_{k-1})^{1+1/r^2}.$$

By taking ε small enough, we have $CT^{1/r^2} \le 1/2$. For such T, we have $T_k \le T_{k-1}/2$. Then, the series $T + T_1 + T_2 + T_3 + \cdots$ is convergent, and its sum is majorized by $2T$. By putting end to end the paths γ, γ_1, γ_2, we thus get a path of d-length $\le 2T$ joining q to q'. Since $T = \widehat{d}_q(q, q') \le \operatorname{const} \|q' - q\| + \operatorname{const} \|q\|^{1-1/r}\|q' - q\|^{1/r}$, we obtain estimate (64). ∎

Theorem 7.31. *Let p any point of M. Then, there exist $\varepsilon > 0$ and $C > 0$ such that for $\|q\| \le \varepsilon$ and $\|q'\| \le \varepsilon$ we have*

$$d(q, q') \le C\|q' - q\| + C\|q\|^{1-1/r}\|q' - q\|^{1/r}. \tag{65}$$

Proof. We have to deal with the case of a singular point p, as the case of regular p has been settled in Lemma 7.30. We shall reduce the singular case to the regular one. Our argument is inspired from the lifting technique of Rothschild and Stein.

Recall from the proof of Proposition 7.26 that the group G_p may be written as a product

$$G_p = H_p \exp(K), \tag{66}$$

where K is the vector space spanned in the Lie algebra of G_p by the vector fields $\widehat{Y}_1, \ldots, \widehat{Y}_n$. Recall from §5.3 the vector fields $\widehat{\xi}_1, \ldots, \widehat{\xi}_m$ on G_p representing the infinitesimal right action of $\widehat{X}_1, \ldots, \widehat{X}_m$ on G_p. Since the action of $\widehat{\xi}_1, \ldots, \widehat{\xi}_m$ preserve the fibration $G_p \to G_p/H_p$, these vector fields can be pushed down to $G_p/H_p = T_pM$, giving back the $\widehat{X}_i$.

The decomposition (66) gives rise to coordinates in G_p, given by

$$(x_1, \ldots, x_n, z_{n+1}, \ldots, z_{\tilde{n}}) \mapsto$$

$$\exp(z_{n+1}\widehat{Z}_{n+1} + \cdots + z_{\tilde{n}}\widehat{Z}_{\tilde{n}}) \exp(x_1\widehat{Y}_1 + \cdots + x_n\widehat{Y}_n).$$

(See again the proof of Proposition 7.26 for the definition of the $\widehat{Z}_j$.) By the above observation we may write

$$\xi_i(x, z) = \widehat{X}_i(x) + \sum_{j=n+1}^{\tilde{n}} \xi_{ij}(x, z)\frac{\partial}{\partial z_j}. \tag{67}$$

Now, we construct an extended manifold $\widetilde{M} = M \times \mathbb{R}^{\tilde{n}-n}$, and an extended system of vector fields on $\widetilde{M}$, by letting

$$\tilde{\xi}_i(x, z) = X_i(x) + \sum_{j=n+1}^{\tilde{n}} \xi_{ij}(x, z)\frac{\partial}{\partial z_j}, \tag{68}$$

with the same ξ_{ij} as in (67).

It is easy to see that $(p, 0)$ is regular point of $\widetilde{M}$, endowed with the sub-Riemannian structure defined by the $\tilde{\xi}_i$. The tangent space to $\widetilde{M}$ at p is (isomorphic to) G_p. Let $\tilde{d}$ the distance defined on $\widetilde{M}$ from vector fields $\tilde{\xi}_1, \ldots, \tilde{\xi}_m$. Clearly, the $\tilde{\xi}_i$ preserves the fibration $\widetilde{M} \to M$, and when pushing them down, one gets back the X_i. This has the effect that every path in M can be lifted to $\widetilde{M}$ by keeping the same length. It follows that the projection $\widetilde{M} \to M$ is distance-decreasing. In particular, its restriction to $M \times \{0\}$ is distance-decreasing: we have

$$d(q, q') \leq \tilde{d}\big((q, 0), (q', 0)\big).$$

Knowing first that $(p, 0)$ is a regular point in $\widetilde{M}$, next that coordinates $x_1, \ldots, x_n,\ z_{n+1}, \ldots, z_{\tilde{n}}$ are privileged coordinates around $(p, 0)$ in $\widetilde{M}$, coordinates $x_1, \ldots, x_n$ keeping the same weights as in M, we can deduce from Lemma 7.30 that

$$\tilde{d}\big((q, 0), (q', 0\big) \leq C\|q' - q\| + C\|q\|^{1-1/r}\|q' - q\|^{1/r}.$$

for q and q' sufficiently near from p. The proof of the theorem is therefore concluded. ∎

The reader will have noticed that we have used, in proving the key Lemma 7.30, a Newton-type method. To reach q' from q, we define successive approximations $q^{(1)}, q^{(2)}, \ldots, q^{(k)}, \ldots$ The definition of point $q^{(k)}$ from the preceding one $q^{(k-1)}$ needs the construction of a tangent space at $q^{(k-1)}$. This allows to get

$$\widehat{d}_{q^{(k)}}(q^{(k)}, q') \leq C'\big(\widehat{d}_{q^{(k-1)}}(q^{(k-1)}, q')\big)^{1+1/r^2}$$

and so to majorize $d\big(q^{(k)}, q'\big)$ by a sequence having zero as a limit.

To use a simple iterative method, based on the use of a single tangent space, namely T_pM, would have led to an induction step like

$$\widehat{d}_p(q^{(k)}, q') \leq C\big(\widehat{d}_p(q^{(k-1)}, q')\big)^{1/r}$$

and it would have been impossible to infer from that that $\lim\limits_{k \to \infty} q^{(k)} = q'$.

The occurring of methods of successive approximations should not come as a surprise. What we have proved is really an open mapping theorem (with bounds), containing Chow's theorem as its corollary, since it says, among other things, that $d(p, q)$ is finite. (By examining our proof, one can check that we have used the conclusion of Chow's theorem only for graded nilpotent Lie groups, where it results from a simple application of Campbell-Hausdorff formula.)

"When the differential of the end-point map E_p is surjective, then E_p is open at 0." This is the actual meaning of Chow's theorem. We only have to compute the differential by using a notion of differential adapted to our context. The differential of $E_p : L^1 \to M$ at 0 should be defined, using the dilations δ_λ, by

$$\widehat{E}_p(u) = \lim_{\varepsilon \to 0} \delta_{\varepsilon^{-1}} E_p(\varepsilon u) \tag{69}$$

and take its values in the Carnot group (or homogeneous space) $T_p M$. In the group case, one even has

$$\widehat{E}_p(u * v) = \widehat{E}_p(u)\widehat{E}_p(v).$$

In this perspective, observe that the effective proofs of Chow's theorem (such as in [14,21]) proceed by assigning to any point q near p a path leading from p to q. On could think this assignment as giving a section of the end-point map. Again, this section is differentiable if one computes the derivative using δ_λ, and it is only Hölderian if one uses the classical notion of derivative for maps $M \to L^1$.

7.4. Comparison of distances in M and in $T_p M$

Here, $\widehat{d} = \widehat{d}_p$.

Theorem 7.32. *There exists constants $\varepsilon > 0$ and $C > 0$ such that for any $q, q' \in B(p, \varepsilon)$, we have*

$$-C\widehat{d}(p,q)d(q,q')^{1/r} \le d(q,q') - \widehat{d}(q,q') \le C\widehat{d}(p,q)\widehat{d}(q,q')^{1/r}. \tag{70}$$

Proof. Let q and q' be such that $\|q\| \le \varepsilon$, $\|q'\| \le \varepsilon$, where ε is small enough. Consider a minimizing geodesic $\gamma : [0, T] \to M$, for the distance d, having velocity 1 and such that $\gamma(0) = q$, $\gamma(T) = q'$. We have $T = d(q,q') \le \mathrm{const}\, \varepsilon$. Denote by $\widehat{\gamma}$ the path defined by using the same starting point and the same control functions, only replacing $X_1, \ldots, X_m$ by their counterparts $\widehat{X}_1, \ldots, \widehat{X}_m$. Proposition 7.29 tells us that

$$\widehat{d}\big(\gamma(T), \widehat{\gamma}(T)\big) \le \mathrm{const}\, \widehat{d}(p,q)T^{1/r} \tag{71}$$

By the triangle inequality, we have

$$\begin{aligned}
d(q,q') = \mathrm{length}(\gamma) = T = \mathrm{length}(\widehat{\gamma}) &\ge \widehat{d}\big(q, \widehat{\gamma}(T)\big) \\
&\ge \widehat{d}(q,q') - \widehat{d}\big(q', \widehat{\gamma}(T)\big) \\
&\ge \widehat{d}(q,q') - \mathrm{const}\, \widehat{d}(p,q)d(q,q')^{1/r}.
\end{aligned}$$

We thus get the left hand part of (70). The second part is proven in the same way by exchanging the roles of d and $\widehat{d}$. Except for one point: to majorize $d\big(\widehat{\gamma}(T), \gamma(T)\big)$ we use the estimate

$$d\big(\gamma(T), \widehat{\gamma}(T)\big) \leq \text{const}\, \widehat{d}(p, q) T^{1/r} \tag{72}$$

which can be proved from (58) in the same way as (59), using only Theorem 7.31 instead of Proposition 7.26. We thus get

$$\widehat{d}(q, q') \geq d(q, q') - \text{const}\, \widehat{d}(p, q)\widehat{d}(q, q')^{1/r},$$

which is the right hand part of (70).   ∎

Corollary 7.33. *For any $\rho > 0$, there exists $\varepsilon > 0$ such that for any $q \in B(p, \varepsilon)$, we have*

$$(1 - \rho)\widehat{d}(p, q) \leq d(p, q) \leq (1 + \rho)\widehat{d}(p, q). \tag{73}$$

We also have

$$C\|q\| \leq d(p, q) \leq C'\|q\| \tag{74}$$

for some positive ε, C, C'.

Proof. The right hand half of (70) gives us

$$d(p, q) \leq \widehat{d}(p, q) + Cd(p, q)\widehat{d}(q, q')^{1/r}$$

Given α, we have $C\widehat{d}(q, q')^{1/r} \leq \alpha$ if q' is near enough from q. We then have $d(p, q) \leq \widehat{d}(p, q) + \alpha\, d(p, q)$ or

$$d(p, q) \leq \frac{1}{1 - \alpha}\, \widehat{d}(p, q) = (1 + \rho)\, \widehat{d}(p, q)$$

if α is chosen so that $1/(1 - \alpha) = 1 + \rho$. The left hand side of (74) is proved in a similar manner.   ∎

We had the following result in mind when defining the notion of privileged coordinates.

Theorem 7.34. *The estimate*

$$d\left(0, (y_1, \ldots, y_n)\right) \asymp |y_1|^{1/w_1} + \cdots + |y_n|^{1/w_n} \tag{75}$$

holds near p if and only if $y_1, \ldots, y_n$ form a system of privileged coordinates at p.

The "only if" part is by definition. The "if" part is a restatement of the preceding corollary.

Corollary 7.35. (The box theorem, without regularity assumptions) *There exists constants C, C' such that (using privileged coordinates)*

$$C\left[-\varepsilon,\varepsilon\right]\times\left[-\varepsilon^{w_2},\varepsilon^{w_2}\right]\times\cdots\times\left[-\varepsilon^{w_n},\varepsilon^{w_n}\right] \subset$$
$$B(p,\varepsilon) \subset C'\left[-\varepsilon,\varepsilon\right]\times\left[-\varepsilon^{w_2},\varepsilon^{w_2}\right]\times\cdots\times\left[-\varepsilon^{w_n},\varepsilon^{w_n}\right]. \tag{76}$$

for ε small enough. (Recall $w_1 = 1$.) This statement holds only if we use privileged coordinates.

Proof. This is a rewriting of Theorem 7.34, using the inequalities

$$\sup_{1\leq j\leq n} |x_j| \leq |x_1| + \cdots + |x_n| \leq n \sup_{1\leq j\leq n} |x_j|. \qquad\blacksquare$$

7.5. Identification of the metric tangent space

Theorem 7.36. *A sub-Riemannian manifold M admits a metric tangent space T_pM, in Gromov sense, at every point p. This space is isometric to the space $\mathbb{R}^n$ endowed with the sub-Riemannian metric associated to the vector fields $\widehat{X}_1,\ldots,\widehat{X}_m$ introduced in Definition 5.15. At regular points, it has a natural group structure.*

Proof. We have only to reexamine the estimates just given. Our estimates (70) imply that

$$\lim_{\varepsilon\to 0} \varepsilon^{-1}\big(d(q,q') - \widehat{d}(q,q')\big) = 0 \tag{77}$$

uniformly for q, q' in $B(p,R\varepsilon)$ (or $\widehat{B}(p,R\varepsilon)$), R being fixed. If $B(p,R\varepsilon)$ and $\widehat{B}(p,R\varepsilon)$ would coincide as subsets of $\mathbb{R}^n$ (recall again that, using a chosen system of privileged coordinates, we identify M and T_pM with $\mathbb{R}^n$), we would invoke Corollary 6.24 to obtain directly

$$\lim_{\varepsilon\to 0} \text{H-dist}\big(\varepsilon^{-1}B(p,R\varepsilon),\varepsilon^{-1}\widehat{B}(p,R\varepsilon)\big) = 0. \tag{78}$$

A little more care is needed. Corollary 7.33 says that, for given ρ, if ε is small enough, any point in q in $\widehat{B}(p,R\varepsilon)$ is within $\widehat{d}$ distance $\leq \rho\varepsilon$ from $B(p,R\varepsilon)$. So, choose a point q' in $B(p,\varepsilon)$ such that $\widehat{d}(q,q') \leq 2\rho\varepsilon$, and call it $f_\varepsilon(q)$. Using (77), we have, for any q and q' in $\widehat{B}(p,R\varepsilon)$,

$$\left|\widehat{d}(q,q') - d\big(f_\varepsilon(q),f_\varepsilon(q')\big)\right| \leq 2\rho\varepsilon + d(q,f_\varepsilon(q)) + \leq 2\rho\varepsilon + d(q,f_\varepsilon(q))$$
$$\leq 2\rho\varepsilon + o(\varepsilon),$$

and

$$\left|\widehat{d}(q,q') - d\big(f_\varepsilon(q), f_\varepsilon(q')\big)\right| \leq 3\rho\varepsilon$$

by taking ε small enough. In the same way, we define $g_\varepsilon : B(p, R\varepsilon) \to \widehat{B}(p, R\varepsilon)$ such that

$$\left|d(q,q') - \widehat{d}\big(g_\varepsilon(q), g_\varepsilon(q')\big)\right| \leq 3\rho\varepsilon$$

if ε is small enough. We now apply Lemma 6.23 to show that

$$\lim_{\varepsilon \to 0} \text{H-dist}\big(\varepsilon^{-1}B(p, R\varepsilon), \varepsilon^{-1}\widehat{B}(p, R\varepsilon)\big) = 0.$$

Since $\varepsilon^{-1}\widehat{B}(p, R\varepsilon)$ is isometric to $\widehat{B}(p, R)$ by virtue of the homogeneity of T_pM, we have thus proved that T_pM is the metric tangent space, in Gromov sense, of M at point p. ∎

7.6. Comparison between the manifold and its tangent spaces. Hölder equivalence and Lipschitz equivalence

Can we use T_pM as a local model for M? In other terms, can we pass from an infinitesimal equivalence between T_pM and M to a stronger, local equivalence?

The map

$$\phi_p : T_pM \to M$$

we have used to compare distances d and $\widehat{d}$ is of Hölder class $C^{1/r}$, by (70), as well as its inverse. (Recall ϕ_p is defined as being the identity in some fixed system of privileged coordinates.) In fact, (70) gives rise to a Lipschitz ϕ_p only when $r = 1$, that is, in the Riemannian case.

This does not prevent the existence of bi-Lipschitz homeomorphisms between M and T_pM. Most likely Lipschitz equivalences between smooth sub-Riemannian manifolds come from differentiable equivalences between the underlying distributions. (As we have said, when the rank of $X_1, \ldots, X_m$ is not constant, the role of the distribution D may be taken up by the submodule generated over the smooth functions by vector fields $X_1, \ldots, X_m$.)

If we replace $X_1, \ldots, X_m$ by $X_1', \ldots, X_m'$ where

$$X_j'(q) = \sum_{i=1}^{m} a_{ij}(q)X_i(q)$$

and $(a_{ij}(q))$ is a smooth invertible $m \times m$ matrix, we get a new sub-Riemannian distance d', which is Lipschitz equivalent to d. So, to each distribution is canonically associated a class of Lipschitz equivalent distances. Conversely, a distribution is determined by the class of distances it defines. Proof in the constant rank case: you can measure the length of a path $c : [0,1] \to M$ using

$$\text{length}(c) = \sup_{0=t_0<t_1<\cdots<t_n=1} \sum d(c(t_i), c(t_{i+1}));$$

once you know which smooth paths have finite length, you know their tangent vectors, hence E_q for every q.

It follows that for smooth sub-Riemannian metrics, the problem of Lipschitz equivalence (as far as these equivalence are realized through diffeomorphisms) is the same as the old problem of equivalence of distributions, which has been studied by Engel, Cartan, Darboux and so many mathematicians. It is certainly true that the distributions on M and T_pM are equivalent when $\dim M = 2k + 1$, $\text{rank } E = 2k$ and $r = 2$, in which case T_pM is isomorphic, for every p, to the Heisenberg group H_{2k+1}. This is a consequence of the well-known theorem of Darboux on normal forms for contact differential forms.

Now, an obstruction to the existence of Lipschitz diffeomorphisms between M and T_pM is the following: since T_pM is a group, all its tangent spaces are isomorphic to it. So, in case such a diffeomorphism would exist, in a neighbourhood of p, all tangent spaces to M should be isomorphic to T_pM. A counter-example to this conclusion has been built by A. Varchenko [36]: For generic regular distributions of dimension 8 on a manifold of dimension 11, the algebraic structure of T_pM varies from point to point. See also [39], pp. 33–34.

One may wonder if this is the only obstruction to the equivalence of distributions (in the regular case). In other words: since the T_pM can be thought as 1-jets of sub-Riemannian metrics, does some form of the h-principle hold, which would assert that if all the T_pM are isomorphic to a given group G then M is locally Lipschitz-equivalent to G?

8. Why is the tangent space a group?

8.1.   Why is the tangent space a group at regular points? Since the first time I realized (after reading Métivier [22]) that Theorem 7.36 should hold, I have been puzzled by this question. Drawing a Lie algebra from the bracket structure of some X_i's did not seem to me the appropriate answer. I remember having, at last, asked M. Gromov about it (1982). The answer came under the form of a little apologue:

Take a map $f : \mathbb{R}^m \to \mathbb{R}^n$. Define its differential as

$$D_x f(u) = \lim_{\varepsilon \to 0} \varepsilon^{-1} \big[f(x + \varepsilon u) - f(x) \big], \tag{79}$$

provided convergence holds. Then $D_x f$ is certainly homogeneous:

$$D_x f(\lambda u) = \lambda D_x f(u),$$

but it need not satisfy the additivity condition

$$D_x f(u + v) = D_x f(u) + D_x f(v).$$

Try $f(x) = |x|$ in dimension one (using $\varepsilon > 0$ only), or $f(x, y) = xy/(x^2 + y^2)$ in $\mathbb{R}^2$ (with ε of arbitrary sign). However, if the convergence in (79) is uniform on some neighbourhood of $(x, 0)$ in $\mathbb{R}^n \times \mathbb{R}^n$, then $D_x f$ is additive, hence linear. So, uniformity was the key. The tangent space at p is a limit, in the Hausdorff sense, of pointed spaces:

$$(T_p M, 0) = \lim_{\varepsilon \to 0} (\varepsilon^{-1} M, p). \tag{80}$$

It certainly is a homogeneous space—in the sense of a metric space having a 1-parameter group of dilations. But when the convergence is uniform with respect to p, which is the case near regular points, in addition, it is a group.

Before giving the proof, I want to tell of another, later, hint, coming from the work of A. Connes. He has made significant use of the following observation: The tangent bundle TM to a differentiable manifold M is, like $M \times M$, a groupoid. In $M \times M$ the composition law is

$$(p, q) * (p', q') \begin{cases} = (p, q') & \text{if } p' = q, \\ \text{not defined} & \text{if } p' \neq q. \end{cases} \tag{81}$$

while TM is a groupoid under the law

$$(p, v) * (p', v') \begin{cases} = (p, v + v') & \text{if } p' = p, \\ \text{not defined} & \text{if } p' \neq p. \end{cases} \tag{82}$$

In fact TM is simply a union of groups. In [8], II.5, it is stated that its structure may be derived from that of $M \times M$ by blowing up the diagonal in $M \times M$. (Technically, the connection between $M \times M$ and TM is by means of a differentiable structure and a compatible groupoid law on

$TM \cup M \times M \times]0,1]$.) This suggests that, putting metrics back into the picture, one should have

$$TM = \lim_{\varepsilon \to 0} \varepsilon^{-1} (M \times M) \tag{83}$$

(which appear to be a strengthening of (80)) in some sense to be made precise.

There is still one question. Since the differentiable structure of our manifold M is the same as in Connes' picture, why do we not get the same abelian group structure? One can answer: The differentiable structure is strongly connected to (the equivalence class of) Riemannian metrics; differentiable maps are locally Lipschitz, and Lipschitz maps are almost everywhere differentiable. There is no such connection between differentiable maps and the metric when it is sub-Riemannian. Put in another way, differentiable maps have good local commutation properties with ordinary dilations, but not with sub-Riemannian dilations δ_λ.

So, one should not be abused by (83) and think that the algebraic structure of $T_p M$ stems from the absolutely trivial structure of $M \times M$! It is concealed in dilations, as we shall now prove.

8.2. A purely metric derivation of the group structure in $T_p M$ for regular p

What distinguishes in a sub-Riemannian manifold singular points from regular ones is that, if p is a regular point, the convergence in

$$\lim_{\varepsilon \to 0} \varepsilon^{-1}(M, q) = (T_q M, q) \tag{84}$$

(it is as well to denote by q, instead of 0, the distinguished point in $T_q M$) is uniform with respect to q in some neighbourhood of p, while it is definitely non-uniform at singular points. By uniform convergence in (84) we mean that, for any $R > 0$,

$$\text{H-dist}\left(B^{\varepsilon^{-1} M}(q, R), B^{T_q M}(q, R)\right)$$

tends to zero as ε tends to zero, uniformly with respect to q.

Let us try to extract the group law in $T_p M$ from the mere groupoid law in $M \times M$: put (p, q) and (q, r) end to end to get (p, r). As we have seen, (84) is expressed by the existence of some mapping

$$\phi_q : T_q M \to M$$

such that

$$\widehat{d^q}(X,Y) - d(\phi_q(X), \phi_q(Y)) = o(\varepsilon) \tag{85}$$

if $\widehat{d^q}(0, X) \le \varepsilon$, $\widehat{d^q}(0, Y) \le \varepsilon$. Since p is regular, the $o(\varepsilon)$ is uniform with respect to q, for q in some neighbourhood of p.

To $(p, X) \in T_pM$ corresponds $(p, q) \in M \times M$, with $q = \phi_p(X)$. Likewise, to $(p, Y) \in T_pM$ corresponds $(p, r) \in M \times M$, with $r = \phi_p(Y)$. To define the product $X * Y$, we would want to have $p = q$, which has not much sense, or Y to belong to T_qM, instead of T_pM.

There is a solution: To let Y pass from T_pM to T_qM, let it grow big, by using some dilation with large ratio λ. Then it will be possible to replace (p, r) by (q, r). The error will be harmless if $d(p, r)$, or $\widehat{d^p}(0, Y)$, is large compared to $d(p, q)$. Then bring back Y to its original size, by using the dilation with ratio λ^{-1}, now in T_qM. This leads us to the formula

$$X * Y = \lim_{\lambda \to \infty} \phi_{pq}^{-1} \delta_{q,\lambda^{-1}} \phi_{pq} \delta_{p,\lambda} Y, \tag{86}$$

where $q = \phi_p(X)$, and ϕ_{pq} denotes $\phi_q^{-1} \phi_p$. The existence of the limit, and the verification of group axioms stem from the uniformity in (85).

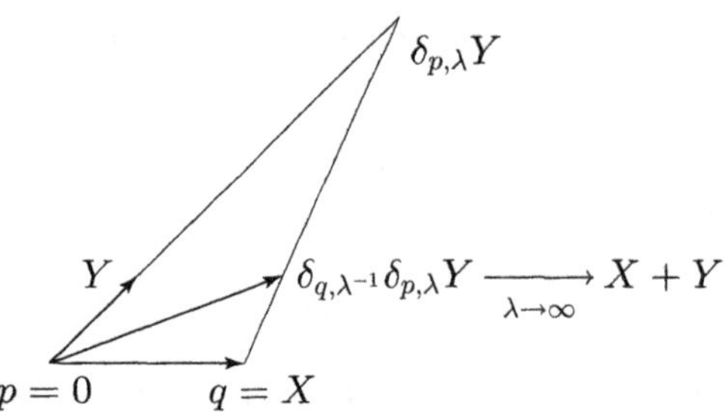

Fig. 5: The group structure of the plane with an origin, defined from dilations

To understand (86), one may apply it to the case of $\mathbb{R}^n$, with $\phi_p(X) = p + X$. It shows how to define the addition of two vectors of same origin, using only dilations and a passage to the limit.

References

[1] A. A. Agrachev and R. V. Gamkrelidze, The exponential representation of flows and the chronological calculus, *Mat. Sbornik* (N.S.), **107** (149) (1978), 467–532, 639. English transl.: *Math. USSR Sbornik*, **35** (1979), 727–785.

[2] A. A. Agrachev, R. V. Gamkrelidze and A. V. Sarychev, Local invariants of smooth control systems, *Acta Appl. Math.* **14** (1989), 191–237.

[3] A. Bellaïche, Métriques de Carnot-Carathéodory sur les variétés, unpublished manuscript, 1981.

[4] A. Bellaïche, J.-P. Laumond and J.-J. Risler, Nilpotent infinitesimal approximations to a control Lie algebra, in *IFAC Nonlinear Control Systems Design Symposium* (NOLCOS 92'),174–181, Bordeaux, France, June 1992.

[5] N. Bourbaki, *Variétés différentielles et analytiques. Fascicule de résultats* (§1 à 7.), Hermann, Paris, 1967.

[6] R. W. Brockett, Control theory and singular Riemannian geometry, in *New directions in applied mathematics*, P. J. Hilton and G. J. Young eds., Springer-Verlag, 1982.

[7] W. L. Chow, Über Systeme von linearen partiellen Differentialgleichungen erster Ordnung, Math. Ann., **117**, (1939) 98–105.

[8] A. Connes, *Non commutative geometry*, Academic Press, San Diego, 1994.

[9] C. Fefferman and D. H. Phong, Subelliptic eigenvalue problems, in *Conference on harmonic analysis in honor of Antoni Zygmund*, (Chicago, Ill., 1981), Wadsworth Math. Ser., Wadsworth, Belmont, Calif., 1983, Vol. II, 590–606

[10] B. Gaveau, Principe de moindre action, propagation de la chaleur et estimées sous-elliptiques sur certains groupes nilpotents, *Acta Math.*, **139** (1977), 95–153.

[11] B. Gaveau, Systèmes dynamiques associés à certains opérateurs elliptiques, *Bull. Sc. Math.*, 2e série, **102** (1978), 203–229.

[12] N. Goodman, *Nilpotent Lie groups*, Springer Lecture Notes in Mathematics, Vol. 562, 1976.

[13] M. Gromov (avec J. Lafontaine, P. Pansu), *Structures métriques pour les variétés riemanniennes*, Cedic-Nathan, Paris, 1981.

[14] M. Gromov, *Carnot-Carathéodory spaces seen from within*, this volume.

[15] V. V. Grušin, On a class of hypoelliptic operators, *Math. Sb.*, **83** (125) (1970), 456–473 (Russian). Engl. trad.: *Math. USSR Sb.* **12** (1970), 458–476.

[16] V. V. Grušin, A certain class of elliptic pseudodifferential operators that are degenerate on a submanifold, *Math. Sb.*, **84** (126) (1971), 163–195 (Russian). Engl. trad.: *Math. USSR Sb.* **13** (1971), 155–185.

[17] B. Helffer et J. Nourrigat, Approximation d'un système de champs de vecteurs, et applications à l'hypoellipticité, *Arkiv för Matematik*, **19** (1979), 237–254.

[18] R. Hermann, Geodesics of singular Riemannian metrics, *Bull. AMS* **79** (1973), 780–782.

[19] H. Hermes, Nilpotent and high-order approximations of vector field systems, *SIAM Review*, **33** (2), 238–264.

[20] L. Hörmander, Hypoelliptic second order differential equations, *Acta Math.*, **119** (1967), 147–171.

[21] C. Lobry, Contrôlabilité des systèmes non linéaires, *SIAM J. Control*, **8** (1970), 573–605. Erratum: "Contrôlabilité des systèmes non linéaires", *SIAM J. Control Optimization*, **14** (1976), 387.

[22] G. Métivier, *Comm. Partial Differential Equations*, **1** (1976), 467–519.

[23] J. Mitchell, On Carnot-Carathéodory metrics, *Journal of Differential Geom.*, **21** (1985) 35–45.

[24] O. A. Oleinik and E. V. Radkevic, *Second order equations with nonnegative characteristic form*, Plenum Press, New York, London, 1973.

[25] P. Pansu, Métriques de Carnot-Carathéodory et quasi-isométries des espaces symétriques de rang un, *Ann. of Math.*, (2) **129** (1989), 1–60.

[26] P. K. Rashevsky, Any two points of a totally nonholonomic space may be connected by an admissible line, *Uch. Zap. Ped. Inst. im. Liebknechta, Ser. Phys. Math.*, **2** (1938), 83–94 (Russian).

[27] Ch. Rockland, Intrinsic nilpotent approximations, *Acta Appl. Math.*, **8** (1987), 213–270.

[28] L. P. Rothschild and E. M. Stein, Hypoelliptic differential operators and nilpotent groups, *Acta Math.*, **137** (1976), 247–320.

[29] M. Spivak, *A comprehensive introduction to differential geometry*, Vol. 1, Sec. edition, Publish or Perish, Inc., Houston, Texas, 1979.

[30] P. Stefan, Accessible sets, orbits, and foliations with singularities, *Proc. London Math. Soc.*, (3) **29** (1974), 699–713.

[31] E. M. Stein, Some problems in harmonic analysis suggested by symmetric spaces and semi-simple groups, in *Actes du Congrès international des mathématiciens*, Nice 1970, Vol. 1, 173–189.

[32] E. M. Stein, *Harmonic Analysis*, Princeton University Press, Princeton, N.J., 1993.

[33] R. S. Strichartz, Sub-Riemannian geometry, *J. Differential Geom.*, **24**, (1986), 221–263.

[34] H. J. Sussmann, V. Jurdjevic, Controllability of nonlinear systems, *J. Differential Equations*, **12** (1972), 95–116.

[35] H. J. Sussmann, Orbits of families of vector fields and integrability of distributions, *Trans. Amer. Math. Soc.*, **180** (1973), 171–188.

[36] A. N. Varchenko, Obstruction to local equivalence of distributions, *Mat. Zametki*, **29** (1981), 939–947 (Russian). Engl. Trad., *Math. notes*, **29** (1981), 479–484.

[37] N. Th. Varopoulos, L. Saloff-Coste and T. Coulhon, *Analysis and geometry on groups*, Cambridge University Press, 1992.

[38] A. M. Vershik and V. Ya. Gershkovich, Nonholonomic problems and the theory of distributions, *Acta Appl. Math.*, **12**, 181–209, 1988.

[39] A. M. Vershik and V. Ya. Gershkovich, Nonholonomic dynamical systems, geometry of distributions and variational problems, in *Dynamical Systems VII*, V. I. Arnold, S. P. Novikov (Eds), Encyclopaedia of Mathematical Sciences, vol. 16, Springer, 1994. (Original Russian edition, VINITI, Moscow, 1987).

Progress in Mathematics, Vol. 144, © 1996 Birkhäuser Verlag Basel/Switzerland

Carnot-Carathéodory spaces
seen from within

MIKHAEL GROMOV[*]

Contents

---

[*]**Acknowledgment**. I thank Richard Montgomery for reading the manuscript and locating a multitude of errors.

0. Basic definitions, examples and problems

Let V be a smooth manifold where we distinguish a subset $\mathcal{H}$ in the set of all piecewise smooth curves c in V. We assume that $\mathcal{H}$ is defined by a *local* condition on curves, i.e. if c is divided into segments $c_1, \ldots, c_k$, then

$$c \in \mathcal{H} \Longleftrightarrow c_i \in \mathcal{H}, \ i = 1, \ldots, k.$$

Next we pick up some Riemannian metric g in V and define

$$\mathrm{dist}(v_1, v_2) = \mathrm{dist}_{\mathcal{H},g}(v_1, v_2), \ v_1, v_2 \in V,$$

as the infimum of the lengths of the *distinguished* curves joining v_1 and v_2 in V. This distance obviously satisfies the usual axioms of a metric, provided every two points in V can be joined by a (distinguished) curve $c \in \mathcal{H}$. Otherwise, dist becomes infinite at the pairs of points in V which admit no distinguished curve joining them.

Example. Let V be the Euclidean plane $\mathbb{R}^2$ and $\mathcal{H}$ consist of piecewise linear curves, where each segment is either vertical or horizontal. Then the corresponding distance between the points $v_1 = (x_1, y_1)$ and $v_2 = (x_2, y_2)$ equals $|x_1 - x_2| + |y_1 - y_2|$, where we use the Euclidean metric of $\mathbb{R}^2$ for g. Notice that this $\mathrm{dist}_{\mathcal{H}}$ is *equivalent* to the ordinary Euclidean distance $\mathrm{dist}_{\mathrm{Eu}}$ in the sense that the identity map $(\mathbb{R}^2, \mathrm{dist}_{\mathcal{H}}) \to (\mathbb{R}^2, \mathrm{dist}_{\mathrm{Eu}})$ is *bi-Lipschitz*, i.e. $C^{-1} \mathrm{dist}_{\mathrm{Eu}} \leq \mathrm{dist}_{\mathcal{H}} \leq C \mathrm{dist}_{\mathrm{Eu}}$ for some $C > 0$. (Here one may take $C = \sqrt{2}$). This equivalence makes $\mathcal{H}$ and $\mathrm{dist}_{\mathcal{H}}$ rather non-interesting from our present point of view.

0.1. Polarizations, horizontal curves and Carnot-Carathéodory metrics. A *polarization* of a manifold V is, by definition, a subbundle of the tangent bundle, say $H \subset T(V)$. One may think of H as a distinguished set of directions (tangent vectors) in V which are called in sequel *horizontal*. (This terminology is motivated by the picture where V is smoothly fibered over some manifold B and H is normal to the fibers.)

A piecewise smooth curve in V is called *horizontal* with respect to H if the tangent vectors to this curve are horizontal.

The metric defined with (the set $\mathcal{H}$ of) the horizontal curves in V is called the *Carnot-Carathéodory metric* associated to H and denoted

dist_H. Notice that the definition of dist_H also involves an auxiliary Riemannian metric g as

$$\mathrm{dist}_H(v_1, v_2) \underset{\mathrm{def}}{=} \inf(g\text{-lengths of } H\text{-horizontal curves between } v_1 \text{ and } v_2),$$

but the effect of g on dist_H is non-essential from our point of view. Namely, two metrics defined with different g_1 and g_2 and some H are bi-Lipschitz equivalent (on each compact subset in V). On the other hand, the role of H is crucial as is seen in the following

0.2. Basic contact example. Let $V = \mathbb{R}^3$ and H be the standard *contact* subbundle, which is the kernel of the (contact) 1-form $\eta = dz + x\,dy$ on $\mathbb{R}^3$. This means that the tangent space (plane) $H_{v_0} \subset T_{v_0}(\mathbb{R}^3) = \mathbb{R}^3$ is given at each $v_0 = (x_0, y_0, z_0) \in \mathbb{R}^3$ by the equation $z + x_0 y = 0$. Notice, that H is generated by the following two independent vector fields, $\partial_1 = \frac{\partial}{\partial x}$ and $\partial_2 = \frac{\partial}{\partial y} - x\frac{\partial}{\partial z}$. These fields do not commute. In fact, their commutator equals $-\frac{\partial}{\partial z}$ and so the three fields ∂_1, ∂_2, and the Lie bracket $[\partial_1, \partial_2]$ span the tangent bundle $T(\mathbb{R}^3)$ at each point $v \in \mathbb{R}^3$.

0.2.A. Connectivity theorem for the contact polarization H.

Theorem. *Every two points in $\mathbb{R}^3$ can be joined by a smooth H-horizontal curve.*

Proof. Take a curve $\underline{c} = (x(t), y(t))$, $t \in [0, 1]$, in the (x, y)-plane joining two given points (x_1, y_1) and (x_2, y_2) and such that the *formal area* "bounded" by $\underline{c}$, defined by the integral $\int_{\underline{c}} x\,dy = \int_0^1 x(t)y'(t)dt$, equals a given number a. (One easily finds such $\underline{c}$, say among curves of constant curvature). Then we take the horizontal lift of $\underline{c} = (x(t), y(t))$ to $\mathbb{R}^3$ by letting $z(t) = z_1 - \int_0^t x(t)y'(t)dt$ for a given value z_1 of z. The lifted curve $c = (x(t), y(t), z(t))$ is indeed horizontal as $dz(t) = z'(t)dt = -x(t)y'(t)dt = -x(t)dy(t)$ and it joins the given points (x_1, y_1, z_1) and $(x_2, y_2, z_2 = z_1 + a)$. $\blacksquare$

Historical Remarks. This result (which seems obvious by the modern standards) appears (in a more general form) in the 1909-paper by Carathéodory on formalization of the classical thermodynamics where horizontal curves roughly correspond to *adiabatic* processes. In fact, the above proof may be performed in the language of Carnot (cycles) and for this reason the metrics dist_H were christened "Carnot-Carathéodory" in [G-L-P].

I suspect that some form of the connectivity theorem was known to Lagrange in the framework of the *nonholonomic mechanics*. (Compare [Ve-Fa] and [Ver-Ger$_{1;2}$]; an instance of a nonholonomic system is given by a billiard ball rolling on the plane, such that the velocity at the lowest point of the ball where it touches the plane must be zero. Here V equals the configuration space, that is $\mathbb{R}^2 \times SO(3)$, and the nonholonomic constraint (on the velocity) is represented by a subbundle $H \subset T(V)$ of rank 3. Now, a child can roll the ball from one given position to another thus proving the connectivity property for H.) Various forms of the connectivity theorem have been persistently appearing in the literature but I have not tried to keep track of them. By tradition, a general commonly used connectivity theorem (see 0.4.) is attributed to Chow (see [Cho], though his paper was neither the first (see, e.g. [Rash]) nor the last on that matter. Finally we notice that the phenomenon of H-connectivity is also seen in the theory of optimal control and in robotics where it is named "controllability" (see [Brock]).

0.2.A'. Contact C-C metric on (V, H). Now, let us look more closely at the C-C metric (C-C = Carnot-Carathéodory) associated to a contact subbundle $H \subset T(\mathbb{R}^3)$. We recall that the H-horizontal curves c in $\mathbb{R}^3$ are the lifts of curves in the (x, y)-plane, such that the z-coordinate of c equals the formal area of the (x, y)-projection $\underline{c}$ of c. If two points v_1 and v_2 in $\mathbb{R}^3$ lie on the same vertical line (or z-line), i.e. have equal (x, y)-coordinates, then the (x, y)-projections $\underline{c}$ of curves c joining these points are *closed* in the (x, y)-plane and so the (formal) area of these curves $\underline{c}$ is bounded by

$$\mathrm{area}\,\underline{c} \leq \mathrm{const}(\mathrm{length}\,\underline{c})^2 \leq \mathrm{const}(\mathrm{length}\,c)^2,$$

where $\mathrm{const} = (4\pi)^{-1}$. It follows that the C-C distance between v_1 and v_2 is bounded from below by the Euclidean distance as follows,

$$\mathrm{dist}_H \geq \mathrm{const}^{-\frac{1}{2}} (\mathrm{dist}_{\mathrm{Eu}})^{\frac{1}{2}}, \qquad\qquad (*)$$

since the Euclidean distance $\mathrm{dist}_{\mathrm{Eu}}$ between our points equals $z_1 - z_2 = \mathrm{area}\,\underline{c}$. One also has the upper bound

$$\mathrm{dist}_H \leq \alpha\,\mathrm{const}^{-\frac{1}{2}}\,\mathrm{dist}_{\mathrm{Eu}}^{\frac{1}{2}}$$

where α is a certain positive function depending on the Euclidean norms of the points v_1 and v_2. (One may take $\alpha = \left(1 + \|v_1\| + \|v_2\|\right)^2$). In fact, one can join v_1 and v_2 by a curve c in $\mathbb{R}^3$ which projects to a circle $\underline{c}$ in $\mathbb{R}^2$, such that

$$\operatorname{dist}_{\mathrm{Eu}} = \operatorname{area}\underline{c} = (4\pi)^{-1}(\operatorname{length}\underline{c})^2$$

while

$$\operatorname{dist}_H \leq \operatorname{length} c \leq \alpha \operatorname{length}\underline{c}.$$

$(\sqrt{\ })$**-Conclusion.** *On every vertical line in $\mathbb{R}^3$ the Carnot-Carathéodory metric is locally equivalent to $\sqrt{\ }$ Euclidean metric.*

On the other hand, the C-C metric is locally equivalent to the Euclidean one on every horizontal curve, for example, on every (straight) x-line in $\mathbb{R}^3$. This makes the C-C metric highly non-isotropic. In fact, a small ε-ball in the C-C metric around each point $v \in \mathbb{R}^3$ looks roughly as a rectangular solid (box) with two ε-edges and one edge of length ε^2 (see 0.3.C). The $\varepsilon\times\varepsilon$-face of this solid is positioned in R^3 approximately tangent to H at v, while the ε^2-edge is approximately normal to H. (Our idea of "approximate" is such that the Euclidean ε-ball is roughly the same thing as an ε-cube). It follows, that the (Euclidean) volume of the ε-C-C ball is about ε^4 and consequently *the Hausdorff dimension of $\mathbb{R}^3$ with the C-C metric* dist_H *equals* 4.

Integrable non-example. If we take the form dz instead of $dz + x\,dy$, then the kernel subbundle H becomes integrable and the C-C metric degenerates: on each horizontal plane dist_H equals the Euclidean distance while every two points with non-equal z-coordinates have $\operatorname{dist}_H = \infty$. This can be remedied by introducing another metric which, by definition, equals $\sqrt{\text{Euclidean}}$ in the z-directions, see [N-S-W] and 1.5.

0.2.B. Internal versus external in C-C geometry. If we live inside a Carnot-Carathéodory metric space V we may know nothing whatsoever about the (external) infinitesimal structures (i.e. the smooth structure on V, the subbundle $H \subset T(V)$ and the metric g on H) which were involved in the construction of the C-C metric. For example, the relations between the C-C metric and the Euclidean one (such as the above equivalence C-C metric $\approx \sqrt{\text{Euclidean}}$ metric on the vertical lines in $\mathbb{R}^3$ and the ε^4-estimates for the *Euclidean* volume of C-C balls) remain invisible for a C-C insider. On the other hand the equality $\dim_{\mathrm{Hau}} V = 4 = \dim_{\mathrm{top}} V + 1$ is an internal feature stated in the intrinsic metric terms.

Main problems

(1) Develop a sufficiently rich and robust internal C-C language which would enable us to capture the essential external characteristics of our C-C spaces. For example, we would like to recognize rank H by a purely C-C metric consideration invariant under a sufficiently large class of transformations, such as bi-Lipschitz transformations and, even better, C^α-Hölder transformations, where $\alpha = 1-\varepsilon$ for a *positive* (albeit small) ε.

(2) Develop external (analytic) techniques for evaluation of internal invariants of V. For example, in order to determine $\dim_{\mathrm{Hau}} V$ we need the external information on the Euclidean volume of small C-C balls.

The purpose of the present paper is to expose what is known in these directions. The basic internal invariants we look at are concerned with subspaces $V' \subset V$ and here the best understood ones are curves ($\dim_{\mathrm{top}} V' = 1$) and hypersurfaces ($\mathrm{codim}_{\mathrm{top}} V' = 1$). On the other hand, the subspaces of intermediate dimensions offer more challenging geometric problems where we are still far from the final solution.

0.3. Heisenberg group view on the contact example. The geometry of the contact C-C metric becomes infinitely more transparent if we replace the Euclidean metric by another Riemannian metric as follows. (As we mentioned earlier, this change does not affect the essential C-C features.) In fact, instead of $\mathbb{R}^3$ we take the three-dimensional *Heisenberg group G* which can be defined as the only simply connected nilpotent non-Abelian Lie group. The Lie algebra $L = L(G)$ of G admits a basis x, y, z, such that $[x, z] = [y, z] = 0$ and $[x, y] = z$. (These relations uniquely define L and, hence, G.) We introduce a polarization $H \subset T(G)$ by taking the left translates of the (x, y)-subspace $H_0 \subset L = T_{\mathrm{id}}(G)$. (One knows, that there exists a diffeomorphism between G and $\mathbb{R}^3$ sending this H to the standard contact subbundle in $\mathbb{R}^3$ but this is not crucial at the present moment.) Next we take a left invariant metric g on G and let dist_H be the C-C metric defined with H and g. First, we must make sure that this is indeed a *metric* by checking the connectivity property for H. This can be done in (at least) two ways. One possibility is to look at the homomorphism (projection)

$$G \to \mathbb{R}^2 = G/\mathrm{center},$$

where the center of G (obviously) equals the 1-parameter subgroup obtained by the exponentiation of the (central) line spanned by z in $L(G)$. The (contact) geometry of this projection of $G = \mathbb{R}^3$ to $\mathbb{R}^2$ is identical to the (x, y)-projection of the basic contact example and the proof of the connectivity theorem 0.2.A applies.

Here is another approach.

0.3.A. Lie group theoretic proof of connectivity. Consider the one-parameter groups G_x and G_y of (right) translations of G corresponding to x and y in $L(G)$. The orbits of these subgroups are obviously tangent to H. On the other hand, G_x and G_y, viewed as subgroups in G, generate G since x and y Lie-generate $L(G)$. It follows, that every two points in G can be joined by a piecewise smooth curve whose every segment is a piece of an orbit of G_x or of G_y. Q.E.D.

0.3.A$'$. Connectivity theorem for general Lie groups. Let G be an arbitrary connected Lie group and H_0 a linear subspace in the Lie algebra of G. This H_0 defines a left invariant polarization $H \subset T(G)$ and then one defines the Carnot-Carathéodory metric dist_H on G. This is an honest metric (nowhere ∞) if and only if the subspace H_0 Lie-generates the Lie algebra $L(G)$. Furthermore, if the auxiliary Riemannian metric used in the definition of dist_H were left invariant, then dist_H is also left invariant on G.

0.3.B. Self-similarity. The C-C metric on the Heisenberg group has an additional remarkable feature which is absent for the left invariant Riemannian metrics on general G. Namely, besides being left invariant and thus admitting a transitive group of isometrics, (which is, of course, a property shared by all left invariant Riemannian metrics) the Carnot-Carathéodory metric on G admits non-trivial *self-similarities*. In fact *there exists a 1-parameter group of diffeomorphisms* $A_t : G \to G, t \in \mathbb{R}_+^\times$, *such that* $A_t \, \mathrm{dist}_H = t \, \mathrm{dist}_H$ *for all* $t \in \mathbb{R}_+^\times$, *which means*

$$\mathrm{dist}\big(A_t(v_1), \, A_t(v_2)\big) = t \, \mathrm{dist}(v_1, v_2)$$

for all v_1 *and* v_2 *in* G.

Proof. Define automorphisms a_t of the Lie algebra $L(G)$ by $(x, y, z) \mapsto (tx, ty, t^2 z)$ for the (x, y, z)-basis, where z is central and $[x, y] = z$. These are, clearly, automorphisms and they exponentiate to automorphisms of G, called $A_t : G \to G$. These A_t preserve H (which is generated by x and y) and they scale the Riemannian metric on H (but not on all of $T(G)$) by t. Therefore, the length of all horizontal (i.e. tangent H) curves is also scaled by t and then dist_H is scaled by t as well.   ■

General self-similar Lie groups. Let G be a simply connected nilpotent Lie group where the Lie algebra L admits a grading, $L = \bigoplus_{i=1}^{d} L_i$, such that $[L_i, L_j] \subset L_{i+j}$. Then the operator $a_t : L \to L$, $t \in \mathbb{R}_+^\times$, defined by $a_t \ell = t^i \ell$ for $\ell \in L_i$ are automorphisms of L. These a_t integrate to a 1-parameter group of automorphisms $A_t : G \to G$, which are similarities for the Carnot metric defined with the polarization corresponding to L_1. (Here as earlier we should assume that L_1 Lie generates L in order to have dist $< \infty$.)

Two-step example. Let G be a two-step nilpotent group. Then L can be graded with $L_2 = [L, L]$ and some subspace $L_1 \subset L$ complementary to L_2. This L_1 obviously Lie-generates L and thus gives us a self-similar C-C metric on G.

Remarks. If V is a homogeneous *Riemannian* manifold which admits a non-trivial self-similarity, then V, clearly, is isometric to $\mathbb{R}^n$ for some n. Analogous non-Riemannian examples are provided by Banach spaces of finite or infinite dimension but the self-similar nilpotent Carnot-Carathéodory manifolds do not spring in one's mind so readily. One knows now-a-days that there are no additional essential examples among *finite dimensional manifolds* (see [Be-Ve]) but there are interesting *infinite dimensional* and/or *disconnected* homogeneous self-similar metric spaces.

Infinite dimensional and totally disconnected examples

(a) Let $L = \bigoplus_{i=1}^{d} L_i$ be an infinite dimensional graded nilpotent Banach-Lie algebra which is (algebraically) Lie-generated by L_1. Then the corresponding nilpotent Banach-Lie group carries a natural homogeneous self-similar C-C metric. The simplest example is that of the infinite dimensional Heisenberg group for $L = L_1 \oplus L_2$, where L_1 is the Hilbert space and $L_2 = \text{center } L = \mathbb{R}$.

(b) Every field with a norm, e.g. the field of p-adic numbers, is homogeneous and self-similar. Furthermore, vector spaces and suitable nilpotent groups over such fields share this property.

(c) Let G be a topological group generated by a family of 1-parameter subgroups. We give a metric to each of these subgroups by choosing parameters and then we define the corresponding C-C metric in G as the *maximal* (or "supremal") left invariant metric for which the inclusions of all our 1-parameter subgroups into G are 1-Lipschitz, i.e. (non-strictly) distance decreasing maps. (This is similar to the definition of the word metric in G with a given generating subset). If G admits a (dilating) automorphism $A : G \to G$, which preserves the distinguished subset of subgroups and dilates the corresponding parameters with a fixed constant $\lambda > 1$, then this A also dilates the C-C metric by λ, which means A is a nontrivial self-similarity. An especially interesting class of such self-similar groups is associated to infinitely graded Lie algebras $L = \bigoplus_{i=1}^{\infty} L_i$ which are Lie-generated by L_1. (This was pointed out to me by I. Babenko.)

0.3.C. Uses of self-similarity: infinitesimal versus asymptotic. If a metric space V admits a non-trivial (i.e. non-isometric) self-similarity fixing a point $v_0 \in V$ then all of geometry of V can be seen in an arbitrary small neighbourhood of v_0. In particular, if V is a C-C manifold then the asymptotic geometry of V (at infinity) can be read in terms of infinitesimal data of the implied polarization at v_0 and vice versa.

Let us apply this to the Heisenberg group G with the dilations A_t (defined by $(x, y, z) \mapsto (tx, ty, t^2 z)$ on the Lie algebra $L = L(G)$) and determine an approximate shape of the Carnot-Carathéodory balls in G. Denote by $\tilde{B}'(\rho) \subset L$ the box defined by the inequalities

$$|x| \leq \rho \, , \ |y| \leq \rho \, , \ |z| \leq \rho^2$$

and let $B'(\rho) \subset G$ be the exponential image of $\tilde{B}'(\rho)$. Clearly, A_t transforms each box $B'(\rho)$ into $B'(t\rho)$ (because A_t commute with $\exp$) and the actual C-C balls in G around $\mathrm{id} \in G$ are transformed by A_t in a similar way, $B(\rho) \mapsto B(t\rho)$ (since A_t is a t-similarity for dist_H). It follows that the boxes $B'(\rho)$ approximate the balls $B(\rho)$ in the sense that

$$B'(C^{-1}\rho) \subset B(\rho) \subset B'(C\rho) \tag{$*$}$$

for all $\rho \geq 0$ and some constant $C > 0$.

The relation $(*)$ for $\rho \to 0$ justifies our earlier claims (see 0.2.A$'$) on the shape of the small C-C balls associated to the contact structure. On the other hand, the same $(*)$ for $\rho \to \infty$ provides an asymptotic information on the ordinary Riemannian left invariant metric in G. Namely the balls in this metric, call them $B^+(\rho) \subset G$, (obviously) have the same asymptotics for $\rho \to \infty$ as the balls $B(\rho)$ (since the two metric involved are both left invariant and determined by length of curves). In fact,

$$B(\rho) \subset B^+(\rho) \subset B(C^+\rho),$$

for all $\rho \geq 1$ and some constant C^+ (where the first constantless inclusion is due to the fact that the Riemannian metric we speak about is the one which underlies the C-C metric and so the corresponding Riemannian distance $\leq$ C-C distance).

It follows that the Riemannian balls $B^+(\rho)$ are asymptotically approximated by the (exponentiated) boxes $B'(\rho)$. In fact, this remains valid for all simply connected nilpotent Lie groups with graded Lie algebras by the above self-similarity argument. (An arbitrary simply connected nilpotent Lie group G is *asymptotic* to a group G_∞ which does admit a self-similarity by a theorem of Pansu cited in Remark (b) below. Thus the large balls in G are box-shaped as well as those in G_∞. Also see [Bass] and [Kari] on this matter.)

0.3.D. Self-similar spaces appearing as tangent cones of equiregular ones. Let $V = (V, \mathrm{dist})$ be a metric space with a reference point v_0. We set $tV = (V, t\,\mathrm{dist})$ for $t \in \,]0, \infty[$ and we want to go to some limits of tV for $t \to 0$ and for $t \to \infty$. An appropriate notion of a limit for our present purpose is the one associated to the *Hausdorff topology* on the "set" of isometry classes of pointed metric spaces (see [Gro$_{\mathrm{GPG}}$] and [G-L-P]). If such a limit exists for $t \to \infty$ it can be thought of as the tangent space (cone) of V at v_0 while the limit of tV for $t \to 0$ looks like the asymptotic cone of V or the tangent space of V at infinity. Notice, that the very definition of the limit makes these tangent cones (spaces) self-similar whenever they exist. On the other hand, if V already admits t-self-similarities fixing v_0 for all $t > 0$ then (tV, v_0) is isometric to (V, v_0) for all $t > 0$ and so the tangent cones to V at v_0 and at infinity exist and are isometric to V.

Examples

(i) If V is a Riemannian manifold then the Hausdorff limit $\lim_{t \to \infty} tV$ does exist and it equals the ordinary tangent space $T_{v_0}(V)$.

(ii) Let G be the Heisenberg group with a left invariant Riemannian metric g and set $g_t = t^{-1} A_t^*(g)$ for the above A_t. It is obvious, that the distance function dist_t associated to g_t converges (in the usual sense) to the Carnot-Carathéodory metric dist_∞ on G for $t \to \infty$. On the other hand the space (G, dist_t) is isometric to $t^{-1}G \underset{\mathrm{def}}{=} (G, t^{-1}\,\mathrm{dist})$. It follows that the *Hausdorff* limit of the spaces tG for $t \to 0$ also exists and is isometric to $(G, \mathrm{dist}_\infty)$. Hence, *the asymptotic* (tangent) *cone of the Riemannian manifold* (G, dist) *equals the Carnot-Carathéodory space* $(G, \mathrm{dist}_\infty)$.

Thus the (infinitesimal data of the) C-C geometry can be recaptured from the asymptotic Ricmannian geometry of G.

Remarks on C-C limits of discrete groups. The above example appears in [Gro$_{\mathrm{GPG}}$] in the surrounding of *discrete* groups G of *polynomial growth*. The polynomial growth ensures the existence of a Hausdorff *sublimit* of tG, $t \to 0$, *an* asymptotic cone, which is a nilpotent Lie group with a C-C metric where the degree of the growth of G translates to the Hausdorff dimension of the limiting C-C metric. In fact, the asymptotic geometry of discrete nilpotent groups provided the major source of inspiration for the initial study of Carnot-Carathéodory spaces.

0.3.D′. Pansu convergence theorem. The existence of an actual Hausdorff limit (not just a sublimit) $\lim_{t \to \infty} tG$ was proven by P. Pansu (see [Pan$_{\mathrm{CBN}}$]) for an arbitrary nilpotent Lie group with a left invariant Riemannian metric. (In fact, Pansu allows in [Pan$_{\mathrm{CBN}}$] a more general class of spaces G including discrete virtually nilpotent groups with word metrics.) Pansu shows that such a limit is isometric to a nilpotent Lie group G_∞ (which is, in general, not isomorphic to G) with a self-similar C-C metric. Thus the asymptotic geometry of every nilpotent group G reduces to the local C-C geometry of G_∞. In particular the asymptotics of the ρ-balls in G for $\rho \to \infty$ is encoded in the (infinitesimal) behavior of ρ-balls in G_∞ as $\rho \to 0$.

0.3.D″. Mitchell cone theorem for equiregular spaces. Consider a C-C metric in V defined with a polarization $H \subset T(V)$ which satisfies the following (genericity) assumption. Choose some tangent vector fields in H which span H and denote by $H_i(v) \subset T_v(V)$, $v \in V$, the subspace in $T_v(V)$ spanned by all commutators of the chosen fields of order $\leq i$. Clearly, this H_i does not depend on the choice of the spanning fields.

Equiregularity Assumption. The dimension $\dim H_i(v)$ is constant in $v \in V$ for each i. proves in [Mit$_{1;2}$] under this assumption that the Carnot-Carathéodory metric in V (for an arbitrarily given underlying Riemannian metric) admits the tangent cone at each point $v_0 \in V$ (i.e. the Hausdorff limit $\lim_{t \to \infty} tV$ exists) and this cone is isometric to some self-similar nilpotent Lie group (see 1.4.).

This shows that the local Carnot-Carathéodory geometry essentially reduces to (asymptotic) geometry of a nilpotent group with a dilation. (One can imagine by looking at Pansu and Mitchell theorems that there is an "inversion $t \longleftrightarrow t^{-1}$" which interchanges local C-C spaces with nilpotent Lie groups, such that the "fixed point set" of this "inversion" consists of the self-similar groups.)

0.4. Chow connectivity theorem. (see 1.1).

Theorem. *Let $X_1, \ldots, X_m$ be C^∞-smooth vector fields on a connected manifold V, such that successive commutators of these fields span each tangent space $T_v(V)$, $v \in V$. Then every two points in V can be joined by a piecewise smooth curve in V where each piece is a segment of an integral curve of one of the fields X_i.*

Remarks and corollaries

(a) *Lie group motivation.* This theorem is quite obvious on the formal level. In fact, let L be the Lie algebra generated by the fields X_i, $i = 1, \ldots, m$, and $G \subset \mathrm{Diff}\, V$ be the subgroup of diffeomorphisms generated by the one-parameter subgroups corresponding to X_i, $i = 1, \ldots, m$. The theorem claims that G is transitive on V provided L spans $T(V)$. This is immediate if L is *finite dimensional* as L can be identified with the Lie algebra of G (which makes G finite dimensional as well) and then the "L spans $T(V)$"-condition amounts to surjectivity of the differential of the orbit map $G \to V$ for $g \mapsto g(v_0)$

for every given point $v_0 \in V$. In fact, this argument applies in the infinite dimensional case as well (see 1.1).

(b) *Polarizations.* If the dimension of the span $H_v \subset T_v(V)$ of the fields X_i at v is independent of v, the span H of these fields is a subbundle in $T(V)$, i.e. a polarization of V in our sense where the orbits of X_i are horizontal. Thus Chow theorem implies the connectivity property for H-horizontal curves mentioned earlier.

(c) *Generic fields.* The assumption of the theorem is satisfied by *generic* C^∞-fields whenever $m \geq 2$ as successive commutators of *two* generic C^∞-fields on V span $T(V)$ as is easy to show. On the other hand, the conclusion of the theorem may be satisfied by some non-smooth continuous fields where the commutators are not even defined. In fact, the connectivity property is valid for a single (!) *generic continuous* field as everybody knows (compare II.5 in [Hart]). Probably (this seems obvious) a generic pair of C^k-fields have the connectivity property for every k.

(d) *Polarizations defined by 1-forms.* As we mentioned in (b) the Chow theorem applies to polarizations $H \subset T(V)$ *viewed as spans of systems of vector fields.* But one also can define H as the common zero set of a system of 1-form on V. This suggests a dual approach to (the proof of) the connectivity property of H which does not directly use orbits of vector fields tangent to H but rather appeals to leaves of 1-dimensional foliations obtained by intersecting H with submanifolds $W \subset V$ with $\operatorname{codim} W = \operatorname{rank} H - 1$. For example, if H is a contact subbundle on a 3-dimensional manifold V and $W_0 \subset V$ is a curve *transversal* to H, one takes 2-dimensional cylinders $W_\varepsilon \subset V$, $\varepsilon > 0$, around W_0 and $H \cap T(W_\varepsilon)$ give us (spiral) curves in W_ε tangent to H which closely follow W_0 for small ε, see Fig 1.

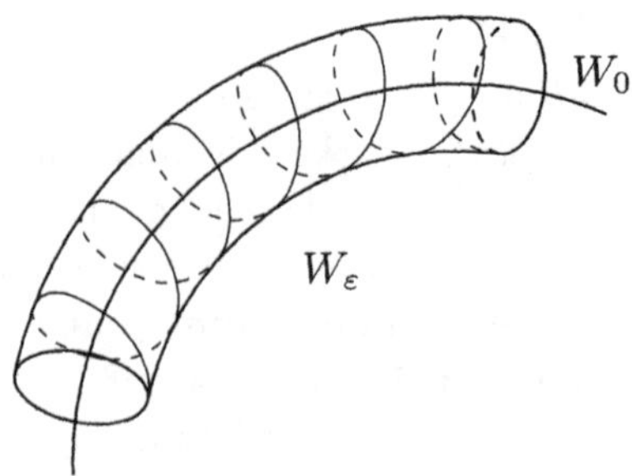

Figure 1

0.5. The shape of C-C balls: Mitchell-Gershkovich-Nagel-Stein-Wainger theorem.

Let $H \subset T(V)$ be a smooth polarization (i.e. sub-bundle) spanned by some vector fields $X_1, \ldots, X_m$. (Every H of rank $H = n_1$ can be spanned by $m \leq n_1 + n$ fields for $n = \dim V$ and locally one needs only $m = n_1$ fields. In fact, our considerations are local and so $m = n_1$ suffices.) We denote successive commutators of our fields by X_i for suitable indices $i > m$ and we assign to each X_i the number $\deg X_i$ which is the degree of the corresponding commutator. Thus, for example, $\deg X_i = 1 \iff i \leq m$, and

$$\deg X_i = 2 \iff X_i = [X_\mu, X_\nu], \ 1 \leq \mu, \ \nu \leq m.$$

(It may happen that $X_i = X_j$ for $i \neq j$ and, moreover, $\deg(X_i) \neq \deg(X_j)$ for $X_i = X_j$, i.e. the degree is assigned to the name (i.e. the subindex) of a field rather than the field itself.)

Now, let X_i, $i = 1, \ldots, M$, be the successive commutators of $X_1, \ldots, X_m$ of degree $\leq d$ and let us assume that these X_i, $i = 1, \ldots, M$, span $T(V)$. We want to characterize the C-C metric in V corresponding to H in terms of the integral curves (1-parameter subgroups) of the fields X_i, $i = 1, \ldots, N$, and of the linear combinations of these fields with constant coefficients. First, for an arbitrary field X on V, we recall the notation $\exp_v X \in V$ which is the result of applying to v the 1-parameter group (flow) corresponding to X at the time $t = 1$. In other words, we obtain $v_1 = \exp_v(X)$ by taking the integral curve of X issuing from v and following it with speed X for time $t = 1$. (If V is non-compact, the field may be globally non-integrable but this is irrelevant at the moment.) Next, we define the exponential map $\exp_v : \mathbb{R}^M \to V$ by

$$(t_1, \ldots, t_M) \mapsto \exp_v(t_1 X_1 + \cdots + t_M X_M).$$

This is, clearly, a smooth map (as we assume X_1 are smooth) and the differential of $\exp_v$ is surjective at the origin $0 \in \mathbb{R}^M$.

Notice that if V is non-compact, the map $\exp_v : \mathbb{R}^M \to V$ need not be defined on all of $\mathbb{R}^M$ (as some field $X = \sum_{i=1}^{M} t_i X_i$ may fail to be globally integrable) but it is always defined in a small neighbourhood of the origin of $\mathbb{R}^M$ and this is all we need for our purpose.

Consider the following box in $\mathbb{R}^M$,

$$\mathrm{Box}(\rho) = \left\{ |t_i| \leq \rho^{\deg X_i}, \ i = 1, \ldots, M \right\} \subset \mathbb{R}^M.$$

0.5.A. Ball-box-theorem. (^{2})

Theorem. *The small C-C balls $B_v(\rho)$ in V around $v \in V$ are uniformly equivalent to the exponential images of the boxes. This means, there are strictly positive continuous functions $C = C(v)$ and $\rho_0 = \rho_0(v)$, such that*

$$\exp_v \mathrm{Box}(C^{-1}\rho) \subset B_v(\rho) \subset \exp_v \mathrm{Box}(C\rho)$$

for all $v \in V$ and $\rho \leq \rho_0(v)$.

Remarks and corollaries

(a) *Hölder equivalence of* C-C *and Euclidean.* The first inclusion $B_v(\rho) \supset \exp_v \mathrm{Box}(C^{-1}\rho)$ gives us a quantitative version of Chow's theorem as it shows that every point v_1 in the exponential image of $\mathrm{Box}(C^{-1}\rho)$ can be joined with v by an H-horizontal curve of length $\leq \rho$. Since the exponential map has surjective differential at the origin, the image of this box contains the Euclidean (or Riemannian) ball in V around v or radius $\approx \rho^d$ (where, recall, $d = \max \deg X_i$). It follows that the identity map

$$(V, \text{ Riemannian metric}) \xrightarrow{\;\mathrm{id}\;} (V, \text{ C-C metric})$$

is C^α-*Hölder* with the exponent $\alpha = d^{-1}$. (Notice that the map

$$(V, \text{C-C metric}) \xrightarrow{\;\mathrm{id}\;} (V, \text{ Riemannian metric})$$

is, obviously, Lipschitz.)

(b) Let us generalize the above and determine the Hölder class of a smooth map of a Riemannian or, more generally, Carnot-Carathéodory manifold W into V. To simplify the matter we assume our polarization H is *equiregular* in the sense that the dimension of the subspace in $T_v(V)$ generated by the commutators of degree $j \leq d$ does not depend on v (compare Mitchell cone theorem cited in 0.3.D). In this case the bundle $T(V)$ is filtered by smooth subbundles

$$H = H_1 \subset H_2 \subset \cdots \subset H_j \subset \cdots \subset H_d = T(V)$$

such that H_j is spanned by the j-th degree commutators of the fields in H. Then we take some polarization $H' \subset T(W)$ whose successive

---

^{2} See [Mit$_{1;2}$], [Ger$_{\mathrm{TSE}}$], [Ger$_{\mathrm{EN}}$], [N-S-W] and 1.1 – 1.3.

commutators span $T(W)$ and turn to the question of the Hölder exponent for a smooth map $f : W \to V$ with respect to the C-C metrics associated to H' and H. Here is the answer.

0.5.B. Hölder exponent evaluation. *The Hölder exponent of f equals j^{-1} for the minimal j, such that the differential $\mathcal{D}$ of f sends H' into H_j. (This includes the case of a Riemannian W for $H' = T(W)$.)*

Notice that this evaluation of the Hölder exponent contains two statements: the fact that f is C^α for $\alpha = j^{-1}$ and that it is *not* C^β for $\beta > j^{-1}$, where the latter is immediate with the inclusion C-C ball $\subset \exp \mathrm{Box}$ in V while the former uses the $\sqrt[j]{\mathrm{Euclidean}}$ property of H_j-horizontal curves in V.

It is not at all clear what is the precise geometrical (infinitesimal) significance of f being C^α with respect to C-C metrics *without* assuming f is smooth. Yet one can think of such maps f as of *generalized solutions* of the P.D.E.-system expressing the inclusion $\mathcal{D}f(H') \subset H_j$ for $j = (\mathrm{ent}\, \alpha^{-1}) + 1$. In fact, one of the general questions concerning C-C manifolds is the following

0.5.C. Hölder mapping problem. Given two C-C spaces V and W and a real number $0 < \alpha \leq 1$, describe the space of C^α-maps $f : W \to V$. For example, when can each continuous map $W \to V$ be uniformly approximated by C^α-maps? When can W be C^α-*embedded* into W? When are V and W C^α-homeomorphic? etc.

These questions can be approached in the smooth case by the P.D.R-techniques (see [Gro$_{\mathrm{PDR}}$] and §4) and what little we can say about general (non-smooth Hölder) maps also uses some P.D.R. (see §4).

0.5.D. Hölder surfaces in contact 3-manifolds. Let V be a 3-dimensional contact C-C manifold (see 0.2) and $f : \mathbb{R}^2 \to V$ a C^α-*embedding*, where $\mathbb{R}^2$ is endowed with the Euclidean metric. Then one can show that $\alpha \leq \frac{2}{3}$ while the natural expectation is $\alpha \leq \frac{1}{2}$ (compare 0.6.C).

0.5.E. Homotopy count of Hölder maps. If V and W are compact Riemannian manifolds, then the number $Nm(\lambda)$, of the homotopy classes of Lipschitz maps $V \to W$ with the Lipschitz constant $L(f) \le \lambda$, satisfies

$$Nm(\lambda) \le C(1 + \lambda)^r, \qquad (*)$$

provided the fundamental group of W is finite and where the exponent r depends only on the homotopy types of V and W, while the constant C may depend on the metrics as well. This is proven in [Gro$_{\mathrm{HED}}$] by appealing to Sullivan's minimal model of the de Rham algebra of differential forms.

Warning. *The argument in* [Gro$_{\mathrm{HED}}$] *uses minimal models only for maps of* spheres *into W and then the general case is obtained by an induction on skeletons of V. This induction, however, contains a gap (pointed out to me by Pansu) which I am still unable to fill in. Yet, one can apply the techniques of minimal models directly to V and obtain by the argument in* [Gro$_{\mathrm{HED}}$] *a polynomial bound on the number of mutually non-equivalent morphisms* $\mathrm{Mod}V \to \mathrm{Mod}W$ *in terms of $L(f)$. Then the rational homotopy theory yields such a bound (i.e. $(*)$) for maps $V \to W$ themselves. (This is one of the creeds of the rational homotopy theory, so I believe.)*

Next we observe that a Hölder C^α-map $f : V \to W$ can be approximated by a Lipschitz map f' such that $L(f') \approx \left(L_\alpha(f)\right)^{-\frac{1}{\alpha}}$. This is done by working on the ε-balls B_ε in V with $\varepsilon \approx \left(L_\alpha(f)\right)^{\frac{1}{\alpha}}$ where $\mathrm{Diam}\, f(B_\varepsilon) \ll 1$. Thus $(*)$ implies a similar inequality for C^α-maps $f : V \to W$ satisfying $L_\alpha(f) \le \lambda$, namely

$$Nm_\alpha(\lambda) \le C(1 + \lambda)^{r\alpha} \qquad (*)_\alpha$$

for $r_\alpha \le r_1^{\frac{1}{\alpha}}$ for the exponent $r = r_1$ in $(*)$.

Now we return to C-C manifolds V and W, invoke the Hölder equivalence between the Riemannian and the C-C metrics and conclude to the bound $(*)$ for compact C-C manifolds. Notice that the exponent r in $(*)$ depends in the C-C case on the polarizations of V and W as well as on their homotopy types.

Problem. Give a precise formula for r in $(*)$ for C-C manifolds.

Notice that this problem is open even in the Riemannian case for most homotopy types of V and W and in the C-C case one would be content to know the answer in simple cases where the homotopy theoretic structure is sufficiently transparent (compare 1.4.E$'$, 2.4.A).

Remark. The Riemannian inequality $(*)$ can be sharpened by replacing the Lipschitz constant $L(f)$ by more precise measures of dilation of f such as the L_p-norm of the differential $\mathcal{D}f$ on the k-th exterior power $\wedge^k T(V)$, for suitable k and p (see [Gro$_{\text{HED}}$] and note that the L_∞-norm of $\mathcal{D}f$ on $\wedge^1 T(V)$ equals $L(f)$). This will be partially extended to the C-C framework in 2.4, 2.5 and §4.

Example. Let V and W be closed connected orientable n-dimensional Riemannian manifolds where W is homeomorphic to the sphere. Then $Nm(\lambda) \sim C\lambda^n$ as the only homotopy invariant of a map $f : V \to W$ is the degree (see [Gro$_{\text{HED}}$] and [G-L-P]). Now let V be C-C associated to a generic subbundle $H \subset T(V)$ of codimension one (e.g. a contact structure). Then, by an easy argument, $Nm(\lambda) \sim C\lambda^{n+1}$ which agrees with the fact that the Hausdorff dimension of such a V is $n+1$ (see below as well as 1.4.E$'$ and 2.5). Notice, that this is sharper than the bound $Nm(\lambda) \lesssim \lambda^{2n}$ provided by the Riemannian estimate $Nm(\lambda) \lesssim \lambda^n$ via the Hölder $C^{\frac{1}{2}}$-equivalence between the contact and Riemannian structures.

0.6. The volume of C-C balls and the Hausdorff dimension. We assume here for simplicity's sake that the polarization $H \subset T(V)$ defining our C-C structure is equiregular and take a frame of vector fields $X_1, \ldots, X_n$, $n = \dim V$, which agrees with the commutators filtration $H = H_1 \subset \cdots \subset H_j \subset \cdots \subset H_d = T(V)$ defined above. (This means the first m_1 fields for $m_1 = \operatorname{rank} H_1$ belong to H_1, the following m_2 fields for $m_2 = \operatorname{rank} H_2/H_1$ belong to H_2, etc.) We denote by $\deg X_i (= \deg i)$ the minimal j such that X_i belongs to H_j. Thus $\deg X_i = 1$ for $i = 1, \ldots, m = \operatorname{rank} H_1$, $\deg X_i = 2$ for $i = m + 1, \ldots, \operatorname{rank} H_2/H_1$, etc. The corresponding box $\mathrm{Box}(\rho) \subset \mathbb{R}^n$ given by the inequalities $|t_i| \le \rho^{\deg X_i}$ (obviously) has the Euclidean volume equal $2^n \rho^D$ for $D = \sum_{i=1}^{n} \deg X_i = \sum_{j=1}^{d} j \operatorname{rank}(H_j/H_{j-1})$, where we assume $H_0 = 0 \subset H$, and by the ball-box theorem the Riemannian volumes of the C-C balls $B_v(\rho)$ in V are roughly the same. Namely, *there* *exist continuous strictly positive functions $C_1(v)$, $C_2(v)$ and $\rho_0(v)$, on V, such that*

$$C_1(v) \leq \left(\mathrm{Vol}\, B_v(\rho)\right)/\rho^D \leq C_2(v) \tag{+}$$

for all $\rho \leq \rho_0(v)$. Consequently, the Hausdorff dimension of V with respect to the C-C metric associated to H equals D. (See [Mit$_{1;2}$] and 1.3.A).

Notice that the inequality $(+)$ partially extends to non-equiregular structures (see [N-S-W]) which allows an evaluation of $\dim_{\mathrm{Hau}} V$ for some non-equiregular H, e.g. for the real analytic ones (see 1.3.A).

0.6.A. On the intrinsic size of C-C balls. If ρ is small, the C-C ball $B_v(\rho)$ approximated by the box with the sides ρ^{d_i}, $i = 1, \ldots, n$, (extrinsically) appears much smaller than the Euclidean ball of radius ρ (or the ρ-cube which is roughly the same as the ρ-ball from our viewpoint) if ρ is small and at least one of the degrees $d_i = \deg X_i$ exceeds one (i.e. if $\mathrm{rank}\, H_1 < n$). Yet we know that both, the Euclidean and Carnot-Carathéodory balls, can be rescaled to the unit size where they look roughly the same and so for each ρ the C-C ball *cannot* be smaller than the Euclidean one of the same radius ρ. In fact, one may imagine the C-C balls as being incomparably larger since their Hausdorff dimensions are greater than n. The following observation justifies this view. *Every C-C ball $B_v(\rho)$, where v is a fixed point and $\rho \leq 1$, admits a surjective distance decreasing map f onto the Euclidean ball $B^{\mathrm{Eu}}(\rho')$, such that f has infinite average multiplicity and $\rho' \geq \delta\rho$ for a fixed (independent of ρ) $\delta > 0$.*

Recall that *the average multiplicity*, also called the *total volume* or the *variation* of a map f with the range in a Euclidean ball $B \subset \mathbb{R}^n$ is defined by the integral

$$\int_B \mathrm{card}\, f^{-1}(x)\,dx.$$

If the domain V of f is an n-dimensional Riemannian manifold and f is Lipschitz, this integral equals the total volume of the induced (singular) Riemannian metric on V and, hence, is finite for *compact* V. But if V is non-Riemannian of the Hausdorff dimension $N > n$, one can only claim the finiteness of the integral

$$\int_B \mathrm{mes}_{N-n}\left(f^{-1}(x)\right)\,dx,$$

as this, for Lipschitz maps f, does not exceed $\mathrm{const}\, \mathrm{mes}_N V$ by the coarea inequality (where $\mathrm{const} \leq C_n L(f)$ for the Lipschitz constant L of f and an universal constant C_n).

Now let us explain how to construct the above claimed map $f : B_v(\rho) \to B^{\mathrm{Eu}}(\rho')$ of infinite average multiplicity. We start with some smooth distance decreasing map f_0 of $B_v(\rho)$ onto $B^{\mathrm{Eu}}(\rho')$ (which is trivial to arrange) and then we use the fact that the C-C metric in $B_v(\rho)$ can be approximated by Riemannian metrics g_t for $t \to \infty$ (see 0.8). When t is large, the map f_0 becomes almost distance decreasing for g_t and it can be easily made honestly g_t-distance decreasing by a small perturbation of f_0. Then f_0 can be uniformly approximated by a g_t-isometric (sic!) map f_t, i.e. preserving the g_t-length of the smooth curves in $B_v(\rho)$ (see 2.4.11 in [Gro$_{\mathrm{PDR}}$]), which has $\int \mathrm{card}\, f_t^{-1}(x)dx = \mathrm{Vol}_{g_t} B_v(\rho) \to \infty$. Finally, one sees that f_t subconverge to a C-C-Lipschitz map $f = f_\infty$ with $\int \mathrm{card}\, f^{-1}(x)dx = \infty$.

Questions

- Can one find a Lipschitz $f : B_v(\rho) \to B^{\mathrm{Eu}}(\rho')$ such that $\mathrm{card}\, f^{-1}(x) = \infty$ for *all* $x \in B^{\mathrm{Eu}}(\rho')$?

- Can one have a Lipschitz f with the *positive* measure $\mathrm{mes}_{N-n}\, f^{-1}(x)$ for all x in $B^{\mathrm{Eu}}(\rho')$ or at least for x in a subset $X \subset B^{\mathrm{Eu}}(\rho')$ of positive measure? (Recently some such f were constructed in [Bat$_3$]).

0.6.B. Hausdorff dimension of submanifolds. Let V' be a smooth submanifold in V and let us evaluate the Hausdorff dimension of V' with respect to the Carnot-Carathéodory metric in $V \supset V'$. We intersect the tangent spaces $T_v(V') \subset T_v(V)$, $v \in V' \subset V$, with H_j and denote these intersections by $H_j'(v)$. These are linear subspaces which filter $T_v(V')$ and we denote by $m_j'(v)$, $v \in V'$, the ranks $\mathrm{rank}\big(H_j'(v)/H_{j-1}'(v)\big)$. Next we let

$$D'(v) = \sum_{j=1}^{d} j\, m_j'(v),\ v \in V',$$

and finally we define $D_H(V')$ by

$$D_H(V') = \max_{v \in V'} D_H'(v).$$

For example, if V' *horizontal*, i.e. is everywhere tangent to H then $D_H(V') = \dim V'$. In general, $D_H(V') \geq \dim V'$ and $D_H(V')$ is maximal for generic submanifolds V'. For instance if V' is a *generic* submanifold of dimension n', then the number $m'_j(v)$ is given for almost all $v \in V'$ by the following obvious formula depending on $\operatorname{codim} V' = n - n'$,

$$m'_j(v) = \begin{cases} m_j = \operatorname{rank} H_j/H_{j-1} & \text{if} \quad \operatorname{codim} V' \leq \operatorname{rank} H_{j-1}, \\ m_j + n' - n & \text{if} \quad \operatorname{rank} H_{j-1} \leq \operatorname{codim} V' \leq \operatorname{rank} H_j, \\ 0 & \text{if} \quad \operatorname{codim} V' \geq \operatorname{rank} H_j. \end{cases}$$

It follows in the case $d = 2$ that

$$D_H(V') - \dim V' = \begin{cases} m_2 = \dim V - \operatorname{rank} H & \text{if} \quad \operatorname{codim} V' \leq \operatorname{rank} H, \\ n' = \dim V' & \text{if} \quad \operatorname{codim} V' \geq \operatorname{rank} H. \end{cases}$$

Evaluation of $\dim_{\mathrm{Hau}} V'$. *The Hausdorff dimension of an arbitrary smooth submanifold V' in a manifold V with a C-C metric associated to H equals the number $D_H(V')$.*

This is an easy corollary of the ball-box theorem for V (see 4.1.A).

Warning. The restricted C-C metric dist_H on V' by no means equals the C-C metric on V associated to the polarization $H' = T(V') \cap H$ on V. In fact, this restricted metric is of more general nature than Carnot-Carathéodory. (But it fits in many cases the definition in §1 of [N-S-W].) Yet there are important examples (e.g. hypersurfaces V' in contact manifolds of dimension ≥ 5) where the restricted C-C metrics $\operatorname{dist}_H |V'$ is Lipschitz equivalent to the C-C metric $\operatorname{dist}_{H'}$ on V' (see 2.4.B).

Let us look at the inequality $\dim_{\mathrm{Hau}} V' \leq N'$ for a given $N' > 0$ as an equation imposed on $V' \subset V$. If V' is a smooth submanifold (and so $\dim_{\mathrm{Hau}} V' = D_H(V')$) this can be indeed represented by a system of partial differential equations on V' expressing some degree of tangency of V' to H_j. For example, the relation $\dim_{\mathrm{Hau}} V' \leq \dim V'$ says, in the case where V' is smooth, that V' is everywhere tangent to H. Thus non-smooth subspaces $V' \subset V$ with $\dim_{\mathrm{Hau}} V' \leq N'$ can be interpreted as generalized solutions of a certain system of P.D.E. (compare the earlier Hölder discussion for mappings). Here is the main question we want to address:

0.6.C. Hausdorff dimension problem for subspaces. Determine the (topological) structure of (the space of) the subsets $V' \subset V$ having $\dim_{\mathrm{Hau}} V' \leq N'$ with respect to a given C-C metric in V. For instance, find the infimum of $\dim_{\mathrm{Hau}} V'$ among all compact subsets $V' \subset V$ of a given topological dimension n'.

Contact example. Every smooth surface in the contact 3-space V has the C-C Hausdorff dimension 3 as follows from the equality $\dim_{\mathrm{Hau}}(V') = D_H(V')$. Then with a little extra effort one can show that *every subset V' of topological dimension 2 has* $\dim_{\mathrm{Hau}} V' \geq 3$ (see 2.1). Notice that this inequality immediately implies non-existence of C^α-embeddings $\mathbb{R}^2 \to V$ for $\alpha > \frac{2}{3}$ (as was claimed in 0.5.D earlier) since the Hausdorff dimension (obviously) agrees with C^α-Hölder maps $f : W \to V$ by the rule $\dim_{\mathrm{Hau}} f(W) \leq \alpha^{-1} \dim_{\mathrm{Hau}} W$.

Exercise. Show that the Euclidean ε-ball in the contact 3-space V can be covered by $\approx \varepsilon^{-1}$ C-C balls of radius ε while every C-C ε-ball needs $\approx \varepsilon^{-2}$ Euclidean ε^2-balls to cover it. Use this to compare the C-C and Euclidean Hausdorff dimensions of subsets in V. State and prove corresponding results for general C-C manifolds (V, H).

0.7. Isoperimetric filling problem. Let S be a k-dimensional cycle in V which is homologous to zero. We want to evaluate the minimal possible Hausdorff dimension of (the supports of) $(k+1)$-chains D in V *filling in S*, i.e. having $\partial D = S$. More specifically we want to bound this minimal $j = \dim_{\mathrm{Hau}} D$ in terms of $i = \dim_{\mathrm{Hau}} S$ and then, moreover, we look for a bound on the (minimal possible) j-dimensional Hausdorff measure of fillings D of S in terms of the i-dimensional Hausdorff measure of S.

0.7.A. Isoperimetric inequality for k = n − 1. If S is a closed hypersurface in V then there is little choice for D : this is a domain in V bounded by S. Thus we exercise no control over the Hausdorff dimension of D but we still may try to bound its Hausdorff measure of an appropriate dimension in terms of the Hausdorff measure of S one dimension less.

A first result of this kind is due to Pansu (see [Pan$_{\mathrm{The3}}$] and [Pan$_{\mathrm{InIs}}$]) who proved the following isoperimetric inequality for the 3-dimensional Heisenberg group V with an equivariant C-C metric.

Pansu isoperimetric inequality. *The 3-dimensional Hausdorff measure of the boundary $S = \partial D$ of a domain $D \subset V$ restricts the 4-dimensional measure of D by*

$$\mathrm{mes}_4\, D \le \mathrm{const}(\mathrm{mes}_3\, S)^{\frac{4}{3}}. \tag{$*$}$$

Notice that mes_4 in this case equals the ordinary (3-dimensional) Haar measure on the group V.

Now we recall that every left invariant *Riemannian* metric on the group V has the same asymptotic geometry as the C-C metric. Thus $(*)$ is essentially equivalent to the following inequality for the ordinary volume and area with respect to a left-invariant Riemannian metric g in V,

$$\mathrm{Vol}\, D \le \mathrm{const}_g(\mathrm{Area}\,\partial D)^{\frac{4}{3}}. \tag{$**$}$$

(This is stronger for large domains D than the Euclidean inequality $\mathrm{Vol}\, D \le \mathrm{const}(\mathrm{Area}\,\partial D)^{\frac{3}{2}}$.)

The inequality $(**)$ was extended by Varopoulos (see [Var-Sa-Co]) to an arbitrary simply connected nilpotent Lie group V,

$$\mathrm{Vol}_n\, D \le \mathrm{const}(\mathrm{Vol}_{n-1}\,\partial D)^{\frac{N}{N-1}},$$

where $n = \dim V$ and N is the *asymptotic Hausdorff dimension* or the exponent of the volume growth of the balls $B(R) \subset V$, that is

$$N = \lim_{R\to\infty}\,\big(\log \mathrm{Vol}\, B(R)\big)/\log R.$$

The isoperimetric inequality for general C-C spaces V of the Hausdorff dimension N reads

$$\mathrm{mes}_N\, D \le \mathrm{const}_V(\mathrm{mes}_{N-1}\,\partial D)^{\frac{N}{N-1}} \tag{$+$}$$

and we shall prove this in 2.3 under suitable assumptions on V. (The inequality $(+)$ in the form of a Soholev inequality for smooth functions on V is due to Varopoulos, see [Var].)

0.7.B. Filling in curves in V. Here $k = 1$ and we look for a "minimal" surface $D \subset V$ which fills in a given closed curve $S \subset V$. We assume for the moment that S is rectifiable (i.e. $\dim_{\mathrm{Hau}} = 1$) and, in fact, we do not loose much by assuming that S is smooth and horizontal (with respect to the polarization underlying the C-C-geometry). But even in this case the evaluation of the (minimal) Hausdorff dimension (and measure) of surfaces D filling in S is a rather subtle matter as is seen in the following

Contact example. Let V be the 3-dimensional Heisenberg group with a left invariant and self-similar C-C metric. We know already (see 0.6.C) that surfaces in V have Hausdorff dimension ≥ 3 and moreover, one can show (using $(*)$ for instance) that every filling D of a simple closed curve S has $\mathrm{mes}_3 D \geq \varepsilon > 0$ for $\varepsilon = \varepsilon(S)$. On the other hand the Riemannian area of a smooth surface bounds mes_3 which implies the (filling) inequality

$$\mathrm{mes}_3 D \leq C(\mathrm{mes}_1 S)^3$$

for a suitable filling D of S.

The situation radically changes if we take the Heisenberg group V of dimension $n > 3$. (Recall that the Heisenberg group of dimension $n = 2m+1$ is characterized by its Lie algebra which admits a basis x_i, y_i, z, $i = 1, \ldots, m$, where z is central, $[x_i, x_j] = [y_i, y_j] = 0$ and the only non-zero commutators between x_i and y_j are $[x_i, y_i] = z$, $i = 1, \ldots, m$.) This V admits an essentially unique left invariant self-similar C-C metric where the underlying (contact) polarization H has codimension one (being spanned by x_i and y_i). Now, for $n > 3$, the filling of closed C-C-rectifiable curves X can be made more efficiently than for $n = 3$. Namely, S can be filled by surfaces D of Hausdorff dimension 2, which means, in effect, that these are horizontal, i.e. are everywhere tangent to H. Moreover, the area of these horizontal surfaces can be made as small as $\mathrm{const}(\mathrm{length}\, S)^2$. *Thus every S admits a filling D for which*

$$\mathrm{mes}_2 D \leq \mathrm{const}(\mathrm{mes}_1 S)^2 \tag{$\star$}$$

(see 3.5 and 4.8 where $(\star)$ generalizes to some non-contact C-C manifolds).

0.7.C. Filling for dim S $>$ 1. Here the situation is rather unsatisfactory, as we lack a non-trivial upper bound on the (best) filling in most cases. For example, we have the following

Open question. Let V be the $2m + 1$-dimensional Heisenberg group with the C-C metric. Does there exist, for every k-dimensional cycle S for $k < m$, a filling D of S satisfying

$$\mathrm{mes}_{k+1} D \leq \mathrm{const}(\mathrm{mes}_k S)^{\frac{k+1}{k}}\,?$$

This is motivated by the fact that the underlying (contact) polarization H admits plenty of $(k+1)$-dimensional submanifolds tangent to it for $k+1 \leq m$. In particular, if S is smooth horizontal, it can be filled in by a horizontal D and thus having $\operatorname{mes}_{k+1}(D) < \infty$. What is unclear at the present moment is how to bound $\operatorname{mes}_{k+1} D$ in terms of $\operatorname{mes}_k S$.

Above middle dimension. If $k = m$ then the "correct" (isoperimetric) filling inequality reads

$$\operatorname{mes}_{k+2} D \leq \operatorname{const}(\operatorname{mes}_k S)^{\frac{k+2}{k}}$$

(which means: "there exists D satisfying ...") and this can be easily derived from an ordinary (Riemannian) filling estimate.

Finally, the k-dimensional cycles S for $k > m$ are expected to admit fillings D satisfying

$$\operatorname{mes}_{k+2} D \leq \operatorname{const}(\operatorname{mes}_{k+1} S)^{\frac{k+2}{k+1}} \qquad (++)$$

as in the (only known) case $k = \dim V - 1$.

Remark

(a) Since the Heisenberg group V admits self-similarities (dilations) $A_t :$ $V \to V$ which scale the r-dimensional Hausdorff measure by t^r for each r, every one of the above inequalities can be reduced to the case where the relevant measure of S equals one and the problem boils down to finding any bound on the measure of a suitable filling D of S.

(b) One knows that the constant in $(++)$ is definitely different from zero. In fact, for every closed smooth k-dimensional submanifold $X \subset V$ with $k \geq m$, there exists a constant $\varepsilon = \varepsilon(S) > 0$, such that *every* filling D of S has

$$\operatorname{mes}_{k+2} D \geq \varepsilon. \qquad (+-)$$

This follows by the argument in 3.1.A.

0.8. Carnot-Carathéodory metrics as limits of Riemannian ones. Consider a smooth manifold V where the tangent bundle is decomposed into the sum of two complementary subbundles, $T(V) = H \oplus H^{\perp}$, and let $A_t : T(V) \to T(V)$ be defined by $(h, h^{\perp}) \mapsto (h, t\,h^{\perp})$ for all $t > 0$. Then we pick up some Riemannian metric g on V and look at the family $g_t = A_t^*(ga)$ as t varies between 1 and ∞.

0.8.A. Riemannian homogeneous spaces and their limits. Let V be a Lie group and $H, H^\perp$ and g be left invariant. Then the metrics g_t are also left invariant and one may naively think that their geometry is fairly simple. But even in the first non-trivial (i.e. non-commutative) case of $V = SU(2)$ and rank $H = 2$, one has insufficient understanding of the asymptotic behavior of the geometry of (V, g_t) for $t \to \infty$. For example, one still does not know if the path (ray) (V, g_t), $t \in [1, \infty[$ is (at least roughly) minimizing in the space $\mathcal{M}$ of Riemannian manifolds with a suitable metric on $\mathcal{M}$. (I have more respect for this problem now than ten years ago when I first faced it, see my note $[\mathrm{Gro_{AGHS}}]$ following a meeting in Torino in 1983 on homogeneous spaces.)

If the complementary bundle $H^\perp$ has rank > 0, then the family of the Riemannian metrics (quadratic forms) g_t *diverges* at each point $v \in V$. Yet the associated distance functions $\mathrm{dist}_t = \mathrm{dist}_{g_t}$ may converge for $t \to \infty$. In fact, if the subbundle H (polarization) Lie generates the tangent bundle (i.e. successive commutators of H-horizontal, fields span $T(V)$) then dist_t (obviously) converges to the Carnot-Carathéodory metric $\mathrm{dist}_\infty = \mathrm{dist}_{H,g}$ on V (compare 1.4.D).

0.8.B. Contact example in the spherical clothing. Let $V = S^3$ and $H \subset T(V)$ be the (2-dimensional horizontal) subbundle normal to the fibers of the Hopf fibration $S^3 \to S^2$. As we take the metrics g_t for t getting larger and larger, the Hopf fibers are becoming longer and longer but the diameter of S^3 with respect to the metric dist_t remains bounded for $t \to \infty$, as every two points in S^3 can be joined by an H-*horizontal* curve whose g_t-length is independent of t. (This drastically contrasts with what happens to the trivial fibration $V = S^2 \times S^1 \to S^2$, where $\mathrm{diam}(V, \mathrm{dist}_t) \to \infty$ for $t \to \infty$.)

Let us formulate two basic problems concerning asymptotic geometry of $V_t = (V, g_t)$ (for $t \to \infty$).

0.8.C. The asymptotic mapping problem. Find the asymptotics for $t \to \infty$ of the *Lipschitz constant* Lip_t of the *homotopy class* of the identity map $(V, g_1) \to (V, g_t)$, where the Lipschitz constant of a class Φ of maps φ is defined as the infimum over $\varphi \in \Phi$ of the Lipschitz constants $L(\varphi) = \sup_{v \in V} \|\mathcal{D}\varphi(v)\|$. More specifically, one wants to know if the Lipschitz constant of the identity map can be significantly decreased (for large t) by homotopying this map. One also asks this question for other dilation characteristics such as the L_p-norms of $\mathcal{D}f$ on $\wedge^k T(V)$ (compare 0.5.E.).

0.8.D. The intermediate volume problem. Take a k-dimensional homology class $\alpha \in H_k(V)$ and evaluate the asymptotics of $\mathrm{Vol}_k(\alpha)$ with respect to g_t as $t \to \infty$, where the volume of a homology class in a Riemannian manifold is defined as the infimum of the volumes of the cycles in this class.

Notice that these two problems are vaguely parallel to the Hölder mapping problem and the Hausdorff dimension problem correspondingly (see 0.5.C and 0.6.B) and we see later clearer relations between this kind of problems.

0.8.E. Families of metrics associated to dynamical systems. Instead of moving g_t by automorphisms $A_t : T(V) \to T(V)$ one may use a family of diffeomorphisms $a_t : V \to V$ and define

$$\mathrm{dist}_t = \sup_{1 \leq \tau \leq t} a_\tau^*(\mathrm{dist})$$

for a fixed (Riemannian) distance function on V. Many dynamical characteristics of a_t, e.g. *the topological entropy*, can be expressed as asymptotic invariants of dist_t. On the other hand, for some dynamical systems (first of all for Anosov systems) one can renormalize dist_t such that the limit for $t \to \infty$ exists and is of Carnot-Carathéodory type. Here is a typical question where these ideas are useful (see 4.10).

0.8.F. The intermediate entropy problem. Recall that for every self-homeomorphism $a : V \to V$ and each compact subset K one can define the topological entropy $\mathrm{ent}(a; K)$ (by using suitable ε-covers of K for the metrics $\mathrm{dist}_i = \sup_{1 \leq j \leq i} (a^{(i)})^* \mathrm{dist}$). Then one defines $\mathrm{ent}_k(a)$ as the infimum of $\mathrm{ent}(a; K)$ over all compact subsets K of topological dimension k. The question is how to evaluate this ent_k for specific (e.g. Anosov) transformations a.

0.8.G. Intrinsic approximation of C-C spaces by Riemannian ones. The approximation of a C-C metric by the Riemannian metric g_t at the beginning of 0.8 makes an essential use of the polarization H underlying the C-C structure, and this is hard to see in purely metric (intrinsic) C-C terms. An alternative intrinsic approximation V_ε to V (where ε corresponds to t^{-1}) appeals to the nerve of a suitable ε-covering of V where each simplex of the nerve is given the metric of the standard

Euclidean ε-simplex. (These V_ε are piecewise Riemannian rather than Riemannian but this hardly matters.) We shall see later on that these V_ε rather closely approximate V for $\varepsilon \to 0$ in a suitable category (albeit V_ε are not necessarily homeomorphic to V) and many geometric invariants of V can be extracted from those of V_ε for $\varepsilon \to 0$ (see 1.4.D and [Ger$_\mathrm{SBRM}$]).

0.9. Conformal C-C geometry and hyperbolic geometry. The most profound geometric applications of the Carnot-Carathéodory structures are centered around the rigidity problems for non-compact symmetric spaces of rank one and are due to Mostow and Pansu (see [Most], [Pan$_\mathrm{QIR1}$]). Recall that every compact metric space V serves as the *ideal boundary* of certain *hyperbolic spaces* CV. Namely, we take $CV = V \times [0, \infty[$ with the maximal (or better to say supremal) metric satisfying the following two conditions.

(i) for each $v \in V$ the embedding $[0, \infty[\to CV$ for $t \mapsto (v, t)$ is (non-strictly) distance decreasing.

(ii) for each $t \in [0, \infty[$ the embedding $V \mapsto CV$ for $v \mapsto (v, t)$ is 2^t-Lipschitz.

Then one can show that the quasi-isometry type of CV determines (suitably defined) quasi-conformal types of V (compare [Mar], [Gro$_\mathrm{HG}$], [Pan$_\mathrm{QCM}$]).

Example

(a) Let $V = S^3$ with the C-C metric associated to the above (contact) polarization (normal to the Hopf fibers). Then the corresponding space CV is quasi-isometric to the complex hyperbolic plane and quasi-isometries of CV induce (possibly non-smooth) contact maps on the ideal boundary $\partial_\infty CV = V$ which are quasi-conformal for the C-C metric (notice that the quasi-conformality is automatic for *smooth* contact maps).

(b) Let $V = S^7$ and $H \subset T(V)$ be horizontal for the Hopf fibration $S^7 \to S^4$. Then the corresponding CV is quasi-isometric to the quaternionic hyperbolic plane. Quasi-isometries of CV induce C-C quasi-conformal transformations of V. But these are quite special (unlike the contact maps in the complex hyperbolic case) by the following

Pansu rigidity theorem. *The above transformations of V are, in fact, conformal and the corresponding quasi-isometrics of CV are equivalent (asymptotic) to actual isometrics of the quaternionic hyperbolic plane.*[3]

Remark. Usually one starts with the complex (or quaternionic) hyperbolic space (e.g. plane) X and reconstructs the C-C structure at the sphere at infinity $\partial_\infty X$ as the limit of normalized Riemannian metrics induced on the concentric spheres $V_t \subset X$ around a fixed point $x_0 \in X$. Namely, one takes g_t on V_t equal to (induced metric)$/\mathrm{Diam}_t$, where Diam_t denotes the diameter of the induced Riemannian metric on V_t (which is about $\exp t$ for $t \to \infty$), and then the distance functions $\mathrm{dist}_t = \mathrm{dist}_{g_t}$ (on $\partial_\infty X$ identified with V_t) converge to the Carnot-Carathéodory metric on the corresponding sphere with the horizontal subbundle (polarization) associated to the Hopf fibration. (If one does all that to the real hyperbolic space of dimension n one gets just the usual round Riemannian metric on S^{n-1} and one does not know what happens for non-symmetric spaces of negative curvature, compare [Gro$_\mathrm{AI}$]).

0.10. Dimension and growth in the asymptotic geometry. We have seen in 0.8. how certain problems for C-C manifolds appear in the limit for families of growing Riemannian metrics on a compact manifold. Then such families can be realized on growing concentric spheres in a single non-compact complete manifold X whose ideal boundary comes along with a C-C geometry which is determined by the asymptotic geometry of X. One can generalize further and take a quite general (not necessarily hyperbolic) complete manifold X and transplant our basic C-C problems to X. For example, one may look for lower bounds on the (volume) growth of subsets $Y \subset X$ in terms of a suitable *asymptotic dimension* of Y in the spirit of [Gro$_\mathrm{AI}$].

1. Horizontal curves and small C-C balls

We analyze the behavior of short horizontal curves issuing from a given point $v \in V$ by a systematic (and somewhat boring) use of the Taylor remainder formula and thus prove several versions of the ball-box theorem (see [Bell] for a more elegant treatment of these problems). We apply this

---

[3] See the original paper [Pan$_\mathrm{QIR1}$] for a more precise and general statement.

to the evaluation of the Hausdorff dimension of V and submanifolds V' in V and also to a count of homotopy classes of Lipschitz maps $V \to W$. Our first step is

1.1. Proof of the Chow connectivity theorem. (compare [Herm]).

Recall that we are given vector fields $X_1, \ldots, X_m$ on a connected manifold V which Lie-generate $T(V)$ and we want to join a given pair of points in V by a piecewise smooth curve where each piece is a segment of an integral curve of some field $X_i, i = 1, \ldots, m$.

Since the problem is local, we may assume the fields X_i integrate to one-parameter groups of diffeomorphisms of V and we must join given points by piecewise orbit curves. In other words we must prove that the group G of diffeomorphisms of V generated by these subgroups is transitive on V. We start with the following

Trivial Lemma. *If G contains one-parameter subgroups, say $Y_1(t)$, $Y_2(t), \ldots, Y_p(t)$, where the corresponding vector fields $Y_1, Y_2, \ldots, Y_p$ span $T(V)$ (without taking commutators), then G is transitive on V.*

Proof. This follows from the implicit function theorem. Namely, for each $v \in V$ we consider the composed action map $E_v : \mathbb{R}^p \to V$ defined by

$$(t_1, \ldots, t_p) \mapsto Y_1(t_1) \circ \cdots \circ Y_p(t_p)(v).$$

The differential of E_v at the origin $0 \in \mathbb{R}^p$ sends $\mathbb{R}^p$ onto the span of the fields Y_i in $T_v(V)$ and, hence, is surjective in our case. Thus the orbit $G(v)$ is open in V for each $v \in V$ by the implicit function theorem and, as V is connected, $G(v) = V$. ∎

Now, let $X(t)$ be a one-parameter group (flow) on V, Y be a vector field and let us look at the transport of Y by $X(t)$, denoted $X_*(t)Y$. We observe that for small $t \to 0$

$$X_*(t)Y = Y + t[X, Y] + o(t)$$

(by the very definition of the Lie bracket) and conclude that, since the commutators of X_i span $T(V)$, there exist vector fields Y_j, $j = 1, \ldots, p \geq m$, on V which span $T(V)$ and such that

(i) $Y_i = X_i$ for $i = 1, \ldots, m$,

(ii) each field Y_j for $j > m$ equals $\big(X_i(t_j)\big)_* Y_{j'}$, i.e. the transport of some $Y_{j'}$ for $j' < j$ by the flow $X_i(t)$ at $t = t_j$, (where i also depends on j).

Finally we note that the one-parameter groups $Y_j(t)$ are contained in G because the transport of a field Y by $X(t)$ corresponds to the conjugation of the one parameter group $Y(\tau)$, i. e.

$$X_*(t)Y = X(t)Y(\tau)X^{-1}(t),$$

and the proof follows by the Trivial lemma.

1.1.A. Quantitative version of the Chow theorem for $\deg X_i \leq 2$ and a preliminary Hölder bound on the C-C metric. Let us try to estimate the length of the piecewise integral curve between given nearby points v and v' in V provided by the above argument, where the length of an orbit is given by the time parameter. (Thus, for example, the length of the orbit for the field $\frac{1}{t}\frac{\partial}{\partial t}$ in $\mathbb{R}$ between the points $l = 1$ and $t = \varepsilon \in \,]0, 1[$ equals $|\log \varepsilon|$). We assume that $T(V)$ is spanned by the commutators of degree ≤ 2 of X_i (i.e. by X_i and $[X_i, X_j]$) and we want to show that v and v' can be joined by a piecewise integral curve, such that

length of the curve $\lesssim$ (Riemannian distance between v and v')$^{\frac{1}{2}}$.

Proof. Take vector fields $Y_1, \ldots, Y_n$ on V such that

(i) The first m_0, for certain $m_0 < m$, among Y_i are some of the fields X_i say $X_1, \ldots, X_{m_0}$, which are linearly independent at v.

(ii) The remaining fields $Y_{m_0+1}, \ldots, Y_n$ are of the form $\big(X_i(\varepsilon)\big)_* X_j$ for some small ε (specified below), such that the corresponding commutators $[X_i, X_j]$ complete $X_1, \ldots, X_{m_0}$ to a full n-frame at v.

Recall, that every field Y_k for $k > m_0$ is of the form $X_j + \varepsilon[X_i, X_j] + o(\varepsilon)$, therefore the vectors $Y_1, \ldots, Y_n$ are linearly independent at v for small ε, and so the differential $\mathcal{D} : \mathbb{R}^n \to T_v(V)$ of the composed orbit map

$$E = E_\varepsilon : (t_1, \ldots, t_n) \mapsto Y_1(t) \circ \cdots \circ Y_n(t)(v)$$

does not degenerate. Moreover, it is clear that the norm of the inverse operator is bounded by

$$\|\mathcal{D}^{-1}\| \leq \mathrm{const}\, \varepsilon^{-1} \quad \text{for small } \varepsilon \to 0.$$

It follows that the E_ε-image of the ε-ball in $\mathbb{R}^n$ around the origin contains the Riemannian ball $B_v(\delta) \subset V$ for $\delta \geq \text{const}' \, \varepsilon^2$, provided $\varepsilon > 0$ is sufficiently small. This follows from the implicit function theorem since the second derivatives of the map E_ε are uniformly bounded in ε (as this is true for the fields Y_i of the form $\big(X_i(\varepsilon)\big)_* X_j$). $\blacksquare$

Reformulation in terms of the C-C metric. The length of piecewise integral curves defines a metric on V which is of a slightly more general type than Carnot-Carathéodory since the fields X_i are not supposed to be independent. But if these fields are independent this metric essentially majorizes the Carnot-Carathéodory one as our curves are (special) H-horizontal for $H = \text{Span}\{X_i\}$. In fact, the two metrics are equivalent as we shall see later and we call our present metric C-C anyway. Now we can reformulate the above proposition in terms of the following Hölder bound on this C-C metric by a Riemannian one,

$$\text{C-C dist} \lesssim (\text{Riem.dist})^{\frac{1}{2}}. \qquad (*)$$

1.1.A$'$. Upper box bound on C-C dist for deg $X_i \leq 2$. Let us refine $(*)$ by taking into account the difference in the behavior of the C-C distance in different directions. Namely, we want to show that in the direction of H the C-C metric is equivalent to the Riemannian one. Namely,

let $c(t)$ be a smooth curve in V issuing from $v \in V$ parametrized by the length parameter t. If $c(t)$ is tangent to H at $t = 0$, then

$$\text{C-C dist}\big(v = c(o), c(t)\big) \lesssim t. \qquad (**)$$

Proof. The tangent vector of $c(t)$ at $v = c(o)$ can be written as a linear combination of X_i at v, say $c'(t) = \sum_{i=1}^m a_i X_i$. Then the value of the composed orbit map

$$E : (t_1, \ldots, t_m) \mapsto X_1(t) \circ \cdots \circ X_m(t)(v)$$

at the point $(ta_1, \ldots, ta_m)$ approximates $c(t)$ for small t as

$$E(ta_1, \ldots, ta_m) = c(t) + O(t^2),$$

by the Taylor remainder theorem (applied to $c(t)$ and to E with respect to some Euclidean structure in V near v). It follows, with $(*)$, that

$$\text{C-C dist}\big(c(t), \, E(ta_1, \ldots, ta_m)\big) \lesssim t$$

and the proof of (∗∗) is implied by the triangle inequality as

$$\text{C-C } \operatorname{dist}\big(v, E(ta_1, \ldots, ta_m)\big) \leq t$$

by the very definition of our C-C distance with piecewise integral curves.

1.1.B. Lower box bound on C-C dist for $\deg X_i \leq 2$. The above shows that every small C-C ball of radius ε around v contains a box of size about ε in the directions tangent to H_v and about ε^2 in the transversal direction (where "about ε" signifies $\operatorname{const}\varepsilon$ for $\operatorname{const} > 0$ independent of ε). Now we want to show that the ε-C-C ball is contained in such a box. Namely, we want to show that along a curve $c_0(t)$ *transversal* to H the C-C distance is $\gtrsim \sqrt{\text{Riemannian distance}}$. Here are two slightly different proofs.

First proof. Assume for the moment that the dimension of the span $\{X_i\}$ is constant at v and let α be a smooth 1-form defined near v, such that

$$\alpha(X_i) = 0, \ i = 1, \ldots, m$$

and

$$\alpha\big(c_0'(t)\big) = 1, \quad \text{for our curve } c(t)$$

issuing from v in the direction $c'(t)$ transversal to H.

Let c be a piecewise smooth curve tangent to H joining v with $v_\varepsilon = c_0(t = \varepsilon)$ and let b denote the closed curve formed by $c_0[0, \varepsilon]$ and c, see Fig. 2 below.

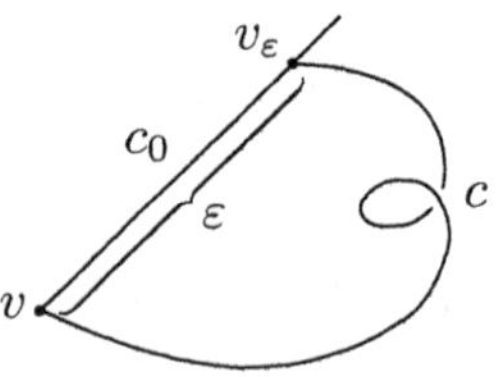

Figure 2

This b bounds a disk D of Riemannian area about ℓ^2 for

$$\ell = \text{length } b = \text{length } c + \varepsilon,$$

and by the Stokes formula

$$\text{Area } D \gtrsim \int_D d\alpha = \int_{c_0[0,\varepsilon]} \alpha \approx \varepsilon,$$

since α vanishes on c (which is tangent to H). It follows that $(\text{length}\,c + \varepsilon)^2 \gtrsim \varepsilon$ and consequently

$$\text{length}\,c \gtrsim \sqrt{\varepsilon}. \qquad\qquad \blacksquare$$

Now, even without assuming that the span of X_i has constant rank, we have a form α which annihilates all X_i at v and has $\alpha\big(c_0'(0)\big) = 1$. Then the integral $\int_D d\alpha = \int_b \alpha$ acquires an extra term, namely $\int_c \alpha$, which is of the order at most $(\text{length}\,c)^2$ since $\alpha(X_i)$ is of order $\leq \varepsilon$ in the Riemannian ε-ball around v. This gives us the relation

$$(\text{length}\,c_1 + \varepsilon)^2 \gtrsim \varepsilon \pm (\text{length}\,c)^2,$$

which implies $\text{length}\,c \geq \sqrt{\varepsilon}$ just the same.

Second Proof. Take a smooth hypersurface $V_0 \subset V$ passing through v and tangent to $H_v \subset T_v(V)$. Then every curve c issuing from v and tangent to H is also tangent to V_0 at v and by the Taylor remainder theorem the Riemannian distance from $c(t)$ to V_0 is bounded by $\approx t^2$. Hence, the Riemannian distance between $v = c(0)$ and $c(t)$ is bounded by t^2 on each curve c_0 transversal to V_0. $\blacksquare$

Remark. Notice that both proofs need the fields X_i to be C^1-smooth and fail for continuous fields where, in fact, the quadratic Hölder bound may be invalid.

1.1.C. Corollary: Ball-box theorem for deg $= 2$. The small ε-C-C ball in V around v is *equivalent* to the following $\varepsilon \times \varepsilon^2$-box. Take a smooth m_0-dimensional submanifold $V_1 \subset V$ through v with $T_v(V_1) = H_v$, take the Riemannian ε-ball in V_1 and then the Riemannian ε^2-neighbourhood of this ball in V. This is our $\varepsilon \times \varepsilon^2$-box. It is equivalent to the ε-C-C ball in the sense that

$$\text{Box}(\delta\varepsilon \times \delta^2\varepsilon^2) \subset \varepsilon\text{-C-C ball} \subset \text{Box}(\Delta\varepsilon \times \Delta^2\varepsilon^2),$$

where δ and Δ are positive constants independent of ε (which can be chosen continuously depending on v in a suitable sense).

1.2. A new proof of the Chow theorem and the Hölder bound on the C-C metric for arbitrary degree. (compare [Mit$_{1;2}$] and [N-S-W]). Our proof of the bound

$$\text{C-C dist} \lesssim (\text{Riem.dist})^{\frac{1}{d}} \tag{+}$$

for the case where $T(V)$ is spanned by the commutators of the fields X_i of degree $d \leq 2$, does not easily extend for $d \geq 3$ but the following more straightforward approach proves out to be more forceful as it yields $(+)$ for all d. The key ingredient is the following (well known) approximate expression for the commutator of one-parameter groups $X_1(t)$ and $X_2(t)$ in Diff V in terms of the 1-parameter group corresponding to the Lie brackets of the fields X_1 and X_2,

$$[X_1(t), X_2(t)]^\circ \overset{\text{def}}{=} X_1(t) \circ X_2(t) \circ X_1^{-1}(t) \circ X_2^{-1}(t) = [X_1, X_2](t^2) + o(t^2), \tag{$*$}$$

where the additive notation refers to some Euclidean structure in a relevant neighbourhood (and where one should note that $X_i^{-1}(t) = X_i(-t)$, $i = 1, 2$).

Proof of $(*)$. We need the following three elementary formulas

$$(1) \qquad\qquad\qquad (\tau Y)(t) = Y(t\tau)$$

(2) $(X+\tau Y)(t)=X(t)\circ Y(\tau t)+o(t\tau)=Y(\tau t)\circ X(t)+o(t\tau)$, for $t, \tau \to 0$.

(3) $\qquad\quad X_1(t)X_2(\tau)X_1^{-1}(t) = (X_2 + \tau[X_1 X_2])t + o(t\tau)$.

Notice, that (1) is obvious, (2) follows from the Taylor remainder theorem for the composition $X(t) \circ Y(\tau t)$, and (3) is implied by (1), (2) and the definition of the Lie bracket (compare 1.1).

Now we obtain $(*)$ in the form

$$X_1(t) \circ X_2(t) \circ X_1^{-1}(t) = [X_1, X_2](t^2) \circ X_2(t) + o(t^2)$$

by applying first (3) and then (2) and (1) to the left-hand side.

Next we observe that $(*)$ by induction implies

$$\big[X_1(t), [X_2(t), X_3(t)]^\circ\big]^\circ = \big[X_1, [X_2 X_3]\big](t^3) + o(t^3)$$

$$\cdot \quad \cdot \quad \cdot \quad \cdot \quad \cdot \quad \cdot \quad \cdot \quad \cdot \quad \cdot \quad \cdot \quad \cdot \quad \cdot \quad \cdot \quad \cdot \quad \cdot$$

$$\big[X_1(t), [\ldots, X_d(t)]^\circ \ldots\big]^\circ = \big[X_1, [\ldots X_d]\ldots\big](t^d) + o(t^d).$$

Proof of (+). We concentrate on the simplest case where $\dim V = 3$ and $T(V)$ is generated by X_1, X_2 and $Y = [X_1, X_2]$. We denote by $Y^\circ(t)$ the one-parameter family (not a subgroup) of diffeomorphisms defined by

$$
Y^\circ(t) = \begin{cases} \left[X_1(|t|^{\frac{1}{2}}), X_2(|t|^{\frac{1}{2}}) \right]^\circ & \text{for } t \geq 0 \\[2mm] \left[X_2(|t|^{\frac{1}{2}}), X_1(|t|^{\frac{1}{2}}) \right]^\circ & \text{for } t \leq 0 \end{cases}
$$

and we observe that the composed map

$$
E^0 : (t_1, t_2, t_3) \mapsto X_1(t) \circ X_2(t) \circ Y^\circ(t)(v)
$$

sends the box $B(\varepsilon) \subset \mathbb{R}^3$ defined by $|t_1| \leq \varepsilon$, $|t_2| \leq \varepsilon$, $|t_3| \leq \varepsilon^2$ into the ε'-C-C ball in V around v for $\varepsilon' \approx \varepsilon$ (in fact, for $\varepsilon' \leq 10\varepsilon$). What remains to show is that the image of this box contains a Riemannian δ-ball around v for $\delta \gtrsim \varepsilon^2$. To see that we compare E^0 with the composed map

$$
E : (t_1, t_2, t_3) \mapsto X_1(t) \circ X_2(t) \circ Y(t)(v),
$$

for which the image of the ε-box is δ-large by the implicit function theorem. In fact, the E-image of the ε^2cube defined by $|t_1| \leq \varepsilon^2$, $i = 1, 2, 3$, contains the required δ-ball. Then we observe with $(*)$ that $E^0 = E + o(\varepsilon^2)$ in the ε^2-cube. It follows by elementary topology (see below) that the E^0-image of the ε^2-cube is essentially as large as the E-image.   ■

Elementary topology lemma. *Let E and E^0 be continuous maps of a compact manifold with boundary into a Riemannian manifold, say $B \to V$ such that*

(i) *$\mathrm{dist}_V(E, E^0) \underset{\text{def}}{=} \sup_{b \in B} \mathrm{dist}\big(E(b), E^0(b)\big) \leq \delta_0$ for some $\delta_0 > 0$.*

(ii) *Every two points in V within distance $\delta \leq \delta_0$ can be joined by a unique geodesic segment of length δ.*

(iii) *The map E is a homeomorphism of B onto its image $E(B)$ in V.*

Then the image $E^0(B)$ contains every point $v \in E(B)$ for which the δ-ball in V around v is contained in $E(B)$.

The (standard) proof of this is left to the reader.

Finally we observe that with the provisions we have made the above proof of (+) extends to the general case (of any number of fields and arbitrary $d = 1, 2, 3, \ldots$) by just adjusting the notations.

1.2.A. Upper box bound on the C-C distance for arbitrary degree. The above argument does not provide decent box-shaped domains inside the C-C balls but the following simple modification of this argument does just that. To see this we concentrate again on the case of three fields X_1, X_2 and $Y = [X_1, X_2]$ spanning $T(V)$ and we observe the following

1.2.A′. C^1-Lemma. *The above family $Y^\circ(t)$ is C^1-smooth.*

Proof. What we know about $Y^\circ(t)$ is the relation

$$Y^\circ(t^2) = [X_1, X_2](t^2) + o(t^2) \qquad\qquad (*)'$$

where $Y^\circ(t^2)$ and $[X_1, X_2](t^2)$ are smooth (as we *assume* our fields X_i are C^∞-smooth). Hence $(*)'$ implies that

$$Y^\circ(t^2) = [X_1, X_2](t^2) + t^3 \varphi(t)$$

where $\varphi(t)$ is smooth, and so

$$Y^\circ(t) = [X_1, X_2](t) + t^{\frac{3}{2}} \varphi(\sqrt{t})$$

is C^1-smooth. ∎

We observe that the differential of $Y^\circ(t)$ at $t = 0$ equals that of $Y(t)$ and so $E(t)$ and $E^0(t)$ also have equal differentials at $t = 0$ (where the C^1-smoothness of E^0 is ensured by that of $Y^0(t)$), and so the E^0-image of the box

$$B(\varepsilon) = \left\{ |t_1| \leq \varepsilon, \ |t_2| \leq \varepsilon, |t_3| \leq \varepsilon^2 \right\}$$

is sent by E^0 onto a box-shaped domain in V by the implicit function theorem (which we now can apply to E^0). Thus the proof of the upper box bound on dist_H is concluded.

1.2.B. Making "smooth" instead of "piecewise smooth" in the Chow connectivity theorem for polarizations H $\subset$ T(V). In the original Chow theorem curves joining given points must necessarily consist of pieces of orbits of different fields and so they cannot be made smooth. But if we have a smooth polarization H, where the commutators of H-horizontal fields span $T(V)$, we may slightly improve the result by showing that

Every two points in V can be joined by a smooth H-horizontal curve in V, i.e. by a smooth immersion $f : [0,1] \to V$ with $f(0) = (v_0)$ and $f(1) = (v)$ and $f'(t) \in H$ for $t \in [0,1]$.

Proof. Either of our two proofs of the Chow theorem provides a smooth family Φ of piecewise smooth curves issuing from v_0, say $\varphi(t) \in \Phi$, $t \in [0,1]$, such that the map $\Phi \to V$ defined by $\varphi \mapsto \varphi(1)$ contains a given point v in its image, i.e. $\varphi(1) = v$ for a certain φ, and this map $\Phi \to V$ is a submersion near φ. The curve φ consists of segments of orbits of certain H-horizontal vector fields $Y_1, \dots, Y_k$ and when these fields are fixed, then φ is uniquely determined by the lengths of the segments. In fact, these lengths serve as coordinates in Φ and so the curves in Φ close to φ are obtained by slightly varying these lengths, called $\ell_i = \ell_i(\varphi)$, $i = 1, \dots, k$. Next, let us smoothly interpolate between Y_i and Y_{i+1} for all $i = 1, \dots, h$. Namely, we introduce a smooth family of fields $Y_t = Y_t(\ell_1, \dots, \ell_k)$, $t \in [0, L_k]$ for $L_k = \sum_{i=1}^{k} \ell_i$, such that

(i) $Y_t = Y_{i+1}$ for $t \in [L_i + \varepsilon, L_{i+1} - \varepsilon]$, for $L_i = \ell_1 + \cdots + \ell_i$ and small $\varepsilon > 0$.

(ii) $\|Y_t\| \leq \mathrm{const}$ for some $\mathrm{const} \geq 0$ independent of ε.

Now we define $f(t)$ as the integral curve of the field Y_t issuing from v_0, i.e. $f(O) = v_0$ and $f'(t) = Y_t$ at $v = f(t)$, and observe that $f \to \varphi$ for $\varepsilon \to 0$. It easily follows (e.g. with Elementary topology lemma) that the map $f \mapsto f(1)$ contains v in its image for a sufficiently small ε.   ■

Acknowledgment. The smoothing problem in Chow's theorem was brought to my attention by Lucas Hsu.

1.3. Lower box bound on the C-C distance for deg ≥ 2. We want to show that on each smooth curve $c_0(t)$ in V issuing from v in the direction $c_0'(t) \in T_v(V)$ transversal to the span of the commutators of given smooth fields X_i, $i = 1, \dots, m$, of degrees $\leq s$, the C-C distance satisfies for small $t > 0$,

$$\text{C-C dist}\big(c(o) = v, c(t)\big) \gtrsim t^{\frac{1}{s+1}}.$$

We shall do it by adopting the second proof from 1.1.B for which we need the following

Definition of tangency. A smooth function f on V is called *constant of order s at $v \in V$* with respect to given fields X_i, if for every differential operator of order r obtained by composing $r \leq s$ of our fields, denoted

$$X_I = X_{i_1} X_{i_2} \ldots X_{i_r} \quad \text{for} \quad I = (i_1, \ldots, i_r),$$

the function $X_I(f)$ vanishes at v, where vector fields are thought of as differential operators of first order acting on functions. Next, a submanifold $V_0 \subset V$ passing through v is called *s-order tangent to X_i*, if every function on V vanishing on V_0 is constant of order s at v.

Examples

(a) If $s = 1$ then this tangency means that the tangent space $T_v(V_0) \subset T_v(V)$ contains the span of the fields X_i at v.

(b) If the system of fields X_i is integrable in the sense that it gives a tangent frame to a foliation of V then each leaf is tangent to X_i of infinite order and the same is true for every submanifold containing a leaf.

The following obvious lemma relates the above definition with our problem.

Lemma

(a) Let f be constant of order s with respect to X_i at v and $c(t)$ be a smooth curve issuing from v and tangent to X_i in the sense that $c'(t) = \sum_{i=1}^{m} a_i(t) X_i$, for some smooth functions $a_i(t)$. (If the span H of X_i has dimension independent of v, then this tangency amounts to the inclusion $c'(t) \in H$ for all t.) Then $f\big(c(t)\big) \lesssim t^{s+1}$.

(b) Let V_0 be tangent to X_i at v with order s then the Riemannian distance from $c(t)$ to V_0 is bounded by

$$\text{Riem.dist}\,\big(c(t), V_0\big) \lesssim t^{s+1}.$$

In other words, the C-C ε-neighbourhood of V near v_0 is contained in the Riemannian ε^{s+1}-neighbourhood. In particular, the C-C distance on each curve $c_0(t)$ transversal to V_0 at v is $\gtrsim$ (Riemannian distance)$^{\frac{1}{s+1}}$.[4]

---

[4] Notice, that the only assumption on the fields X_i is C^∞-smoothness.

In order to use this lemma and exhibit actual boxes in small C-C balls, we need sufficiently many submanifolds tangent to X_i with prescribed order. These are provided by the following

Infinitesimal lemma. *Let $T_0 \subset T_v(V)$ be a linear subspace containing the commutators of X_i of orders $\leq s$ at v. Then there exists a submanifold $V_0 \subset V$ passing through v, having $T_v(V_0) = T_0$ and tangent to X_i at v with order s.*

One may prove this by a straightforward linear algebra in the space of d-th order jets of submanifolds (and functions) at $v \in V$. But one can also make a short-cut with the following

Exponential lemma. *Let $Y_1, \ldots, Y_{m_1}, \ldots, Y_{m_2}, \ldots, Y_{m_r}, \ldots, Y_{m_s}$ be linearly independent vector fields on V, where for each $r = 1, \ldots, s$ the fields $Y_{m_{r-1}+1}, Y_{m_{r-1}+2}, \ldots, Y_{m_r}$ are taken among commutators of X_i of degree r, such that the fields $Y_1, \ldots, Y_{m_r}$ have the same span at v as the commutators of X_i of degree $\leq r$. (Obviously, such Y_j exist.) Then the image V_0 of the exponential map $\exp_v : \mathbb{R}^{m_s} \to V$ corresponding to Y_j is tangent to X_i at v with order s.*

Proof. To make it simple we start with the case $s = 2$ and assume for the moment that the fields X_i are linearly independent and so $m_1 = m$ and $Y_j = X_j$ for $j = 1, \ldots, m$. Let f be a smooth function vanishing on V_0 (or rather on a germ of V_0 at v) and observe that

(1) $[X_i, X_j]f(v) = 0$, $i, j = 1, \ldots, m$, since $T_v(V_0)$ contains the second degree commutators of X_i.

(2) For every field $X = \sum_{i=1}^{m} a_i X_i$, the operator X^2 satisfies $X^2 f(v) = 0$. In fact $X^p f(v) = 0$ for all $p = 1, 2, 3, \ldots$, as f vanishes on the orbit $X(t)(v)$ (which is contained in the exponential image of $\mathbb{R}^m$ corresponding to X_i).

It follows that

$$(X_i X_j + X_j X_i)f(v) = \big((X_i + X_j)^2 - X_i^2 - X_j^2\big)f(v) = 0$$

which implies with (1) that $(X_i X_j)f(v) = 0$ as well. This gives us the required tangency of V_0 to X_i in the case where X_i are independent and the dependent case is taken care of as follows. Every field among X_i, say X_{i_0}, can be written as a sum, $X_{i_0} = Y'_{i_0} + X^0_{i_0}$, where Y'_{i_0} is some combination of Y_j, $j = 1, \ldots, m_0$, with constant coefficients, such that Y'_{i_0} equals X_{i_0} at v and so $X^0_{i_0} = X_{i_0} - Y'_{i_0}$ vanishes at v. The relations $[X_i, X_j]f(v) = 0$ imply that $[X^0_{i_0}, X_i]f(v) = 0$ since $X^0_{i_0}$ is a combination of X_i. Therefore the (obvious) vanishing $X^0_{i_0}X_i f(v) = 0$ implies $X_i X^0_{i_0} f(v) = 0$, and consequently

$$X_i X_j f(v) = (Y'_i + X^0_i)(Y'_j + X^0_j)f(v) =$$
$$Y'_i Y'_j f(v) + Y'_i X^0_j f(v) + X^0_i Y'_j f(v) + X^0_i X^0_j f(v) = 0,$$

where the vanishing of the first term is ensured by (the argument in) the independent case.

Case $s = 3$. Now we start with the relations $\big[X_i, [X_j, X_k]\big]f(v) = 0$ and we use $Y^p f(v) = 0$ for every combination Y of Y_j with constant coefficients. Thus we get $Y_i Y_j\, f(v) = 0$ for $i = 1, \ldots, m_1$ and $j = m_1 + 1, \ldots, m_2$ as well as for $i = m_1 + 1, \ldots, m_2$ and $j = 1, \ldots, m_1$, by using $[Y_i, Y_j]f(v) = 0$ and $Y^2 f(v) = 0$ for $Y = Y_i, Y_j$ and $Y_i + Y_j$. This implies that the (second order) operators $X_i[X_j, X_k]$ and $[X_j, X_k]X_i$ also vanish at $f(v)$ where dependencies among X_i and $[X_j, X_k]$ are taken care of as above for $s = 2$.

Next we show that the symmetrization of $Y_i Y_j Y_k f(v)$, over all (six) permutations of the three indices i, j and k, vanishes for $i, j, k = 1, \ldots, m_0$. This is done with the identity $(aY_i + bY_j + cY_k)^3 f(v) = 0$ for $a, b, c = 0, \pm 1$. This implies with the above that $Y_i Y_j Y_k f(v) = 0$ for $i, j, k = 1, \ldots, m_0$. What remains is to pass from Y_i to X_i, where X_i may be dependent, but these dependencies are easily taken care of as earlier.   ■

Case $s \geq 4$. The proof of this is clear by now.

We conclude by explicitly describing a box-shaped domain in V containing the C-C ball around v provided by the above lemmas.

Step 1. For every $r = 1, 2, \ldots$ we take a submanifold $V_r \subset V$ passing through v such that $T_v(V_r)$ equals the span of the commutators of X_i of degrees $1, \ldots, r$. A specific such V_r is provided by the exponential map corresponding to $Y_1, \ldots, Y_{m_r}$ from the exponential lemma.

Step 2. We take the Riemannian ε^{r+1}-neighbourhood of V_r for each r and intersect these over $r = 0, 1, \ldots$ (with the convention $V_0 = \{v\}$). This intersection is, indeed, box-shaped if we take $V_0 \subset V_1 \subset \cdots \subset V_2 \subset \cdots$, (which is possible with the exponential lemma) and it contains the ε'-C-C ball around $v \in V$ for $\varepsilon' \approx \varepsilon$.

Matching upper and lower box bounds. Our (families of) boxes inside (see 1.2.A) and outside (see above) the C-C balls are slightly different but this does not bring any confusion into the geometric picture as these families are equivalent in an obvious sense as a simple argument shows. (See [N-S-W] for further information.)

1.3.A. Doubling and covering properties for balls; equisingularity and the Hausdorff dimension.

The ball-box theorem (as stated in 0.5.A) immediately implies the following universal bound on the Riemannian volume of concentric C-C balls in a compact manifold V,

$$\operatorname{Vol} B_v(2\rho) \leq C \operatorname{Vol} B_v(\rho) \qquad (*)$$

for all $v \in V$ and real ρ and some constant $C = C(V)$. (This is one of the major applications of the ball-box theorem indicated in [N-S-W].) Consequently we obtain, as an obvious corollary, the following (purely internal) metric property of V.

Every ball $B_v(2\rho)$ can be covered by at most k balls of radii ρ for some k depending only on V (but not on v or ρ).

Now, suppose H is equiregular. Then, clearly, the ball-box theorem shows that small ρ-balls have volumes $\approx \rho^D$ (see $(+)$ in 0.6) and thus $\dim_{\mathrm{Hau}} V = D$ with $0 < \mathrm{mes}_D < \infty$, where D is computed in terms of the commutator filtration $0 \subset H = H_1 \subset H_2 \subset \cdots \subset H_d = T(V)$ by $D = \sum_{i=1}^{d} i \operatorname{rank}(H_i/H_{i-1})$. Furthermore, let $V' \subset V$ be a submanifold in V such that the intersections $H_i' = H_i \cap T(V')$ have constant ranks for all $i = 1, \ldots, d$. Then in the (proof of the) ball-box theorem one may use frames *adapted* to V': if some vector from the frame is tangent to V' at a given point v', then the corresponding vector field is tangent to V' near v'. Thus we extend the ball-box theorem to the restricted metric $\mathrm{dist}_H | V'$, and see that a small ρ-ball in this metric is approximated by a box in V' having k_i' (among his $\sum_{i=1}^{d} k_i'$) sides of length $\approx \rho^i$ for $i = 1, \ldots, d$, where k_i' denotes $\operatorname{rank} H_i'/H_{i-1}'$. Now we see as earlier that the ρ-balls in V' have volumes $\approx \rho^{D'}$ for $D' = \sum_{i=1}^{d} i k_i'$ and so satisfy the same properties as those in V. In particular we see that $\dim_{\mathrm{Hau}}(V', \mathrm{dist}_H) = D'$.

Non-equiregular fields. First, let us try to understand how generic those frames of fields $X_1, \ldots, X_m$ are which Lie-generate $T(V)$. As the Lie generation condition involves jets of arbitrary large orders, one might think that for $m \geq 2$ the (jets of) non-generating frames have infinite codimension. In fact this is true modulo the following trivial

Observation. If C^∞-fields X_i, $i = 1, \ldots, m$ on V vanish at some point $v \in V$, then so do their commutators of all orders. Thus the jets of a non Lie-generating frame at a fixed point $v \in V$ have codim $\leq mn$ for $n = \dim V$.

On the other hand let X be a non-vanishing field on V. Then for generic fields Y the successive commutators

$$Y_1 = Y, \ Y_2 = [X, Y_1], \ Y_3 = [X, Y_2] \ldots,$$

generate $T(V)$, where *the jets of exceptional Y's have infinite codimension*. In particular, *generic polarization of* rank ≥ 2 *Lie-generate* $T(V)$.

Proof. Let $X_1 = X$, $X_2 \ldots, X_n$ be a frame of commuting fields at a point $v \in V$ and $Y = \sum_{i=1}^n a_i X_i$. Then generically, up to infinite codimension, all but finitely many iterated Lie derivatives $L_X L_X \ldots L_X a_i$ do not vanish at a given point and so the fields $Y_1, Y_2, \ldots$ are not contained in a given hyperplane $H' \subset T_v(V)$. This hyperplane, for the above frame, is given by the span of $X_1, \ldots X_{i-1}, X_{i+1}, \ldots, X_n$ for some i and as we are free to change our frame it may be arbitrary. This implies our assertion.

Now we see with the above two facts that the space of jets of m-frames of fields for $m \geq 2$ contains two strata Σ_0 and Σ_1 where codim $\Sigma_0 = mn$ and codim $\Sigma_1 = \infty$ such that a frame Lie-generates $T(V)$ at v if and only if its jet at v misses $\Sigma = \Sigma_0 \cup \Sigma_1$.

Questions. Let $\Sigma^r \supset \Sigma$ correspond to the frames whose commutators of order $\leq r$ do not generate $T_v(V)$. What is the structure of Σ^r and $\Sigma^r - \Sigma^{r-1}$? What is the possible geometry of the subset $V^{(r)} \subset V$ where the r-th order commutators (of some fields) fail to generate $T_v(V)$ for $v \in V^{(r)}$ while the following commutators do generate $T(V)$?

Example. Let $n = \dim V = 2$ and one of the fields in the frame, say X, does not vanish at v. Then the intersection of $V^{(r)}$ with every orbit of X is discrete (as all commutators vanish at an accumulation point) and so $\Sigma^{(r)}$ is locally contained in a smooth curve.

Our interest in the above questions stems from the problem of evaluation of the Hausdorff dimension of non-equiregular (V, H) where we should know the (discontinuity) structure of the functions $n_i(v) = \operatorname{rank} H_i(v)$ on V.

Let us look what happens in the above plane example where we have two fields $X = \frac{\partial}{\partial x}$ and $Y = a(x,y)\frac{\partial}{\partial y}$ on the (x,y)-plane $\mathbb{R}^2$ such that the function a and the derivatives $\frac{\partial a}{\partial x}$, $\frac{\partial^2 a}{\partial^2 x}, \ldots, \frac{\partial^s a}{\partial^s x}$ vanish on some closed subset A in the line $\{x = 0\}$ and $a \neq 0$ outside A. We give $\mathbb{R}^2$ the maximal (or supremal) metric for which all X- and Y-orbits $\mathbb{R} \to \mathbb{R}^2$ are (non-strictly) distance decreasing (where one should use partial orbits if the fields are not integrable) and determine $\dim_{\mathrm{Hau}} \mathbb{R}^2$ with this metric $\mathrm{dist}_{X,Y}$ as follows. This metric, outside A, is equivalent to the Euclidean metric dist and so

$$\dim_{\mathrm{Hau}}(\mathbb{R}^2 - A, \ \mathrm{dist}_{X,Y}) = 2.$$

On the other hand, $\mathrm{dist}_{X,Y}$ on A is equivalent to $(\mathrm{dist})^{\frac{1}{s+1}}$ and thus

$$\dim_{\mathrm{Hau}}(A, \mathrm{dist}_{X,Y}) = \big(\dim_{\mathrm{Hau}}(A, \mathrm{dist})\big)^{s+1}.$$

Consequently

$$\dim_{\mathrm{Hau}}(\mathbb{R}^2, \mathrm{dist}_{X,Y}) = \max\Big(2, \big(\dim_{\mathrm{Hau}}(A, \mathrm{dist})\big)^{s+1}\Big),$$

and we see that *any* real number ≥ 2 may appear as $\dim_{\mathrm{Hau}}(\mathbb{R}^2, \mathrm{dist}_{X,Y})$.

A similar example can be arranged in $\mathbb{R}^3$ with a *polarization* spanned by two *independent* fields, e.g. $X = \frac{\partial}{\partial x}$ and $Z = a(x,y)\frac{\partial}{\partial y} + \frac{\partial}{\partial z}$, where the function a is as above, and the commutator $Y = [X, Z] = \frac{\partial a}{\partial x}\frac{\partial}{\partial y}$ is of the same kind as earlier.

Thus, in the general case, one cannot say much more about $\dim_{\mathrm{Hau}}(V, \mathrm{dist}_H)$ than

(1) $\dim_{\mathrm{Hau}} V \leq D = \max\limits_{v \in V} \sum_{i=1}^{d} i k_i$, for $k_i = \mathrm{rank}\big(H_i(v)/H_{i-1}(v)\big)$.

(2) There is an open dense set $U \subset V$ where $D \leq \frac{n(n-1)}{2} + 1$ and so $\dim_{\mathrm{Hau}} U$ is also bounded by $\frac{n(n-1)}{2} + 1$.

But the picture appears more regular for sufficiently generic H admitting so-called *equisingular stratifications*. Such a stratification is a partition of V into locally closed submanifolds (strata) V_ν, such that

(i) the ranks $n_i(v)$ of $H_i(v) \subset T_v(V)$ are constant on each V_ν,

(ii) the numbers $n_{i,\nu} = \mathrm{rank}\big(H_i(v) \cap T(V_\nu)\big)$ are also constant on each V_ν.

If this is the case we see as earlier that

$$\dim_{\mathrm{Hau}} V_\nu = \sum_{i=1}^{d} i(n_{i,\nu} - n_{i-1,\nu})$$

and $\dim_{\mathrm{Hau}} V = \max_\nu \dim_{\mathrm{Hau}} V_\nu$.

It is clear that real analytic H admits equisingular stratifications and the same seems to be true for the generic C^∞-polarizations H by virtue of (E_1) on p. 34 in [$\mathrm{Gro_{PDR}}$]. Similarly one defines equisingularity of $V' \subset (V, H)$ and obtain, in particular, the integrality of $\dim_{\mathrm{Hau}}(V', \mathrm{dist}_H)$ in the real analytic and C^∞-generic cases.

1.4. Canonical coordinates, almost Lie groups, nilpotent tangent cones and a sharp version of the ball-box theorem for equiregular polarizations. Suppose we are given a frame of smooth vector fields $Y_1, \ldots, Y_n$ on V, $n = \dim V$, where each Y_i is assigned an integer $\deg Y_i = \deg i \geq 1$, such that commuting fields at most add degrees, i.e.

$$[Y_i, Y_j] = \sum_k c_{ijk}(v) Y_k, \qquad (*)$$

where $c_{ijk} = 0$ for $\deg k > \deg i + \deg j$ (compare 0.5). Then for each $v \in V$ one defines the following Lie algebra L_v with a preferred basis, denoted Y_i^v where the Lie brackets are given by the formulae

$$[Y_i^v, Y_j^v] = \sum_k \delta_{ijk} c_{ijk}(v) Y_k^v, \qquad (*)^v$$

where $\delta_{ijk} = 1$ for $\deg k = \deg i + \deg j$ and $\delta_{ijk} = 0$ otherwise. To comprehend the meaning of $(*)^v$ we ε-scale the fields Y_i according to their degrees, denote ${}^\varepsilon Y_i = \varepsilon^{\deg Y_i} Y_i$, and express the multiplication table $(*)$ in terms of ${}^\varepsilon Y_i$. This gives

$$[{}^\varepsilon Y_i, {}^\varepsilon Y_j] = \sum_k \varepsilon^{d_{ijk}} c_{ijk} {}^\varepsilon Y_k, \qquad {}^\varepsilon(*)$$

for $d_{ijk} = \deg i + \deg j - \deg k$. Now we see that $(*)^v$ equals the limit of ${}^\varepsilon(*)$ at v for $\varepsilon \to 0$ which shows that L_v is indeed a Lie algebra which is nilpotent of degree at most $\max_i \deg Y_i$.

Definition. (compare [Good], [Bell]). The simply connected (nilpotent) Lie group N_v corresponding to L_v with distinguished left invariant fields corresponding to Y_i^v is called the *nilpotent tangent* cone of the frame $\{Y_i\}$ at v.

We still denote the distinguished fields on N_v by Y_i^v and we want to show that the formal limit relation $^\varepsilon(*) \to (*)^v$ for $\varepsilon \to 0$ implies an actual convergence $^\varepsilon Y_i \to Y_i^v$ in suitable local coordinates in V and N_v. In fact, one can use for this purpose any system of coordinates in V near v which is made up in a canonical way out of the fields $^\varepsilon Y_i$ but we shall stick to the coordinates t_i defined with the composition of the one parameter groups $Y_i(t)$, namely, with the map

$$E_v : (t_1, \dots, t_n) \mapsto Y_1(t) \circ Y_2(t) \circ \cdots \circ Y_n(t)(v),$$

defined on a certain cube

$$B(\rho) = \{|t_i| \le \rho\} \subset \mathbb{R}^n.$$

We identify the Lie algebra $L_v = L(N_v)$ (which comes along with the basis Y_i^v) with $\mathbb{R}^n$ and we denote by $E_0 : L_v = \mathbb{R}^n \to N_v$ the map defined by composing the one-parameter subgroups $Y_i^v(t)$ corresponding to the fields Y_i^v on N_v. Now we assume the "radius" $\rho > 0$ is so small that the maps E_v and E_0 are diffeomorphisms of $B(\rho)$ onto their respective images in V and N_v and we transport the (left invariant) fields Y_i^v from N_v to V (or rather to the image $E_v(B(\rho)) \subset V$) by (the differential of) the map

$$E_v \circ E_0^{-1} : E_0(B(\rho)) \to E_v(B(\rho)).$$

(Notice that the map E_0^{-1} is defined for all ρ as E_0 is, in fact, a diffeomorphism of L_v onto N_v, but this is irrelevant for our local discussion.) We denote the transported fields by $\bar{Y}_i^v$ and we want to compare them with the fields Y_i in smaller and smaller "boxes" around v in V obtained by the following ε-scaling of a fixed (cubical) box $\bar{B}(\rho) = E_v(B(\rho)) \subset V$. We denote by $a_\varepsilon : \mathbb{R}^n \to \mathbb{R}^n$ the linear operator sending the Euclidean basis $\{e_i\}$ to $\{\varepsilon^{\deg i} e_i\}$ and we apply this notation to our systems of vector fields Y_i, Y_i^v, etc. For example, we write $a_\varepsilon(Y_i)$ for $^\varepsilon Y_i = \varepsilon^{\deg i} Y_i$. Then the ε-scaled box $\bar{B}_\varepsilon(\rho) \subset V$ is defined as the E_v-image of $a_\varepsilon B(\rho) \subset \mathbb{R}^n$. Equivalently, $\bar{B}_\varepsilon(\rho)$ can be defined as the image of $B(\rho)$ under the E_v-map corresponding to the fields $a_\varepsilon Y_i = {}^\varepsilon Y_i$.

Next we observe that the operators a_ε on $L_v = \mathbb{R}^n$ are automorphisms of the Lie algebra L_v and we denote by $A_\varepsilon : N_v \to N_v$ the corresponding automorphisms (self-similarities) of N_v. Then we transport these for $\varepsilon \leq 1$ to V via $E_v E_0^{-1}$ and denote the transported maps by

$$\bar{A}_\varepsilon : \bar{B}(\rho) \to \bar{B}_\varepsilon(\rho) \subset \bar{B}(\rho),$$

where the "transport" is defined by

$$\bar{A}_\varepsilon = (E_v E_0^{-1}) A_\varepsilon \left(E_v E_0^{-1} \right)^{-1}.$$

Let us summarize what we have obtained so far. We have chosen a small "curved cube" $\bar{B}(\rho) \subset V$ around v, that is the E_v-image of an actual ρ-cube in $\mathbb{R}^n$. We have on this cube two systems of fields, Y_i and $\bar{Y}_i^v$ which are related as follows.

(1) The two systems of fields coincide at v,

$$\bar{Y}_i^v(v) = Y_i(v), \quad i = 1, \ldots, n.$$

(2) The multiplication table for $\bar{Y}_i^v$ has constant coefficients,

$$\left[\bar{Y}_i^v, \bar{Y}_j^v \right] = \sum_k \bar{c}_{ijk} \bar{Y}_k^v \qquad (\bar{*})^v$$

where $\bar{c}_{ijk}$ are the structure constants of the Lie algebra L_v with the distinguished basis Y_i^v.

(3) The multiplication table $(\bar{*})^v$ is obtained from that for the fields $^\varepsilon Y_i = a_\varepsilon(Y_i)$ (see $^\varepsilon(*)$) by sending $\varepsilon \to 0$ and by evaluating the limit at v.

(4) We have diffeomorphisms $\bar{A}_\varepsilon$ of $\bar{B}(\rho)$ onto smaller box-like domains $\bar{B}\varepsilon(\rho) \subset \bar{B}(\rho)$, such that (the differentials of) $\bar{A}_\varepsilon$ on $\bar{Y}_i^v$ commute with a_ε, i.e. $\bar{A}_\varepsilon(\bar{Y}_i^v) = a_\varepsilon(\bar{Y}_i^v)$, or equivalently, $\bar{A}_\varepsilon^{-1}\left(a_\varepsilon(\bar{Y}_i^v)\right) = \bar{Y}_i^v$.

Now we want to understand what happens to the fields $\bar{Y}_i^\varepsilon = \bar{A}_\varepsilon^{-1}\left(a_\varepsilon(Y_i)\right)$ on $\bar{B}(\rho)$ (where the corresponding fields Y_i need be defined on $\bar{B}_\varepsilon(\rho)$) for $\varepsilon \to 0$.

1.4.A. Convergence Proposition. *If the fields Y_i are C^1-smooth (which we assume all along) then the fields $\bar{Y}_i^\varepsilon$ uniformly converge to $\bar{Y}_i^v$ on $\bar{B}(\rho)$.*

Proof. The multiplication table for the fields $\bar{Y}_i^\varepsilon$ on $\bar{B}(\rho)$ is obtained from that of $^\varepsilon Y_i$ on $\bar{B}_\varepsilon(\rho)$ (see $^\varepsilon(*)$ above) with the map $\bar{A}_\varepsilon : \bar{B}(\rho) \to \bar{B}_\varepsilon(\rho)$. Namely,

$$[\bar{Y}_i^\varepsilon, \bar{Y}_j^\varepsilon] = \sum_k \bar{c}_{ijk}^\varepsilon \, \bar{Y}_k^\varepsilon, \qquad (\bar{*})^\varepsilon$$

where

$$\bar{c}_{ijk}^\varepsilon(v') = \varepsilon^{d_{ijk}} c_{ijk}\big(\bar{A}_\varepsilon(v')\big), \ \ v' \in \bar{B}(\rho),$$

for $d_{ijk} = \deg i + \deg j - \deg k$.

It follows, by the continuity of c_{ijk} at our point v (in the "center" of $\bar{B}(\rho)$), that the oscillations of c_{ijk} on $\bar{B}_\varepsilon(\rho)$ go to zero for $\varepsilon \to 0$ and so the same is true for the oscillation of $\bar{c}_{ijk}^\varepsilon$ on $\bar{B}(\rho)$. Thus we have uniform convergence on $\bar{B}(\rho)$,

$$\bar{c}_{ijk}^\varepsilon \to \bar{c}_{ijk} \quad \text{for } \varepsilon \to 0.$$

Now we are going to prove the required convergence $\bar{Y}_i^\varepsilon \to \bar{Y}_i^v$ by observing that $\bar{Y}_i^\varepsilon$ and $\bar{Y}_i^v$ satisfy similar systems of ordinary differential equations (in the coordinates $t_1, \ldots, t_n$) with the coefficients $\bar{c}_{ijk}^\varepsilon$ and $\bar{c}_{ijk}$ playing identical roles. To do this we lift all our objects to the cube $B(\rho) \subset \mathbb{R}^n = L_v$ where we use the coordinates $t_1, \ldots, t_n$ and we introduce the following notations $\tilde{c}_{ijk}^\varepsilon$, the lifts of the functions $\bar{c}_{ijk}^\varepsilon$ to $B(\rho)$ via the map $E_v : B(\rho) \to \bar{B}(\rho)$.

$\tilde{Y}_i^\varepsilon$, the lifts of the fields $\bar{Y}_i^\varepsilon$ to $B(\rho)$ by $(E_v)^{-1}$. (This makes sense since E_v^ε is a diffeomorphism of $B(\rho)$ onto $\bar{B}(\rho)$ by our assumption.)

$\tilde{Y}_i^v$, the lifts of the fields Y_i^v to $B(\rho) \subset \mathbb{R}^n = L_v$ via the map $E_0 : L_v \to N_v$. Notice that the map $E_v : B(\rho) \to \bar{B}(\rho)$ sends $\tilde{Y}_i^v$ to $\bar{Y}_i^v$.

First observation. The fields $\tilde{Y}_i^\varepsilon$ are related to the fields $\partial_i = \frac{\partial}{\partial t_i}$ on $B(\rho) \subset \mathbb{R}^n$ by the following identities,

$$
\begin{aligned}
\tilde{Y}_1^\varepsilon &= \partial_1 \text{ on } B(\rho) \\[2mm]
\tilde{Y}_2^\varepsilon &= \partial_2 \text{ on the subspace } \{t_1 = 0\} \subset B(\rho), \\[2mm]
\tilde{Y}_3^\varepsilon &= \partial_3 \text{ on} \{t_1 = 0, t_2 = 0\} \subset B(\rho) \\[2mm]
&- - - - - - - - - - - \\[2mm]
\tilde{Y}_n^\varepsilon &= \partial_n \text{ on the } t_n\text{-line } \{t_i = 0, \ i = 1, \ldots, n-1\} \subset B(\rho).
\end{aligned}
\qquad (A)
$$

Furthermore, the fields Y_i^v satisfy the same system of relation on $B(\rho)$.

To see that, we first observe that the lifts $\tilde{Y}_i$ to $B(\rho)$ of the original fields $Y_i = \bar{Y}_i^1$ satisfy (A) as immediately follows from the definition of the map $E_v : \mathbb{R}^n \to V$ via the composition of the one-parameter subgroups $Y_i(t)$. Similarly, the fields $\tilde{Y}_i^v$ satisfy (A) since these are the lifts of Y_i^v from N_v to $L_v = \mathbb{R}^n$ via the map $E_0 : L_v \to N_v$ obtained by composing $Y_i^v(t)$. What remains to show is that the passage from $Y_i = \bar{Y}_i^1$ to $\bar{Y}_i^\varepsilon = \bar{A}_{\varepsilon^{-1}}\big(a_\varepsilon(Y_i)\big)$ does not change the fields $\tilde{Y}_i^\varepsilon$ on those parts of $B(\rho)$ where the relations (A) apply. Namely

$$\tilde{Y}_1^\varepsilon \;=\; \tilde{Y}_1 \;=\; \partial_1 \;(= \tilde{Y}_1^v) \text{ on all of } B(\rho)$$

$$\tilde{Y}_2^\varepsilon \;=\; \tilde{Y}_2 \;=\; \partial_2 \;(= \tilde{Y}_2^v) \text{ on the subspace } \{t_1 = 0\} \subset B(\rho)$$

and so on. To see this we shall bring $\bar{A}_\varepsilon$ from V to $\mathbb{R}^n$ by taking

$$\tilde{A}_\varepsilon = E_v^{-1}\bar{A}_\varepsilon E_v : B(\rho) \to B_\varepsilon(\rho) = a_\varepsilon\big(B(\rho)\big).$$

Then $\tilde{Y}_i^\varepsilon = \tilde{A}_{\varepsilon^{-1}}\big(a_\varepsilon(\tilde{Y}_i)\big)$ and (A) for $\tilde{Y}_i^\varepsilon$ follows from the relations (A) for $\tilde{Y}_i^v$, which are

$$\tilde{Y}_1^v = \partial_1 \text{ on } B(\rho),$$
$$\tilde{Y}_2^v = \partial_2 \text{ on } \{t_1 = 0\} \subset B(\rho),$$

etc., and the commutation between $\tilde{A}_\varepsilon$ and a_ε on the fields $\tilde{Y}_i^v$, i.e. $\tilde{A}_{\varepsilon^{-1}}\big(a_\varepsilon(\tilde{Y}_i^v)\big) = \tilde{Y}_i^v$, which follows from the corresponding relation for $\bar{A}_\varepsilon$ (and eventually for A_ε). Q.E.D. (This extra argument for $\tilde{Y}_i^\varepsilon$ was needed as the fields $\tilde{Y}_i^\varepsilon$ were defined with E_v rather than with the map corresponding to $\bar{Y}_i^\varepsilon(t)$.)

Second observation. The fields $\tilde{Y}_i^\varepsilon$ satisfy the following linear differential equations,

$$\partial_1 \tilde{Y}_i^\varepsilon \;=\; \sum_k \tilde{c}_{1,i,k}^\varepsilon \tilde{Y}_k, \; i = 2, 3, \ldots, n, \text{ on } B(\rho)$$

$$\partial_2 \tilde{Y}_i^\varepsilon \;=\; \sum_k \tilde{c}_{2,i,k}^\varepsilon \tilde{Y}_k, \; i = 3, \ldots, n \text{ on } \{t_1 = 0\} \subset B(\rho)$$

$$- \; - \; - \; - \; - \; - \; - \; - \; - \; - \; - \; - \; - \; -$$

$$\partial_{n-1} \tilde{Y}_n^\varepsilon \;=\; \sum_k \tilde{c}_{n-1,n,k}^\varepsilon \tilde{Y}_k \text{ on the } (t_{n-1}, t_n)\text{-plane in } B(\rho). \tag{B}$$

Furthermore, the fields $\tilde{Y}_i^v$ satisfy an identical system with (the constants) $\bar{c}_{ijk}$ instead of $\tilde{c}_{ijk}^2$.

The equations (B) follow from the commutation relations for $\tilde{Y}_i^\varepsilon$, that are

$$[\tilde{Y}_i^\varepsilon, \tilde{Y}_j^\varepsilon] = \sum_k \tilde{c}_{\varepsilon_{ijk}} \tilde{Y}_k^\varepsilon, \qquad (\tilde{*})^\varepsilon$$

the identities (B) and the obvious formula $[\partial_i, Y] = \partial_i Y \underset{\text{def}}{=} \frac{\partial Y}{\partial t_i}$ for all fields Y on $\mathbb{R}^n$. Similarly, we see the validity of these relations for $\tilde{Y}_i^v$ with $\bar{c}_{ijk}$ in place of $\tilde{c}_{ijk}^\varepsilon$.

To understand the meaning of (A) and (B) let us read these equations from bottom to top. The last identity in (A), i.e. $\tilde{Y}_n^\varepsilon = \partial_n$, should be thought of as an initial value datum for the (last among (B)) equation $\partial_{n-1}\tilde{Y}_n^\varepsilon = \sum_{k=1}^n \tilde{c}_{n,n-1,k}^\varepsilon \tilde{Y}_k$ on the (t_{n-1}, t_n)-plane (given by the equations $t_i = 0$, $i = 1, \ldots, n-2$). Notice that this equation also involves the fields $\tilde{Y}_k$ for $k \leq n-1$ on the (t_{n-1}, t_n)-plane but these are equal to ∂_k on this plane according to (A). Next, we take $\tilde{Y}_n^\varepsilon$ on the (t_{n-1}, t_n)-plane obtained by solving our initial value problem and we also take $\tilde{Y}_{n-1}^\varepsilon = \partial_{n-1}$ on this plane as given by the second from the bottom relation (A). Then the pair $(\tilde{Y}_n^\varepsilon, \tilde{Y}_{n-1}^\varepsilon)$ serves for initial values for the second from bottom equations in (B) on the (t_{n-2}, t_{n-1}, t_n)-space which are

$$\partial_{n-2}\tilde{Y}_{n-1}^\varepsilon = \sum_k \tilde{c}_{n-2,n-1,k}^\varepsilon \, \tilde{Y}_k^\varepsilon, \ \ \partial_{n-2}\tilde{Y}_n^\varepsilon = \sum_k \tilde{c}_{n-2,n,k}^\varepsilon \, \tilde{Y}_k^\varepsilon,$$

where the field $\tilde{Y}_k^\varepsilon$ on the right hand side for $k \leq n-2$ equals ∂_k according to (A).

Conclusion. The fields $\tilde{Y}_i^\varepsilon$ on $B(\rho)$ are uniquely determined by their values at v and by the functions $\tilde{c}_{ijk}^\varepsilon$ via the equation (A) and (B). It follows, by an elementary theorem on dependence of solutions of linear O.D.E. upon initial conditions and coefficients, that the fields $\tilde{Y}_i^\varepsilon$ are continuous in $\tilde{c}_{ijk}^\varepsilon$. In particular, $\tilde{Y}_i^\varepsilon \to \tilde{Y}_{ijk}^v$ for $\varepsilon \to 0$ as $\tilde{c}_{ijk}^\varepsilon \to \bar{c}_{ijk}$. Consequently, the fields $\bar{Y}_i^\varepsilon$, which are images of $\tilde{Y}_i^\varepsilon$ under E_v, converge to $\bar{Y}_i^v$. Q.E.D.

1.4.A′. Uniformity of the convergence. Recall that the objects appearing in 1.4.A are constructed with the use of a distinguished point $v \in V$. These are the "curved cube" $\bar{B}(\rho) = \bar{B}(v, \rho)$, the fields $\bar{Y}_i^v$ coming from N_v and the fields $\bar{Y}_i^\varepsilon$ obtained by some rescaling and "homotopies" $\bar{A}_{\varepsilon^{-1}} = \bar{A}_{\varepsilon^{-1}}^v$ of Y_i. Now, we claim that the norm $\|Y_i^v - \bar{Y}_i^\varepsilon\|$ can be bounded independently of v. More precisely, we have the following

Uniform version of 1.4.A. *If the fields Y_i are C^1-smooth, then, for each compact subset $V_0 \subset V$, there exists a positive number ρ and a function $\delta(\varepsilon) \to 0$ for $\varepsilon \to 0$, such that for each $v \in V_0$ the fields $\bar{Y}_i^v$ and $\bar{Y}_i^\varepsilon$ are well defined on $\bar{B}(\rho) = \bar{B}(v, \rho)$ and $\|\bar{Y}_i^v - \bar{Y}_i^\varepsilon\| \leq \delta(\varepsilon)$, where the norm refers to the Riemannian metric on V which makes the frame Y_i orthonormal.*

This is immediate by observing that the proof of 1.4.A is "uniform in v".

1.4.A″. On C^r-convergence. *If the fields Y_i are C^{r+1}-smooth then $\bar{Y}_i^\varepsilon$ converge to $\bar{Y}_i^v$ in the C^r-topology and this convergence is uniform in v in the above sense.*

Proof. If Y_i are C^{r+1} then the coefficients c_{ijk} are C^r and so for $r \geq 1$ the Lie derivatives of the functions $\bar{c}_{ijk}^\varepsilon$ with respect to the fields $\bar{Y}_\mu^\varepsilon$ go to zero as fast as ε, i.e.

$$|\bar{Y}_\mu^\varepsilon \, \bar{c}_{ijk}^\varepsilon| \leq \mathrm{const}\,\varepsilon.$$

This estimate lifts to $B(\rho)$ where it reads

$$|\tilde{Y}_\mu^\varepsilon \tilde{c}_{ijk}^\varepsilon| \leq \mathrm{const}\,\varepsilon,$$

and since the fields $\tilde{Y}_\mu^\varepsilon$ are close to a fixed frame (namely $\tilde{Y}_\mu^v$) this implies

$$|\partial_\mu c_{ijk}| \leq \widetilde{\mathrm{const}}\,\varepsilon.$$

Then by going through (A) and (B) one obtains a C^1-bound on $\tilde{Y}_i^\varepsilon$, namely $|\partial_\mu \tilde{Y}_i^\varepsilon| \leq \mathrm{const}'\,\varepsilon$. With this, if $r \geq 2$, one can pass from the bound on the Lie derivatives $\bar{Y}_\mu^\varepsilon \bar{Y}_\nu^\varepsilon \, \bar{c}_{ijk}^\varepsilon$ to the bound on $\partial_\mu \partial_\nu \tilde{c}_{ijk}^\varepsilon$ which yields, in turn, a bound on $\partial_\mu \partial_\nu \tilde{Y}_{ijk}^\varepsilon$ and so on. ∎

1.4.B. Approximation of equiregular Carnot-Carathéodory spaces by self-similar nilpotent groups. Let $H \subset T(V)$ be an equiregular polarization on V which means that the subsets of tangent vectors $H = H_1 \subset H_2 \subset \cdots \subset H_i \subset \cdots$ spanned by the commutators of H-horizontal fields of degrees $\leq i$ are actual subbundles in $T(V)$, i.e. their fibers $(H_i)_v \subset T_v(V)$ have dimensions constant in $v \in V$ (compare 0.3.D). We assume, moreover, that the commutators of order $\leq d$ span all of $T(V)$, i.e. $H_d = T(V)$, and then we take a frame of vector fields $Y_1, \ldots, Y_n$ adapted to H_i, i.e. $Y_1, \ldots, Y_{n_1}$, for $n_1 = \operatorname{rank} H_1$, belong to H_1, then $Y_{n_1+1}, \ldots, Y_{n_2}$, $n_2 = \operatorname{rank} H_2$, belong to H_2 etc. This frame comes along with a deg-function, where the first n_1 vectors Y_i have deg $= 1$, the following $n_2 - n_1$ have deg $= 2$ etc. We invoke the nilpotent tangent cone N_v associated to Y_i and we recall our diffeomorphism $E_v E_0^{-1}$ which sends a small neighbourhood of the identity element in N_v onto some neighbourhood of a given point v in V. We fix some Riemannian metric in V (in order to have the C-C metric) and we endow N_v with a corresponding left invariant metric for which the differential of $E_v E_0^{-1}$ at id $\in N_v$ is isometric. Now both, V and N_v, have C-C metrics, say dist in V and dist* in N_v and the diffeomorphism $E_v E_0^{-1}$ brings dist* to V (or rather to our small neighbourhood in V around v where the action takes place). This transported metric is denoted dist$^*_v$ on V. Denote by $B^*(v, \varepsilon) \subset V$ the ε-ball in V around v with respect to dist* and observe that 1.4.A implies the following

Local approximation theorem. ([5]) *If the subbundle $H \subset T(V)$ is sufficiently smooth then the difference between the metrics* dist *and* dist$^*_v$ *on $B^*(v, \varepsilon)$ is $o(\varepsilon)$, i.e. $\varepsilon^{-1}\big(\operatorname{dist}(v_1, v_2) - \operatorname{dist}^*_v(v_1, v_2)\big) \xrightarrow[\varepsilon \to 0]{} 0$, for all pairs of points $v_1, v_2 \in B^*(v, \varepsilon)$.*

Proof. To see the picture clearly for $\varepsilon \to 0$, we rescale our metrics and neighbourhoods by ε^{-1} using the (expanding) diffeomorphism $\bar{A}_{\varepsilon^{-1}}$ acting in V near v (see 1.4.A). To simplify the matter, we assume for the moment that our Riemannian metric in V comes from the left invariant metric in N_v via $E_v E_0^{-1}$. (This assumption, in fact, does not restrict the generality as our old metric agrees with the new one at v.) Then we observe that $\bar{A}_\varepsilon$ scales dist$^*_v$ by ε, i.e. $\bar{A}^*_\varepsilon \operatorname{dist}^*_v = \varepsilon \operatorname{dist}^*_v$, wherever $\bar{A}^*_\varepsilon$ and dist$^*_v$ are defined. In particular, $\bar{A}^*_{\varepsilon^{-1}}$ transform $B^*(v, \varepsilon)$ to $B^*(v, 1)$, provided E_v^{-1},

---

and hence $\bar{A}_{\varepsilon}^{*}$ are defined on the ball $B^{*}(v,1)$. In fact, this can be always achieved by multiplying the underlying Riemannian metric by a fixed large constant and we assume from now on that the unit ball $B^{*}(v,1)$ is small enough for our game.

Now we compare $\varepsilon^{-1}\operatorname{dist}$ and $\varepsilon^{-1}\operatorname{dist}^{*}$ in $B^{*}(v,\varepsilon)$ by bringing them to $B^{*}(v,1)$ by $A_{\varepsilon^{-1}}$. We observe that $A_{\varepsilon^{-1}}(\varepsilon^{-1}\operatorname{dist}^{*}) = \operatorname{dist}^{*}$ and so we must prove the uniform convergence on the unit ball,

$$\left(A_{\varepsilon^{-1}}(\varepsilon^{-1}\operatorname{dist}) - \operatorname{dist}^{*}\right) \to 0 \ \text{ for } \ \varepsilon \to 0.$$

To prove this we use 1.4.A which shows that the polarization $A_{\varepsilon^{-1}}(H)$ on $B^{*}(v,1)$ converges to the polarization H^{*} corresponding to dist^{*} and the Riemannian metric (i.e. quadratic form) on this polarization transported from the original metric on H converges to the metric in H^{*}, since the vectors $Y_1, \ldots, Y_{n_1}$ spanning H satisfy, according to 1.4.A,

$$\bar{A}_{\varepsilon^{-1}}(\varepsilon Y_i) \underset{\varepsilon \to 0}{\longrightarrow} \bar{Y}_i^{v}, \ i = 1, \ldots, n_1,$$

where $\bar{Y}_i^{v}$ are certain vector fields spanning H^{*}.

To conclude the proof we would need the following continuity of the Carnot-Carathéodory metrics defined by (H,g) where $H \subset T(V)$ is a polarization and g is a Riemannian metric on H,

if $(H_\varepsilon, g_\varepsilon)$ converges to (H^{*}, g^{*}), then the C-C metrics also converge, i.e. $\operatorname{dist}_\varepsilon \to \operatorname{dist}^{*}$.

This is indeed so if the convergence $H_\varepsilon \to H^{*}$ is understood in the C^{d-1}-topology (where d is the bound on the degrees of the commutators of the fields in H^{*} spanning $T(V)$) due to a uniform bound on the metrics $\operatorname{dist}_\varepsilon$ (see below) but, in general, e.g. for C^{0}-convergence $H_\varepsilon \to H^{*}$, the functions $\operatorname{dist}_\varepsilon$ for arbitrarily small $\varepsilon > 0$ may be, a priori, infinite on certain pairs of points in V and so one cannot speak of the ordinary (uniform) convergence $\operatorname{dist}_\varepsilon \to \operatorname{dist}^{*}$. However we do have the following weak convergence defined with the Hausdorff distance between subsets in V with respect to dist^{*}.

Weak convergence lemma. *If $(H_\varepsilon, g_\varepsilon)$ uniformly converge to (H^*, g^*) then every $\mathrm{dist}_\varepsilon$-ball $B^\varepsilon(v, \rho)$ Hausdorff-converges to the corresponding dist^*-ball $B^*(v, \rho)$ and this convergence is uniform on compact subsets in $V \times \mathbb{R}_+$, i.e.*

$$\mathrm{dist}^*_{\mathrm{Hau}}\big(B^\varepsilon(v, \rho), B^*(v, \rho)\big) \leq \delta(\varepsilon)$$

for $\delta(\varepsilon) \to 0$ for $\varepsilon \to 0$, where one may use a fixed function $\delta(\varepsilon)$ for each compact subset of points (v, ρ).

Proof. Let H_1 and H_2 be mutually close polarizations with close quadratic forms. Then, at least locally, there exist mutually close frames of orthonormal vector fields spanning H_1 and H_2 and with these fields one establishes a correspondence between H_1- and H_2-horizontal curves as follows. The curves $c_1(t)$ and $c_2(t)$ parametrized by arc length (coming from the quadratic forms in H_1 and H_2 respectively) *correspond* to each other if at each moment t their derivatives,

$$c_1'(t) \in (H_1)_{v_1 = c_1(t)} \ \text{ and } \ c_2'(t) \in (H_2)_{v_2 = c_2(t)}$$

have identical decompositions with respect to the frames in H_1 and H_2. Clearly, corresponding curves issuing from nearby points remain close for a certain time which implies the required closeness of the C-C balls. (We suggest the reader would check that this "close" talk can be made rigorous and uniform.)

Corollary. *If the metrics $\mathrm{dist}_\varepsilon$ are uniformly bounded, i.e. if each $\mathrm{dist}_\varepsilon$-ball of radius ρ around v contains the dist^*-ball around v of radius ρ^* for some strictly positive function $\rho^*(\rho)$, $\rho > 0$, then $|\mathrm{dist}_\varepsilon - \mathrm{dist}^*| \to 0$ uniformly on compact subsets in V.*

This is obvious by the triangle inequality.

Conclusion of the proof of the approximation theorem for C^{2d-2}-smooth polarizations $H \subset T(V)$. If H is C^r-smooth, then the corresponding frame $Y_1, \ldots, Y_n$ obtained by taking commutators of degree $\leq d$ are C^{r-d+1}-smooth and then H_ε converges to H^* in C^{r+d}-topology according to 1.4.A$''$. In particular, the convergence $H_\varepsilon \to H^*$ is C^{r+d}. Now we use the fact that $T(V)$ is generated by commutators of H^*-horizontal fields of degree $\leq d$ which, according to the Chow connectivity theorem, makes $\mathrm{dist}^* < \infty$. Since the Chow theorem appeals to the derivatives of H^* of order $\leq d - 1$, it is stable under small C^{d-1}-perturbations of H^* which implies a uniform bound on $\mathrm{dist}_\varepsilon$ whenever H_ε is sufficiently C^{d-1}-close to H^*. $\blacksquare$

The case of H being C^d-smooth. We start by giving a bound on a metric in pure "Hausdorff terms". To grasp the idea, imagine we have two metrics on V, say dist and dist^*, such that *every* dist-ball $B(v,\rho)$ is sufficiently dist^*-Hausdorff close to the corresponding dist^*-ball $B^*(v,\rho)$. Say,

$$\text{dist}^*_{\text{Hau}}\big(B(v,\rho), B^*(v,\rho)\big) \le \rho/10.$$

Then the triangle inequality implies that the metrics are close as functions on $V \times V$. In particular (and most importantly)

$$B(v,\rho) \subset B^*(v, \alpha\rho), \text{ for a fixed } \alpha > 0,$$

(where one can take $\alpha = \frac{3}{5}$).

Now we turn to the proof of the approximation theorem and observe that the weak convergence lemma makes every small dist-ball around v quite close to the corresponding dist^*_v-ball. But here, unlike the above discussion, the metric dist^*_v depends on v. To remedy that we exclude v by setting $\text{dis}^*(v,v') = \text{dis}^*_v(v,v')$. Of course, dis^* is not, in general, a metric, but the above argument also works for *quasi-metrics*, that are positive functions on $V \times V$, vanishing exactly on the diagonal and satisfying the following *approximate triangle inequality*,

$$\text{dis}(v,v'') \le \text{const}\big(\text{dis}(v,v') + \text{dis}(v',v'')\big), \tag{+}$$

which must be uniform on compact subsets in $V \times V \times V$ (i.e. for each $K \subset V \times V \times V$ there exists const, so that $(+)$ holds for all $v,v',v'' \in K$).

In order to prove $(+)$ for dis^* we recall that the function dist^* comes from the nilpotent group N_v with self-similarities $A_\varepsilon : N_v \to N_v$ defined with $a_\varepsilon : L_v \to L_v$ on the Lie algebra $L_v = L(N_v)$. Since A_ε and a_ε commute with the (composed orbit) map $E_0 : L_v \to N_v$, (i.e. $E_0 a_\varepsilon = A_\varepsilon E_0$) the ε-balls in N_v around $\text{id} \in N_v$ are equivalent to the E_0-images of the ε-boxes B_ε in $L_v = \mathbb{R}^n$ defined by $|t_i| \le \varepsilon^{\deg i}$. (We have already seen this picture for the exponential map in 0.3.C). Therefore, the inequality $(+)$ for dis^* reduces to the corresponding property of the E_v-images of the boxes $B_\varepsilon \subset \mathbb{R}^n = T_v(V)$. Namely, we need to show, that if two such images, say $E_v(B_\varepsilon)$ and $E_{v'}(B_\delta)$ in V intersect, then $E_{v'}(B_\delta)$ is contained in $E_v(B_{\varepsilon'})$ for $\varepsilon' \le \text{const}(\varepsilon + \delta)$. (Warning: the commutation relations $\deg[Y_i, Y_j] \le \deg i + \deg j$ is crucial for this property.) To see this we recall that $E_v(B_\varepsilon)$ consists of the second ends of piecewise smooth curves issuing from v which are built of n segments $c_1, \ldots, c_n$, where c_1

is a piece of orbit of Y_1 of length $\leq \varepsilon^{\deg 1}(= \varepsilon)$, c_2 such a piece for Y_2 of length $\leq \varepsilon^{\deg 2}$ and so on. Thus the problem reduces to showing that if we add to such curve a new piece c_{n+1} of the orbit of Y_j of length $(\delta)^{\deg j}$, then there exists a curve of n pieces $c'_1, \ldots, c'_i, \ldots, c'_n$ with lengths $\leq (\varepsilon')^{\deg i}$ for $\varepsilon' \leq \mathrm{const}(\varepsilon + \delta)$, such that the second end of the new curve equals the free end v'' of c_{n+1}, see Fig. 3 below.

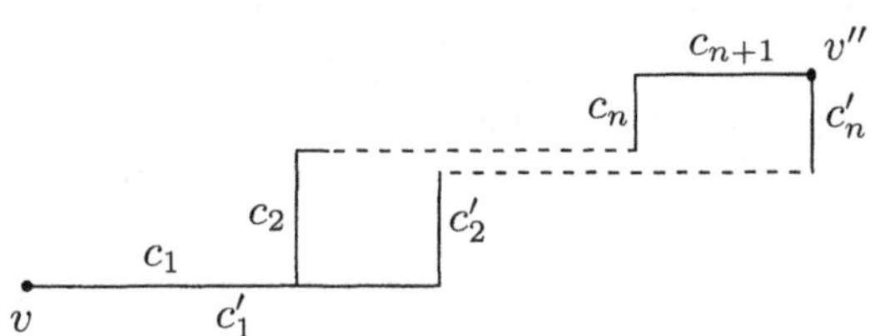

Figure 3

In other words, we must compensate for changing the order of the orbits and this can be achieved with the relation $(*)$ in 1.2. (compare [N-S-W]). This is straightforward and we leave it to the reader.

Finally, the weak convergence lemma shows that the "Hausdorff distance" between dist and dis* on $B^*(v, \varepsilon) \subset V$ is $O(\varepsilon)$ uniformly in v (because of 1.4.A$'$). This yields a bound on the metric dist by the above argument (but now, of course, with a constant different from $3/5$) which concludes the proof of the approximation theorem for C^d-smooth H.

Remarks and corollaries

(a) It seems, the conclusion of the theorem should stand for H being C^{d-1}.

(b) If H is C^{d+1}, it is easy to show that $o(\varepsilon)$ in the approximation theorem can be replaced by $O(\varepsilon^2)$.

(c) The ball-box theorem is an immediate corollary of the approximation theorem as the balls in N_v *are* (obviously) box-like as we have mentioned several times.

(d) The Mitchell theorem concerning the tangent cones of C-C manifolds (see (iii) in 0.3.D) immediately follows from the approximation theorem. In fact, one only needs here the (weak) dis*-Hausdorff convergence of dist_ε rather than the final result on the uniform convergence (see [Mit$_{1;2}$]).

(e) The ball-box theorem implies (at least in the equiregular case) that small balls in V are "essentially contractible", i.e. *each small ε-ball is contractible within the concentric ball of radius $C\varepsilon$ for a fixed $C \geq 1$.* In fact, one can squeeze a *topological* ball (box) between the ε and the $C\varepsilon$-balls. Furthermore, the balls in nilpotent groups with self-similarities are honestly (and obviously) contractible (i.e. the above C equals one). It follows by the approximation theorem that for all V one has $C \to 1$ for $\varepsilon \to 0$ and it is likely (for sufficiently smooth C-C data) that $C = 1$ for small ε, i.e. small balls are probably contractible.

1.4.C. Pinching and related problems for C-C metrics. The approximation of a frame of fields Y_i with almost constant coefficients in the multiplication table $[Y_i, Y_j] = \sum_k c_{ijk} Y_k$ by a frame where the corresponding coefficients are truly constant (see 1.4.A) represents a simplest instance of the *stability phenomenon* for (homogeneous) geometric structures. In general, one looks for a weakest possible local or infinitesimal criterion for a given geometric structure to be homogeneous or almost homogeneous in a suitable sense. More specifically, when dealing with a Riemannian metric, one makes up such a criterion in terms of curvature (and covariant derivatives of the curvature) sometimes by *pinching* the curvature between two constants. (This explains the "pinching" terminology.) In our case the structure was given by a frame of vector fields which form *an almost Lie group* in the terminology of [Ru] and there are several results due to Min-Oo and Ruh allowing an approximation of an almost Lie group by an actual Lie group which goes infinitely deeper than our proposition 1.4.A. In fact, our argument with choosing O.D. equations (A) and (B) parallels the initial phase of the proof of Rauch's comparison and pinching theorems for Riemannian manifolds. (See [Gro$_{SAP}$] for an exposition of these techniques and ideas.)

Now we notice that the study of a general geometric structure, say σ on V, can be often reduced to that of an auxiliary Riemannian metric $\bar{g}$. For example, every frame σ of vectors $Y_1, \ldots, Y_n$ on V defines a unique metric $\bar{g} = \bar{g}(\sigma)$ on V for which this frame becomes orthonormal and so the almost Lie group problem can be, in principle, viewed as a special case of the Riemannian pinching problem.

The essential feature of σ which allows one to make the step $\sigma \mapsto \bar{g}$ is the *compactness* of the (isotropy) groups $\mathrm{Aut}(V, v, \sigma)$ for all $v \in V$. So we may expect that the C-C metrics given by pairs (H, g), where g is a positive quadratic form on the polarization H, should give rise to Riemannian metrics $\bar{g}$ naturally (or at least canonically) associated to (H, g). Here is an example where everything is perfectly nice.

Contact C-C manifolds. Let $H \subset T(V)$ be contact, i.e. H is (locally) given by a 1-form η on V such that the differential $d\eta$ is non-singular on H. Notice that in this case $n - 1 = \mathrm{rank}\, H$ is even, say $2m$. Now, using g on H we can specify η by requiring the $2m$-form $(d\eta)^m$ on H to be equal up to $\pm$ sign to the volume element of g on H which defines η up to $\pm$ sign and consequently $d\eta$ on V. This $d\eta$ defines a 1-dimensional subbundle ℓ transversal to H, namely $\ell = \mathrm{Ker}\, d\eta$, and we have a metric g' on ℓ defined by the condition $\|\eta\|_{g'} = 1$. Finally, the pair (g, g') defines the required metric $\bar{g} = g \oplus g'$ on V.

Let us indicate a general (and rather ugly) construction which provides $\bar{g}$ for a C-C structure (H, g) on V which is everywhere infinitesimally close to a fixed (model) homogeneous structure (H_0, g_0) on V_0. We fix some Riemannian metric $\bar{g}_0$ on V_0 invariant under automorphisms of (H_0, g_0) (which is possible since the isotropy group $\mathrm{Aut}(V_0, v_0, H_0, g_0)$ is compact as we assume our C-C is a *metric* on V_0) and then for a large i and each point $v \in V$ we consider the i-jets of maps $(V, v) \to (V_0, v_0)$ which are "ε-isometric" for a fixed small $\varepsilon > 0$, i.e. for which the image of the jet of (H, g) at v is sent ε-close to the jet of (H_0, g_0) at v_0. Using these jets we induce $\bar{g}_0$ at each point $v \in V$ and thus obtain a family $\bar{g}_\varepsilon$ of metrics on V. Finally, one extracts a single metric $\bar{g}$ out of $\bar{g}_\varepsilon$ by some averaging or envelope construction, e.g. by taking $\bar{g}$ which has the maximal volume element among all metrics smaller than all g_ε. (The reason why we first bring in some ε and then smooth it out is due to the fact that the infinitesimal symmetries of (H_0, g_0) are usually destroyed by small perturbations. Yet we want our $\bar{g}$ on V to remember these symmetries.)

The above indicates an approach to an infinitesimal stability (pinching) problem for C-C structures but we are attracted by more interesting (and more difficult) purely metric stability and/or homogeneity problems. Here are examples of these.

Metric criteria for homogeneity. Let (V, dist) be a Carnot-Carathéodory manifold. Suppose that the small ε-balls in V are mutually *δ-isometric* in an appropriate sense where $\delta = \delta(\varepsilon)$ fast goes to zero for $\varepsilon \to 0$. For example, let the Hausdorff distance between these balls, thought of as abstract metric spaces, satisfy

$$\text{dist}_{\text{Hau}}\big(B(v_1, \varepsilon), B(v_2, \varepsilon)\big) \leq \delta = \text{const}\, \varepsilon^p \tag{$*$}$$

for a sufficiently large p. Does this imply that (V, dist) is locally homogeneous?

Example. If (V, dist) is a C^p-smooth Riemannian manifold then $(*)$ implies that (V, dist) is infinitesimally homogeneous of order $p - 1$ (see below) and then the positive answer is provided by a theorem of Singer (see [Sin], p. 165 in [Gro$_{\text{PDR}}$] and [D-G]). The key step here is the implication $(*) \Rightarrow$ infinitesimal homogeneity which says, in effect, that the curvature of V and its covariant derivatives can be read of the distance (properties) on finite subsets in small balls in V. This is done by observing that $(*)$ implies the existence of a (possibly discontinuous) δ-isometric map $\varphi : B(v_1, \varepsilon) \to B(v_2, \varepsilon)$ which can be smoothed e.g. using the Riemannian center of mass construction (see [Kar]). Thus we obtain a smooth δ'-isometric map φ' where $\delta' \approx \delta$ and where we hold control over the derivatives of the map so that the metrics dist on $B(v_2, \varepsilon)$ and $\varphi'(\text{dist})$ brought from $B(v_1, \varepsilon)$ become close with many (about p) derivatives. (We leave the details to the reader. Notice that even in the Riemannian framework one does not know how to get rid of the smoothness assumption on dist.)

A somewhat easier version of the problem appears if we are already given a homogeneous (model) C-C space (V_0, dist_0) and replace $(*)$ by the corresponding distance inequality between the balls $B(v_0, \varepsilon) \subset V_0$ and $B(v, \varepsilon) \subset V$. (Here, for example, the low smoothness of V does not cause serious problems.)

Almost homogeneity and pinching problems. Now we require the inequality $(*)$ for a *fixed* small ε, or for ε in a fixed interval $[\varepsilon_1, \varepsilon_2]$ where $\varepsilon_1 > 0$. The expected conclusion is the existence of a locally homogeneous C-C metric on V which is close to the original C-C metric. Of course, one needs here extra topological and local geometric assumptions on V (see below).

One arrives at a more traditional pinching problem if one replaces $(*)$ (now for a fixed ε) by the corresponding relation between the balls in V and in a homogeneous model space V_0. The desired conclusion is the existence of an almost isometric covering map $V_0 \to V$.

Necessary restriction on V. Usually, one insists while "pinching" V that it should be compact or at least metrically complete in order to avoid irrelevant complications. (One should separately consider the case where V_0 is incomplete and, moreover, admits no complete homogeneous manifold locally isometric to V_0). Another group of extra conditions should take care of the possibility that the "injectivity radius" of V becomes small of order ε. One can rule this out by insisting on "essential contractibility" of the ε-balls (compare (e) preceding 1.4.C). Yet one may wish to allow Inj.Rad. $\to 0$. Then one should either replace the comparison between balls by that between appropriate coverings of the balls (e.g. universal coverings of $C\varepsilon$-balls restricted to ε-balls for some $C > 1$), or to work out a more subtle comparison between the balls themselves that would allow $\varepsilon \to 0$. (Notice that the traditional pinching condition on the curvature, i.e. $K \in [\kappa_1, \kappa_2]$, can be detected by looking metrically at arbitrarily small balls but conditions imposed on the derivatives of the curvature become invisible on very small balls around $v \in V$ unless K vanishes at v.)

There are two basic approaches to the above problems. The first uses special curves (e.g. geodesics) in V satisfying certain O.D.E.'s similar to the system (B) we met in 1.4.A. Some results in this direction, allowing one to recapture the smooth structure of V out of the metric via the geodesics, appear in [Ham]. (Notice that O.D.E.'s may govern not only curves but also some functions on V, such as $v \mapsto \operatorname{dist}(v_0, v)$, and these functions may be used for embeddings $V \to \mathbb{R}^q$ (with large q) which sometimes serve almost as good as canonical coordinates on V.) The second approach uses some P.D.E.'s. In the Riemannian case the best results were obtained with non-linear P.D.E.'s (see [Ru]) but for C-C manifolds these are not yet available. Yet, we do have the (linear) Hörmander-Laplace operator Δ on V and the corresponding diffusion which can be reconstructed in many (all?) cases out of dist as follows. Define the *energy-density* $ef(v)$ of a Lipschitz function $f : V \to \mathbb{R}$ at $v \in V$ by

$$\lim_{\varepsilon \to 0} \varepsilon^{-N} \int_{B(v,\varepsilon)} |f(v) - f(v')|^2 dv',$$

where N denotes the Hausdorff dimension of (V, dist) and dv' refers to the Hausdorff measure. Then the energy of f on V is

$$E(f) = \int_V ef(v)dv.$$

Equivalently we can define $E(f)$ as the limit of the integrals $\int_{V \times V} |f(v) - f(v')|^2 \psi_\varepsilon dv\, dv'$ for $\varepsilon \to 0$ with appropriate weights $\psi_\varepsilon(v, v')$ which localize at the diagonal for $\varepsilon \to 0$. Then the extremal functions f for $E = E(f)$ satisfy the Hörmander-Laplace equation $\Delta f = 0$ and therefore are smooth (see [Hör]). These can be used to build up the smooth structure on V and one may try with this structure to reduce the metric problems (of pinching and homogeneity) to the corresponding infinitesimal problems. (Notice that the diffusion can be also constructed purely geometrically as the limit of convolutions of some kernels $\Phi_\varepsilon = \Phi_\varepsilon(v, v')$, namely

$$\lim_{i \to \infty} \underbrace{\Phi_{\varepsilon_i} * \Phi_{\varepsilon_i} * \cdots * \Phi_{\varepsilon_i}}_{i}$$

where $\Phi_\varepsilon(v, v')$ are suitably chosen functions of the form $\Phi_\varepsilon = \varphi_\varepsilon\big(\mathrm{dist}(v, v')\big)$ which localize at the diagonal in $V \times V$ for $\varepsilon \to 0$ and where ε_i goes to zero at an appropriate rate for $i \to \infty$.)

1.4.D. Riemannian and piecewise Riemannian approximation of C-C metrics.

Let V be an equiregular C-C manifold as in 1.4.B with the polarization $H \subset T(V)$ and the subbundles

$$H = H_1 \subset H_2 \subset \cdots \subset H_i \subset \cdots \subset H_d = T(V)$$

generated by the commutators of orders $1, 2, \ldots, i, \ldots, d$. We assume V is compact and we want to approximate the C-C metric dist on V by suitably controlled Riemannian metrics $\mathrm{dist}_\varepsilon$ on V (compare 0.8).

Approximation theorem. *There exist Riemannian metrics* $\mathrm{dist}_\varepsilon$ *on* V *for positive* $\varepsilon \to 0$, *such that*

(1) $|\,\mathrm{dist} - \mathrm{dist}_\varepsilon\,| = O(\varepsilon)$ *uniformly on* $V \times V$.

(2) *The curvatures of (the Riemannian metric tensors underlying)* $\mathrm{dist}_\varepsilon$ *are* $O(\varepsilon^{-2})$ *and the injectivity radii of* $\mathrm{dist}_\varepsilon$ *are* $O(\varepsilon^{-1})$.

Proof. Fix a Riemannian metric g on V, let $H_i^\perp = H_i \ominus_g H_{i-1}$, and let A_ε be the linear operator $A_\varepsilon : T(V) \to T(V)$ fixing H and acting by $\tau \mapsto \varepsilon^{-(i-1)}$ on the vectors $\tau \in H_i^\perp$ for $i = 2, \ldots, d$. Then the distance functions $\mathrm{dist}_\varepsilon$ of the Riemannian tensors (metrics) $A_\varepsilon^*(g)$ on $T(V)$ satisfy the requirement of the theorem. This is immediate if V is a self-similar Lie group and the general case follows from 1.4.A - 1.4.A''. (The details are left to the reader.)

Remarks

(a) The metrics $\mathrm{dist}_\varepsilon$ we constructed increase as $\varepsilon \to 0$ and their total volumes grow as ε^{-m} for $m = \dim_{\mathrm{Hau}} V - \dim_{\mathrm{top}} V$.

(b) The weaker bound $O\!\left(\varepsilon^{-2(d-1)}\right)$ on the curvature of g_ε is obvious and needs no special (commutator) relations between H_i.

(c) The conditions (2) of the theorem says that $\mathrm{dist}_\varepsilon = \varepsilon \, \mathrm{dist}_\varepsilon^*$ where the metrics $\mathrm{dist}_\varepsilon^*$ have uniformly bounded (local) geometries as $\varepsilon \to 0$.

(d) It is unclear if one can improve (1) to $|\,\mathrm{dist} - \mathrm{dist}_\varepsilon\,| = o(\varepsilon)$ (for a suitable $\mathrm{dist}_\varepsilon$ satisfying (2)).

Uniqueness of the approximation. Let $\mathrm{dist}_\varepsilon$ and $\mathrm{dist}_\varepsilon'$ be two families of Riemannian metrics on V approximating dist with the above conditions (1) and (2). Then, ideally, one would like to have a bi-Lipschitz map $f_\varepsilon : (V, \mathrm{dist}_\varepsilon) \to (V, \mathrm{dist}_\varepsilon)$, such that

(1) $\displaystyle \sup_{v \in V} \mathrm{dist}\big(v, f_\varepsilon(v)\big) \leq \varepsilon$ for the original C-C metric dist in V.

(2) The Lipschitz constants $L(f_\varepsilon)$ and $L(f_\varepsilon^{-1})$ are bounded independently of ε.

This is likely to be true but may be hard to prove and also appears too refined at the present crude state of the study of C-C geometries. On the other hand the following more natural equivalence between $\mathrm{dist}_\varepsilon$ and $\mathrm{dist}_\varepsilon'$ comes quite effortlessly.

Equivalence Proposition. *There exist Lipschitz maps $f_\varepsilon : (V, \mathrm{dist}_\varepsilon) \to (V, \mathrm{dist}_\varepsilon')$ and $f_\varepsilon' : (V, \mathrm{dist}_\varepsilon') \to (V, \mathrm{dist}_\varepsilon)$ with the implied Lipschitz constants independent of ε such that the composed maps $f_\varepsilon' \circ f_\varepsilon$ and $f_\varepsilon \circ f_\varepsilon'$ are ε-close to the identity with respect to the metrics $\mathrm{dist}_\varepsilon$ and $\mathrm{dist}_\varepsilon'$ correspondingly. Moreover, these composed maps can be joined with the identity by the Lipschitz homotopies $V \times [0, \varepsilon] \to V$ where again the Lipschitz constants do not depend on ε.*

Proof. Since $(V, \mathrm{dist}'_\varepsilon)$ has ε^{-1}-bounded geometry one can smooth the identity map $(V, \mathrm{dist}_\varepsilon) \to (V, \mathrm{dist}'_\varepsilon)$ by using the center of mass construction in $(V, \mathrm{dist}'_\varepsilon)$ for the ε-balls coming from $(V, \mathrm{dist}_\varepsilon)$. Alternatively, one may triangulate $(V, \mathrm{dist}_\varepsilon)$ into standard (fat geodesic) ε-simplices and then approximate the identity map by a piecewise "linear" map. In either way one arrives at a Lipschitz map $f_\varepsilon : (V, \mathrm{dist}_\varepsilon) \to (V, \mathrm{dist}'_\varepsilon)$ which is ε-close to Id and f'_ε is constructed in the same way. Then the required ε-homotopies are constructed in a similar fashion with the additional ingredient of the local contractibility of C-C balls proved in 1.4.B. The details are left to the reader.

Exercise. Generalize the above to approximations $\mathrm{dist}_\varepsilon$ and $\mathrm{dist}'_{\varepsilon'}$ where ε' may be non-equal to ε.

Open problem. Characterize C-C spaces in terms of their (controlled) approximation by Riemannian manifolds with respect to the Hausdorff distance between metric spaces. Study other metric spaces admitting sufficiently nice approximations by Riemannian ones.

1.4.D′. Approximation by nerves. Every compact metric space can be ε-approximated by the nerve of a suitable ε-covering and in the C-C case this approximation enjoys some particularly nice features due to the following two facts.

    I. *The doubling property for C-C balls:* $\mathrm{mes}\, B(2\varepsilon) \le \mathrm{const}\,\mathrm{mes}\, B(\varepsilon)$.

    II. *Contractibility of $B(\varepsilon)$ within $B(2\varepsilon)$.*

    We shall use the following standard construction of a "good" covering by ε-balls. Take a maximal δ-separated net $\Sigma \subset V$ and cover V by the ε-balls with the centers in Σ for $\varepsilon = 3\delta$. Denote by V_ε the nerve of this cover which we equip with the piecewise Riemannian metric $\mathrm{dist}_\varepsilon$ where each simplex is isometric to the regular Euclidean ε-simplex of the corresponding dimension. One sees with I that $\dim V_\varepsilon$ is bounded by a constant independent of ε, and, in fact, bounded, for small ε, by const_n, $n = \dim V$. Moreover, the local geometry of V_ε is bounded by ε^{-1} in the following strong sense: every ε'-ball in $(V_\varepsilon, \mathrm{dist}_\varepsilon)$ for $\varepsilon \le \varepsilon' \le 1$, contains at most $M \le \mathrm{const}_n(\varepsilon'/\varepsilon)^N$ simplices for $N = \dim_{\mathrm{Hau}} V$.

    Next, we introduce a partition of unity of V inscribed into our cover where all functions can be made λ-Lipschitz with $\lambda \le \mathrm{const}_n \varepsilon^{-1}$. Then the corresponding continuous map $f_\varepsilon : V \to V_\varepsilon$ is Lipschitz with $L(f_\varepsilon) \leq$ const_n. Finally, using II, we construct a continuous map $\overline{f}_\varepsilon : V_\varepsilon \to V$ which sends each 0-simplex to the center of the corresponding ball and such that the image of each simplex has diameter $\leq \mathrm{const}_n$. The composed maps $f_\varepsilon \circ \overline{f}_\varepsilon$ and $\overline{f}_\varepsilon \circ f_\varepsilon$ are both $C\varepsilon$-close to the identities for $C \leq \mathrm{const}_n$ and by II $\overline{f}_\varepsilon \circ f_\varepsilon : V \to V$ is homotopic to Id by a homotopy moving all points by at most $C\varepsilon$. This is almost true for $f_\varepsilon \circ \overline{f}_\varepsilon : V_\varepsilon \to V_\varepsilon$. Namely we invoke simplicial maps $M_{\varepsilon'}^\varepsilon : V_\varepsilon \to V_{\varepsilon'}$ for all $\varepsilon' > 5\varepsilon$ where each 1-simplex of V_ε goes to a 1-simplex of $V_{\varepsilon'}$ such that the corresponding ball $B(\varepsilon) \subset V$ is contained in $B(\varepsilon')$. Then we (obviously with I and II) have a homotopy of $M_{\varepsilon'}^\varepsilon \circ f_\varepsilon \circ \overline{f}_\varepsilon$ to Id in $V_{\varepsilon'}$ whenever ε'/ε is sufficiently large, i.e. $\geq \mathrm{const}_n$. Moreover, this homotopy can be realized by a Lipschitz map $V \times [0, \varepsilon] \to V_{\varepsilon'}$ with the implied Lipschitz constant independent of ε.

Now one can see that the above properties of V_ε make this approximation equivalent to the one considered earlier where the equivalence used in the above Equivalence Proposition should be argumented by stabilizing maps $M_{\varepsilon'}^\varepsilon$.

Remark. One can avoid $M_{\varepsilon'}^\varepsilon$ by slightly modifying V_ε to ensure their sufficient local contractibility (and thus the homotopy equivalence to V). But this is somewhat artificial and one should not be worried by $M_{\varepsilon'}^\varepsilon$, albeit they complicate the notations. In what follows we just pretend that $M_{\varepsilon'}^\varepsilon$ are not there.

Example. Pretend V_ε is homotopy equivalent to V and represent the cohomology classes in $H^*(V; \mathbb{R})$ by simplicial cocycles in V_ε. Then for each $h \in H^*(V; \mathbb{R})$ we minimize the ℓ_p-norm of these cocycles, denote this minimum by $\|h\|_{\ell_p}^\varepsilon$, and observe that the asymptotic rate of growth of $\|h\|_{\ell_p}^\varepsilon$ for $\varepsilon \to 0$ gives us an interesting geometric invariant of $(V, \mathrm{dist}; h)$ similar to the norm on the Alexander-Spanier cochains considered in 3. Furthermore, the ε-asymptotics of the spectrum of the combinatorial Laplace operator in V_ε (as well as the ℓ_{pq}-spectra) gives us further invariants of V.

Combinatorial Hyperbolization. Take $\varepsilon_i = 5^{-i}$, $i = 1, 2, \ldots$ and let W be the joint infinite mapping cylinder of the maps $M_{\varepsilon_{i-1}}^{\varepsilon_i} : V_{\varepsilon_i} \to V_{\varepsilon_{i-1}}$. This W carries a natural hyperbolic metric where the (formerly ε) simplices are given the unit size, such that $V_i \subset W$ appears as a sphere of radius i (if we add to W the mapping cone of $V_1 \to \{$point$\}$) and V serves as the ideal hyperbolic boundary of W. Then W harbours many geometric invariants of V. In fact, all quasi-isometry invariants of W are quasi-conformal invariants of V. Most prominent of these are ℓ_p and ℓ_{pq}-cohomology of W (where W must be slightly regularized, i.e. made contractible and having all R-balls contractible within concentric λR-balls for a fixed large λ).

Finally, observe that W is quasi-isometric to the hyperbolic "cone" CV in 0.9 and one can think of W as a $p.\ell.$ regularization of CV which is essentially equivalent to the Rips complex of CV or the nerve of a covering of CV by unit balls.

1.4.D''. Anisotropic blow-up of $(\mathbf{V}, \mathbf{H})$. Let us return to the Riemannian metric $\mathrm{dist}_\varepsilon$ constructed in the proof of approximation theorem and look more closely at the rescaled (blown up) metric $\mathrm{dist}_\varepsilon^* = \varepsilon^{-1} \mathrm{dist}_\varepsilon$ (compare Remark (c)). This $\mathrm{dist}_\varepsilon^*$ (or, rather, the corresponding g_ε^*) is obtained from a given Riemannian metric g on V by scaling (blowing up) by ε^{-1} in the direction of $H = H_1$, by ε^{-2} in the direction of $H_2 \ominus_g H_1$, and so on up to ε^{-d} in $H_d \ominus_g H_{d-1}$. The metric $\mathrm{dist}_\varepsilon^*$, as we have seen, has uniformly bounded (local) geometry for $\varepsilon \to 0$ and if $Y_1, \ldots, Y_{n_1}, Y_{n_1+1}, \ldots, Y_{n_2}, \cdots, Y_{n_d=n}$, where $n_i = \mathrm{rank}\, H_i$, is an orthonormal frame of fields on V near some $v \in V$ which agree with the splitting $T(V) = \bigoplus_{i=1}^d (H_i \underset{g}{\ominus} H_{i-1})$, then the frame

$$a_\varepsilon\{Y_i\} = \varepsilon Y_1, \ldots, \varepsilon Y_{n_1}, \varepsilon^2 Y_{n_1+1}, \ldots, \varepsilon^2 Y_{n_2}, \ldots, \varepsilon^d Y_{n_{d-1}}, \ldots, \varepsilon^d Y_n,$$

becomes unitary (orthonormal) in the rescaled Riemannian metric g_ε^*. Now, as $\varepsilon \to 0$, the family $V_{\varepsilon^{-1}} = (V, v, \mathrm{dist}_\varepsilon^*, a_\varepsilon\{Y_i\})$ converges (in the pointed Lipschitz topology, see [G-L-P]) to a Riemannian manifold, say V_∞, with a distinguished orthonormal frame of fields, say Y_i^∞, whose commutators obviously satisfy

$$[Y_i^\infty, Y_j^\infty] = \sum_k c_{ijk} Y_k^\infty$$

where c_{ijk} are defined at the beginning of 1.4. Now, one clearly sees that this V_∞ can be identified with the (tangent) nilpotent Lie group N_v (see the beginning of 1.4.) which appears this time as a geometric limit of anisotropically rescaled copies of V rather than an abstract Lie group with a given Lie algebra.

Question. (Compare 4.11). What is the behavior of the standard analytic objects (fields), such as harmonic forms and spinors on $V_{\varepsilon^{-1}}$, for small $\varepsilon > 0$? What happens in the (anisotropic adiabatic) limit as $\varepsilon \to 0$. (These kinds of questions are often asked for fibrations over $V_{\varepsilon^{-1}}$ where the geometry of the fibers does not change as $\varepsilon \to 0$.)

1.4.E. Smoothing Lipschitz functions on C-C manifolds. Let (V, H) be a compact equiregular C-C manifold as in 1.4.D. with the commutator subbundles $H_1 = H \subset H_2 \subseteq \cdots \subset H_i \subset \cdots \subset II_d = T(V)$ and let $f : V \to \mathbb{R}$ be a Lipschitz function with the implied Lipschitz constant bounded by some number $\lambda > 0$.

Smooth approximation theorem. *There exists a family of smooth functions* $f_\delta : V \to \mathbb{R}$ *for small positive* δ*, say for* $\delta \in]0, 1]$*, satisfying the following two inequalities.*

(1) $|f - f_\delta| \leq \delta$ *everywhere on* V,

(2) $\|\mathcal{D}f_\delta|H_i\| \leq C \left|\lambda\delta^{-1}\right|^i$, $i = 1, \ldots d$, *everywhere on* V,

where $\mathcal{D}f_\delta|H_i$ *denotes the differential of* f_δ *restricted to* H_i *and where* C *is a positive constant depending on (the geometry of)* V *but not on* f.

Proof. Let V_ε be the Riemannian approximation to V with $\varepsilon = \delta/\lambda$ and let $K(v, v')$ be a standard Riemannian smoothing kernel on V on the scale slightly smaller than ε. Namely $K(v, v')$ is supported in the $C_1^{-1}\varepsilon$-neighbourhood of the diagonal in $V \times V$ for a fixed large constant C_1 and the Riemannian covariant derivatives of K are bounded by $C_2\varepsilon^{-1}$. (For example, for each v one takes the Riemannian ball $B_v \subset V$ of radius $C_1^{-1}\varepsilon$ and defines $K(v, v')$ to be $\mu_v \operatorname{dist}(v', \partial B_v)$ for $v' \in B$ and $K(v, v') = 0$ for $v' \in V - B$, where the constant μ_v is chosen so that $\int_{B_v} K(v, v')dv' = 1$.)

Then the Riemannian smoothing of f with K, i.e.

$$f_\varepsilon(v) = \int f(v')K(v, v')dv',$$

is the desired approximation as an obvious argument shows.

Corollary. *Let W be a compact Riemannian manifold or, more generally, a complete manifold with bounded local geometry. Then every λ-Lipschitz map $f : V \to W$ is homotopic to a smooth map $f_1 : V \to W$, such that $\|\mathcal{D}f_1|H_i\| \leq C\,\lambda^i$, $i = 1, \ldots, d$.*

1.4.E′. On the homotopy count of Lipschitz maps. Let W be the sphere S^n for $n = \dim V$ and let us estimate the degree of a λ-Lipschitz map $V \to W$ f in terms of λ. We replace f by the above f_1 and use the bound of $\deg f_1$ by the norm of the differential on the n forms. Thus we get

$$\deg f = \deg f_1 \leq \mathrm{const}\,\| \wedge^n \mathcal{D}f\| \leq \mathrm{const}_1 \prod_{i=1}^{d} \lambda^{im_i}$$

for $m_i = \mathrm{rank}\, H_i/H_{i-1}$. This agrees with the formula for $N = \dim_{\mathrm{Hau}} V$, i.e.

$$\deg f \leq \mathrm{const}_1\, \lambda^N, \text{ for the Lipschitz constant } \lambda = \lambda(f).$$

Notice that this bound is sharp as V contains $\approx \lambda^N$ disjoint balls of radii λ^{-1} where each can be λ-Lipschitz mapped onto S^n with degree 1.

The above argument also allows an improvement of the estimate in 0.5.E of the number $Nm(\lambda)$ of the homotopy classes of maps $f : V \to W$ in terms of $\lambda = L(f)$, where W is an arbitrary compact simply connected Riemannian manifold. Let us spell it out for maps of the contact C-C sphere S^3 to S^2. Here the number $Nm(\lambda)$ is controlled by the Hopf invariant $h(f)$ which is defined as follows (compare 2.5). Fix the normalized area form ω on S^2, pull it back to S^3 by f, i.e. take the 2-form $f^*(\omega)$ on S^3. This "integrates" to some 1-form α on S^3, where "integrates" means $d\alpha = \omega$. Then

$$h(f) = \int_{S^3} \alpha \wedge f^*(\omega).$$

If S^3 were Riemannian, this would imply $h(f) \leq \mathrm{const}\,\lambda_{\mathrm{Ri}}^4$ since $\|f^*(\omega)\| \leq \mathrm{const}_1\,\| \wedge^2 \mathcal{D}f\| \leq \mathrm{const}_1\,\lambda_{\mathrm{Ri}}^2$ and since (this is a key point) one could find α with $\|\alpha\| \leq \mathrm{const}_2\,\|f^*(\omega)\|$. Since the contact C-C metric is $C^{\frac{1}{2}}$-equivalent to the Riemannian one this implies the bound $h(f) \leq \mathrm{const}\,\lambda^8$ for C-C Lipschitz constant λ as was indicated in 0.5.E. But now the above corollary yields the bound $\|f_1^*(\omega)\| \leq \mathrm{const}\,\lambda_{\mathrm{Ri}}^3$ for the Riemannian norm of a suitable smooth f_1 approximating f which leads to the improved bound

$$h(f) = h(f_1) \leq \mathrm{const}_3\,\lambda^6.$$

Question. What is the exponent r in the true bound of $h(f)$ by $C\lambda^r$ for $\lambda \to \infty$? (The above shows that $r \le 6$ and it is easy to see that $r \ge 4$). A closely related question concerns an estimate of $h(f)$ of an f which is λ-Lipschitz with respect to the Riemannian $(SU(2)$-invariant) metric g_t approximating the C-C metric on S^3 (see 0.8.B), where the asymptotics of h now depend on $t \to \infty$ as well as on $\lambda \to \infty$ (as $d = 2$, this g_t equals the above g_ε^* for $\varepsilon = t^{-1}$).

1.5. Anisotropic metrics beyond Carnot-Carathéodory. Let us indicate a class of metrics on V where the ball-box theorem is built-in into the definition. We consider a Euclidean vector bundle X over V endowed, besides the projection $X \to V$, with a smooth ("exponential") map $E : X \to V$ which fixes the zero section and such that the differential $\mathcal{D}E$ maps the tangent space of each fiber X_v at the origin, say $T_0(X_v)$, onto $T_v(V)$. Then we equip X with a fiber preserving and fiberwise linear automorphism $A : X \to X$ with norm ≤ 1 (i.e. distance decreasing) on each fiber X_v, and we say that a metric dist on V agrees with (X, E, A) if for some $\rho = \rho_i$ and all $i = 1, 2, \ldots$ the balls $B_v(\rho) \subset V$ are equivalent to the E-images of the ellipsoids $A^i(B_0(1)) \subset X_v$ for the unit (Euclidean) ball $B_0(1)$ in the fiber X_v, for all $v \in V$, where the equivalence means, as earlier, the inclusion

$$E\, A^{i+c}\big(B_0(1)\big) \subset B_v(\rho) \subset E\, A^{i-c}\big(B_0(\rho)\big).$$

Then a metric on V is called (pleasantly) *anisotropic* if it agrees with some (X, E, A). This generalizes C-C metrics as well as those defined in [N-S-W]. In particular, the restriction of C-C metrics to submanifolds are anisotropic in our sense.

It is unclear if there are anisotropic metrics significantly different from those defined in [N-S-W]. To figure this out one should be able to decide for which (X, E, A) there exists a metric (or quasi-metric) on V agreeable with these data.

Let us indicate another generalization by first recalling that the ball-box theorem applies to the metrics associated to systems of vector fields on V. Such fields, say $X_1, \ldots, X_m$, viewed as linear functions on the cotangent bundle $T^*(V)$, define a semipositive quadratic form, namely $h = \sum\limits_{i=1}^{m} (X_m)^2$ on $T^*(V)$. Then one observes that every suitably generic smooth semidefinite form h on $T^*(V)$, which does not have to be a sum of squares of *smooth* linear functions (vector fields), also defines a metric on V, where the geometry of the small balls is somewhat similar to that for $h = \Sigma(X_i)^2$, as was proven by Fefferman and Phong (see [Fe-Ph] and [Je-Sa]), but the argument is significantly harder than that for $h = \Sigma(X_i)^2$. (All that was explained to me by N. Varopoulos.) We conclude by suggesting a study of more general infinitesimally defined classes of horizontal curves and associated metrics, e.g. corresponding to semipositive forms on $T^*(V)$ of (even) degree > 2. (A geometric aspect of the problems is related to the structure of the convex hull of the spaces of maps $V \to S \subset \mathbb{R}^n$ for a smooth submanifold $S \subset \mathbb{R}^n$, where one of the questions reads

$$\text{conv.hull}(\text{maps}(V \to S)) \overset{?}{=} \text{maps}\,(V \to \text{conv.hull}\,(S)).$$

(Compare pp. 170, 205 and 206 in [Gro$_\text{PDR}$].)

2. Hypersurfaces in C-C spaces

Much of the classical analysis in $\mathbb{R}^n$ (and in Riemannian manifolds) deals (explicitly) with functions and (often implicitly) with hypersurfaces which may serve as levels of functions. A typical example is provided by the Sobolev inequalities reflecting the isoperimetry of hypersurfaces in $\mathbb{R}^n$. We shall see in this section that many classical Euclidean results extend to C-C spaces where the main technical tools are provided by H-horizontal curves, often appearing as orbits of H-horizontal fields, and by certain closed $(n-1)$-forms vanishing on H. Our major applications concern Sobolev-type spaces of maps $V \to W$ associated to H, where we establish C-C counterparts of corresponding Riemannian results due to Karen Uhlenbeck.

2.1. A lower bound on the Hausdorff dimension of a hypersurface $V' \subset V$. Let V' be a compact subset in an equiregular C-C manifold V of topological dimension $\dim V' = n - 1$ for $n = \dim V$. We show below that the Hausdorff dimension of V' with respect to the induced C-C metric satisfies

$$\dim_{\text{Hau}} V' \geqslant \dim_{\text{Hau}} V - 1, \tag{$*$}$$

as was stated in §0.6.

The first result of this kind was obtained by P. Pansu who observed that such lower bound on $\dim_{\mathrm{Hau}}$ of hypersurfaces follows from his (and Varopoulos) isoperimetric inequality (see [Cor] and p.187 in [Gro$_{\mathrm{AI}}$]). In fact, the lower bound $(*)$ on $\dim_{\mathrm{Hau}} V'$ is a more basic and elementary fact than the isoperimetric inequality and we shall establish it by a simple direct argument. Then we shall elaborate this argument in order to obtain a lower bound on the Hausdorff measure of V', needed for the isoperimetric inequality.

Proof of $(*)$ for equiregular polarizations H. We shall use the following

Characteristic property of dim $= n - 1$. *If V' is a compact subset of topological dimension $n-1$ in an n-dimensional manifold V, then there exists an $\varepsilon > 0$ and a simple curve c in V i.e. an embedding $c : [0,1] \to V$, such that every continuous curve $f : [0,1] \to V$ which is ε-close to c (in the sense of the inequality*

$$\mathrm{dist}_V(c(t), f(t)) \leqslant \varepsilon , \quad t \in [0,1],$$

for a fixed distance in V) meets V', i.e. $f([0,1]) \cap V' \neq \varnothing$.

This follows from the elementary homological dimension theory (see [Nag] and 4.5; a non-fastidious reader may just use the above intersection property as a definition of the dimension).

We use an approximation of c by piecewise horizontal curves and then make them smooth as in 1.2.B. In fact the argument in 1.2.B provides a smooth family of smooth horizontal curves close to c, say $f(t, x)$ for $t \in [0,1]$ and x running over a (small) ball $B \subset \mathbb{R}^{n-1}$, such that the global map $f : [0,1] \times B \to V$ is smooth generic with a (possible) folding along some smooth hypersurface $W \subset [0,1] \times B$ as the only singularity (where f is not an immersion). If V' contains an "essential piece" of W, i.e. if $f^{-1}(V') \cap W$ contains a non-empty open subset $W' \subset W$, then $f(W') \subset V'$, being a *smooth* hypersurface in V must be almost everywhere *transversal* to H (by the connectivity property for H-horizontal curves) which makes

$$\dim_{\mathrm{Hau}} f(W') \geqslant \dim_{\mathrm{Hau}} V - 1$$

by the discussion in 0.6.A.

Now we turn to the essential case where W has nothing to do with V' and then we may assume that f is an immersion, and moreover, a smooth embedding $[0,1] \times B \subset V$, where all curves $[0,1] \times b \subset V$ are H-horizontal and intersect V'. Thus the projection $p : [0,1] \times B \to B$ is surjective on $V'_0 = V' \cap [0,1] \times B$ and the levels of p are horizontal. Now we recall that every small ε-ball in $[0,1] \times B \subset V$ looks as a box with the edges $\varepsilon^{\deg i}$, $i = 1, \ldots, n$ where $\deg i = 1$ for $i = 1, \ldots, n_1 = $ rank H, $\deg i \geqslant 2$ for $i = n_1 + 1, \ldots, n$, where the first edge of the box may be assumed equal to a segment of some level of p (see 1.3.A). It follows that the p-image of this box is a box with the edges $\approx \varepsilon^{\deg i}$, for $i = 2, \ldots, n$. In particular, the volume of this image satisfies

$$(n-1)\text{-volume } (p(\text{box})) \lesssim \varepsilon^{-1}(n\text{-volume (box)}).$$

Now, if a system of ε_j-boxes B_j, $j = 1, 2, \ldots$, cover X_0, then their projections cover B and so $\Sigma_j \, \varepsilon_j^{-1} \operatorname{Vol} B_j \geqslant$ const. This implies the desired inequality

$$\dim_{\mathrm{Hau}} V' \geqslant \dim_{\mathrm{Hau}} V'_0 \geqslant \dim_{\mathrm{Hau}} V - 1,$$

by the discussion in 0.6.

2.1.A. A systolic bound on $\operatorname{mes}_{N-1} V'$ of $(*)$. Suppose V' is $(n-1)$-dimensionally *essential* in V, i.e. it can not be homotoped to a subset of dimension $\leqslant n - 2$. This is the case, for example, where V' supports a (Čech) cycle of dimension $n - 1$ non-homologous to zero in V. Then, if V is compact, *the $(N-1)$-dimensional Hausdorff measure of V' for $N = \dim_{\mathrm{Hau}} V$ is bounded from below by a constant $s = s(V, \text{C-C dist})$,* (called the $(n-1)$-*systole* of V, see [Gro$_{\mathrm{SISI}}$] for a survey of systoles). This is proven with an obvious modification of the above argument.

Exercise. Show that every closed curve in V can be approximated by a smoothly immersed closed horizontal curve. (This is somewhat more than needed for the proof of our claim.)

Remark. If V' supports an essential $(n-1)$-cycle in V modulo some subset $V_0 \subset V$, then again we have a bound on the measure of V' by some $s = s(V, V_0, \text{dist})$ and such a bound was, in fact, instrumental in our proof of $(*)$.

Exercise. Assume V admits an equisingular stratification by V_ν (see 1.3.A), such that $\dim(V' \cap V_\nu) = \dim V_\nu - 1$ for certain ν and bound from below the Hausdorff dimension of $V' \cap V_\nu$.

2.2. Lower systolic bounds for families of metrics on compact manifolds V. Let $A : T(V) \to T(V)$ be a smooth automorphism (fixing V) and let us look at the Riemannian metrics $A^t(g)$, $t = 1, 2, 3, \ldots$, for a fixed g as $t \to \infty$ (compare 0.8). To simplify the matter we assume the absolute values of the eigenvalues λ_i of A and their multiplicities are constant on V. Then the volume growth of (V, g_t) for $g_t = (A^t)^* g$ can be expressed in terms of the eigenvalues of A by

$$\lim_{t \to \infty} \; (\mathrm{Vol}(V, g_t))^{\frac{1}{t}} = \prod_{i=1}^{n} |\lambda_i|.$$

Now, in order to measure the volume growth of submanifolds $X \subset V$ of positive codimension we recall the following

Definitions. The *volume of a k-dimensional* (integral) *homology class* is the infimum of the volumes of the cycles realizing this class. Then the ($\mathbb{Z}$-homological) *systole* $\mathrm{sys}_k(V, g)$ is defined as the infimum of these volumes over all non-zero homology classes. Finally, the *absolute* (homotopy) *systole* absys_k is defined as the infimum of the volumes of k-dimensional subsets in V which can not be contracted to $(k - 1)$-dimensional subsets in V. (Here "subset" means a piecewise smooth sub-polyhedron in V.) It is clear that $\mathrm{absyst}_k \leqslant \mathrm{syst}_k$.

Let $m \leqslant n$ be the first integer such that $\lambda_{m+1} > \lambda_m$ and the (eigen) subbundle H_m corresponding to $\lambda_1, \ldots, \lambda_m$ Lie-generates $T(V)$.

Theorem. *The absolute systole of codimension one of (V, g_t) for $t \to \infty$ is bounded from below according to the inequality*

$$\lim_{k \to \infty} \; \infty \, (\mathrm{absys}_{n-1}(V, g_t))^{\frac{1}{t}} \geqslant |\lambda_m^{-1}| \prod_{i=1}^{n} |\lambda_i|. \tag{$\star$}$$

The proof of $(\star)$ is the same as that of the above inequalities $(*)$ and $(+)$.

Remark. The inequality $(\star)$ is by no means sharp. For example if V is the 2-torus and $H_1 \subset T(V)$ is the standard 1-foliation with slope α, then the asymptotics of $\mathrm{sys}_1(V, g_t)$ depends on the continuous fraction expansion of α.

Problem. Find the actual asymptotics of $\mathrm{sys}_{n-1}(V, g_t)$ for *generic* A.

2.2.A. A lower systolic bound via closed horizontal $(n-1)$-forms.
Our lower bounds on hypersurfaces in V used families of H-horizontal
curves (for a given $H \subset T(V)$ which Lie spans $T(V)$) or, equivalently
smooth maps $p : V \to B^{n-1}$ with H-horizontal fibers. Notice that such a
p pulls back (necessarily closed) $(n-1)$-forms on B^{n-1} to closed $(n-1)$-
forms ω on V which *vanish* on H, i.e. vanish on every (local) hyperplane
field containing H, and such forms, called *H-horizontal*, can be then used
to study hypersurfaces in $X \subset V$. In fact, one can produce sufficiently
many such forms by a purely algebraic argument. Namely we have the
following

Linear Lemma. *If the subbundle $H \subset T(V)$ Lie generates $T(V)$ then
every $(n-1)$-dimensional de Rham cohomology class in V can be represented by a closed horizontal $(n-1)$-form ω.*

First (not quite linear) proof. We may assume (by passing to the
double cover of V if necessary) that V is oriented and so the cohomology
$H^{n-1}(V; \mathbb{R})$ is dual to $H_1(V; \mathbb{R})$. Then every integral class in $H_1(V; \mathbb{R})$
can be realized by a closed horizontal curve c (see the above exercise)
which gives us a closed $(n-1)$-current, called c^*, representing the class
$[c]^* \in H^{n-1}(V; \mathbb{R})$ where the latter "$*$" denotes the Poincaré duality.
Now, in order to pass from currents to forms one needs some smoothing
or diffusion of currents preserving H-horizontality. This is easy if V admits
a transitive action of a connected group G preserving H as one can diffuse
the current c^* by taking $\int_G c^* d\mu$, where $d\mu$ is a smooth measure with a
compact support on G (localized near $\mathrm{id} \in G$). For example, this diffusion
is available if our polarization H is a constant structure. In the general
case, the diffusion is achieved with a smooth family of horizontal curves,
say $c_b \subset V$, $b \in B$, such that the corresponding map $S' \times B \to V$ (for
c_b parametrized by the circle S') is a submersion. The existence of such
family is proven in the same manner as of an individual c (the details are
left to the reader).

Second (purely linear) proof. We still assume V is oriented and take
a non-vanishing oriented volume form Ω on V. Then the *interior product*
with Ω establishes an isomorphism between vector fields X and exterior
$n-1$ forms, i.e.
$$X \leftrightarrow X \cdot \Omega,$$

and similarly bivectors correspond to $(n-2)$-forms,

$$X \wedge Y \leftrightarrow (X \wedge Y) \cdot \Omega.$$

Closed $(n-1)$-forms correspond in this picture to *divergence free* vector fields, where the divergence δX of X is the function defined by the equality

$$L_X \, \Omega = (\delta X)\Omega,$$

where L_X denotes the Lie derivative. We recall the formula

$$d((X \wedge Y) \cdot \Omega) = [X, Y] \cdot \Omega + Y \cdot L_X \, \Omega - X \cdot L_Y \, \Omega$$

which implies that the field

$$[X, Y] + \delta(X)Y - \delta(Y)X$$

has zero divergence and, moreover, corresponds to an *exact* $(n-1)$-form. It follows, that for all functions a, the field

$$a[X, Y] + (Xa + a\delta(X))Y - \delta(aY)X$$

corresponds to an exact $(n-1)$-form, or in other words $a[X, Y]$ equals $(Xa + a\,\delta(X))Y - \delta(aY)X$ modulo (the fields corresponding to) exact forms. (The latter expression is antisymmetric in X and Y, as it should be, since $Xa + a\,\delta(X) = \delta(aX)$.)

Next, we look at $H \subset T(V)$ and observe that $(n-1)$-forms vanishing on H correspond to vector fields sitting in H. Thus, to prove the lemma, we must find a divergence free H-horizontal field in a given cohomology class. We pick up some fields $X_1, \ldots, X_s$ spanning H (here s may be greater than rank H) add to these X_i their successive commutators, say $X_j, j = s+1, \ldots, r$, which span $T(V)$ and observe that every cohomology class in $H^{n-1}(V; \mathbb{R})$ can be represented by a divergence free field of the form $\Sigma_{i=1}^r \, a_i \, X_i$. But the above formulae allow us to replace every (commutator) term in this sum with $i > s$ by (cohomologically) equivalent lower terms and thus we obtain a desired divergence free representative of the form $\Sigma_{i=1}^s \, a_i' \, X_i$. $\blacksquare$

Remarks

(a) The above algebraic discussion can be neatly expressed with the differential operator d^H which is obtained by composing the exterior differential $d : \Lambda^{n-2}(V) \rightarrow \Lambda^{n-1}(V)$ with the quotient homomorphisms $\Lambda^{n-1}(V) \rightarrow \Lambda^{n-1}(V)/H^{\perp}$ where $H^{\perp}$ denotes the (sheaf of sections of the) subbundle of $(n-1)$-forms vanishing on H. Namely, we have shown that this d^H is a *surjective* operator. Moreover, our argument provides another differential operator, say $\delta^H : \Lambda^{n-1}(V)/H^\perp \to \Lambda^{n-2}(V)$ (which order equals the degrees of commutators needed to span $T(V)$) such that $d^H \delta^H = \mathrm{Id}$. Furthermore, the "coefficients" of δ^H are expressed by some universal rational functions in the components of some jet of H represented by a section $V \to \mathrm{Gr}_{n_1}(T(V))$ for $n_1 = \mathrm{rank}\ H$. (This is not surprizing as the P.D.E. system $d^H \varphi = \omega$ is underdetermined in our case and the theorem proven in 2.3.8 (E) in [$\mathrm{Gro_{PDR}}$] predicts the existence of δ for generic H.)

Exercise. Show that the existence of a differential operator δ^H making $d^H \delta^H = \mathrm{Id}$ is not only implied by the commutator generation property of H but also implies this property of H.

(b) Observe that our second proof is purely algebraic (modulo de Rham theorem), technically trivial and self-contained unlike the first proof based on an elaborated version of Chow connectivity theorem (paradoxally, the first proof is geometrically obvious while the second one makes no lasting impression on a geometrically moulded mind).

(c) The abundance of closed $(n-1)$-forms vanishing on H is dual to what happens to 1-forms: every closed 1-forms vanishing on H (obviously) equals zero. In fact, every closed 1-current φ satisfying $X\varphi = 0$ for all H-horizontal fields X necessarily vanishes. Furthermore, there exists a differential operator δ'_H acting from $\Lambda^2(V)$ to the subspace $H_\perp \subset \Lambda^1(V)$ of 1-forms vanishing on H, such that

$$(\delta'_H\ \omega) \wedge \lambda = \omega \wedge (\delta_H\ \lambda) \ \mathrm{mod}\ d\Lambda^{n-1}(V),$$

where ω is an arbitrary 2-form and λ is an $(n-1)$-form modulo $H^\perp$. In fact, this formula (essentially) defines δ'_H as a formal adjoint of δ^H which makes $\delta'_H\ d = \mathrm{id}$ since the differential d on 1-forms is a formal adjoint of d on $(n-2)$-forms and $d^H \delta^H = \mathrm{id}$.

(d) The operators d_H and d^H make sense for forms of all degrees, see 3.3 and 4.1.E.

Now let us return to the metrics $g_t = (A^t)^* g$ on V and notice that the g_t-norms of the forms ω vanishing on H satisfy for the same m as in $(\star)$,

$$\|\omega\|_{g_t} \leqslant \mathrm{const}\ |\lambda_m|^t \ / \ \prod_{i=1}^{n} |\lambda_i|^t. \qquad (-)$$

If ω is closed then the volume of every homology class $C \in H_{n-1}(V; \mathbb{R})$ is bounded from below by

$$(\mathrm{Vol}_{g_t} C)\, \|\omega\|_{g_t} \geqslant \mathrm{const}' \left| \int_C \omega \right|.$$

Since for each C there is a closed ω vanishing on H_m with $\int_C \omega \neq 0$ by the linear lemma, we obtain once more the bound

$$\mathrm{Vol}_{g_t} C \gtrsim \left(\prod_{i=1}^{n} |\lambda_i| \right)^t \Big/ |\lambda_m|^t.$$

2.3. Isoperimetric inequality. We show in this section that the $(N-1)$-dimensional measure of a closed hypersurface S is a C-C manifold V of Hausdorff dimension N is minorized under certain restrictions on V, by the N-dimensional measure of the domain $D \subset V$ bounded by S as follows

$$\mathrm{mes}_{N-1} S \geqslant \mathrm{const}_V (\mathrm{mes}_N D)^{\frac{N-1}{N}}. \qquad (*)$$

For example, we shall see that $(*)$ is valid for all simply connected nilpotent Lie groups with left-invariant C-C metrics as well as for all compact equiregular manifolds V.

This result for the Heisenberg group is due to Pansu (see $[\mathrm{Pan_{InIs}}]$) and in the general case to Varopoulos (see [Var]) who uses a rather elaborate random walk argument and proves $(*)$ in the (equivalent) form of a Sobolev inequality for smooth functions on V. Our argument presented below is more direct, using only the ball-box theorem and the issuing Vitali covering lemma, and it does not presupposes any a priori regularity of S. The basis for a relation between mes_N and mes_{N-1} is established by the following lemma which is implied by the ball-box theorem.

2.3.A. Flow tube estimate for small C-C balls. *Let X be a smooth H-horizontal vector field on an equiregular C-C manifold (with the underlying polarization $H \subset T(V)$), take a C-C ball $B = B(v, \varepsilon) \subset V$ and consider the $X(t)$-orbit $T = T(v, \varepsilon, \ell)$ of B under the flow $X(t)$ for $t \in [0, \ell]$, i.e. $T = \{X(t)(v), t \in [0, \ell], v \in B\} \subset V$. Then the top-dimensional Hausdorff measure of T satisfies in the limit for the radius of B going to zero the following inequality*

$$\lim_{\varepsilon \to 0} (\mathrm{mes}\, T \,/\, \varepsilon^{-1} \mathrm{mes}\, B) \leqslant \ell \, \mathrm{const}_X (v, \ell),$$

where the function $\mathrm{const}_X(v, \ell)$ *is uniformly bounded on compact subsets in* $V \times \mathbb{R}_+$.

Proof. The ball-box theorem implies that the Hausdorff measure is equivalent to the Riemannian (Lebesgue) measure and that the transversal section to the tube T at v has $(n-1)$-dimensional Riemannian measure $O(\varepsilon^{-1} \operatorname{mes} B)$. ∎

2.3.A′. Corollary: Flow tube estimate for hypersurfaces. *Let* X *be as above and let* S *be a hypersurface contained in a compact subset* $V_0 \subset V$. *Then the* $X(t)$-*orbit* $T(S, \ell)$ *of* S *for* $t \in [0, \ell]$, *satisfies*

$$\operatorname{mes} T(S, \ell) \leqslant \ell \operatorname{const}_{X, V_0}(\ell) \operatorname{mes}_{N-1} S,$$

for $N = \dim_{\mathrm{Hau}} V$, *where* $\mathrm{const}(\ell)$ *is bounded on each segment* $[0, \ell_0] \subset \mathbb{R}$.

 Proof. Use the definition of the Hausdorff measure of S recall the relation $\operatorname{mes} B(\varepsilon) \approx \varepsilon^N$ and apply 2.3.A.

Using 2.3.A′ one may bound from below the measure $\operatorname{mes}_{N-1} S$ of a closed hypersurface S in V in terms of the volume (or, equivalently of mes_N) of the domain $D \subset V_0$ bounded by S as follows. Suppose that every orbit of $X(t)$ starting from a point $v \in V_0$ for $t = 0$ leaves V_0 at some moment $t_0 = t_0(v) \leqslant \ell_0 = \ell_0(V, X)$. Then, D is contained in the flow tube $T(S, \ell_0)$ and so

$$\operatorname{mes}_N D \leqslant \operatorname{const} \ell_0 \operatorname{mes}_{N-1} S$$

for $\mathrm{const} = \mathrm{const}(V_0, X)$.

We want a similar bound in the relative case where S may have a boundary which is contained in the boundary of V_0. Here one should take proper care in choosing X so that sufficiently many orbits of X intersect S which now constitutes only a part of the boundary of D. For example if D is a flow tube for X starting and terminating on ∂V_0, then X is useless for our purpose, but we shall see below how to find finitely many fields X_i, $i = 1, \ldots, k$ such that at least one of them will serve our purpose. We denote by $T(U, \ell, \{X_i\})$ the *iterated orbit* of a given subset $U \subset V$ defined as the union of piecewise smooth curves in V which issue from U and which consist of k segments where the i-th segment equals a piece of the orbit of X_i of length $\leqslant \ell$. It is clear (by Chow connectivity theorem) that there exist smooth H-horizontal fields X_i, $i = 1, \ldots, k$, such that the iterated orbit of every point v in our compact subset V_0 is sufficiently large, i.e. for each positive $\ell \leqslant \ell_0 = \ell_0(V_0) > 0$ the "orbit" $T(v, \ell, \{X_i\})$ contains the C-C ball B in V around v of radius ℓ. (Notice that $T(v, \ell, \{X_i\})$ is contained in the ball $B = B(v, R_+)$ for $R_+ = \text{const}\,\ell$.)

Now take some point $v \in V_0$ and suppose that some R-ball $B = B(v, R)$ is contained in V_0. Assume, moreover, that the "orbit" $T(B, \ell, \{X_i\})$ is also contained in V_0 for $\ell = 2R$ (for which it suffices to assume that the concentric ball of radius $R' = \text{const}\,R$ is contained in V_0). We claim that for every sufficiently small subset $D \subset V_0$ a definite percentage of the measure of $D \cap B$ can be moved away from D by some field X_i in time $\equiv \ell = 2R$. Namely, we have the following trivial

2.3.B. Measure moving lemma. *There exist positive constants ℓ_0 and α (depending on V_0 via X_i), such that for every $R \leqslant \ell_0/2$ and every measurable subset $D \subset V_0$ with $\mathrm{mes}\, D \leqslant \frac{1}{2}\,\mathrm{mes}\, B$ for $B = B(v, R)$, there exist some field among X_i, say X_{i_0} and a measurable subset $D_0 \subset D$, such that*

(1) $\mathrm{mes}\, D_0 \geqslant \alpha\,\mathrm{mes}\, D \cap B$

(2) *every X_{i_0}-orbit of length $\ell = 2R$ issuing from D_0 is contained in V_0 but is not contained in D.*

Proof. If otherwise, the "orbit" $T(D \cap B, \ell, \{X_i\})$ would be "almost contained" in D which is impossible as this orbit contains $B(v, R)$ which has significantly greater volume than D.

2.3.B′. Corollary: Local isoperimetric inequality. *The part S of the boundary of D strictly inside V_0, i.e. $S = \partial D \cap \mathrm{Int}\, V_0$, is bounded from below by*
$$\mathrm{mes}_N(D \cap B) \leqslant \beta\, R\, \mathrm{mes}_{N-1} S, \tag{$\star$}$$
for some constant $\beta = \beta(V_0)$.

Proof. Apply 2.3.A′ to X_{i_0}.

Important remark. Notice that all action takes place in the concentric ball $B' \supset B$ of radius $R' \leqslant \mathrm{const}_{V_0} R$ and so the inequality $(\star)$ may be strengthened to

$$\mathrm{mes}_N(D \cap B) \leqslant \beta \, R \, \mathrm{mes}_{N-1}(S \cap B'). \qquad (\star)'$$

Now we want to prove the global scale invariant inequality $\mathrm{mes}_N D \lesssim (\mathrm{mes}_{N-1} S)^{\frac{N}{N-1}}$ by covering a substantial part of D by balls of various size to which $(\star)'$ applies. This is achieved with the following version of Vitali covering lemma which is quite standard in this framework and follows from the doubling property for the concentric balls (this was clarified to me by N. Varopoulos),

$$\mathrm{mes}\, B(v, 2R) \leqslant \mathrm{const}_{V_0} \mathrm{mes}\, B(v, R)$$

for all $v \in V_0$ and $R \leqslant \mathrm{diam}\, V_0$, where this property is immediate with the ball-box theorem.

2.3.C. Vitali covering lemma. *For each $\lambda > 1$ there exist positive numbers $\mu > 0$ and $\delta > 0$, such that for every measurable subset $D \subset V_0$ of measures μ there exist balls $B_i = B(v_i, R_i) \subset V$, $i = 1, \ldots, m$ around some points $v_i \in V_0$ satisfying the following properties.*

(1) The balls B_i are mutually disjoint; moreover, the concentric balls $B(v_i, \lambda R_i)$ are also mutually disjoint.

(2) The balls B_i contain at least δ-part of the total measure of D, i.e.

$$\sum_{i=1}^{m} \mathrm{mes}(B_i \cap D) \geqslant \delta \, \mathrm{mes}\, D.$$

(3) The intersection $B_i \cap D$ is δ-substantial in each ball, i.e.

$$\mathrm{mes}(B_i \cap D) \geqslant \delta \, \mathrm{mes}\, B_i \, , \quad i = 1, \ldots, m.$$

(4) The intersections of D with the (larger) λR_i-balls $B(v_i, \lambda R_i)$ are somewhat smaller than B_i, i.e.

$$\mathrm{mes}(D \cap B(v_i, \lambda R_i)) \leqslant \frac{1}{2} \, \mathrm{mes}\, B(v, R_i) \quad \text{for} \quad i = 1, \ldots, \tau.$$

Proof. Consider concentric balls $B(v, R_j)$ for $v \in D$ of radii $R_j = 2^{-j} R_0$ for $R_0 = \mathrm{Diam} V_0$ and $j = 1, 2, \ldots$. If v is a density point of D, then $\mathrm{mes}(D \cap B(v, R_j)) \geqslant \frac{1}{2} \mathrm{mes} B(v, R_j)$ for large j. If $\delta > 0$ is small and $\mu < \delta \, \mathrm{mes} B(m, R_1)$, then there exists first j, say j_0, such that $\mathrm{mes}(B(v, R_{j_0}) \cap D) \geqslant \delta \, \mathrm{mes} B(v, R_{j_0})$. Furthermore, by making μ and δ smaller, we arrive at the situation where $\lambda R_{j_0} < R_1$ and the intersection of D with $B(v, \lambda R_{j_0})$ is somewhat smaller than B_j in the sense of (4). Thus for each density point $v \in D$ we constructed a ball $B(v, R = R(v))$ satisfying the above (3) and (4) and now we select the required B_i among them. We start with the ball $B_1 = B(v_1, R(v_1))$ for the point v_1 where the function $R(v)$ assume its maximum on D (notice that $R(v)$ takes finitely many values). Then we take the point $v_2 \in D$ outside $B(v_1, 2\lambda R(v_1))$ where again $R(v)$ is maximal on $D - B(v_1 \ 2\lambda R(v_1))$. Clearly the ball $\lambda B_2 = B(v_2, \lambda R(v_2))$ does not intersect $\lambda B_1 = B(v_1, \lambda R(v_1))$. Then we take the maximal ball outside $2\lambda B_1 \cup 2\lambda B_2$ for B_3 and so on. The resulting balls B_i satisfy (1), (3) and (4). Furthermore, the concentric balls $2\lambda B_i$ cover D, (this is obvious) and so by the doubling property these B_i contain definite part of D, i.e. satisfy (2) with some $\delta' > 0$ which may be somewhat smaller than the one used above.

To conclude the proof we need to ensure that all (or almost all) points $v \in D$ are density points with respect to the C-C distance. In fact, the proof of that follows by the above covering argument presented in a proper light. On the other hand, the subset D we work with is open (or at least contains an open dense set of full measure) and so the density problem becomes irrelevant.

Remark. The above proof does not need equiregularity of V if "mes" is understood in the Riemannian (Lebesgue) sense.

2.3.D. Isoperimetric inequality in compact regions $V_0 \subset V$. *There exist constants $\mu > 0$ and $C \geqslant 0$ (depending on V_0) such that every domain D inside V_0 with $\mathrm{mes}\, D \leqslant \mu$ bounded by a closed hypersurface S satisfies*

$$\mathrm{mes}\, D \leqslant C(\mathrm{mes}_{N-1} S)^{\frac{N}{N-1}}, \qquad (**)$$

where $N = \dim_{\mathrm{Hau}} V$ and $\mathrm{mes} = \mathrm{mes}_N$.

Proof. Take B_i from the covering lemma with a sufficiently large λ and apply $(\star)'$ to each intersection $D \cap B_i$. Then we observe that each (small) ball $B = B(R)$ has $\operatorname{mes} B \gtrsim R^N$ and since the measure of $D \cap B$ is compatible with $\operatorname{mes} B$ the inequality $(\star)'$ gives us the bound $\operatorname{mes}_{N-1}(S \cap B') \gtrsim R^{N-1}$, which implies, in turn, that

$$R \operatorname{mes}_{N-1}(S \cap B') \lesssim (\operatorname{mes}_{N-1}(S \cap B'))^{\frac{N}{N-1}}.$$

Thus we can free $(\star)'$ of R and obtain

$$\operatorname{mes}_N(D \cap B_i) \leqslant C_{V_0} \operatorname{mes}_{N-1}(S \cap \lambda B_i) \qquad (\star)_i$$

for all i. These $(\star)_i$ add up to $(\ast\ast)$ became the balls B_i exhaust an essential part of D (property (2)) while the balls λB_i do not intersect according to (1).

Remarks

(a) If V is a compact manifold to start with we do not need any V_0 but we still have to restrict D in size. For example, if V is a closed connected manifold, then $(\ast\ast)$ holds true with $C = C(V)$ for all $D \subset V$ with $\operatorname{mes} D \leqslant \frac{1}{2} \operatorname{mes} V$. But in general, removing V_0 from the picture is possible only under special favourable circumstances as indicated in 2.3.E below.

(b) The inequality $(\ast\ast)$ and its proof can be transplanted to the asymptotic framework of 2.2 where this can be used for evaluating the Sobolev constant and the first eigenvalue of (V, g_t) for $t \to \infty$.

(c) The inequality $(\ast\ast)$ implies a similar inequality (with the same constant C) for *multiple* domains D over V which are *immersions* of D to V ore more generally (ramified) maps $D \to V$ without foldings. This is proven by applying $(\ast\ast)$ to each subset $D_i \subset V$ consisting of the points covered by D at least i times. Furthermore, this works for continuous families of domains $D_t \subset V$ represented by levels of positive functions $f : V \to \mathbb{R}_+$ (for $D_t = f^{-1}[0, t]$) and gives one a Sobolev inequality for f. (This argument is due to Mazia.)

2.3.D$'$. Excluding V_0 from the game. If V is a nilpotent Lie group with a self-similarity, then every bounded domain can be scaled into a fixed compact subset $V_0 \subset V$ and so the constant C can be assumed independent of V_0. In particular, $(\ast\ast)$ implies the ordinary isoperimetric inequality in $\mathbb{R}^n$ (with $N = n$ and non-sharp constant). In fact, since an arbitrary simply connected nilpotent Lie group is asymptotic to a group V with a self-similarity, the inequality $(**)$ holds true in such V with $C = C(V)$. In general, one looks for inequalities like $(**)$ and $(*)$ with constants controlled by some easily computable invariants of V. In the Riemannian category one likes inequalities with the constants depending on the curvature. Such inequalities are available in two somewhat opposite cases, namely $K \leqslant 0$ and Ricci $\geqslant -\kappa$, see [B-Z] and one wishes to distinguish appropriate C-C manifolds in a similar manner.

Examples of C-C manifolds with "$K < 0$". Let V_0 be nilpotent Lie group with a polarization H_0 a metric g_0 on H_0 and with self-similarities $A_t : V_0 \to V_0$. Take $V = V_0 \times \mathbb{R}$, with the polarization $H \subset T(V)$ obtained as the pull-back of H_0 under the projection $V \to V_0$ and with the metric g on H given by $g = A_t\, g_0 + dt^2$. (This correspond to the horospherical coordinates in the ordinary hyperbolic space $H^n = \mathbb{R}^{n-1} \times \mathbb{R}$ with $g = e^t g_0 + dt^2$ for $g_0 = \Sigma_{i=1}^{n-1} dx_i^2$.) One can show (we leave it to the reader) that this V satisfies $(**)$ with some $C = C(V)$. Another class of negatively curved C-C manifolds appears in [Ge$_{\mathrm{GG3d}}$] who studies horizontal contact structures in S^1-bundles over surfaces of genus $\geqslant 2$ where the condition $K < 0$ refers to *Webster* curvature. Probably, some of Ge's results may be extended to (S^1 and more general) bundles over Riemannian (e.g. Kählerian) manifolds of negative curvature.

Examples of C-C manifolds with Ricci $\geqslant 0$. The round sphere S^{2n-1} with the standard contact structure and metric looks as a good candidate for "$K > 0$" (compare [Hsu$_{\mathrm{SRCT}}$]). The same can be said about S^{4n-1} (with the polarization normal to the Hopf fibers) and S^{15}. Here one expects sharp isoperimetric inequalities in the spirit of Paul Levy (see [Gro$_{\mathrm{PLI}}$]). Further examples of positive (or at least bounded from below) curvature (may) appear in the constructions parallel to those in the Riemannian geometry. These are

(i) *Cartesian products and some non-trivial fiber bundles.*

(ii) *Factors by compact isometry groups* (which may act non-freely if one allows spaces with singularities).

(iii) *Special "cone" and "joint" constructions.*

(iv) *Convex hypersurfaces.* We do not know what corresponds to these in C-C manifolds but the idea of convexity suggests looking at strictly convex (and pseudoconvex) hypersurfaces in $\mathbb{C}^n$ with the natural C-C structures on them.

Examples of $|K| \leqslant$ const. The standard examples are of the form $V = \widetilde{V}/\Gamma$ where $\widetilde{V}$ is a nilpotent Lie group and Γ a discrete free isometry group of $\widetilde{V}$. One probably can show with our argument that every $D \subset V$ satisfies

$$\mathrm{mes}_N D \leqslant C(\mathrm{Diam} D)\, \mathrm{mes}_{N-1} \partial D$$

with C depending on (the C-C metric on) $\widetilde{V}$ but not on Γ.

The above $\widetilde{V}/\Gamma$ correspond to flat manifolds in the Riemannian geometry. The "small variable curvature" can be expressed in term of systems of (locally defined) vector fields on V compatible with H and g (in an appropriate sense) and having "multiplication table" for the Lie bracket with small coefficients. (The difficulty which emerges here is the same as in the pinching discussion in §1, that is the absence of canonical coordinates in V and/or of a parallel transport.)

2.3.D″. Isoperimetric inequality in non-equiregular spaces. The key role of equiregularity appears in the equivalence of the N-dimensional Hausdorff measure with the Riemannian (Lebesgue) measure on V. In the general case, the Hausdorff measure can be supported on a proper subset $A \subset V$, such as a Cantor sets in the example in, and then, of course, the inequality $(**)$ fails to be true. On the other hand if V admits an equisingular stratification, one, probably, has a meaningful isoperimetric inequality on each stratum V_ν of dimension $\geqslant 2$. Observe that the metric on such a stratum is not, in general, Carnot-Carathéodory and the corresponding inequality for N_ν-dimensional measure for $D_\nu \subset V_\nu$ may involve the N'_ν-dimensional measure of ∂D_ν for $N'_\nu < N_\nu - 1$. For example, V_ν may have such metric as $(\mathrm{dist}_{\mathrm{Riem}})^{\frac{1}{d}}$ and then the isoperimetric inequality bounds $\mathrm{mes}_{N_\nu} D$ by $\mathrm{mes}_{N_\nu - d} \partial D$. It seems, inequality of this kind are satisfied by all equiregular submanifold $V' \subset V$ (not only on $V' = V_\nu$).

On the other hand, there is a meaningful version of the isoperimetric inequality $(**)$ which needs no equiregularity assumption. To formulate this we denote by mes the *Riemannian* measure in V and then define the *Minkowski volume* $\mathrm{mes}' S$ by

$$\mathrm{mes}' S = \liminf_{\varepsilon \to 0} f\varepsilon^{-1}\, \mathrm{mes}\, U_\varepsilon(S)$$

where $U_\varepsilon(S)$ denotes the Carnot-Carathéodory ε-neighbourhoods of S.

Theorem. *If D, C and S satisfy the assumptions of 2.3.D then*

$$\operatorname{mes} D \leqslant C(\operatorname{mes}' S)^{\frac{N}{N-1}} \qquad (**)'$$

where N is defined in terms of the (variable) $n_i(v) = \operatorname{rank} H_i(v)$ by

$$N = \max_{v \in V} \sum_{i=1}^{d} i(n_i(v) - n_{i-1}(v))$$

(and H is not assumed equiregular anymore).

The proof is essentially identical to that of $(**)$.

Notice that this theorem follows from a result by Varopoulos in [Var] where the isoperimetric inequality is directly linked to the volume behaviour of (small) balls in V. Also see [F-G-W]$_{1,2}$, [F-L-W], [Haj$_{\text{SSAM}}$], [Ha-Ko] and [Co-SC].

2.3.E. Green forms, pencils of curves and an integral geometric proof of the isoperimetric inequality. We start with a nilpotent group V with a self-homotopy $A : V \to V$ and we call a closed $(n-1)$-form ω_0 on $V - \{0\}$, where 0 stands for the identity element, a *Green form* if it is

 (i) H-horizontal.

(ii) A-invariant.

(iii) closed and non-exact.

       Notice that the A-invariance implies that

(iv) $\|\omega_0\|_v \leqslant \operatorname{const} \operatorname{dist}^{-(N-1)}(0, v)$, $v \in V$.

Also observe that the non-exactness of ω_0 makes

$$\int_S \|\omega_0\|_s \, ds \geqslant \left| \int_S \omega_0 \right| = c_0 \neq 0$$

for a fixed c_0 and all smooth closed hypersurfaces S around the origin, where ds refers to the $(N-1)$-dimensional Hausdorff measure on S.

Example. If $V = \mathbb{R}^n$ such an ω_0 may be obtained as the radial pull-back of the volume form on $S^{n-1} \subset \mathbb{R}^n$.

Lemma. *Every V admits a Green form.*

Proof. Divide $V - \{0\}$ by the (infinite cyclic) group $\{A^i\}$ generated by A, take some H-horizontal closed non-exact $(n-1)$-form $\bar{\omega}$ on the quotient space $(V - 0)/\{A^i\}$ (which exists according to Linear Lemma in 2.2.A) and pull $\bar{\omega}$ back to V for the quotient map

$$V - \{0\} \to V - \{0\}/\{A^i\}.$$

Remarks

(a) A slight readjustment of the proof of 2.2.A yields a Green form invariant under the one parameter group $\{A^t\}$, $t \in \mathbb{R}$.

(b) A standard ω_0 comes as the dual of the horizontal gradient of the fundamental solution of the Hörmander-Laplace operator on V.

Santalo-type proof of the isoperimetric inequality for a compact domains $D \subset V$ with a smooth boundary S. If D contains 0 we have the following lower bound on the integral of $\mathrm{dist}^{-(N-1)}$ with respect to the Hausdorff measure on S,

$$\int_S \mathrm{dist}^{-(N-1)}(0, s)ds \geqslant \varepsilon > 0, \tag{+}$$

since $\int_S \omega_0 = c_0 \neq 0$. On the other hand

$$\int_D \mathrm{dist}^{-(N-1)}(0, v)dv \leqslant \mathrm{const}_0 (\mathrm{mes}\, D)^{\frac{1}{N}}, \tag{++}$$

as the left hand side integral is obviously bounded by $r^{N-1} \mathrm{mes}\, B$ for the ball $B = B_0(r) \subset V$, where $r = r(D)$ is chosen so that $\mathrm{mes}\, B = \mathrm{mes}\, D$. We apply (+) to $\mathrm{dist}(v, s)$, for all $v \in D$ and integrate over D. Thus we get

$$I = \int_D dv \int_S \mathrm{dist}^{-(N-1)}(v, s)ds \geqslant \varepsilon\, \mathrm{mes}\, D.$$

Then we change the order of integration and obtain with (++),

$$I = \int_S ds \int_D \mathrm{dist}^{-(N-1)}(s, v)dv \leqslant \mathrm{const}_0\, \mathrm{mes}_{N-1} S (\mathrm{mes}\, D)^{\frac{1}{N}}.$$

Thus

$$(\mathrm{mes}\, D)^{\frac{N-1}{N}} \leqslant \varepsilon^{-1}\, \mathrm{const}_0\, \mathrm{mes}_{N-1} S.$$

Remarks. The original argument by Santalo proceeds slightly differently and leads to the *sharp* isoperimetric inequality on $\mathbb{R}^2$ as well as on S^2 and H^2. Also this argument provides the basic (non-sharp) isoperimetric inequality on all complete simply connected Riemannian manifolds with $K \leqslant 0$ which suggests a similar extension for C-C manifolds.

Green forms for equiregular V. Now a Green form ω_v is defined in a punctured neighbourhood of some point $v \in V$ and it must satisfy the above (i), (iii) and (iv). Such an ω_v is constructed by using the approximation of V at v by a nilpotent group $\overline{V}$ (called N_v in 1.4). Namely, according to 1.4.A$''$ our polarization on each sufficiently small annulus $\mathrm{An}_\rho = B_v(3\rho) - B(\rho)$ can be smoothly approximated by that on the corresponding annulus in a certain nilpotent Lie group $\overline{V} = \overline{V}(V, v)$, such that when An_ρ and $\overline{\mathrm{An}}_\rho$ are scaled to the unit size and brought together, the polarization H becomes ε-close to $\overline{H}$ in C^n-topology (as we assume H as smooth as we need) with $\varepsilon = \varepsilon(\rho) \to 0$ for $\rho \to 0$. Then we cover $B_v(\rho) - \{v\}$ by slightly overlapping annuli, say

$$\mathrm{An}_i = B(\lambda 2^i) - B(2^{i-1}/\lambda) \quad \text{for } \lambda = \sqrt{\frac{3}{2}}$$

and large $i = i_0,\, i_0 + 1, \ldots$, and we construct a closed horizontal form ω_i on each An_i representing the generator of $H^{n-1}(A_i; \mathbb{Z}) = \mathbb{Z}$ (where we ignore possible (?) minor nuisance at $\partial \mathrm{An}_i$) such that on the rescaled unit annulus this form become ε_i-close to a fixed form $\overline{\omega}_1$ on $\overline{\mathrm{An}}_1$. This is done by slightly adjusting the proof of Linear Lemma which also allows us to match ω_i with ω_{i-1} and ω_{i+1} on the overlaps of the annuli. Everything is quite simple and left to the reader. We also trust the reader to reprove, using ω_v, the inequality $(**)'$.

Remark. The new proof can be made "softer" and thus brought closer to our first proof by observing that the form ω_v does not need to be closed but rather having the differential $d\omega_v$ growing no faster near v than $(\mathrm{dist}(v, \cdot))^{-(N-1)}$. This makes $\int_D d\omega_v$ small for small D and does not affect our estimates. Then a suitable ω_v can be constructed by just prescribing several first terms of its (Laurent type) asymptotic expansion at v. (Our $\omega_v = \omega_v(v')$ can be found in the form $\omega_v = d^{-(N-1)}(v, v')\, \omega_v'(v')$, where d is a properly smoothed distance function and $\omega_v'(v')$ is a smooth form for all v' near v. Here the relevant properties of ω_v are ensured by first Taylor terms of $\omega_v'(v')$ at v.)

Pencils of curves emanating from v. We saw in 2.2.A how a closed horizontal $(n-1)$-form comes along with (a measure on) a family of closed horizontal curves in V. In our present situation we need families (pencils) of non-closed horizontal curves issuing from a fixed point v. (For example, the standard form ω_0 on $\mathbb{R}^n$ corresponds to the family of the straight rays coming from the origin with the obvious spherical measure on these rays.) Every such family provides certain information on hypersurfaces S around v since the measure on S is recorded by the curves in the pencil as they intersect S leaving the domain $D \ni v$ bounded by S. The best pencils are those where the corresponding ω_v is a true form (not just a current) which grows no faster than $(\mathrm{dist})^{-(N-1)}$ at v. Such pencils can be constructed as follows. Take sufficiently many horizontal fields $X_0, \ldots, X_k$ on V in general position and let

$$X_\mu = \sum_{i=1}^{k} \mu_0 \, X_i$$

for $\mu = (\mu_0, \ldots, \mu_k)$ being the points in the standard k-simplex Δ. Then for each $v \in V$ we have the family of integral curves $c_{v,\mu}$, $\mu \in \Delta$, issuing from v with the natural measure $d\mu$ on this family. We suggest the reader would follow the integral geometric proof with ω_v associated to this $c_{v,\mu}$ and observe how close this comes to our first proof. In fact much of the standard analysis on (V, H) can be performed with suitable measures on the space of horizontal curves and we shall encounter later on similar measures on higher dimensional horizontal submanifolds.

2.4. Singular integrals and Sobolev inequalities.

As we mentioned earlier the isoperimetric inequality yields, by Mazia argument, the following Sobolev inequality for the L_p-norm, $p = \frac{N}{N-1}$, of a function f on V in terms of the L_1-norm of the *differential of f restricted to H*,

$$\int_V |f(v)|^{N/N-1} dv \leqslant \mathrm{const} \left(\int_V \|df(v) \mid H\| \, dv \right)^{N/N-1}. \qquad (*)$$

This implies the bound

$$\|f\|_{L_p} \leqslant \mathrm{const}_q \|df \mid H\|_{L_q} \qquad (*)_q$$

for all q in the interval $1 \leqslant q < N$ and $\frac{1}{p} = \frac{1}{q} - \frac{1}{N}$ as follows. Apply $(*)$ to $|f|^a$ for $a = \frac{p(N-1)}{N}$ and get

$$\|f\|_{L_p}^a \lesssim \| |f|^{a-1} df \mid H\|_{L_1}$$

where "$\underset{\sim}{\lesssim}$" means "$\leqslant \text{const}_q$". Then use the Hölder inequality

$$\||f|^{a-1}df \mid H\|_{L_1} \lesssim \||f|^{a-1}\|_{L_b} \cdot \|df \mid H\|_{L_q}$$

for $b = \left(1 - \frac{1}{q}\right)^{-1}$, and observe that $\||f|^{a-1}\|_{L_b} = \|f\|_{L_p}^c$ for $c = pb^{-1}$. Thus $\|f\|_{L_p}^{a-c} \lesssim \|df \mid H\|_{L_q}$, which yields $(*)_q$ since $a - c = 1$ with our choice of a, b and c.

The inequality $(*)$ for $q > 1$ can be also derived from the following estimates for convolution integrals. Let V be a nilpotent group and $K(v)$ be a function (convolution kernel), such that

$$|K(v)| \leqslant (\text{dist}(0,v))^{-(N-1)} , \ v \in V.$$

Then

$$\|K * f\|_{L_p} \leqslant \text{const}_q \|f\|_{L_q}, \qquad\qquad (**)_q$$

for all q in the interval $1 < q < N$ and $\frac{1}{p} = \frac{1}{q} - \frac{1}{N}$.

This is classical for $V = \mathbb{R}^N$ and the Euclidean proof (see, e.g. Ch.V in [Ste]) extends to the general V (see [Fol]).

Finally, we recall the Green form $\omega_0(v)$ of 2.3.E and take the associated divergence free vector fields $X(v)$ which clearly has

$$\|X(v)\| \leqslant \text{const}(\text{dist}(0,v))^{-(N-1)}.$$

We observe that every function f on V decaying at ∞ can be reconstructed from $df \mid H$ by "convolution" with $X(v)$, as $f(0) = \int_V df(X(v))dv$, and so

$$(**)_q \Rightarrow (*)_q \quad \text{for } q > 1.$$

2.4.A. Dilation and homotopy. The above inequality $(*)_q$ fails to be true for $q = N$ and $p = \infty$ (where $\|f\|_{L_\infty}$ refers to $\sup_{v \in V} |f(v)|$) but the proof of $(*)_q$ implies a bound on $\|f\|_{L_\infty}$ by $\|df \mid H\|_{L_q}$ for all $q > N$. In fact, one may reconstruct $f(v_1) - f(v_2)$ from the differential df on H for given v_1 and v_2 using a divergence free vector field X_{v_1,v_2} on $V - \{v_1, v_2\}$ which vanishes away from a ball containing v_1 and v_2 and has non-zero fluxes through small spheres around v_1 and v_2. One sees instantaneously with such a field (using the Hölder inequality) that the norm $\|df \mid H\|_{L_q}$ for $q = N + \alpha$ bounds the Hölder constant $L_\beta(f)$ for the exponent $\beta = \alpha/N + \alpha$ for all $\alpha > 0$ (compare 3.6.B$_1$) which implies a similar bound on the Hölder constant of a map f of V into a Riemannian manifold by the L_q-norm of the differential $\mathcal{D}f$ restricted to H.

Corollary. *Let V be a compact equiregular C-C manifold and W be a compact Riemannian one. Then, for every $\alpha > 0$ and $C > 0$, the space of maps $f : V \to W$ with $\|\mathcal{D}f \mid H\|_{L_{N+\alpha}} \leqslant C$ is compact in the uniform topology and, in particular, contains at most finitely many homotopy classes of maps. Furthermore, if C is sufficiently small, then f is null-homotopic.* (Here, as earlier, $N = \dim_{\mathrm{Hau}} V$.)

Remarks

(a) The fields X_{v_1,v_2} in the above argument can be replaced by a pencil of curves c_μ between v_1 and v_2 (e.g. the orbits of X_{v_1,v_2}). In fact, the difference $f(v_1) - f(v_2)$ is given by the integral of df on c_μ over c_μ for each μ. On the other hand the μ-average of the latter integrals can be bounded in term of $\|df \mid H\|_{L_q}$ for sufficiently "rich" (or "thick") pencils (compare 3.6.B$'$, 3.6.D and 5.4.B).

(b) The above corollary (and its proof) remains valid for non-equiregular V with $N = \max\limits_{v \in V} \Sigma_{i=1}^{a}\, i(n_i(v) - n_{i-1}(v))$ as in 2.3.D$''$.

2.4.B. High levels of functions f with $\|df \mid H\|_{L_N} < \infty$ on (codim 1)-stable C-C manifolds of (formal) Hausdorff dimension N. The above compactness property fails to be true for $\alpha = 0$ as a suitable positive function $\varphi = \varphi(\rho)$ defined for $\rho > 0$ and satisfying $\lim_{\rho \to 0} \varphi(\rho) = \infty$ may have $\int_V \|d(\varphi(\mathrm{dist}(v_0, v)))|H\|^N dv < \infty$. Yet one can sometimes understand pretty well the ways a functions f with $\|df \mid H\|_{L_q} < \infty$, where $q \leqslant N$, goes to infinity. For example, if V is compact Riemannian (i.e. $H = T(V)$) and $q > N - 1$, then f is "zero-dimensional at infinity" in the following sense. *Every connected component of the ε^{-1}-level $\{v \in V \mid |f(v)| > \varepsilon^{-1}\}$ has diameter $\leqslant \delta$ where $\delta \to 0$ for $\varepsilon \to 0$* (and, in fact, one may estimate δ in terms of ε and $q - N + 1$). More generally if $q > N - i$, $i = 1, \ldots N - 1$, then f is "at most i-dimensional at infinity". Namely, *the i-width of the ε^{-1}-level goes to zero with ε*, i.e. this level can be retracted to an $(i - 1)$-dimensional subset by a δ-move in V.

This is well known and, I guess, is due to Karen Uhlenbeck. The proof follows by observing that the restriction of f to the $(N - i)$-skeleton of a generic δ-triangulation of V is bounded in a controlled way which allows a δ-push of our ε^{-1}-level to the complementary $(i - 1)$-skeleton (compare 3.4).

(codim 1)-*Stability*. In order to apply the above argument to a C-C manifold (V, H) and $i = 1$ one needs sufficiently many hypersurfaces $V' \subset V$, such that the Hölder estimates from 2.4.F' apply to $f \mid V'$ with $N-1$ in place of N. For this it would suffice to have $(V', H' = H \cap T(V'))$ equiregular of Hausdorff dimension $N - 1$ which is essentially the same as Lipschitz equivalence of the metrics $\mathrm{dist}_{H'}$ and dist_H on V'.

Definition. We say that V is (codim 1)-*stable* if for each $v \in V$ and every hyperplane $T' \subset T_v(V)$ there exists (a germ of) a smooth hypersurface $V' \subset V$ passing through v with $T_v(V') = H'$ such that the *formal Hausdorff dimension* $N' = \max_{v' \in V'} \Sigma_{i=1}^d i(n_i'(v) - n_{i-1}'(v))$, for n_i' being the ranks of the spaces $H_i' \subset T'(V')$ spanned by k-th order commutators of H'-horizontal fields on V', satisfies $N' \leqslant N-1$ for the formal Hausdorff dimension N of (V, H).

Examples. The Riemannian manifolds (where $H = T(V)$) are, obviously, (codim 1)-stable. Contact manifolds of dimension $n = 2m + 1$ are stable for $n \geqslant 5$ and are (codim 1)-unstable for $n = 3$. In fact, generic H of rank $n - 1$ on n-dimensional manifolds V are (codim 1)-stable starting from $n = 4$. On the other hand, H's of rank 2 are unstable. In general, a generic H with rank $H \gg$ corank H is (codim 1)-stable but if corank $\gg$ rank then H is typically unstable.

Now we see with this definition that *smooth functions f on* (codim 1)-*stable manifolds with* $\|df \mid H\|_q < \infty$ *for* $q > N - 1$ *have high levels of "zero dimension at infinity"* as in the Riemannian case.

Conformal interpretation. The connectedness results for ε^{-1}-levels can be equivalently expressed in the conformal language. Namely, let dist_φ be conformal to the original C-C metric dist on (V, H) with the conformal factor $\varphi = \varphi(v)$ (corresponding to $\|df(v) \mid H\|$) which is a positive Borel function on V with $\|\varphi\|_{L_q} < \infty$. If $q > N$ the metric dist_φ is equivalent to dist by the above discussion but for $q \leqslant N$ dist_φ may become infinite on certain pairs of points. However there is a well defined maximal subset $V_\varphi \subset V$ where dist_φ is finite and one can reformulate the above properties of $f(v)$ (corresponding to $\mathrm{dist}_\varphi(v, v_0)$) as certain connectedness at infinity of the metric space $(V_\varphi, \mathrm{dist}_\varphi)$. Here the case $q = N$ appears especially attractive as the condition $\|\varphi\|_{L_N} < \infty$ is equivalent to $\mathrm{mes}_N(V_\varphi, \mathrm{dist}_\varphi) < \infty$.

2.5. Homotopy bounds by $\|\mathcal{D}f \mid H\|_{L_N}$, taut maps and the bubbling phenomenon. Despite the failure of the compactness, the homotopy finiteness conclusion may hold true for certain $q \leqslant N$ as the homotopy invariants of a map $f : V \to W$ may be estimated, under favorable topological assumptions on V and W, in terms of the L_q-norm of the differential $\mathcal{D}f$ on H. For example, if W is a compact aspherical manifold, then the homotopy class of f is determined by the restriction of f to the 1-skeleton S_1 of a triangulation of V. If we choose this S_1 horizontal and sufficiently generic, then the length of $f(S_1)$ will be (obviously) bounded in terms of $\|\mathcal{D}f \mid H\|_{L_1}$ and so we conclude that there are *at most finitely many homotopy classes of maps $f : V \to W$ with $\|\mathcal{D}f \mid H\|_{L_1} \leqslant$ const. Furthermore, if const $\leqslant c_0$ for some sufficiently small $c_0 > 0$, then f is null-homotopic.* (All this is standard for V Riemannian.) Similarly, one may treat the case where $\pi_i(W) = 0$ for $i > i_0$, and V contains sufficiently many i_0-dimensional submanifolds $V' \subset V$ having $\dim_{\mathrm{Hau}} V' < q$ with respect to the polarization $H' = H \cap T(V)$.

A somewhat different argument shows, in the case V is Riemannian, that the number of the homotopy classes of maps $f : V \to W$ for an arbitrary W is bounded by $\|\mathcal{D}f\|_{L_N}$ for $N = \dim V$. For example, $\|\mathcal{D}f\|_{L_N}$ obviously bounds the topological degree of f in the case $\dim W = \dim V$ and there are similar bounds for the other (rational) homotopy invariants of f (compare 1.4.E$'$). However, such a "bound on degree" consideration does not immediately yield null homotopy of an f with small $\|\mathcal{D}f\|_{L_N}$ (e.g. for maps $f : S^4 \to S^3$ with small $\|\mathcal{D}f\|_{L_4}$); nor it extends to general C-C manifolds. But there is an alternative approach borrowed from the (quasi)conformal geometry and based on a suitable form of the Schwarz lemma (compare 2.6). The Schwarz lemma does not directly apply to general (non-quasi-conformal) maps $f : V \to W$ but the following modification (probably, first pointed out by Uhlenbeck) works just as well. First we observe that in homotopy problems the map f can be assumed *taut* in the following sense.

Definition. A map f of a smooth manifold into a Riemannian one is called *taut* if no homotopy of f can make the induced (possibly singular) Riemannian metric strictly smaller.

Example. A taut map into $\mathbb{R}^n$ is necessarily (and obviously) constant. Similarly, every taut map which lands inside a small (and hence convex) ball in a Riemannian manifold is also constant.

Now, if V and W are compact, then every map f is obviously homotopic to a taut one, say f_{tau}, such that $\|\mathcal{D}f_{\text{tau}}(v)\| \leqslant \|\mathcal{D}f(v)\|$ everywhere on V and, in particular, $\|\mathcal{D}f_{\text{tau}}\|_{L_N} \leqslant \|\mathcal{D}f\|_{L_N}$.

Proposition. *Let V be compact and W has locally bounded geometry. Then there exists $c_0 > 0$, such that every taut map $f : V \to W$ with $\|\mathcal{D}f\|_{L_N} \leqslant c_0$ is constant.*

Proof. Normalize W so that Conv Rad $W = 1$. Then if some small ball $B_\varepsilon \subset V$ lands in a unit ball in V then, the tautness of f obviously yields the following

Key relation. $\mathrm{Diam} f(B_\varepsilon) = \mathrm{Diam} f(S_\varepsilon)$ for the boundary sphere S_ε of B_ε.

The second fact we are going to use is that for every small sphere $S_\varepsilon \subset V$ there is a concentric sphere S_δ for $\varepsilon \leqslant \delta \leqslant 2\varepsilon$ such that $\mathrm{Diam} f(S_\delta) \leqslant d_0 = d(c_0)$ where $d_0 \to 0$ for $c_0 \to 0$. (This follows from the bound on $f \mid S^{N-1}$ by $\|\mathcal{D}f \mid S^{N-1}\|_{L_N}$, as we have already seen.) If we knew that $\mathrm{Diam} f(B_\delta) \leqslant 1$, we could conclude that, in fact, $\mathrm{Diam} f(B_\delta) \leqslant d_0$ which for small c (and hence d_0) would make $\mathrm{Diam} f(V) = 1$ and yield the desired constancy of f. Thus we are lead to look at the ball $B_\varepsilon \subset V$ of the minimal radius ε which has $\mathrm{Diam} f(B_\varepsilon) = 1/2$. Then the concentric ball B_δ for every $\delta \leqslant 2\varepsilon$ has $\mathrm{Diam} f(B_\delta) \leqslant 1$ as for each $v \in B_\delta$ there is a ball B'_ε containing v and intersecting B_ε (which, as we know, has $\mathrm{Diam} f(B'_\varepsilon) \leqslant 1/2$). But then for a suitable $\delta \geqslant \varepsilon$ we have $\mathrm{Diam} f(B_\delta) \leqslant \delta$ and so $\mathrm{Diam} f(B_\varepsilon) \leqslant d_0$. If $d_0 < \frac{1}{2}$ this leads to a contradiction and thus proves the desired constancy of f.   ■

Relative case. Let us try to generalize the above to the case where f is *taut relative* to a certain subset $V_0 \subset V$, i.e. the homotopy involved in the definition of tautness is required to be constant on V_0. If V_0 consists of a single point $v_0 \in V$ then such an f with small $\|\mathcal{D}f\|_{L_N}$ is still constant and maps all of V to $f(v_0) \in W$, but for several points the situation is different as for every finite subset V_0 in V there obviously exists a map f with prescribed values on V and having arbitrarily small energy

$\|\mathcal{D}f\|_{L_N}$. (A similar property is enjoined by all closed subsets V_0 in V of conformal capacity zero.) However, if such an f is taut relative to V_0 it must have "one-dimensional shape". Namely, a *small perturbation of f factors through a map of V into a tree with j extremities for $j = \operatorname{card} V_0$, and the implied map of this tree to W sends the i-th extremity to $f(v_i)$ for $V_0 = \{v_1, \dots, v_i, \dots, v_j\}$.*

Proof. One sees as earlier that there are small disjoint balls $B_{\rho_i}(v_i) \subset V$, $i = 1, \dots, j$, such that the map is nearly constant on the complements of these balls, i.e. has small $\operatorname{Diam} f(V - \bigcup_i B_{\rho_i}(v_i))$, and also f is nearly constant on the concentric spheres $S_\varepsilon(v_i) = \partial B_\varepsilon(v_i)$ for $i = 1, \dots, j$ and $\varepsilon \leqslant \rho_i$.

Remark. The "one-dimensionality" conclusion remains valid for V_0 consisting of the union of small disjoint spheres $S_{\rho_i}(v_i) = \partial B_{\rho_i}(v_i)$, provided the map f is already known to be nearly constant on these spheres.

The case where c is large. Now we do not expect that f with $\|\mathcal{D}f\|_{L_N} \leqslant c$ is constant but we still can control it on a domain $U \subset V$ if the integral of $\|\mathcal{D}f\|^N$ over (some ε-neighbourhood of) U is sufficiently small. Then we may try to cover V by a finite (depending over c) number of such domains and thus show f is piecewise controlled in a suitable sense.

Here is the relevant (trivial)

Covering lemma. *Let V be a metric space and μ a Borel measure of total mass $c < \infty$ on V without atoms. Then for arbitrary positive constants c_0, ρ and λ there exist a positive integer $k \leqslant k_0(c, c_0, \rho, \lambda)$ and $\varepsilon \geqslant \varepsilon_0(c, c_0, \rho, \lambda) > 0$, such that V can be partitioned into k pieces, $V = \cup_{i=1}^k V_i$ with the following three properties.*

(1) *The (major) piece V_1 is obtained from V by removing at most k disjoint balls of radii $\leqslant \rho$. Each of the remaining pieces V_i, $i = 2, \dots, k$, is either a ball or a ball B minus several balls which are strictly contained in B. The total number of balls involved is bounded by k and their radii are bounded by ρ. (As we insist that the number of pieces V_i is exactly k some of them may be taken empty.)*

(2) *For a sphere $S \subset V$ let $[1, \lambda]S \subset V$ denote the annulus consisting of concentric spheres of radii $\mu \operatorname{rad} S$ for $\mu \in [1, \lambda]$. Then every two among the above balls, say B_1 and $B_2 \neq B_1$, have the boundary spheres S_1 and S_2 λ-disjoint in the sense that $[1, \lambda]S_1 \cap [1, \lambda]S_2 = \varnothing$.*

(3) *For every point $v \in V_1$ the ball of radius ε in V_1 has mass $\leqslant c_0$, i.e. $\mu(B_\varepsilon(v) \cap V_1) \leqslant c_0$ and for each $i \geqslant 2$, every ball $B_\delta(v)$, $v \in V_i$, of radius $\delta = \varepsilon \operatorname{diam} V_i$ has $\mu(B_\delta(v) \cap V_i) \leqslant c_0$.*

Now we apply this lemma to the measure $\|\mathcal{D}f\|^N \, dv$ on V. By moving the boundary spheres within their respective λ-annuli, we can make the integrals of $\|\mathcal{D}f\|^N$ over these spheres small which makes small the diameters of the f-images of these spheres. Then, on each connected component $V_i \subset V$ of the complement to these spheres, the map $f_i = f \mid V_i \to W$ falls into one of the two categories.

(1) *Prebubbles.* f_i is (c_0, ε)-controlled on V_i endowed with the metric $\operatorname{dist}_i = \operatorname{dist}/\operatorname{Diam}V_i$ where dist denotes the original metric in V and where the control means that the f_i-image of the ε-balls from V_i have diameters $\leqslant c_0$ in W, where c_0 is given beforehand and $\varepsilon = \varepsilon(c_0, \|\mathcal{D}f\|_{L_N}) \to 0$ for $c_0 \to 0$.

(2) *Bridges.* f_i has small energy, i.e. $\int_{V_i} \|\mathcal{D}f\|^N \leqslant c_0$.

(Notice that (1) and (2) are not mutually exclusive.) We apply the previous remark to the bridges and see what can be called a *prebubbling decomposition* of f (see Fig. 4) as the actual (Uhlenbeck's) bubbling occurs in the limit for sequences of maps $f_\nu : V \to W$ with $\|\mathcal{D}f_\nu\|_{L_N} \leqslant c$ where $\operatorname{diam} f_\nu(S_{\nu,i}) \to 0$ for the boundary spheres $S_{\nu,i}$ of the $V_{\nu,i}$'s and the bubbles appear as the limits of the maps $f_{\nu,i} : (V_{\nu,i} \operatorname{dist}/\operatorname{Diam}V_{\nu,i}) \to W$ for the prebubbles $f_{\nu,i}$ and $\nu \to \infty$. (Sometimes one excludes from the ranks of bubbles the mainland piece $V_{\nu,1}$ where $\operatorname{Diam}V_{\nu,1}$ does not go to zero for $\nu \to \infty$.)

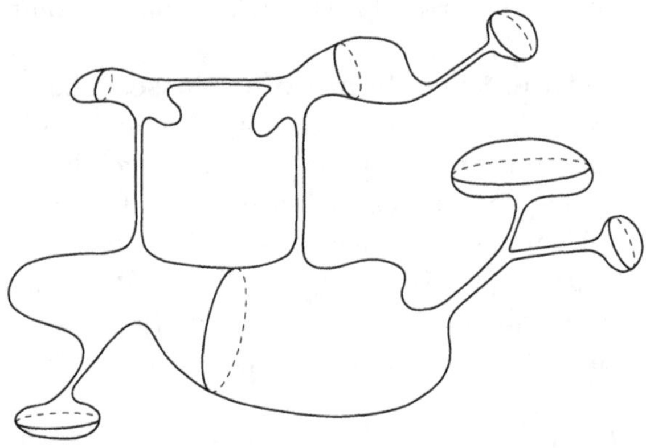

Figure 4

As a consequence of the prebubbling decomposition we see again that the number of the homotopy classes of maps $f : V \to W$ is bounded in terms of $\|Df\|_{L_N}$ provided V and W are compact and W is *simply connected* or rather $\pi_1(W)$ *trivially* acts on $\pi_n(W)$ for $n = N = \dim V$. Furthermore, *all this extends to* (codim 1)*-stable C-C manifolds such as contact manifolds V of dimension ≥ 5, for example.* So we conclude for such V's, that *the maps $f : V \to W$ with small norms $\|\mathcal{D}f \mid H\|_{L_N}$ are null-homotopic and if $\pi_1(W)$ acts trivially on $\pi_n(W)$, $n = \dim V$, then there are at most finitely many homotopy classes of maps with $\|\mathcal{D}f \mid H\|_{L_N} \leq c$ for every $c \geq 0$* (where the implied number of the classes of maps depends on c).

Question. If V is Riemannian then all rational homotopy invariants of maps $f : V \to W$ (e.g. $\deg f$ for $N = n = \dim W = \dim V$) admit polynomial bounds by $\|\mathcal{D}f\|_{L_N}$ (see 3.6 and compare 1.4.E$'$). We want to know if this is also true for the general C-C case. For example, what is the actual bound on $\deg f$, for $\dim W = \dim V < N = \dim_{\mathrm{Hau}} V$ in terms of $\|\mathcal{D}f \mid H\|_{L_N}$? Is it $\approx \|Df \mid H\|_{L_N}^N$?

A natural approach to this problem consists in seeking a "nice" covering of V by standard domains V_i, $i = 1, \ldots, k$, (e.g. balls $B(r_i)$ of some radii r_i) such that the number k is bounded by something like const $\int_V \|\mathcal{D}f(v) \mid H\|^N \, dv$ and the f-images of all V_i have $\mathrm{Diam}(f(V_i)) \leq \varepsilon$ for a fixed small positive ε. The latter inequality would follow if somewhat enlarged V_i, say $2V_i$, (e.g. doubled concentric balls $B(2r_i)$) satisfied $\int_{2V_i} \|\mathcal{D}f \mid H\|^N \leq c_0$ for a fixed small $c_0 > 0$. The difficulty here stems from the possibility of fast variation of the integrant $\|\mathcal{D}f(v) \mid H\|^N$ on V (compare [Kor]).

2.5.A. Weak stability of homotopy classes. Let a sequence f_i of smooth maps converges to a smooth map f in some weak topology, e.g. almost everywhere and we want to have all f_i for large i homotopic to f. Then the above shows this is indeed so if the norm $\|\mathcal{D}f(v) \mid H\|$ is bounded on V in the following sense. There is $\rho > 0$, such that the integral of $\|\mathcal{D}f(v) \mid H\|^N$ over every ρ-ball B in V is universally bounded by a sufficiently small positive constant, i.e. $\int_B \|\mathcal{D}f(v) \mid H\|^N \, dv \leq c_0$, (where we assume as earlier that V is (codim 1)-stable.

Corollary. *The homotopy class of f is stable under small perturbation in the F_N^H-topology (defined in 2.5.E).*

2.5.B. On the topology of the space F_c of maps f with $\|\mathcal{D}f \| H\|_{L_N} \leqslant c$. The above discussion only touches the zero-homotopies (i.e. connected components) of F_c or rather the image of these in the space of all maps, $F = \cup_{c<\infty} F_c$. Yet the full homotopy structure of F_c and its image in F is an equally interesting matter and our argument provides some information on the higher dimensional homotopy of F_c. Namely, if V is (codim 1)-stable we have the following

Homotopy finiteness proposition. *The homotopy image of F_c in F is bounded in terms of c provided W is compact simply connected.* [6]

2.5.C. Surface maps of small area and related questions. The above is well known in the Riemannian case. For example, the space F_{c_0} of smooth maps f of a compact surface Σ into a complete Riemannian manifold W with bounded geometry, defined by $F_{c_0} = \{f \in F \mid \|\mathcal{D}f\|_{L_2} \leqslant c_0\} \subset F$, contracts to (the space of) constant maps provided $c_0 > 0$ is small enough.

Let us take the sphere S^2 for the above Σ and observe that a generic smooth map $f : V \to W$, for $\dim W \geqslant 2$, can be made *conformal* onto its image by composing with a self-homeomorphism of S^2. Since $\|\mathcal{D}f\|_{L_2}^2 = $ Area $f(S^2)$ for such f, we obtain the above contractibility property for Area f in place of $\|\mathcal{D}f\|_{L^2}$. Similarly, we see that *the space of based maps $f : S^2 \to W$ with bounded area, $\{f \mid f(s_0) = w_0$, Area $f \leqslant c\}$ has bounded homotopy image in the space of all based maps (i.e. with $f(s_0) = w_0$), provided W is compact simply connected, and is contractible for sufficiently small $c \leqslant c_0 > 0$.* (Here the area of f is counted with due multiplicity, i.e. Area $f \underset{\mathrm{def}}{=} \int_{S^2} |\text{Jacobian } f| = \int_{S^2} \|\Lambda^2 \mathcal{D}f\|$.)

If Σ is a general compact surface ($\neq S^2$), it seems not hard to show that the maps f of small area simultaneously contract to maps with 1-dimensional images. Furthermore, one should be able to understand the (space of) maps with Area $f \leqslant c$ properly taking into account $\pi_1(W)$ and the action of $\pi_1(W)$ on $\pi_i(W)$ for $i \geqslant 2$.

---

[6] The boundedness means that the inclusion $F_c \subset F$ factors up to a homotopy through a map of F_c into a finite polyhedron with the number of cells bounded by some function of c.) *Furthermore, if $c < c_0$, then the space F_c homotopy retracts (within itself, not only in $F \supset F_c$) to the subspace of the constant maps.* (Here we allow non-trivial $\pi_1(W)$. Moreover, W may be non-compact, but yet with bounded geometry.

Questions. Let $n = \dim V \geqslant 3$, e.g. $V = S^3$. What is the homotopy structure of (the space of) maps $V \to W$ with $\mathrm{Vol}_n f \leqslant c$? Do these maps for small c (and with suitable homotopy restrictions on V and W) simultaneously contract to $(n-1)$-dimensional maps? If so, what is the algebro-topological structure of spaces of such maps?

2.5.D. On the homotopy role of the L_q-norm of the differential $\mathcal{D}f$ on H for $q < N$. Let us look at the maps f with $\|\mathcal{D}f \mid H\|_{L_q} \leqslant c$ for some $q < N$. Now, we do not have the full finiteness-contractibility result but one obtains a weaker homotopy conclusion by restricting f to the k-skeleton $V^k \subset V$ brought to a sufficiently general position with respect to f. Then the bound $\|\mathcal{D}f \mid H\|_{L_q} \leqslant c$ implies a similar bound on V^k, namely

$$\|\mathcal{D}f \mid T(V^k) \cap H\|_{L_q} \leqslant c', \qquad (*)'$$

where $c' = \mathrm{const} \cdot c$, where the "tangent bundle" $T(V^k)$ is understood as the set of the vectors tangent to the (smooth!) simplices in V^k and where the L_q-norm of the differential $\mathcal{D}f$ on the intersection $T(V^k) \cap H$ is obtained by integration over V^k. Now, if the intersection $T(V^k) \cap H$ induces a C-C structure of formal Hausdorff dimension $N' \leqslant q$ on each k-face of V^k, then the bound $(*)'$ has non-trivial homotopy effect on $f' = f \mid V^k$ and hence on f. Namely, if $\|\mathcal{D}f \mid H\|_{L_q}$ is sufficiently small, then the restriction $f \mid V^k$ is contractible whenever V^k can be *stably* brought into a position where $\dim_{\mathrm{Hau}}(V^k, T(V^k) \cap H)) \leqslant q$. Here the stability means the existence of a measure μ on the space of embeddings (i.e. positions) V^k in V having $\dim_{\mathrm{Hau}} \leqslant q$ and such that the push-forward of the measure $\mu \times dv^k$ to V is absolutely continuous with respect to Lebesgue (or equivalently C-C Hausdorff) measure in V.

If V is Riemannian (i.e. $\dim V = \dim_{\mathrm{Hau}} V$) then the above stability is trivially satisfied with $N' = k$ and so every smooth map $f : V \to W$ with $\|\mathcal{D}f\|_{L_q}$ small is null-homotopic on some (and hence every) k-skeleton $V^k \subset V$ with $k \leqslant q$. In fact the converse is also known to be true. Every smooth map $f : V \to W$ sending a small neighbourhood of V^k to a point can be composed with a suitable diffeotopy $\varphi_t : V \to V, 0 \leqslant t < \infty$, such that $\|\mathcal{D}f \circ \varphi_t\|_{L_q} \underset{t \to \infty}{\to} 0$ for every $q < k + 1$. (If $V = S^n$ and $V^0 \subset V$ is the south pole, one uses the north pole south pole push for φ_t. In general one uses such pushes in the cells in $V - V^k$. First every n-cell B is radially pushed from the center $b_0 \in B$ toward the boundary, so that in the limit for $t \to \infty$ all of $B - \{b_0\}$ goes to ∂B. Then one composes the above push toward V^{n-1} with a similar push of a small neighbourhood of V^{n-1} toward V^{n-2} and so on, see p.388 in [E-L] and references therein. Probably, a similar construction can be carried over for certain C-C manifolds, e.g. the contact ones.

Now, let us look at the maps f with $\|\mathcal{D}f \mid H\|_{L_q} < c$ with a possibly large c and observe that there are at most finitely many homotopy classes of restrictions of f to V^k in the following two cases, (i) $q > N'$ and (ii) $q = N'$ and $\pi_1(W)$ acts trivially on $\pi_k(W)$, where N' is, as earlier, the minimal integer so that V^k can be made stably of formal Hausdorff dimension N' for the C-C metric associated to $T(V^k) \cap H$. For example, $N' = k$ in the Riemannian case. If V is a contact C-C manifold, then $N' = k$ for $2k < \dim V$ and $N' = k+1$ for $2k \geqslant \dim V$ as follows from the discussion in 3.4.B, and see 4.? for the general C-C case.

The space $F_{q,c}$. Let us look at the homotopy property of the space $F_{q,c}$ of the maps f with $\|\mathcal{D}f \mid H\|_{L_q} \leqslant c$. We have just seen that the zero-dimensional homotopy (i.e. connected components) of $F_{q,c}$ are strongly affected by q and c but, probably, there is no additional link between the geometry of maps f (encoded into q and c) and higher homotopies of spaces of these f for $q < N$. Namely, if $f_a : V \to W$ is a family of smooth map parametrized by a compact polyhedron $A \ni a$ such that each f_a can be individually contracted to $F_{q,c}$ then, conjecturally, the whole family can be continuously moved to $F_{q,c}$ (or, possibly, to $F_{q,c'}$ for $c' = c'(c, A)$) in the case where $q < N$. This appears easy in the Riemannian case. For example, the above diffeotopy φ_t of V works for families of maps $f_a : V \to W$ which send a *fixed* (i.e. independent of a) skeleton V^k to a point. In general, if $\|\mathcal{D}f_a\|_{L_q}$ is small for all a, we can only have f_a almost constant on V_a^k depending on a. In fact we can make V_a^k constant in a on each simplex of a suitable subdivision of A and then $\varphi_{a,t}$ can be probably build using some induction on skeletons (or partition of unity) in A.

Exercise. Determine the homotopy structure of the space of maps $f : S^1 \times [0,1] \to S^2$ with $\|\mathcal{D}f\|_{L_q} \overset{\text{def}}{=} \left(\int \|\mathcal{D}f\|^q \right)^{\frac{1}{q}} \leqslant c$ for given $c > 0$ and $1 \leqslant q < 2$.

2.5.E. The space F_q^H of measurable maps f with $\|\mathcal{D}f \mid H\|_{L_q} <$ ∞. The norm $\|\mathcal{D}f \mid H\|_{L_q}$ makes sense for those measurable (possibly discontinuous) maps $f : V \to W$ where the derivative $\partial_\tau f \in T(W)$ exists for almost all $\tau \in H$. Then $\|\mathcal{D}f \mid H\|_{L_q}$ may be defined by integrating $\|\partial_\tau f\|^q$ over the (unit sphere) bundle of the unit vectors in H and the space of maps with $\|\mathcal{D}f \mid H\|_{L_q} < \infty$ is denoted by F_q^H. If $q > N$, then our earlier argument works equally well for maps $f \in F_q^H$ and show these are, in fact, continuous and even C^β for $\beta = \frac{q-N}{q}$. It is not so, of course, for $q \leqslant N$ and, moreover, continuous maps in F_q^H are not necessarily dense in this space (see (c_1) below). In fact, one know exactly for which q continuous maps are dense when V is Riemannian (see [Beth]) but the corresponding result in the C-C category remains conjectural. A closely related (apparently more global) question is that of the homotopy content of F_q^H and the homotopy structure of the inclusions $C^\infty \cap F_q^H \subset F_q^H$ and $C^\infty \cap F_{q,c}^H \subset F_{q,c}^H \subset F_q^H$ where $F_{q,c}^H = \{f \in F_q^H \mid \|\mathcal{D}f \mid H\|_{L_q} \leqslant c\}$, where the space F_q^H is given L_q^H-topology arizing from the norm $\|\mathcal{D}f \mid H\|_{L_q}$ via an embedding of W into some Euclidean space. One asks in this regard, for example, which homotopy invariants of smooth maps extends to F_q^H and are F_q^H-continuous (or continuous in some weaker topology on F_q^H).

2.5.E$'$. Examples of F_q^H-non-density of smooth maps in F_q^H. ([7]) Let f_0 be the radial projection of the n-ball $V = B^n$ to the boundary $S^{n-1} = \partial B^n$. This map is in F_q^H for $H = T(V)$ and all $q < n$ but for $q > n-1$ it is *not* an F_q^H-limit of smooth maps $f : V \mapsto S^{n-1}$ since the F_q^H-closeness between maps, say $\mathrm{dist}_{L_q^H}(f, f_0) \leqslant \varepsilon$, implies, for $q > n-1$, the uniform closeness of the (continuous!) maps on a sphere $S_\rho^{n-1} \subset B^n$ concentric to ∂B^n and, hence, contractibility of the map $f_0 \mid S_\rho^{n-1}$. The same applies to composed maps $f = \varphi \circ f_0$ for non-null-homotopic maps $\varphi : S^{n-1} \to W$. Moreover, this is also valid for $q = n - 1$ as follows from corollary in 2.5.A applied to concentric spheres in the ball $B = B^n$.

Now let V be a smooth (topological) ball around the origin in a nilpotent Lie group with a one-parametric self-similarity such that each orbit of this self-similarity transversally meets $S^{n-1} = \partial V$ at a single point. Then the radial projection $f_0 : V \to S^{n-1}$ along the orbits is in F_q^H for all $q < N = \dim_{\mathrm{Hau}} V$ and neither this f_0 nor any $f = \varphi \circ f_0$ for a non-null-homotopic φ can be approximated by smooth maps in F_{N-1}^H (under the standing assumption of (codim 1)-stability of V).

---

[7] Compare [Sh-Uhl], [Beth] and [Haj$_{\mathrm{ASM}}$].

2.5.F. Space F_N^H. Let us look at the homotopy structure of a map $f \in F_N^H$. We start with the (well known) Riemannian case (where $H = T(V)$) and observe that if $f \in F_q^H$ then the restriction $f \mid V'$ has $\|\mathcal{D}f \mid TV'\|_{L_q(V')} < \infty$ for a generic hypersurface V' in V and so for $q > \dim V - 1$ this restriction in continuous. However, the homotopy class of $f \mid V'$ may jump under small perturbations of V' in V for $q < \dim V$. Let us show this does not happen for $q = n = \dim V$. We observe that for each point $v \in V$ and every small positive ε there is a sphere S_δ around v of radius δ in the interval $\varepsilon \leqslant \delta \leqslant 2\varepsilon$, such that the L_n-norm of f on this sphere S_δ with the normalized metric ($= \mathrm{dist}\,/\delta$) is bounded by the L_n-norm of f on the ball $B_{2\varepsilon}$ (by integrating $\|\mathcal{D}f\|^n$ over the annulas between S_ε and $S_{2\varepsilon}$). It follows, that $\mathrm{Diam}f(S_\delta)$ is bounded in terms of $\int_{B_{2\varepsilon}} \|\mathcal{D}f(v)\|^n \, dv$, as we, in fact, have seen earlier and therefore for every $\varepsilon > 0$ one can cover V by balls of radii between ε and 2ε, such that every ball among these at most $\nu = \nu(n)$ neighbours. We assume without loss of generality that the union Σ of the boundary spheres of these balls is connected and we partition V into the connected components of the complement $V - \Sigma$, say $V = \cup U_i$, where the boundary of each U_i has $\mathrm{Diam}f(\partial U_i) \leqslant \nu \mathrm{Diam}f(S_\delta)$ which uniformly (in i) goes to zero as $\varepsilon \to 0$. Then one can regularize f by using some standard continuous extension of $f \mid \partial U_i$ for each i to all of U_i within the (small) ball of radius $\rho = \mathrm{Diam}f(\partial U_i)$ in W. Furthermore, given a hypersurface $V' \subset V$, one can do the same to the (finer) partition into the connected components U_i' of $V - (\Sigma \cup V')$ as all $\mathrm{Diam}f(U_i')$ are necessarily small for small (now, depending also on V') ε. Thus the (continuous) restriction $f \mid V' : V' \to W$ admits a *continuous* extension to all of V which, in particular, imply the contractibility of $f \mid V'$ for small spheres in V (which are contractible in V). Then, obviously, the homotopy type of $f \mid V'$ is invariant under the deformations of V' in V and also under homotopies of f in the space F_n^H for $H = T(V)$.

The same reasoning applies to (codim 1)-stable C-C manifolds V. For example, *if V is a contact C-C manifold, then every map $f \in F_N^H$, $N = n + 1$, has a well defined homotopy class of the restriction of f to the $(n - 1)$-skeleton of V, provided $n = \dim V \geqslant 5$.* (It is unclear what happens for 3-dimensional contact manifolds.)

2.5.F′. Regularization of F_N^H-maps. The above process of filling small U_i-holes by continuous maps (extending $f \mid \partial U_i$) allows us to approximate every $f \in F_N^H$ by a continuous map, say $f' : V \to W$, such that any two such approximation to f are mutually homotopic. More precisely, we have the following proposition (which is well known in the Riemannian case and is due, I believe, to K. Uhlenbeck).

Let V be a compact (codim 1)-stable C-C manifold (i.e. with sufficiently many $(N-1)$- dimensional hypersurfaces for $N = \dim_{\mathrm{Hau}} V$, e.g. V is Riemannian or contact of dimension $\geqslant 5$) and $f : V \to W$ be a map with $\|\mathcal{D}f \mid H\|_{L_N} \leqslant c < \infty$. Then for every $\varepsilon > 0$ there is a decomposition of V, say $V = V_\varepsilon \cup V_{1-\varepsilon}$ with the following three properties.

(1) *V_ε is an open subset in V of $\mathrm{mes}_N V_\varepsilon \leqslant \varepsilon$ and $V_{1-\varepsilon} = V - V_{1-\varepsilon}$; furthermore, each connected component U of V_ε has $\mathrm{Diam}(U) \leqslant \delta$ where $\delta \to 0$ for $\varepsilon \to 0$.*

(2) *The restriction $f \mid V_{1-\varepsilon}$ is a continuous, moreover, C^β-Hölder map for $\beta = 1/N$ (where the implied Hölder constant may depend on ε). Furthermore the image $f(\partial U)$, of the boundary of every component U of V_ε, has $\mathrm{Diam} f(\partial U) \leqslant \delta'$ where $\delta' \to 0$ for $\varepsilon \to 0$.*

(3) *If W has locally bounded geometry (e.g. compact) and ε is sufficiently small, then $f \mid V_{1-\varepsilon}$ admits a continuous extension say $f_\varepsilon : V \to W$ which is contained in F_N^H and, moreover, has $\|\mathcal{D}f_\varepsilon \mid H\|_{L_N} \leqslant \|\mathcal{D}f \mid H\|_{L_N}$. (In fact this f_ε may be chosen taut on V_ε relative to the boundary $\partial V_\varepsilon \subset V_{1-\varepsilon}$.)*

The proof follows by our earlier argument and is left to the reader.

Notice that the maps f converge to f in F_N^H for $\varepsilon \to 0$ and so they are all mutually homotopic for small ε by the weak homotopy stability observed in 2.5.A. Also notice that the regularization $f \mapsto f_\varepsilon$ applies to families of maps and shows that *the space F_N^H is homotopy equivalent to the space of continuous maps $V \to W$* (where V and W are compact and V is (codim 1)-stable, i.e. has many "nice" hypersurfaces as we always assume). In fact *the inclusion of the space C^1 of smooth maps $V \to W$ into F_N^H is a homotopy equivalence. Furthermore the space F_N^H (as well as $C^1 \subset F_N^H$ with the induced topology) is locally contractible.* (Of course, this all is well known for Riemannian manifolds V.)

Example. Let V be homeomorphic to the sphere S^n and W be an n-dimensional Riemannian manifold with locally bounded geometry. Suppose there exists a L_N^H-map $f : S^n \to W$ which has degree 1 in the following strong sense: there is an open subset $U \subset S^n$ such that the map is one-to-one on U and moreover is a homeomorphism of U onto some $U' \subset W$ with $f^{-1}(U') = U$. Then f can be approximated by a continuous map of degree 1 and so W is a homotopy sphere.

Remarks

(a) The degree 1 condition can be more succinctly expressed if V is Riemannian by $\int_V f^*(\omega) = 1$ for the normalized oriented volume form ω on W but I do not know how to do this in the general C-C case.

(b) Here and in future N refers to the Hausdorff dimension of V if it is *equiregular* and to the *formal dimension* $\max \Sigma i(n_i - n_{i-1})$ otherwise. But in fact many of our results hold true for $N = \dim_{\mathrm{Hau}}$ under milder (genericity) assumptions than equiregularity.

2.5.G. Restriction of L_q^H-maps to k-dimensional submanifolds in codim-stable manifolds for $q < N$ and $k < q$. If V is Riemannian then the restriction $f \mid V$ is obviously continuous for generic $V' \subset V$ of dimension $k < q$ (where $f \in F_q^{T(V)}$). Furthermore if $k \leqslant q - 1$, then the homotopy type of this restriction is well defined. This is derived from the case codim $V' = 1$ as follows. First, let codim $V' = 2$, take two generic hypersurfaces V_1' and V_2' in V and observe that the restrictions $f \mid V_i' \in F_q^{T(V)}$, $i = 1, 2$, are continuous and if $q > n - 2 = \dim V_i' - 1$, then the restriction of f to the (transversal!) intersection $V' = V_1' \cap V_2'$ is also continuous. Furthermore, if $q \geqslant n - 1$ then the homotopy type of $f \mid V'$ is stable under deformations of V_1' and of V_2' by (c_2) applied to V_2' and V_1' correspondingly. Next, this intersection argument applies to all V' which are transversal intersections of hypersurfaces, e.g. to those with trivial normal bundles and then we localize in a standard fashion to make the intersection trick work for all V'. Moreover, one easily shows with a properly localized intersection argument that f defines the homotopy type of the restriction of f to the k-skeleton of V for $k = \mathrm{ent}(q - 1)$ (i.e. k is biggest integer $\leqslant q - 1$).

Codim-stability. Call V *stable in codimension* $n - k$ if for each tangent k-plane $T' \subset T(V)$ there exists a germ of smooth submanifold V' tangent to T' such that the formal Hausdorff dimension of V' at most equals that of V minus $n - k$ (compare 2.4.B). This means, intuitively, that generic k-dimensional submanifolds V' have $\mathrm{codim}_{\mathrm{Hau}} V' = \mathrm{codim}_{\mathrm{top}} V'$.

Example. Contact manifolds are $(n - k)$-stable for $k \geqslant 3$.

If V is $(n - k)$-stable then the restriction of $f \in F_q^H$ to generic V' of dimension $k < q$ is continuous and if $k \leqslant q - 1$ the homotopy class of this restriction is well defined by the above Riemannian argument. This applies in particular to k-dimensional submanifolds in contact manifolds for $k \geqslant 3$ (where the situation for $k \leqslant 2$ remains unclear).

Restriction to horizontal submanifolds V' in a contact V. *Such V' are plentiful for $k < n/2$, $n = \dim V$, and the restriction $f \mid V'$ is continuous for generic horizontal V' provided $q > k$ (compare 3.1). Now we claim that if $k \leqslant q - 1$ and $k + 1 < \dim V/2$, then the homotopy class of this (continuous) restriction $f \mid V'$ is stable under (horizontal!) deformations of V'. This implies (via the horizontal triangulation of V, see 3.4.B) that each $f \in F_q^H$ has well defined homotopy class of the restriction of f to the k-skeleton of V for every $k < (\dim V/2) - 1$ and $q \geqslant k + 1$.*

Proof. To grasp the idea we start with the Riemannian case and indicate another way of reducing the case codim $V' > 1$ to that of codim $V' = 1$. For example, let V' be a small k-dimensional sphere in V and let us show that, generically, the restriction $f \mid V'$, which is continuous for $q > k$, is contractible for $q \geqslant k + 1$. We make this sphere V' the boundary of a small $(k + 1)$-ball $V_1' \subset V$ and observe that the restriction $f \mid V_1'$ is genericly contained in the space $F_q^{T(V_1')}$ on V_1'. Then we restrict further, to a generic sphere in V_1' concentric to $\partial V_1'$, and obtain contractible (as well as continuous) f on a *generic* small k-sphere in V. This argument easily generalizes to all (generic) $V' \subset V$ and then it extends into the contact framework with the provisions of 3.1, 3.4.A.

2.5.G$'$. On singularities of $f \in F_q^H$. One can imagine, following Karen Uhlenbeck, every map $f \in F_q^H$ as being regular (e.g. continuous) away from a certain (pole-like) singularity $\Sigma_f \subset V$ which, in the case where V is Riemannian, has "dim" $\Sigma_f \leqslant \mathrm{ent}(n - q)$. This is justified by the following three facts

(1) generic V' of dimension $k < q$ misses Σ_f as $f \mid V'$ is continuous;

(2) if $k \leqslant q - 1$, then generic 1-parameter families of V's miss Σ_f and so the homotopy class of $f \mid V'$ is well defined;

(3) the blow-up construction in 2.5.E$'$ applied to the balls $B^{k+1} \times b \in B^{k+1} \times B^{n-k-1} = V$, $b \in B^{n-k+1}$, gives us $f \in F_q$, with $q = k + 1 - \varepsilon$ for all $\varepsilon > 0$, and with $(n - k - 1)$-dimensional singularity.

Now, in the C-C case, one should think of Σ_f as some virtual subset (in V or in some auxiliary jet space over V) so that generic submanifolds V' and generic *horizontal* submanifolds of the same topological dimension have different chances to meet Σ_f. (Notice that both, "generic" and "generic horizontal" submanifolds V' of *Hausdorff* dimension ℓ, miss Σ_f, $f \in F_q^H$, if $\ell < q$, and the restriction $f \mid V'$ is homotopically sound if $\ell \leqslant q - 1$.)

We conclude by observing that the restriction problem is still not solved in full generality for C-C manifolds. Namely, when does every map $f \in F_q^H$ restrict to a continuous map on a generic submanifold $V' \in V$ of dimension k? When is the homotopy class of the restriction of f to the k-skeleton of V well defined? When does there exist at least one submanifold $V' \subset V$ of dimension k for which $f \mid V'$ is continuous?

2.5.H. On local estimates for taut L_N^H-maps. Consider a taut map f on an ε_0-ball $B(\varepsilon_0) \subset V$ where $\int_B \|\mathcal{D}f(v) \mid H\|^N \, dv \leqslant c_0$ for a small $c_0 > 0$ and let us evaluate the diameters of the f-images of the concentric ε-ball $B(\varepsilon)$ for $\varepsilon \leqslant 1/2 \, \varepsilon_0$. Since f is taut, this diameter does not exceed the infimum of those for the concentric spheres $S(\rho)$ for $\rho \in [\varepsilon_0, \varepsilon]$ and Diam $f(S(\rho))$ is bounded by const $\rho^{-1} \int_{S(\rho)} \|\mathcal{D}f(v) \mid H\|^N \, dv$, provided the induced C-C geometry on these spheres has formal Hausdorff dimension $N - 1$. (In fact, the C-C spheres are usually non-smooth and so the formal dimension makes no sense. The true condition we need is, of course, (codim 1)-stability which allows us to approximate the spheres by piecewise smooth hypersurfaces of formal dimension $N - 1$.) Then we integrate over $\rho \in [\varepsilon_0, \varepsilon]$ and conclude to the uniform continuity of f on $B(\varepsilon^0/2)$ with the logarithmic modulus of continuity,

$$\mathrm{dist}(f(v), f(v')) \leqslant \mathrm{const}_0 \ \varepsilon_0^{-1}(-\log \mathrm{dist}(v, v'))^{-\frac{1}{N}}$$

for all v, v' in $B(\varepsilon^0/2)$.

2.5.H′. Hölder estimates. Let us additionally assume that our f minimizes the energy $f \mapsto \|Df \mid H\|^N_{L_N}$ and prove that then f is Hölder. It is well known (and obvious by the previous discussion) that the Hölder bound for taut maps issues from the following monotonicity inequality

$$\int_{B(2\varepsilon)} \|\mathcal{D}f(v) \mid H\|^N \, dv \geqslant \lambda \int_{B(\varepsilon)} \|\mathcal{D}f(v) \mid H\|^N \, dv \qquad (*)$$

for $\varepsilon \leqslant \varepsilon_0/2$ and some $\lambda > 1$ independent of ε. Now, since f is minimizing, no extension of f from the sphere $S(2\varepsilon) = \partial B(2\varepsilon)$ to a map f' on $B(2\varepsilon)$ may have $\int_{B(2\varepsilon)} \|\mathcal{D}f'(v) \mid H\|^N \, dv < \int_{B(2\varepsilon)} \|\mathcal{D}f(v) \mid H\|^N \, dv$, and so the following lemma yields $(*)$.

Modification lemma. Every taut map $f_0 : B(2\varepsilon) \to W$ which lands in a (small) ball within the range of the convexity radius of W can be modified to a map $f' : B(2\varepsilon) \to W$ agreeing with f_0 on $S(2\varepsilon)$, landing in the same (small) ball in W and having

$$\int_{B(2\varepsilon)} \|Df'(v) \mid H\|^N \, dv \leqslant C \int_{A(\varepsilon)} \|Df_0(v) \mid H\|^N \, dv, \qquad (*+)$$

where $A(\varepsilon)$ denotes the annulus $B(2\varepsilon) - B(\varepsilon)$ in V.

Proof. Let B_0 be the minimal (convex!) ball in W containing the image $f_0\left(B\left(\frac{3}{2}\varepsilon\right)\right)$ and w_0 be the center of W_0. Observe that the radius of B_0 is bounded by $R_0 = \mathrm{Rad}\, B_0 \leqslant C_0 \left(\int_{A(\varepsilon)} \|\mathcal{D}f_0(v) \mid H\|^N \, dv\right)^{\frac{1}{N}}$ by our earlier argument (using (codim 1)-stability and the tautness of f_0). Then we use the notation $w \mapsto \delta w$, $\delta \in [0, 1]$, for the geodesic scaling of the ball B_0 toward the center (i.e. δw stands for the geodesic convex combination $\delta w + (1 - \delta) w_0$) and define f' on $B(2\varepsilon)$ by "compressing" f_0 by means of the cut-off distance function. Namely, take

$$d(v) = \begin{cases} 1 \text{ for } v \in B(2\varepsilon) - B\left(\frac{3}{2}\varepsilon\right), \\[2mm] (\varepsilon/2)^{-1} \mathrm{dist}(v, B(\varepsilon)) \text{ for } v \in B(\tfrac{3}{2}\varepsilon) \end{cases}$$

and $f'(v) = d(v)\, f_0(v)$. The horizontal differential of f' clearly satisfies

$$\|\mathcal{D}f'(v) \mid H\| \leqslant \varepsilon^{-1} R_0 + \|\mathcal{D}f_0(v) \mid H\|$$

all $v \in A(\varepsilon)$ and $\mathcal{D}f'(v) = 0$ on $B(\varepsilon)$. Thus $(*+)$ follows from the above bound on R_0 and the volume bound $\mathrm{mes}_N B(2\varepsilon) \lesssim \varepsilon^N$.

Remarks

(a) The minimizing property of f can be replaced by "quasi-minimizing".
This means that, for every relative compact domain $U \subset V$ and every
map $f' : V \to W$ obtained by a homotopy of V fixed outside U, the
total energy of f' on U can not significantly smaller than that of f,
i.e.

$$\int_U \|\mathcal{D}f'(v) \mid H\|^N \geqslant C \int_U \|\mathcal{D}f(v) \mid H\|^N \, dv$$

for a fixed constant $C > 0$ given beforehand (and defining our class of
C-quasi-minimizing maps). Notice that quasi-minimizing maps need
not to be taut but the Hölder bounds holds true all the same by a
simple additional argument. Also observe that quasi-conformal (sometimes called "quasi-regular") maps between Riemannian manifolds
are taut as well as quasi-minimizing for the energy $E_n = \|\mathcal{D}f\|^n_{L_n}$,
$n = N = \dim_{\mathrm{Hau}} V$ and, similarly, contact quasi-conformal maps are
quasi-minimizing for E_{n+1}. Thus we recapture the (well known, compare [Ko-Re]) Hölder estimate for quasi-conformal maps.

(b) One knows, thanks to K. Uhlenbeck, much more than mere Hölder for
minimizing maps in the Riemannian case but the corresponding analysis is yet to be developed on C-C manifolds. In fact, our monotonicity
argument is borrowed from the (well known) Riemannian situation,
where, in fact, one can use a straightforward radial extension of maps
from the spheres $S(\rho) = \partial B(\rho)$ to the balls $B(\rho)$. Such an extension
implicitly uses the fact that the radial projection of the ball $B(\rho)$ to
the sphere $S(\rho)$ is Lipschitz away from the center. Nothing of the kind
will work for general C-C manifolds where smooth hypersurfaces are
not neighbourhood Lipschitz retracts.

(c) Our estimates became global in certain cases, e.g. for *non-constant*
quasi-minimizing maps f of a nilpotent group V with a natural C-C
geometry into a Riemannian manifold W with non-positive curvature.
Here the monotonicity argument provides a lower bound

$$\int_{B(R)} \|\mathcal{D}f \mid H\|^N \, dv \geqslant \mathrm{const}\, R^\alpha \qquad\qquad (**)$$

for some const > 0 and $\alpha > 0$ and all $R \geqslant 1$, where $B(R)$ stands
for the R-ball around the origin. (Probably, honest *minimizing* maps
satisfy $(**)$ with $\alpha = N = \dim_{\mathrm{Hau}} V$.) Furthermore, if f is bounded
one can easily replace const R^α by const $\exp \alpha R$, but it is unclear if

non-constant bounded (quasi)minimizing maps exist at all (Liouville problem).

(d) Let us indicate an integrated version of our "compressing" argument in the proof of the modification lemma which sometimes applies to the energy $E_q = \int \|\mathcal{D}f \mid H\|^q$ for $q < N$. Now we use variable $w_0 = f_0(v_0)$, where v_0 runs over the annulas $A(\varepsilon) = B(2\varepsilon) - B(\varepsilon)$. Then we estimate the average of the energies of the extended maps $f'_{v_0}(v)$ obtained by "compression" toward $w_0 = f_0(v_0)$. We write this as if W were a Euclidean space, namely

$$f'_{v_0}(v) = d(v)(f_0(v) - f_0(v_0))$$

and have, after the normalization making $\varepsilon = 1$,

$$\|\mathcal{D}f'_{v_0}(v)\| \leqslant 2(f_0(v) - f_0(v_0)) + \|\mathcal{D}f_0\|.$$

Then our average is bounded by,

$$(\mathrm{Vol}\, A(\varepsilon))^{-1} \int_{A(\varepsilon)} dv_0 \int_{B(2\varepsilon)} \|\mathcal{D}f'_{v_0}(v) \mid H\|^q \, dv \leqslant$$

$$\mathrm{const} \left(\int_{A(\varepsilon)} \|Df_0 \mid H\|^q \, dv + \iint_{A(\varepsilon) \times A(\varepsilon)} |f_0(v) - f_0(v_0)|^q \, dv \, dv_0 \right).$$

In order to make this work we need an estimate for the second integral on the right hand side by the first one (where $|f_0(v) - f_0(v_0)|$ should be understood as $\mathrm{dist}_W(f_0(v, f_0(v_0))$. If $\dim V \geqslant 2$ and $A(\varepsilon)$ is connected, the desired estimate is a trivial case of the Sobolev inequality (which is discussed in sharper form below) and so the monotonicity comes along if f_0 lands within the convexity radius of W, for example if W is complete simply connected with non-positive curvature. Unfortunately, this does not yield Hölder for $q < N$ although this likely to be true for many E_q-minimizing maps for $q > 1$. (One may try here the standard Riemannian trick of infinitesimal deformations of f along radial fields similar to our "compression".)

2.5.I. On semicontinuity of the energy $E_q(f)$. We want to know that if f_i weakly converge to f for $i \to \infty$ then

$$E_q(f) \leqslant \liminf_{i \to \infty} E_q(f_i),$$

which is important for the calculus of variations. If the limit map is
a.e. smooth this semicontinuity can be sometimes derived from the E_q-
minimizing property of linear maps between Euclidean spaces or some
other standard maps between relevant spaces approximating our f at the
regular points. Here we indicate another approach based on a regularization of energy functionals (compare [Jo]). First we introduce a class
of energies similar (and somewhat more general) than E_q. Such an energy will be constructed with some auxiliary space M endowed with the
following structures

(1) Projection $p : M \to V$

(2) A measure $d\mu$ on M

(3) A vector field X on M.

Our main requirement is that each vector X_m, $m \in M$, is sent
by p to a *horizontal* vector at $v = p(m) \in V$. Here we require just
enough regularity of M and p make sense of this. For example, M
may be a smooth manifold and p a smooth map, but, in general, p
should be smooth only along the orbits of X and just measurable in
the transversal directions. Now we can define the energy $E_q^\mu(f)$ by

$$E_q^\mu(f) = \int_M \|X(p \circ f)\|^q \, d\mu$$

where $X \varphi \underset{\mathrm{def}}{=} (\mathcal{D}\varphi)(X)$. If the push-forward of the measure $d\mu$
under the map $M \to H$ for $m \mapsto X_m$ p dominates in an obvious way
the measure of the unitary subbundle of H, then E_q^μ is equivalent to
E_q, i.e.

$$C^{-1} E_q \leqslant E_q^\mu \leqslant C \, E_q,$$

and for suitable $(M, d\mu)$ one may have $E_q^\mu = E_q$.

In what follows we add an extra assumption on X, and $d\mu$

(4) X integrates to a flow $X(t)$ on M and this flow preserves the measure
$d\mu$.

Examples

(a) If V is Riemannian (i.e. $H = T(V)$), then the unit tangent bundle
with the Liouville measure and the geodesic flow provides a model
example.

(b) Fix a measure on V (e.g. the Hausdorff measure), let A be a family
of measure preserving horizontal fields X_a on V (these, as we know,

are plentiful) and let $M = V \times A$ with the product measure for some measure on A and the field $(v, a) \mapsto X_a$ at v in $V = V \times a$.

Now we introduce the following ε-regularization of the energy.

$$E_q^{\mu,\varepsilon} = \varepsilon^{-1} \int_M (\mathrm{dist}_W(p \circ f(m), p \circ f(X(\varepsilon)m)))^q \, d\mu.$$

It is clear (without (4)) that $E_q^{\mu,\varepsilon} \to E_q^\mu$ for $\varepsilon \to 0$. What is more interesting (albeit obvious), is the inequality

$$E_q^{\mu,\varepsilon} \leqslant E_q^\mu \quad \text{for all } \varepsilon > 0$$

which follows by integration and use of (4) from the corresponding inequality on a single X-orbit in M.

Finally we observe that every energy $E_q^{\mu,\varepsilon}(f)$ for $\varepsilon > 0$ is continuous with respect to the uniform topology in the space of maps f and therefore, E_q^μ is semicontinuous in this topology. Now we can apply it to E_N^μ when it is equivalent to E_N and prove *the existence of the energy minimizing Hölder map in every homotopy class of maps $V \to W$, provided V is (codim 1)-stable (e.g. being contact of dimension $\geqslant 5$) and $\pi_n(W) = 0$.* (If $\pi_n(W) \neq 0$ one realizes the homotopy classes modulo the action of $\pi_1(W)$ on these classes as is seen by looking at the bubbling picture.)

On V's which are not (codim 1)-*stable.* If rank $H \geqslant 3$, then, genericly, the polarization $H' = H \cap T(V')$ Lie-generates $T(V')$ (at least, apart from a stratified subset of positive codimension) and so the (formal) Hausdorff dimension of (V', H') is finite, say N'. Then our regularity arguments for $f \in F_N^H$ remain valid with N replaced by $N' + 1$ but this does not seem to do us any good as we have Hölder estimates for $F_{N+\varepsilon}^H$ anyway.

2.6. Isoperimetric inequalities and quasi-conformal mappings.
The theory of quasi-conformal (= quasi-regular) mappings between *Riemannian* manifolds, say $f : W \to V$, relies on isoperimetric inequalities in V for large domains $D \subset V$ which for the nilpotent groups V are governed by the corresponding Carnot-Carathéodory isoperimetric inequalities. (In fact, this motivated the study of such inequalities, see [G-L-P] and [Pan$_{\mathrm{InIs}}$].) Let us spell it out in details. Suppose every domain D in a Riemannian manifold V of volume $\geqslant \mu_0$ satisfies the isoperimetric inequality with the exponent $\alpha = \frac{N}{N-1}$ for $N > n = \dim V$, i.e.

$\mathrm{Vol}_n\, D \leqslant C(\mathrm{Vol}_{n-1}, \partial D)^{\alpha}$, provided $\mathrm{Vol}_r\, D \geqslant \mu_0$. Then *quasi-conformal mappings* $f : W \to V$ are *uniformly continuous (and even Hölder)* if $|K(W)| \leqslant \mathrm{const} < \infty$ by the following standard argument. Consider concentric R-balls in W mapped into V, say $B(R) \subset W$, and let $p(w)$ denotes the Jacobian of f. Then $\mathrm{Vol}_n\, f(B(R)) = \int_{B(R)} p(w)\, dw$ and by quasi-conformality, the volume of the sphere $S(R) = \partial B(R)$ in the image satisfies

$$\mathrm{Vol}_{n-1}\, f(S(R)) \leqslant C_f \int_{S(R)} p^{\frac{n-1}{n}}(w)\, dw.$$

If for some $R_0 < R$ the volume $\mu(R) = \mathrm{Vol}_n\, f(B(R))$ satisfies $\mu(R_0) \geqslant \mu_0$, then the isoperimetric inequality applied to $f(B(R))$ provides a lower bound on $\sigma(R) = \int_{S(R)} p^{\frac{n-1}{n}}(w)\, dw$, that is

$$\sigma(R) \geqslant C_1(\mu(R))^{\beta} \quad , \quad \beta = \frac{N-1}{N}.$$

On the other hand, since

$$\mu(R) = \int_{B(R)} p(w)\, dw = \int_0^R dR \int_{S(R)} p(w)\, dw,$$

the derivative $\mu'(R) = \int_{S(R)} p(w)\, dw$ satisfies by Hölder inequality,

$$\mu'(R) \geqslant (\sigma(R))^{\frac{n}{n-1}}\, (\mathrm{Vol}_{n-1}\, S(R))^{-\frac{1}{n-1}},$$

which implies with the above that

$$\mu'(R) \geqslant C_2(\mu(R))^{\gamma}\, \mathrm{Vol}_{n-1}\, S(R)^{-\frac{1}{n-1}} \ , \ \text{for } \gamma = \frac{\beta n}{n-1} > 1.$$

Now, if $\mathrm{Vol}_{n-1}\, S(R)$ grows no faster than $C_0\, R^{n-1}$, the ratio R/R_0 must be bounded by

$$R/R_0 \leqslant \mathrm{const} = \mathrm{const}(\mu_0, C_2, \gamma, C_0) = \mathrm{const}(\mu_0, N, C, C_0),$$

since the above differential inequality for $\mu(R)$ with the initial condition $\mu(R_0) \geqslant \mu_0$ predicts the blow-up of $\mu(R)$ at some moment after R_0. Thus we obtain a bound on $\int_{B(R_0)} p(w)\, dw$ and, hence, by quasi-conformality, a similar bound on the L_n-norm of the differential,

$$\int_{B(R)} \|\mathcal{D}f(w)\|^n\, dw = \int_0^R dR \int_{S(R)} \|\mathcal{D}f(w)\|^n\, dw.$$

If the spheres $S(R)$ have (roughly) the standard geometry, the bound on the integral $\int_{S(R)} \|\mathcal{D}f(w)\|^n$ implies, by the Sobolev inequality, the bound on Diam $f(S(R))$, and therefore (by openness of the map f) the required bound on Diam $f(B(R_0))$. Finally, in order to use all this for bounded curvature $|K(W)| \leqslant$ const, we notice that up to the last moment we only needed the bound on $\mathrm{Vol}_{n-1} S(R) \lesssim R^{n-1}$ ensured by Ricci $W \geqslant$ $-$ const, while the standardization of the geometry of $S(R)$ is achieved for $|K(W)| \leqslant$ const by using small balls in $T_w(W)$ immersed into W by the exponential map (which may be non-injective).

Remarks and corollaries

(a) Since the isoperimetric inequality applies to multiple domains (see 2.3.D (c)) the above argument allows non-injective maps f which may, moreover, have ramification points.

(b) The uniform continuity property remains valid whenever V satisfies an isoperimetric inequality which is asymptotically (for large domains) stronger than the Euclidean one). In this case one obtains a differential inequality $\mu' \geqslant \psi(\mu, R)$ which makes μ grow faster than R^n though not forcing a blow up of $\mu(R)$ for $R < \infty$. Then one uses the (obvious in this case) fact that the ball $B(R)$ can be covered by at most $k = c(R/R_0)^n$ of balls of radius $R_0/2$ and therefore some of these balls, say $B_1(R_0/2)$ around some $v_1 \in B(R)$ satisfies

$$\mathrm{Vol}\, f(B_1(R_0/2)) \geqslant 2\, \mathrm{Vol}\, f(B(R_0)),$$

since $\mu(R) = \mathrm{Vol}\, f(B(R))$ grows faster than R^n and we choose R/R_0 large. Then we find next ball $B_2(R_0/4)$ with center $v_2 \in B_1$ and so on. These balls B_i, $i = 1, 2, \ldots$, have infinite volume altogether and they are contained in $B(2R)$. Thus $\mathrm{Vol}\, f(B(R))$ does become infinite in finite time unless $\mathrm{Vol}\, B(R_0)$ was rather small, exactly as in the case considered earlier. We conclude by noticing that even the Euclidean inequality $\mathrm{Vol}_n D < C(\mathrm{Vol}_{n-1}\, \partial D)^{\frac{n}{n-1}}$ may be used for this purpose if the constant C is sufficiently small with respect to the quasi-conformality constant of f.

(c) The key use of the quasi-conformality in our argument is the inequality between the norms of the differential $\mathcal{D}f$ on the forms of degree n and $n-1$, i.e.

$$\left(C_f^{-1}\|\wedge^{n-1}\, \mathcal{D}f(w)\|\right)^{\frac{n}{n-1}} \leqslant \|\wedge^n\, \mathcal{D}f(w)\| \underset{\mathrm{def}}{=} p(w).$$

This suggests introducing more general classes of *pinched* maps satisfying

$$\| \wedge^{n-1} \, \mathcal{D}f(w)\| \leqslant A\| \wedge^{n} \, \mathcal{D}f(w)\|^{a} + B\| \wedge^{n} \, \mathcal{D}f(w)\|^{b}$$

for certain a and b close to $\frac{n-1}{n}$. One knows that such maps do share some quasi-conformal properties (see [Deg], and §7.C in [Gro$_{\mathrm{FPP}}$]) and they seem to be relevant in the Hölder geometry of C-C spaces. (Another avenue of generalizing quasi-conformality is suggested by taut E_N-quasi-minimizing maps mentioned earlier in 2.5.H′.)

(d) If V is a simply connected non-Abelian nilpotent groups it satisfies, as we know, an asymptotically N-dimensional isoperimetric inequality for $N > n = \dim V$ and so quasi-conformal maps $f : W \to V$ are uniformly continuous if $|K(W)| \leqslant \text{const} < \infty$. Furthermore, every quasi-conformal map $f : \mathbb{R}^n \to V$ is constant. Now, if W also such a non-Abelian nilpotent group and f is bijective as well as quasi-conformal than f is *quasi-isometric*, i.e. bi-Lipschitz on the large scale. Then the corresponding *asymptotic tangent cones* $V^\infty = \lim\limits_{\varepsilon \to 0} \varepsilon V$ and $W^\infty = \lim\limits_{\varepsilon \to 0} \varepsilon W$, which are certain nilpotent groups with self-similarities, are bi-Lipschitz equivalent with respect to their (limit) C-C metrics and by Pansu theorem they are isomorphic as Lie groups (see [Pan$_{\mathrm{QIR1}}$]). In particular, if V and W admit dilating automorphisms (e.g. being of nilpotency degree two) then the existence of a quasi-conformal homeomorphism $V \leftrightarrow W$ makes them isomorphic.

(d') The above is probably not hard to generalize to some "pinched" homeomorphisms $V \leftrightarrow W$ but it seems more difficult to decide when there exists a non-injective non-constant quasi-conformal map of a non-Abelian nilpotent group into another (possibly Abelian) nilpotent group (compare [Hol], [Hol-Rick]).

(e) According to Varopoulos, every discrete group Γ which grows faster than $\mathbb{Z}^n$ satisfies the isoperimetric inequality with the exponent $\alpha = \frac{N}{N-1}$ for $N \geqslant n + 1$. It follows, that if Γ serves as the fundamental group of a closed manifold $\overline{V}$ of dimension $n < N = N(\Gamma)$, then quasi-conformal maps $W \to \overline{V}$ which lift to the universal covering V of $\overline{V}$ are uniformly continuous for $|K(W)| \leqslant \text{const} < \infty$. Furthermore, if W also appears as the universal covering of a closed manifold whose fundamental group Γ' grows faster than $\mathbb{Z}^n$, then the existence of a quasi-conformal homeomorphism makes Γ and Γ' quasi-isometric. Probably, this remains true with no extra assumption on the dimension.

Conjectures

(i) *If the universal coverings of two closed manifolds are quasi-conformally homeomorphic then the fundamental groups are quasi-isometric.*

(ii) *If the group $\Gamma = \pi_1(\overline{V})$ is not virtually Abelian, then quasi-conformal maps $W \to \overline{V}$ are uniformly continuous under the standing assumptions $|K(W)| \leqslant \text{const} < \infty$.*

(iii) **Example.** Let $V = V_1 \times V_2$, where V_1 is a simply connected non-Abelian nilpotent Lie group and V_2 is a closed manifold. Then (ii) implies that every quasi-conformal map $\mathbb{R}^n \to V$ is constant. (This example as well as the above conjectures can be probably solved with the techniques developed in [Hol-Rick].

(iv) The notion of a quasi-conformal map makes sense for Carnot-Carathéodory spaces. Such maps, however, are rather rare species for general C-C spaces (see [Pan$_{\text{QIR1}}$] where quasi-conformal maps are shown to be conformal in many cases) but they are plentiful in the contact case due to the abundance of contact maps. Our uniform continuity proof easily extends to the C-C category and all applications have C-C counterparts. (The reader may replace everywhere "manifold" by "contact manifold" and ponder over the significance of resulting theorems and conjectures. Then we suggest the paper [Kor-Rei] for a more systematic and profound study of contact quasi-conformal maps. Notice that these maps were discovered by Mostow [Mos] in his remarkable work on the rigidity of locally symmetric spaces of rank one.) Finally, we want to bring reader's attention to "pinched" maps in the C-C category which are (at least) as abundant as Hölder maps and to which some quasi-conformal techniques still apply.

3. Carnot-Carathéodory geometry of contact manifolds

Recall that a contact structure on V is given by a codimension one polarization $H \subset T(V)$ with non-degenerate *curvature form* $\Omega : H \wedge H \to T(V)/H$ which can be defined in the following two equivalent fashions.

(1) Represent H locally as the kernel of a 1-form, say η on V, identify $T(V)/H$ with the trivial line bundle and then define Ω as $d\eta|H$. Notice that this can be done globally on V if H is *coorientable*, i.e. if the line bundle $T(V)/H$ is trivial.

(2) Define $\Omega(X, Y)$ on pairs of vector fields tangent to H by $\Omega(X, Y) = [X, Y] \bmod H$ and verify this is indeed a 2-form, i.e. $\Omega(aX, bX) = ab\,\Omega(X, Y)$ for arbitrary smooth functions a and b on V.

To simplify the matter we assume below that H is coorientable, we fix η and write ω for $d\eta|H$. This is an ordinary 2-form on the bundle H and "non-degenerate" applies to this form in the usual sense. Notice that the non-degeneracy of ω makes rank H even and so $n = \dim V$ is odd, say $n = 2m + 1$. Also observe that if $\omega(= \Omega)$ does not vanish (which is the case for contact structures and $n \geq 3$) then the commutators of degree ≤ 2 span $T(V)$ and so C-C balls in V look as $(\underbrace{\varepsilon \times \varepsilon \times \cdots \times \varepsilon}_{2m} \times \varepsilon^2)$-boxes, where the $(\varepsilon \times \varepsilon \times \cdots \times \varepsilon)$-face is (roughly) tangent to H and the ε^2-edge is transversal to H. In particular, the Hausdorff dimension of V equals $n + 1$ for $n = \dim_{\text{top}} V$ and this remains valid for more general (non-contact) H where the non-degeneracy of ω is weakened to mere non-vanishing of ω.

The equality $\dim_{\text{Hau}} V = n + 1$ implies that there exists no C^α-Hölder homeomorphism (or just surjective map)

$$f : (V, \text{Riem}\,.\,\text{metric}) \to (V, \text{C-C metric})$$

for $\alpha > \frac{n}{n+1}$. On the other hand, the identity map is C^α for $\alpha = \frac{1}{2}$ and one may suspect that there are no V^α-homeomorphisms for $\alpha > \frac{1}{2}$. We shall prove below the inequality $\alpha \leq \frac{m+1}{m+2}$ with $m = (n-1)/2$ for C^α-homeomorphisms $(V, \text{Riem}) \to (V, \text{C-C})$ which improves the above bound $\alpha \leq \frac{n}{n+1}$. Here is the basic C-C feature of contact manifolds that makes this possible.

3.1. Abundance of contact horizontal submanifolds. *If $k \leq m = \frac{n-1}{2}$, then there are plenty of k-dimensional H-horizontal submanifolds in V. In particular, every continuous map $\varphi_0 : \mathbb{R}^k \to V$ can be uniformly approximated by smooth immersions φ everywhere tangent to H.*

This is proven in full generality in § 3.4.3 of [Gro$_{\text{PDR}}$] on the basis of a suitable h-principle but the abundance of H-horizontal spaces can be seen quite elementarily as follows. Every contact manifold admits by Darboux' theorem local coordinates (near each point), say, $x_1, \ldots, x_m, y_1, \ldots, y_m, z$, such that $H = \text{Ker}\,\eta$ for

$$\eta = dz + x_1 dy_1 + \cdots + x_m dy_m.$$

Now every function f on $\mathbb{R}^m$ with coordinates $y_1, \ldots, y_m$ defines the following (jet) map $J_f : \mathbb{R}^m \to \mathbb{R}^{2m+1}$,

$$J_f : (y_1, \ldots, y_m) \mapsto (x_i = \frac{\partial f}{\partial y_i}, y_i = y_i, \ z = -f),$$

which is H-horizontal, since

$$J_f^*(\eta) = -df + \sum_{i=1}^{m} \frac{df}{dy_i} dy_i = 0.$$

The images $J_f(\mathbb{R}^m) \subset V$ for various f and their contact transforms in V constitute the bulk of H-horizontal manifolds in V needed for C-C geometry, but it is faster to refer to the general results in $[\mathrm{Gro_{PDR}}]$.

3.1.A. A lower bound on the Hausdorff dimension of k-dimensional subsets in V for k > n − 1/2. The above "horizontal abundance" has two somewhat opposite C-C uses. First it tells us that there are plenty of injective Lipschitz maps $\mathbb{R}^k \to V$ for $k \leq m = (n-1)/2$ which provide k-dimensional submanifolds having

$$\text{C-C-}\dim_{\text{Hau}} = k = \dim_{\text{top}}.$$

Secondly, one easily shows with "horizontal abundance" that *every compact k-dimensional subset $V' \subset C$ for $k \geq m+1$ satisfies* C-C-$\dim_{\text{Hau}} V' \geq m + 2$.

Proof. Since $\dim_{\text{top}} V' = k$ there exists a continuous map $\mathbb{R}^{n-k} \to V$, such that every nearby map meets V (compare Alexandroff theorem in 4.5). Now, by 3.1, we may assume that this map is H-horizontal and moreover, we can easily arrange a "parallel family" of such maps that define a submersion of some neighbourhood $U \subset V$ onto $\mathbb{R}^k$, say $\psi : U \to \mathbb{R}^k$, such that the levels $\psi^{-1}(y) \in U$ are H-horizontal for all $y \in \mathbb{R}^k$ and such that $\psi(U \cap V') = \mathbb{R}^k$. Then we use the ball-box theorem as in 2.1 and conclude that $\dim_{\text{Hau}} V' \geq k + 1$. ∎

Corollary. *Every C^α-embedding $\mathbb{R}^k \to V$ for $k \geq m + 1$ has $\alpha \leq \frac{m+1}{m+2}$. In particular there is no C^α-homeomorphism (and even no C^α-map of locally non-zero degree) $\mathbb{R}^n \to V$ for $\alpha > \frac{m+1}{m+2}$.*

Question. Can one improve upon this bound on α?

3.2. Polarizations with degenerate curvature ω. Suppose the (curvature) form $\omega(=\Omega)$ on H has constant rank $2r > 0$. Then every field X in $\operatorname{Ker}\omega$ satisfies $[X, Y] = 0 \bmod H$ for all Y in H. This implies that the subbundle $\operatorname{Ker}\omega \subset H \subset V$ is integrable of dimension $2m - 2r$ and so, locally, the polarization H is induced by a smooth map $V \to V_0^{2r+1}$ from some contact structure H_0 on V_0^{2r+1}. Now we see that V contains (quite a few) of H-horizontal submanifolds of dimension $n - r - 1$ which are pull-backs of r-dimensional H_0-horizontal manifolds in V_0. These give us Lipschitz embeddings $\mathbb{R}^k \to V$ for $k \leq n - r - 1$ and topological k-dimensional submanifolds $V' \subset V$ with

$$\dim_{\text{top}} V' = \dim_{\text{Hau}} V' = k.$$

But such V' are rather exceptional for $k > r$, because if the projection of V' to V_0 is not totally degenerate, i.e. has $\dim_{\text{top}} = k > r$, then $\dim_{\text{Hau}} V' \geq k + 1$. It (obviously) follows that V receives no C^α map $\mathbb{R}^n \to V$ of locally positive degree for $\alpha > \frac{r}{r+1}$. It is also clear that the $r = \operatorname{rank}\omega$ metrically distinguishes the corresponding C-C manifolds. Moreover, the same argument shows that

every proper C^α-map $V_{r_1} \to V_{r_2}$ of non-zero degree has

$$\alpha < \frac{n - r_2}{n - r_2 + 1}, \text{ if } r_1 < r_2,$$

and

$$\alpha < \frac{r_2 + 1}{r_2 + 2}, \text{ if } r_1 > r_2(> 0).$$

(Here r_1 and r_2 denote the $\frac{1}{2}$ranks of the implied curvature forms on V_{r_1} and V_{r_2} while $\dim V_{r_1} = \dim V_{r_2} = n$ and basic examples of maps of locally positive degree are given by homeomorphisms $V_{r_1} \to V_{r_2}$.)

In fact, the first inequality is detected with $(n - r_2)$-dimensional horizontal submanifolds in V_{r_1}: these have $\dim_{\text{Hau}} = n - r_2$, in V_{r_1} but the images (of some of them) in V_{r_2} must have $\dim_{\text{Hau}} = n - r_2 + 1$. In the second case, where $r_1 > r_2$, one uses horizontal $(r_2 + 1)$-dimensional manifolds. These are "dense" in V_{r_1} but not in V_{r_2} and so the Hausdorff dimension (of some of them) jumps from $r_2 + 1$ to $r_2 + 2$ under our map $V_{r_1} \to V_{r_2}$.

Question. Let $r_1 = r_2 = r$. Then one expects that C^α-homeomorphisms $V_r \to V_r$ with α close to one preserve the foliation defined by Ker ω. Intuitively, this foliation can be visualized in the C-C geometry of V_r by observing that the leaf through a given point $v \in V_r$ equals the intersection of all "sufficiently regular" (or "generic") submanifolds $V' \subset V_r$ passing through v and having

$$\dim_{\mathrm{Hau}} V' = \dim_{\mathrm{top}} V' = n - r - 1.$$

(Another characteristic property of this foliation is the existence of "many" Lipschitz homeomorphisms of V_2 preserving the leaves.) The question is how to make a *rigorous* description of this foliation in C-C C^α-Hölder terms for some $\alpha = 1 - \varepsilon$.

3.3. Differential forms and straight Alexander-Spanier cocycles.

Instead of H-horizontal submanifolds in V one may use H-horizontal exterior forms a on V where "horizontality" means "vanishing on H" or, equivalently, representability by $\eta \wedge b$ for the 1-form η defining H (which we assume coorientable in this section). Such forms can be either obtained with measures on (sufficiently rich) families of H-horizontal submanifolds (viewed as currents, compare 2.2.A.) or by a purely algebraic consideration based on the following well-known property of *non-degenerate* forms ω on H.

Lefschetz lemma. *The operator $\Lambda^k H \to \Lambda^{k+2} H$ defined by the exterior product with ω (where $h \mapsto \omega \wedge h$) is injective for $k \leq m-1$ and surjective for $k \geq m - 1$ (where $m = \frac{1}{2}\operatorname{rank} H$).*

Using this lemma we obtain a complete description of closed H-horizontal k-forms on V. Namely

Every closed H-horizontal k-form vanishes if $k \leq m$ On the contrary, if $k \geq m + 1$, then such forms are plentiful, e.g. *every k-dimensional de Rham cohomology class can be represented by a closed H-horizontal form.* This is proven by inverting the exterior differential d on horizontal forms. More explicitly, let $H\Lambda^k(V) \subset \Lambda^k(V)$ denote the bundle (as well as the sheaf) of horizontal forms on V and observe that $\Lambda^k H = \Lambda^k(V)/H\Lambda^k(V)$. Then we denote by $\bar{d} : \Lambda^k(V) \to \Lambda^{k+1} H$ the composition of d with the projection (restriction homomorphism) $\Lambda^{k+1}(V) \to \Lambda^{k+1}(H)$ and let d' stand for the restriction of d to $H\Lambda^k(V) \subset \Lambda^k(V)$.

Algebraic inversion lemma. ([8]) *There exist homomorphisms (i.e. differential operators of order zero)*

$$\bar{\delta} : \Lambda^k(H) \to \Lambda^{k-1}(V) \quad \text{for} \quad k \geq m+1$$

and

$$\delta' : \Lambda^{k+1}(V) \to H\Lambda^k(V), \quad \text{for} \quad k \leq m,$$

such that

$$\bar{d}\,\bar{\delta} = \mathrm{Id} \quad \text{and} \quad \delta' d' = \mathrm{Id}.$$

Proof. First we construct $\bar{\delta}$ by defining $x = \bar{\delta}\,\bar{a}$ as a (canonically chosen) solution of the equation $\bar{d}x = \bar{a}$ which is equivalent to the equation $dx|H = \bar{a}$. We solve this by taking $x = \eta \wedge y$ for the form η defining H and observe that $d(\eta \wedge y)|H = d\eta \wedge y|H$ where the 2-form $\omega = d\eta|H$ is nonsingular and the equation $d\eta \wedge y|H = \bar{a}$ is solvable for $\deg y \geq \frac{1}{2}\operatorname{rank} H - 1$ by the Lefschetz lemma. In fact, since the operator $y \mapsto d\eta \wedge y$ is surjective it admits a right inverse, say ω_r^{-1}, such that $d\eta \wedge \omega_r^{-1} y = y$, and then the operator $\bar{\delta} : \bar{a} \mapsto \eta \wedge \omega_r^{-1}\,\bar{a}$ serves as the required right inverse for $\bar{d}$. Next, we turn to d' where we have to show that the equation $d'x' = 0$ implies $x' = 0$. This equation is equivalent to $d(\eta \wedge x) = 0$ or $d\eta \wedge x = \eta \wedge dx$ which says that $d\eta \wedge x|H = 0$, and so $x|H = 0$ according to the Lefschetz lemma. Hence, $x' = \eta \wedge x = 0$ and the required left inverse δ' of d' may be given by $\delta'(a) = \eta \wedge (\omega_{\mathrm{left}}^{-1} a|H)$.

Remarks. (a) These operators $\bar{d}$ and d' are mutually formally adjoint and so are their inverses $\bar{\delta}$ and δ'.

3.3.A. Rumin complex. The above properties of differential forms on contact manifolds trivially follow from the elegant *contact de Rham theorem* discovered by Michel Rumin in his theses (see [Rum$_{1;2}$]). To state this theorem we denote by $I^* \subset \Lambda^*(V)$ *the differential ideal* generated by η, i.e. I^* consists of the differential forms on V representable as $\eta \wedge x + d\eta \wedge y$ for some forms x and y. Then we observe that the exterior differential d sends I^k to I^{k+1} for $i = 0, \ldots, 2m$ which gives us operators $\Lambda^k(V)/I^k \to \Lambda^{k+1}(V)/I^{k+1}$, denoted by d_H. Notice that these operators (and the spaces they act upon) depend only on $H = \ker \eta$ and so they are contact invariant. The operators d_H work well below the middle dimension while above the middle dimension one uses d^H, that is the

---

restriction of d to the *annihilator* J^* of I^* with respect to the exterior product. (*Closed* forms in J^* are the same as closed H-horizontal forms). The following operator crosses the middle line.

Rumin operator. $D : \Lambda^m(V)/I^m \to J^{m+1}$. *There exists a unique operator* $\tilde{D} : \Lambda^m(H) \to H\Lambda^{m+1}$, *such that* $\tilde{D}(\alpha) = d\tilde{\alpha}$ *for every* m-*form* $\tilde{\alpha}$ *on* V *satisfying* $\tilde{\alpha}|H = \alpha$ *and* $d\alpha \in H\Lambda^{m+1}$. *Furthermore* $\tilde{D}$ *passes to the quotient* $\Lambda^m(V)/I^m = \Lambda^m(H)/\{\omega \wedge x \mid x \in \Lambda^{m-2}H\}$, $\omega = d\eta|H$, *i.e.* $\tilde{D}(\omega \wedge x) = 0$ *and* D *is defined as the resulting operator*

$$\Lambda^m(V)/I^* \to H\Lambda^{m+1} \subset J^{m+1}.$$

The proof is straightforward with the Lefschetz lemma. Notice, following Rumin, that D is a second-order differential operator, since the lift $\alpha \mapsto \tilde{\alpha}$ is a first-order operator as is clear from the computation needed to define $\tilde{D}$.

Rumin-de Rham theorem. *The sequence*

$$0 \longrightarrow \mathbb{R} \longrightarrow \Lambda^0(V) \xrightarrow{d_H} \Lambda^1(V)/I^1 \xrightarrow{d_H} \cdots \longrightarrow$$

$$\Lambda^m(V)/I^m \xrightarrow{D} J^{m+1} \xrightarrow{d^H} \cdots \longrightarrow J^{2m+1} \longrightarrow 0$$

is a locally exact complex. Its cohomology is isomorphic to the usual de Rham cohomology.

In fact, Rumin's proof (essentially explained above) provides a chain homotopy equivalence between Rumin's and de Rham complexes where all homomorphisms (especially chain homotopies) are given by *differential* operators. (See [Vin$_{1;2;3}$], [Br-Gr] and [Ge$_{\mathrm{BNCC}}$] for further algebraic results of this kind.)

3.3.B. Construction of straight Alexander-Spanier cocycles with a controlled growth at the diagonal. Recall that the real k-dimensional cohomology of V can be represented by *straight* (Alexander-Spanier) cochains that are functions $c(v_0, v_1, \ldots, v_k)$ defined arbitrarily near the principal diagonal in $\underbrace{V \times V \times \cdots \times V}_{k+1}$. If V is given a metric then we may restrict c to the ε-neighbourhood U_ε of the diagonal for small $\varepsilon > 0$ (here V is compact) and the supremum of $|c|$ on U_ε for $\varepsilon \to 0$

is a relevant characteristic of c. (I have picked up this idea from Alain Connes).

Example. Let V be Riemannian and c represent a non-zero class. Then $\|c\|_\varepsilon \underset{\mathrm{def}}{=} \sup |c| \big| U_\varepsilon \gtrsim \varepsilon^k$. To see this we fix a homology class c' on which $[c]$ does not vanish and then, for each $\varepsilon > 0$, represent c' by a cycle built of $N_\varepsilon \approx \varepsilon^{-k}$ geodesic Riemannian simplices of diameter $\leq \varepsilon$. Now the lower bound on $\|c\|_\varepsilon$ follows from the obvious inequality which is valid for all sufficiently small $\varepsilon > 0$,

$$[c](c') \leq \mathrm{const}\, N_\varepsilon \|c\|_\varepsilon.$$

Now let V be a manifold with an equiregular polarization H of arbitrary codimension. Then every $(k+1)$-tuple of points $(v_0, \ldots, v_k)$ in the C-C ball $B(v, \varepsilon) \subset V$ can be canonically spanned by an actual smooth simplex Δ_ε with vertices $v_0, \ldots, v_k$ of C-C diameter $\leq \mathrm{const}\,\varepsilon$. This can be done locally, for example, by using (local) exponential maps corresponding to (locally defined) full frames of vector fields on V formed by the commutators of suitable H-horizontal fields. If ω is a smooth k-form on V which vanishes on H, then

$$\int_{\Delta_\varepsilon} \omega \leq \varepsilon^{k+1}.$$

In fact, the ball-box theorem allows one to reduce this inequality to the obvious special case where $V = \mathbb{R}^n$ and Δ_ε is spanned by k vectors contained in the box

$$B_\varepsilon = \{|x_i| \leq \varepsilon,\ i = 1, \ldots, n_1,\ |x_i| \leq \varepsilon^r,\ i = n_1 + 1, \ldots, n\} \subset \mathbb{R}^n$$

for $n_1 = \mathrm{rank}\, H$ and $\dot\omega = dx_{i_1} \wedge dx_{i_2} \wedge \cdots \wedge dx_{i_k}$, where $i_k \geq n_1 + 1$. (See 4.1.C' for a more general result.)

Corollary. *Let V be a contact $(2m + 1)$-dimensional Carnot-Carathéodory manifold. Then every k-dimensional cohomology class with compact support for $k \geq m + 1$ can be represented by a straight cochain c satisfying $\|c\|_\varepsilon \leq \mathrm{const}\, \varepsilon^{k+1}$ for all $\varepsilon > 0$.*

Notice that if $H^k(V; \mathbb{R}) = 0$ this Corollary may be applied to open subsets $U \subset V$ with $H^k(U; \mathbb{R}) \neq 0$. Thus one proves once more that the identity map $(V, \mathrm{Riem}.) \to (V,\ \text{C-C})$ cannot be uniformly approximated by C^α-maps with $\alpha > \frac{m+1}{m+2}$.

One can get more mileage from Alexander Spanier cochains by observing that the norm $\|c\|_\varepsilon$ is semimultiplicative under the cup-product

$$\|c_1 v c_2\|_\varepsilon \leq \text{const}\,\|c_1\|_\delta\,\|c_2\|_\delta\,, \qquad \delta = 2\varepsilon.$$

It follows that certain classes in the product of contact C-C manifolds $V = V_1 \times \cdots \times V_i$ in the dimension $k = m_1 + 1 + \cdots + m_i + 1$ are representable by cochains c with $\|c\|_\varepsilon \lesssim \varepsilon^{i+k}$. Therefore, *every proper C^α-map of non-zero degree*

$$(V, \text{Riem.metric}) \to (V, \text{product C-C metric})$$

must have $\alpha \leq \frac{k}{k+1}$. In particular, if $\dim V_1 = \dim V_2 = \cdots = \dim V_i = 3$ (i.e. $m_1 = m_2 = \cdots = m_i = 1$), then $\alpha \leq \frac{2}{3}$.

Question. Does the latter estimate $\alpha \leq \frac{2}{3}$ remain valid for $i = \infty$? (Of course, one should be more specific here about the geometry and topology of infinite products.)

3.4. Width and filling radius. Let V be the $(2m + 1)$-dimensional Heisenberg group with the standard (contact) C-C metric. We want to bound the width (and the filling radius) of subsets (and cycles) $V' \subset V$ in terms of their Hausdorff measures where the width $\text{wid}_k(V' \subset V)$ is defined as the infimum of the numbers $\delta > 0$ for which there exists a continuous map $f : V' \to V$ having the topological dimension of the image at most k and $\text{dist}_V(v', f(v')) \leq \delta$ for all $v' \in V'$.

3.4.A. A bound on wid by mes. *Every closed subset $V' \subset V$ satisfies*

$$\text{wid}\, V'_{k-1} \leq \text{const}_m (\text{mes}_k V')^{\frac{1}{k}} \quad \text{for } k = 1, \ldots, m, \qquad (*)$$

$$\text{wid}_{k-1} V' \leq \text{const}_m (\text{mes}_{k+1} V')^{\frac{1}{k+1}} \quad \text{for } k = m + 1, \ldots, 2m + 1. \, (**)$$

Proof. The inequality $(*)$ follows from the corresponding Riemannian inequality. Namely, we first scale V' (by a C-C self-similarity of V) to have $\text{mes}_k V' = 1$, observe that C-C $\text{mes}_k \geq \text{Riem-mes}_k$ and then use the implication

$$\text{Riem-mes}_k V' \leq 1 \Rightarrow \text{wid}_{k-1} V' \leq \text{const}_V$$

which is valid for all contractible Lie groups V with left-invariant Riemannian metrics by the Federer-Fleming isoperimetric argument (compare p. 17 in [Gro$_{\text{FRM}}$]). Now we turn to the (more interesting) inequality $(**)$ where we shall use a contact version of the method of Federer-Fleming. Namely we need the following

3.4.B. Contact triangulation. *There exists a triangulation* Tr *of* V *into piecewise smooth simplices which is invariant under some discrete cocompact subgroup* Γ *(in the Heisenberg group* V*) and such that all simplices of dimensions* $\leq m$ *are* H*-horizontal, i.e. piecewise tangent to the implied contact structure* $H \subset T(V)$.

This can be derived from the h-principle for Legendre maps (see [$\mathrm{Gro_{PDR}}$]) or proved by an elementary (albeit cumbersome) contact argument which we leave to the reader (compare 3.5 and 4.2). Next we come to the following

3.4.B′. Integral-geometric intersection inequality. *Let* $V'' \subset V$ *be a compact smooth* H*-horizontal submanifold of dimension* $\ell = 2m + 1 - k$ *and* $V' \subset V$ *be a closed* k*-dimensional subset with finite C-C Hausdorff measure* mes_{k+1}. *Then for almost all* $v \in V$ *the Heisenberg translate (written additively)* $V'' + v$ *intersects* V' *at finitely many points and the integral of the intersection number over* V *is bounded by*

$$\int_V \#(V' \cap V'' + v)dv \leq \mathrm{const}_{V''} \, \mathrm{mes}_{k+1} V'.$$

In fact,

$$\int_V \#(V' \cap V'' + v)dv \leq \mathrm{const}_V \, \mathrm{mes}_\ell V'' \, \mathrm{mes}_{k+1} V'$$

as follows by our argument in 2.1 and in 3.1.A.

Corollary. *For every contact* Γ*-invariant triangulation* Tr *of* V *and every* $k = m + 1, \ldots, 2m$ *there exists a positive number* $\varepsilon = \varepsilon(Tr) > 0$, *such that for every* $V' \subset V$ *with* $\mathrm{mes}_{k+1} V' \leq \varepsilon$ *there exists a translate* $V' + v$ *of* V' *which does not intersect the* ℓ*-skeleton* $\mathrm{Tr}^\ell \subset \mathrm{Tr}$ *for* $\ell = 2m + 1 - k$.

Proof. In fact, for every fundamental domain $U \subset V$ of Γ, the integral over U of the intersection number between Tr^ℓ and the translates of V' is bounded by

$$\int_U \#(\mathrm{Tr}^\ell \cap V' + u)du \leq \mathrm{const} \, \mathrm{mes}_{k+1} V'$$

which makes this number zero for some $u \in U$ for small $\mathrm{mes}_{k+1} V'$.

Now the proof of $(**)$ is immediate. We rescale V' to have small fixed $\mathrm{mes}_{k+1} = \varepsilon > 0$ and then bring it to the position where it misses Tr^ℓ.

Then it can be mapped to the $(k-1)$-skeleton of the dual partition of V (as $k-1 = 2m+1-\ell-1$ for our $\ell = 2m+1-k$) by a map f within a fixed distance from the identity. ∎

3.4.C. Asymptotic Riemannian version of 3.4.A. *Let the Heisenberg group V be equipped with a left-invariant Riemannian metric. Then every smooth k-dimensional submanifold $V' \subset V$ for $k \geq m+1$ satisfies*

$$\text{Riem wid}_{k-1} V' \leq \text{const}_V (\text{Vol}_k V')^{\frac{1}{k+1}}. \qquad (++)$$

Proof. Indeed,
$$\text{Riem-Vol}_k \geq \text{C-C mes}_{k+1},$$
which reduces $(++)$ to $(**)$. (Notice that $(++)$ is truly interesting for large $\text{Vol}_k V' \to \infty$.)

Exercise. State and prove a $(+)$-version of $(*)$.

Remarks

(a) There are two other asymptotic versions of 3.4.A: one concerns families of Riemannian metrics on (compact) manifolds approximating C-C metrics (as in 0.8.G and 1.4.D) and the other deals with certain infinite subsets in the complex hyperbolic space where the ideal boundary carries a contact C-C geometry (compare 7.C in $[\text{Gro}_{\text{AI}}]$).

(b) For every contractible Lie group V with a left-invariant Riemannian metric one has certain bound on the widths of k-dimensional subsets $V' \subset V$ in terms of their volumes,

$$\text{wid}_{k-1} V' \leq \psi(\text{Vol}_k V)$$

for some function $\psi = \psi_V(X)$, but one knows little about the (asymptotic) behaviour of this ψ for general contractible groups V. (Notice that this ψ is essentially a quasi-isometric invariant which can be also defined for discrete groups in the spirit of $[\text{Gro}_{\text{AI}}]$.)

Filling Radius. This is defined for k-dimensional *cycles* V' in V as the infimum of those $\delta \geq 0$, such that V' becomes homologous to zero in the δ-neighbourhood of (the support of) V'. It is obvious that $\text{Fill Rad} \leq \text{wid}_{k-1}$ and so the theorem 3.4.A yields a bound on Fill Rad in terms of the measure of (the support of) V'. Unfortunately, the constant in the inequalities $(*)$ and $(**)$ depends on $\dim V$ while one is inclined to allow this constant to depend only on $\dim V'$. Such a bound on the filling radius in $\mathbb{R}^n$, namely

$$\operatorname{Fill Rad} V' \leq \operatorname{const}_k (\operatorname{Vol}_k V')^{\frac{1}{k}},$$

is even known with the *sharp* constant $\operatorname{const}_k = \big(\operatorname{Vol} B(1)\big)^{-\frac{1}{k}}$ for the unit Euclidean ball $B(1) \subset \mathbb{R}^k$ thanks to a result by Bombiery and Simon proven by the variational techniques (see [Bom], and p. 106 in [Gro$_{\text{FRM}}$]). It seems plausible that the k-cycles V' in the Heisenberg group V with the *standard* C-C metric (which is Hermitian with respect to the curvature from ω on H) also satisfy the filling inequality *independent* of $\dim V$, i.e.

$$\operatorname{Fill Rad} V' \leq \operatorname{const}_k (\operatorname{mes}_k V')^{\frac{1}{k}}, \quad \text{for } k \leq m,$$

and

$$\operatorname{Fill Rad} V' \leq \operatorname{const}_k (\operatorname{mes}_{k+1} V')^{\frac{1}{k+1}}, \quad \text{for } k \geq m+1,$$

but the determination of the best const_k does not appear realistic.

3.5. Lipschitz maps of Riemannian manifolds into contact C-C ones. Let V be a contact C-C manifold of dimension $n = 2m+1$ and W be a Riemannian manifold of dimension k. If $f : W \to V$ is a C^1-smooth Lipschitz map then f is (obviously) horizontal, i.e. tangent to the implied contact subbundle $H \subset T(V)$ (of rank $2m$) and moreover, if f is C^2-smooth it is Ω-*isotropic*, i.e. the curvature form Ω of H pulls back to zero by the differential $\mathcal{D}_f$, as

$$\mathcal{D}_f^* \Omega = \mathcal{D}_f^* d\eta = d\mathcal{D}_f^* \eta = 0$$

where η is the 1-form defining H by $\operatorname{Ker} \eta = H$ and the equality $\mathcal{D}_f^* \eta = 0$ expresses the H-horizontality of f. (All this remains valid for C^1-maps f if the differential $d\mathcal{D}_f^* \eta$ is understood as a distribution.) Since the 2-form Ω is non-singular on H, the map f has $\operatorname{rank} \mathcal{D}_f \leq m$ and so it is everywhere singular if $k = \dim W > m$. On the other hand if $k \leq m$, the map f can be very well *regular* which means "immersion", i.e. the injectivity of $\mathcal{D}_f$ all over W. Such a map is, locally, a diffeomorphism onto its image which is a horizontal k-dimensional submanifold in V. (Horizontal submanifolds are called *Legendre* for $k = m$.) The local geometry of horizontal manifolds in V is well understood. If V' is such a submanifold in V, then for each point $v \in V'$ there are local coordinates x_i, y_i, z, $i = 1, \ldots, m$, in V at v, such that $H = \operatorname{Ker}(\eta = dz - \sum_{i=1}^m x_i dy_i)$ and the projection Y of V' to $\mathbb{R}^m$ given by the y_i's coordinates is regular at v. Then, if $k = m$, the map $Y : V' \to \mathbb{R}^m$ is locally onto some neighbourhood $U \subset \mathbb{R}^m$ and V' is represented by the (graph of the) 1-jet of the function $\zeta : u \mapsto z \circ Y^{-1}(u)$

on U as the relation $dz - \sum_i x_i dy_i = 0$ makes $x_i = \frac{\partial \zeta}{\partial y_i}$, $i = 1, \ldots, m$. Thus we obtain a 1-1 correspondence between (germs of) H-horizontal m-dimensional submanifolds in V close to V' at v and (germs of) functions on $\mathbb{R}^m$. In general, if $p \leq m$, we represent horizontal submanifolds in V by 1-jets of functions in $\mathbb{R}^m$ along p-dimensional submanifolds in $\mathbb{R}^m$. It follows that the local geometry in the space of horizontal immersions $W \to V$ is essentially the same as in the space of functions on W. In particular, the sheaf of horizontal immersions $f : W \to V$ is *microflexible* in the sense of [$\mathrm{Gro_{PDR}}$]. This means, roughly speaking, that every deformation of f initially defined near some compact subset $W_0 \subset W$, say ψ_t, $t \in [0,1]$, where $\psi_{t=0}$ equals f on some neighbourhood $U_0 \supset W_0$ and where ψ_t is H-horizontal and regular on U_0 for each t, can be extended to an H-horizontal deformation f_t of $f = f_{t=0}$ for $t \in [0, \varepsilon]$ for some *positive* $\varepsilon \leq 1$ (depending on ψ_t) and where "extension" means $f_t \mid W_0 = \psi_t \mid W_0$ for $t \in [0, \varepsilon]$.

Moreover, one knows that the horizontal immersions are, in fact, *flexible* which means one can take the above ε equal to 1, but this is a rather difficult theorem (see [$\mathrm{Gro_{PDR}}$]) which is not truly needed for our present purpose. What we need, however, is a *piecewise* smooth extension f_t for $t \in [0,1]$ which is horizontal and regular on each piece for all t. The existence of such an extension is achieved with the *Poenaru pleating* (or *folding*) *lemma* which delivers such homotopy where f_1 for $\dim W = 1$ looks as in Fig. 5 below (compare 4.4, also see pp. 51 and 112 in [$\mathrm{Gro_{PDR}}$]).

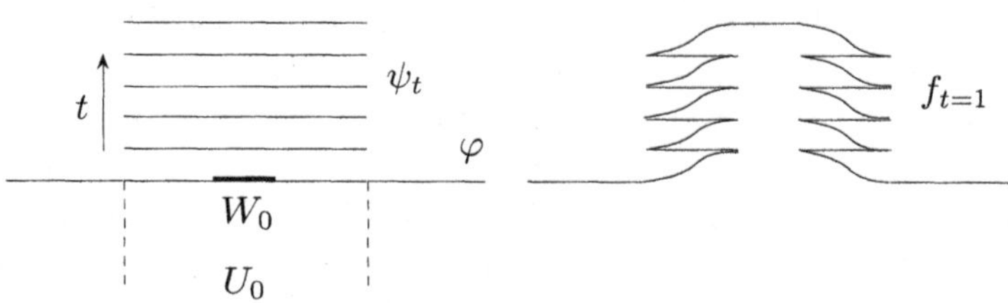

Figure 5

Notice that piecewise smooth horizontal maps are Lipschitz as well as the smooth ones and, if one wishes, one can make them smooth by applying a smooth self-mapping $W \to W$ collapsing a neighbourhood of the non-smoothness locus. But the regularity cannot be recaptured once it is lost. In fact, piecewise regular maps are not difficult to deal with as we know how to deform them (using microflexibility) over the regular pieces. On the other hand, smooth horizontal maps without any a priori control over the domain of regularity, admit no nice 1-jet representation and they cannot be so easily deformed.

Now we state (and indicate the proofs of) several theorems which ensures sufficiently many piecewise smooth Lipschitz maps as in the case of maps into a Riemannian (rather than C-C) manifold V.

3.5.A. First Lipschitz approximation theorem. *Every continuous map $f : W \to V$ admits a fine C^0-approximation (which amounts to the uniform approximation for compact V) by piecewise smooth and piecewise H-horizontal (and hence locally Lipschitz) maps (where we assume throughout $k \leq m$, for $k = \dim W$ and $2m + 1 = \dim V$).*

Sketch of the proof. First we recall the necessary and sufficient condition for an approximation of f by *smooth* horizontal maps f_{hor}. This reads,

there exists a continuous injective homomorphism $\varphi : T(W) \to f^*(H)$ with an Ω-isotropic image (i.e. with $\Omega|\varphi\big(T(V)\big) = 0$ for the curvature form Ω on H corresponding to $\omega = d\eta$ for the 1-form η defining H).

The necessity is easy as one can make φ out of the differential $\mathcal{D}f_{\mathrm{hor}} : T(W) \to f^*_{\mathrm{hor}}(H)$ for f_{hor} close to f. The sufficiency follows from the (dense) h-principle for horizontal maps (see p. 339 in [Gro$_{\mathrm{PDR}}$]). Notice that the proof of the h-principle is based upon flexibility of the sheaf of horizontal immersions. As we are content here with *piecewise* horizontal maps, we may use Poenaru's pleating (or folding) trick and thus obtain an alternative (more elementary) construction of piecewise horizontal maps approximating f under the assumption of the existence of φ. This works, for example, in the case where W is homeomorphic to $\mathbb{R}^k$ and so φ exists being a section of a certain bundle over $W(= \mathbb{R}^k)$.

Now, to grasp the idea suppose the map f decomposes as

$$W \xrightarrow{\ e\ } W' \xrightarrow{\ f'\ } V$$
$$\underbrace{\qquad\qquad\qquad}_{f} \uparrow$$

where f' is a smooth horizontal immersion and e is some continuous map. Then, assuming $\dim W' \geq k = \dim W$, the required approximation of f

can be achieved by approximating e by piecewise smooth and piecewise regular maps $W \to W'$. These are constructed by using sufficiently fine triangulations of W, approximating e on the vertices and then "linearly" interpolating to the simplices in W.

Next suppose there exists a vector bundle T^* over W of rank $\ell \geq k = \dim W$ which admits an injective Ω-isotropic homomorphism $\varphi : T^* \to f^*(H)$ (where Ω-isotropic means $\varphi^*(\Omega) = 0$). Then φ on each fiber T_v^*, $v \in W$, can be turned into a germ, say T_v', of a ℓ-dimensional H-horizontal submanifold in V through $f(v) \in V$ continuously depending on v. Then, by microflexibility of such germs, one can, after slightly perturbing them, construct, for each $\varepsilon > 0$, a non-Hausdorff (or branched as in 4.4) manifold W_ε' horizontally immersed into V, say by $f_\varepsilon' : W_\varepsilon' \to V$, such that f can be ε-approximated by the composed map $f_\varepsilon' \circ e$ for a suitable $e : W \to W_\varepsilon'$. Compare D - D'' in 2.2.7 of [Gro$_{\mathrm{PDR}}$] and see Fig. 6 below.

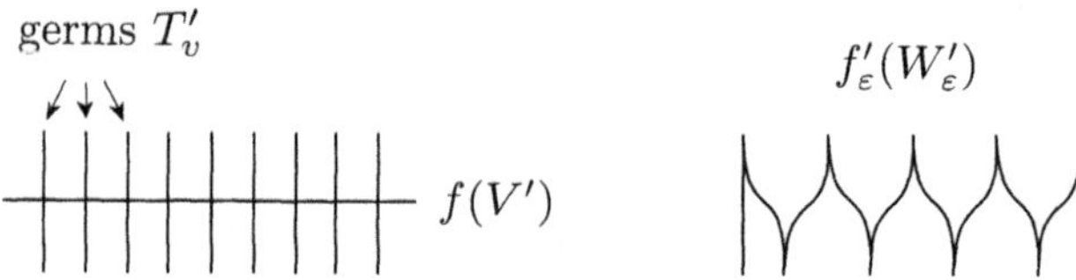

Figure 6

Then e is made piecewise "linear" as in the case of an ordinary W'.

Example. If V is contractible, then the required T^* and φ obviously exist and the above applies.

Notice that we have used so far only the microflexibility of horizontal immersions and no specific contact geometry.

General case. Suppose W is triangulated and we want to approximate $f : W \to V$ by continuous maps which are smooth horizontal and regular on all simplices. The obvious necessary condition requires the existence of Ω-isotropic injective homomorphisms $\varphi_\Delta : T(\Delta) \to f^*(H)$ for all simplices Δ in the triangulation of W, such that $\varphi_{\Delta'}$ over each face Δ' of Δ is obtained by restricting φ_Δ to $T(\Delta') \subset T(\Delta)$. Then the h-principle in 3.4.3 of [Gro$_{\mathrm{PDR}}$] implies that this condition is necessary as well as sufficient and the approximation theorem reduces to finding a triangulation of W for which the homomorphism φ_Δ exist. In fact, these φ_Δ exist for every sufficiently fine triangulation of W which is most clearly seen if

$$(V, H) = \left(\mathbb{C}^m \times \mathbb{R}, \ H = \ker \left(\sum_{i=1}^{m} x_i dy_i - dz \right) \right).$$

Here one may use any triangulation of W and use a generic piecewise linear map $W \to V$ whose projection to $\mathbb{C}^m$ is regular and *totally real* on all simplices of W, where an affine subspace in $\mathbb{C}^m$ is called totally real if it contains no $\mathbb{C}$-line. The space of totally real subspaces in $\mathbb{C}^m$ contracts (by an easy argument) to the space of the ω-isotropic subspaces for the form $\omega = \sum_i dx_i \wedge dy_i$ which essentially equals the curvature Ω of H. Furthermore, this contraction can be made compatible with the inclusions between totally real (respectively, ω-isotropic) subspaces of different dimension. Thus the differential of our piecewise linear map contracts to the desired piecewise linear Ω-isotropic homomorphism $\{\varphi_\Delta\}$. (A more logical proof of the existence of $\{\varphi_\Delta\}$ should use certain connectivity of the *Tits' building* which is a polyhedron with the vertices corresponding to linear ω-isotropic subspaces in $\mathbb{C}^m$ and where the simplices correspond to flags of such subspaces in $\mathbb{C}^m$.)

3.5.A$'$. Lipschitz approximation of families of maps $\mathbf{W \to V}$.

Suppose we start with a family of continuous maps $f_p : W \to V$ where p runs over some compact space and we want an approximation by a P-family of maps where all members are Lipschitz. This can be done again with piecewise horizontal maps but now the implied triangulation of W must depend on P. (A generic family of piecewise linear maps of a fixed triangulation into $\mathbb{R}^m$ may become non-regular on some simplices for certain values $p \in P$ if $\dim P$ is large.) For example, if P is a manifold, one may use some triangulation of $W \times P$ and use generic piecewise linear maps $W \times P \to \mathbb{C}^m \times \mathbb{R}$. This idea can be extended to general compact (and even locally compact) spaces Q foliated by k-dimensional smooth manifolds which allows for $k \leq m$ an *approximation of every continuous map $f : Q \to V$ by leafwise Lipschitz maps*. This conclusion is by no means deep (not to say plain trivial). Yet, I do not see a short elementary proof of it.

3.5.B. Extension of piecewise horizontal maps. Take a submanifold or, more generally, a piecewise smooth subcomplex $W_0 \subset W$ and let $f_0 : W_0 \to V$ be a piecewise regular horizontal map. Then the above arguments allow an extension of f_0 to a similar map of all of W to V provided f_0 admits a continuous extension to W. In the case where W_0 is a smooth submanifold in W and f_0 is smooth this can be done by a more or less straightforward application of the flexibility (or microflexibility + pleating) of the sheaf of smooth horizontal maps. In general, one should note that even a local extension of f_0 at a non-smooth point is a global problem. For example, if $f_0(W_0)$ is a piecewise horizontal curve in V, then an extension to an ambient $W \supset W_0$ of dimension two at a breaking point v of f_0, amounts to joining two points (corresponding to two tangent vectors τ_1 and τ_2, see Fig. 7) in the standard contact sphere $S_v^{2m-1} \subset H_v = \mathbb{R}^{2m}$ by a horizontal path in S_v^{2m-1}.

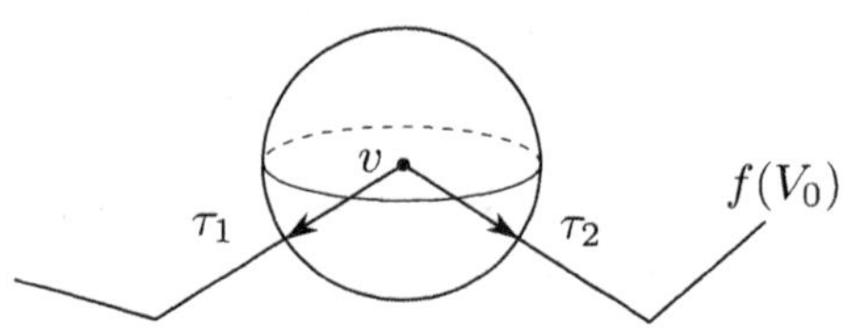

Figure 7

Compactness and complexity. We want to introduce a suitable notion of *complexity* (or size) of a piecewise smooth map so that the complexity of the above extension of f_0 from W_0 to W would admit a bound in terms of the complexity of f_0 on W_0. An equivalent formulation can be achieved with some topology in the space of our maps so that for each precompact subset F_0 of maps $W_0 \to V$ there would exist of a precompact set F of maps $W \to V$, where each $f_0 \in F_0$ could be extended to an $f \in F$. (Eventually, we want to reduce complexity to the Lipschitz constant but this will be done later on the basis of more complicated preliminary notion of complexity.) In what follows we assume the manifolds V and W are compact and we fix Riemannian metrics in both of them. Then we may speak of the norms of the differentials $\|\mathcal{D}^r f\|$ for C^r-maps $f : W \to V$, and of the norm of $(\mathcal{D}f)^{-1}$. The latter is finite iff f is regular (i.e. an immersion). Now we introduce the r-complexity $\|f\|_r$ of an immersion $f : W \to V$ as the supremum over W of the sum of the norms

$$\|(\mathcal{D}f)^{-1}\| + \|\mathcal{D}f\| + \|\mathcal{D}^2 f\| + \cdots + \|\mathcal{D}^r f\|.$$

Warm-up exercise. *If* $\dim V > \dim W$ *and an immersion* $f_0 : W_0 \to V$ *admits an extension to an immersion* $W \to V$, *then, for each* $r = 1, 2, \ldots$, *there is such an extension* f *whose complexity* $\|f\|_{r-1}$ *is bounded by* $(\Phi\|f_0\|_r)$ *for some function* $\Phi(c) = \Phi_{V,W,W_0}(c)$.

The main ingredient of the proof is the flexibility of immersions (Smale-Hirsch theory) which implies that if some C^1-immersion $f_\infty : W_0 \to V$ is a C^1-limit of immersions $f_i : W_0 \to V$, extendable by immersions $\tilde{f}_i : W \to V$, then f_∞ also extends to a C^1-immersion $\tilde{f}_\infty : W \to V$. (This is the only place where we need $\dim W < \dim V$.) Next, assuming f_∞ is C^{r-1}, the extension $\tilde{f}_\infty$ can be easily smoothed to C^{r-1} and then, if the convergence $f_i \to f_\infty$ is C^{r-1}, each f_i for $i \geq i_0$ obviously extends to $\bar{f}_i : W \to V$ which is C^{r-1}-close to $\tilde{f}_\infty$ and thus have $\|\bar{f}_i\|_{r-1} \leq \text{const}$. All this yields what we need as the set of immersions $W_0 \to V$ with a given bound on $\|\cdot\|_r$ is C^{r-1}-precompact.

Remark. One could easily recapture the loss of one derivative by a simple smoothing argument but this is unnecessary for our ultimate purpose (where such a smoothing becomes problematic).

Horizontal version. The above discussion extends to the case where V is a contact manifold and "immersion" is everywhere replaced by "horizontal immersion" and where we assume as earlier that $\dim V \geq 2 \dim W + 1$. (This inequality is necessary for the existence of a horizontal immersion $W \to V$.) In fact, horizontal immersions have the same flexibility and approximation properties as ordinary immersions. Here, as earlier, one could estimate $\|f\|_r$ by $\|f_0\|_r$ rather than by $\|f_0\|_{r+1}$ but we allow the loss of smoothness to make the present discussion suitable for a generalization to non-contact C-C manifolds V.

Complexity made piecewise. If W is smoothly triangulated then each simplex comes along with a smooth map $\sigma : \Delta \to V'$ where Δ is the unit Euclidean simplex of dimension $i = 0, 1, \ldots, k = \dim W$. Then the C^r-complexity of the triangulation is defined as the sum of all $\|\sigma\|_r$ and the C^r-complexity of a piecewise regular map f which is actually smooth on each simplex in W is defined as the sum of that of the triangulation with the complexities $\|f\|_r$ on all simplices Δ of the triangulation. In general, if no triangulation is specified, the complexity of f refers to the infimum of those with respect to all triangulation of W for which f is smooth and regular on the simplices.

Admission. This definition suffers from a variety of defects, where the major one is the lack of the scale invariance. This will be taken into account later on.

Extension Lemma. *If a horizontal piecewise regular map $f_0 : W_0 \to V$ admits a continuous extension to $W \supset W_0$ then it also admits a horizontal piecewise regular extension f whose piecewise complexity $\|f\|_r$ can be bounded by some function of $\|f_0\|_{r+r_0}$, where $r_0 \geq 0$ depends on $\dim W$ (where, as earlier, V is contact of dimension $2m + 1$ and $\dim W \leq m$).*

Idea of the proof. The construction of f is essentially the same as that of the approximation in the previous section. It is useful to keep the following points in mind.

(a) The standard induction by the skeletons of a suitable triangulation of W reduces the general problem to the case where W is a unit Euclidean simplex Δ and W_0 is the boundary of Δ. This explains why the total loss of regularity, i.e. the number r_0 depends only on $\dim W$. (In fact, one can, with some extra effort, make $r_0 = 0$.)

(b) As we are allowed to subdivide W as much as needed, we do not have to use the flexibility of horizontal maps but only microflexibility augmented by Poenaru's pleating (folding) construction. This becomes especially important for more general (non-contact) C-C geometries where the flexibility is less available.

(c) The *local* extension of f_0 at the faces of dimension i leads to the corresponding *global* extension problem with the dimension shifted down by $i + 1$.

(d) The steps of the extension are completely constructive in nature which ensures the required bound

$$\|f\|_r \leq \Phi(\|f\|_{r+r_0}). \tag{$*$}$$

(e) All ingredients of the extension construction are geometrically rather trivial with the exception of pleating (or the flexibility which contains the pleating as the major geometric component) but the overall proof becomes lengthy and boring if one tries to write down the details. (This is a good excuse for us not to do it here.)

(f) Even if the original map f_0 was everywhere smooth and regular, on W_0 it does not always admit a regular horizontal extension to W' and the division into pieces is unavoidable.

Scale invariance of Φ. If we replace V and W by λV and $\lambda' W$, i.e. we multiply the underlying metrics in V and W by the constants λ and λ', then $\Phi = \Phi_{V,W}$ in $(*)$ may change; however, if $\lambda, \lambda' \geq 1$ then these $\Phi_{\lambda,\lambda'} = \Phi_{\lambda V, \lambda' W}$ are bounded independently of λ and λ'. This follows from the fact that λV converges to the Heisenberg group for $\lambda \to \infty$ and $\lambda' W$ converges to $\mathbb{R}^k$, $k = \dim W$.

3.5.C. Smoothing Lipschitz maps. Let f_0 be a Lipschitz map of a Riemannian manifold W into a C-C contact manifold V. We want to *smooth* f_0, i.e. to approximate it by smooth Lipschitz maps $f_\varepsilon : W \to V$ whose Lipschitz constants $L(f_\varepsilon)$ converge to $L(f_0)$ for $\varepsilon \to 0$. (This is easy and well known when V is Riemannian.) Unfortunately, we are able to do it only for $m \geq k$, where $2m + 1 = \dim V$ and $k = \dim W$, and also we cannot achieve the sharp bound of $L(f_\varepsilon)$ by $L(f_0)$. Yet we have the following

3.5.C$'$. Second Lipschitz approximation theorem. *If $f_0 : W \to V$ is Lipschitz, then, for $m \geq k$, it admits a uniform approximation by smooth horizontal maps $f : W \to V$ whose Lipschitz constants $L(f) = \sup_{v \in V} \|\mathcal{D}f\|(v)$ are bounded by*

$$L(f) \leq CL(f_0)$$

for some universal constant $C = C_m$.

Proof. First enlarge V and W by rescaling $V \mapsto \varepsilon^{-1} V$ and $W \mapsto L(f_0)\varepsilon^{-1} W$ for a small $\varepsilon > 0$. Then triangulate $L(f_0)\varepsilon^{-1} W$ into simplices Δ with the complexities bounded by a universal constant $\mathrm{const}'_k > 0$ (which is possible for small ε as everybody knows). Restrict f_0 to the 0-skeleton of this triangulation and then extend it into each Δ inductively by the skeletons inside of a ball in $\varepsilon^{-1} V$ of radius $R \leq \mathrm{const}_m$. This extension may be chosen with the complexity bounded by a universal constant C_m according to the scale invariance of Φ mentioned earlier. Thus we obtain a map $f_\varepsilon : L(f_0)\varepsilon^{-1} W \to \varepsilon^{-1} V$ whose complexity and, hence, the Lipschitz constant are bounded by C_m. This very map f_ε is $CL(f_0)$-Lipschitz for the original (unscaled) metrics in W and V and it converges to f_0 for $\varepsilon \to 0$ for these metrics. This gives us a piecewise smooth approximation of f_0 which can be made smooth by composing f_ε with a self-mapping of W retracting a small neighbourhood of the singular locus of f_ε onto this locus.   ■

Remarks

(a) Our argument applies, strictly speaking, only to *compact* manifolds W but an obvious modification of the above yields, for all W, a fine C^0-approximation of f_0 by smooth maps f with $\|\mathcal{D}f\|(w)$ controlled at each $w \in W$ by the Lipschitz constant of f_0 on a small ball in W around w.

(b) The above argument also yields certain information when f_0 is a C^α-Hölder map, for example if f_0 is C^1-smooth and hence $C^{\frac{1}{2}}$-Hölder. Namely, *such an f_0 can be ε-approximated (with respect to the C-C metric in V) by f_ε with*

$$L(f_\varepsilon) \le C_m \varepsilon^{\alpha-1} L_\alpha(f_0),$$

where L_α is the Hölder constant of f_0.

(c) The second approximation theorem, as the first one, should extend to families and to maps of foliations into V.

3.5.D. Construction and extension of non-piecewise smooth Lipschitz maps. We start with the following

Trivial Example. Let $v_0, v_1, \ldots$ be a sequence of points in V (where V is contact Carnot-Carathéodory as earlier) such that

$$\operatorname{dist}(v_0, v_1) < \frac{1}{2}, \qquad \operatorname{dist}(v_1, v_2) < \frac{1}{4}, \qquad \operatorname{dist}(v_2, v_3) < \frac{1}{8}, \ldots.$$

Then we join v_0 and v_1 by a smooth horizontal curve of length $\le \frac{1}{2}$, we join v_1 and v_2 by such a curve of length $\frac{1}{4}$ and so on. Thus we obtain in the limit a 1-Lipschitz map $f : [0,1] \to V$ with $f(0) = v_0$, $f(\frac{1}{2}) = v_1$, $f(\frac{3}{4}) = v_3, \ldots f(1) = \lim_{i \to \infty} v_i$.

One can think of the points $1, \frac{1}{2}, \frac{3}{4}, \frac{7}{8}$ as vertices of an infinite triangulation of $[0,1]$ and interpret the above construction as a Lipschitz extension of a map from the zero-skeleton $= \{1, \frac{1}{2}, \ldots\}$ to all of $[0, 1]$. In fact, such an extension is possible according to 3.5.B for higher dimensional (finite or infinite) triangulations of Riemannian manifolds where all simplices Δ are *small and fat*. This means that the diameters $d(\Delta)$ are uniformly bounded and the rescaled simplices $d^{-1}\Delta$ for $d = d(\Delta)$ have uniformly bounded complexities.

Here, in Fig. 8, is an example of such a triangulation of the unit square $\square$ which is obtained by refining a (more natural) partition of $\square$ into (small and fat) pentagons.

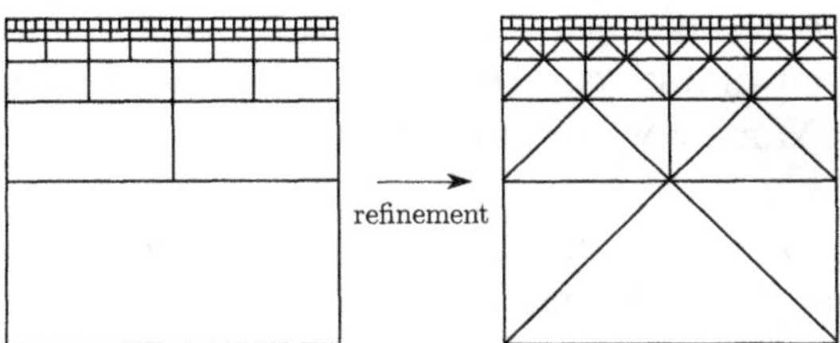

Figure 8

Every Lipschitz map of the set of the vertices of such a triangulation into V extends to all of the square $\square$ provided $\dim V \geq 5$. This is done by induction on skeletons where at each step we use the (rescaled) Extension Lemma of 3.5.B. Notice that this lemma provides at each stage maps of bounded complexity (on the scale of each simplex) which makes the induction possible. For example, suppose we are given some Lipschitz map f_0 of the top (limit) segment I of $\square$ into V. This map, composed with the orthogonal projection of $\square$ to I gives us a Lipschitz map of the whole square $\square$, and hence of the zero skeleton into V. Then the above gives a new map of $\square$ to V which is smooth and regular on each triangle and which equals (by continuity) f_0 on the top segment I of $\square$. We conclude that for every Lipschitz map $f_0 : I \to V$ there exists a map $f : \square = [0,1] \times I \to V$ which is piecewise smooth and regular on $\square - I$, which extends f_0 from $I = 1 \times I \subset \square$ and whose Lipschitz constant can be bounded by that of f_0 (provided V is compact of dimension at least 5).

Now let us do the same to the partition of the square $\square$ in Fig. 9 below where the whole boundary plays the role of the limit segment I.

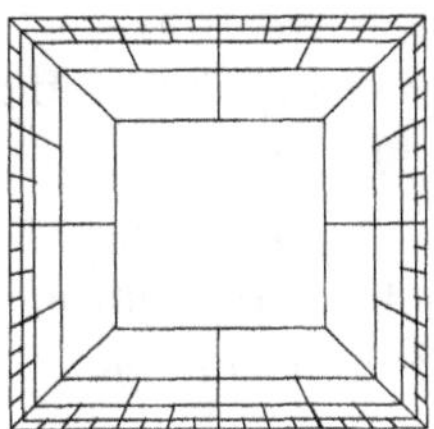

Figure 9

Here again, every Lipschitz map f_0 of the boundary $\partial\square$ into V extends to the zero skeleton with the central projection $\square \to \partial\square$ and then f_0 admits a Lipschitz extension f to $\square$, provided it admits just a continuous extension (where $\dim V$ is assumed ≥ 5 as earlier). What is most important here is the control of the Lipschitz constant of f by that of f_0 which makes our infinite construction relevant even if the original map f_0 were smooth. Namely we have the following

Disk extension theorem. *Let V be a compact simply connected contact C-C manifold of dimension ≥ 5. Then every Lipschitz map f_0 of the boundary circle S^1 of the unit disk $D \subset \mathbb{R}^2$ (which is bi-Lipschitz to $\square$) extends to a Lipschitz map $f : D \to V$, where the Lipschitz constant of f is bounded by*

$$L(f) \leq CL(f_0) \text{ for some } C = C(V).$$

As an immediate corollary we have the following

Isoperimetric inequality. *Every closed C-C rectifiable curve in V bounds a disk of the area (i.e. 2-dimensional C-C Hausdorff measure) satisfying*

$$\text{Area} \leq C_V (\text{length})^2. \qquad (*)$$

Remarks

(a) If length $\gg 1$ the above inequality can be (obviously) improved to

$$\text{Area} \leq C_V \text{ length}.$$

(b) The inequality $(*)$ as well as the Disk extension theorem remain valid for certain non-compact manifolds V. For example, this is the case for the Heisenberg groups H^n for $n \geq 5$. In fact, everything for H^n reduces to the compact case by the obvious use of the self-similarities of H^n.

(c) The present proof of $(*)$ simplifies and conceptualizes the argument in $5.A_3$ of [Gro$_{AI}$]. Also Thurston has been claiming a variety of isoperimetric inequalities in the spirit of $(*)$ and I suspect the above proof must be close to what he had in mind as he insisted (hearsay) on the notion of the *Lipschitz homotopy groups* in the metric and asymptotic geometries. The inequality $(*)$ also appears in [Lee] in the study of minimal Lagrange surfaces in Kähler manifolds.

Our extension argument obviously generalizes to higher dimensions and yields the following

Lipschitz extension theorem. *Let W be a compact k-dimensional Riemannian manifold, $W_0 \subset W$ be a submanifold and let V be a compact contact C-C manifold of dimension $n \geq 2k + 1$. Then, if V is $(k-1)$-connected, every Lipschitz map $f_0 : W_0 \to V$ extends to a Lipschitz map $f : W \to V$ where the Lipschitz constant of f satisfies*

$$L(f) \leq CL(f_0) \quad \text{for} \quad C = C(V, W, W_0).$$

Remarks and questions

(a) It seems not difficult to generalize the above theorem to families and to leafwise Lipschitz maps of foliations into V.

(b) Most results in this section probably extend to the case where the source manifold W is C-C rather than Riemannian. The first case to check is where W is C-C *contact* of dimension $< \dim V$. (It is less clear what are *equidimensional non-injective* Lipschitz maps between contact manifold. For example, one does not know if and how such maps fold. On the other hand, such maps may easily collapse domains to points and to more general horizontal submanifolds. Then, using the collapse to points one may, for instance, construct equidimensional Lipschitz maps of a given degree between closed contact manifolds where the receiving one is the sphere S^{2k+1} with the standard contact structure.)

(c) Recall that the Nash-Kuiper theorem (see [Gro$_{\text{PDR}}$], for instance) allows one to approximate under favorable topological conditions, $(1-\varepsilon)$-Lipschitz maps between Riemannian manifolds by isometric C^1-maps where "isometric" means preservation of the length of the curves. The Nash part of the theorem was transplanted by D'Ambra to the contact C-C manifolds (see [DA$_{\text{C1}}$]). On the other hand, the Riemannian immersion theory has a Lipschitz (non-C^1) counterpart (which allows, in particular, equidimensional isometric maps, see 2.4.11 in [Gro$_{\text{PDR}}$]). This suggests similar results for (Lipschitz) isometric maps into contact C-C manifolds.

(d) Our Lipschitz extension theorem does not seem to yield any higher dimensional isoperimetric inequality. Nobody knows yet (except, possibly, Bill Thurston) whether every (closed horizontal) surface S in the

Heisenberg group H^n for $n \geq 7$ bounds something 3-dimensional of the volume (i.e. the 3-dimensional C-C Hausdorff measure) satisfying

$$\text{Volume} \leq C(\text{Area})^{\frac{3}{2}}.$$

(e) There are many other dilation characteristics of maps $f : W \to V$ besides the Lipschitz constant and the volume, such as the *p-energy* $\int_W \|\mathcal{D}f(w)\|^p dw$, which is defined whenever f is almost everywhere horizontal. If the receiving space V is Riemannian rather than C-C, one knows how to control the energy of extension of f_0 from $W_0 \subset W$ to W by a suitable energy of f_0 on W_0 (see [G-E] and p. 388 in [E-L]). Some of these extension results may be valid with values in contact C-C manifolds.

Besides the p-energies which measure the dilation of f along curves one may measure the dilation of the volumes of k-dimensional submanifolds by using the action of the differential on the k-th exterior power of $T(W)$. Then one defines the (k,p)-energies $\int_W \|\Lambda^k \mathcal{D}f(w)\|^p dw$ and raises the extension problem for these. Most generally, one may try an extension with a control over several such energies which amounts to controlling the joint distribution of the functions $T_k(w) = \text{Trace}\,\Lambda^k \mathcal{D}^* \mathcal{D}(w)$, for $\mathcal{D} = \mathcal{D}f$, with respect to the Riemannian measure dw.

3.6. Controlled integration of differential forms and bounds on the rational homotopy invariants of maps.

We start with a recollection of some known Riemannian facts which we then will extend to contact C-C manifolds.

Let V be a compact Riemannian manifold. Then every exact k-form α on V can be *integrated* to a $k-1$ form β, which means $d\beta = \alpha$, such that

$$\|\beta\|_{L_p} \leq C\|\alpha\|_q \qquad\qquad (*)_q$$

for all q in the interval $1 < q < N$, $\frac{1}{p} = \frac{1}{q} - \frac{1}{\dim V}$ and some constant $C = C(V,p)$ for $k = 1$. This is just the Sobolev inequality of 2.4 and the proof for $k \geq 2$ is as follows. Choose β with $d^*\beta = 0$ as well as $d\beta = \alpha$ and write, using the elliptic theory, $\beta = K\alpha$, where K is a singular integral operator with the kernel $K(v, v')$ satisfying $\|K(v, v')\| \leq \text{const}\big(\text{dist}(v, v')\big)^{-(\dim V - 1)}$. (In fact, the general case can be reduced to that of $V = \mathbb{R}^n$ where K is *the same* convolution kernel as in 2.4 applied to each scalar component of α.) Then the above $(*)_q$ follows from $(**)_q$ in 2.4. (Probably $(*)_1$ is not true for $k \geq 2$.)

Remarks

(a) The above proof shows, in fact, that

$$\|J^1\beta\|_q \leq C_1\|\alpha\|_q \tag{$*$}'_q$$

where J^1 denotes the 1-jet of β, and $(*)'_q$ implies $(*)_q$ via the Sobolev inequality.

(b) Inequalities like $(*)_q$ are dual to *filling inequalities* evaluating the minimal k-chain filling in a given k-cycle, but the precise relation between the two classes of inequalities is not quite clear apart from the cases $k = 1$ and $k = \dim V$ (compare 2.4 and 3.6.B).

The inequality $(*)_q$ is useful for bounding rational homotopy invariants of smooth maps $f : V \to W$ in terms of a suitable $L_q\Lambda^k$-energies, $E_{k,q}(f) = \int_V \|\Lambda^k \mathcal{D}f(v)\|^q dv$, mentioned earlier. If W is simply connected then every such invariant is obtained, according to D. Sullivan, by the following procedure.

Step 1. Take some closed forms $\alpha_1, \ldots, \alpha_i, \ldots$ in W and let $\alpha_i^* = f^*(\alpha_i)$. Notice that the ($L_p$-norm of α_i^*)^p can be bounded by $\mathrm{const}\, E_{k,q}(f)$ for each $k \leq \deg \alpha_i$ and $q = (k^{-1} \deg \alpha_i)p$.

Step 2. Take among the closed forms α_i^* the non-exact ones and evaluate (i.e. integrate) them against some cycles in V. The resulting numbers are the invariants of the first level (encoding the action of f on the real cohomology).

Step 3. Integrate the exact forms among α_i^* and call the integrals β_i. Consider the products $\gamma_{ij} = \beta_j \wedge \alpha_i^*$ and observe that γ_{ij} is closed whenever $\alpha_j \wedge \alpha_i = 0$. Then the values of the closed non-exact γ_{ij} on the cycles in V constitute the set of the invariants of the second level.

Step 4, etc. Integrate exact γ_{ij}. Consider their products with forms obtained at the previous stages and go on. Notice that the integrated forms are estimated by $(*)_q$ while the L_1-norms of products are controlled by the Hölder inequality. Finally the value of a closed k-form ω on a k-cycle V' in V, that is $\int_{V'} \omega$ is bounded by

$$\int_{V'} \omega = \mathrm{const}\, \|\omega\|_{L_1}$$

for const depending on V and the homology class of V'.

Example. Let $W = S^2$ and $f : V \to W$ send the fundamental cohomology class of S^2, represented by the oriented area form α, to zero. Then α^* integrates to some 1-form β on V and the set of the values $\int_{V_i} \beta \wedge \alpha^*$ over the 3-cycles V_i in V constitutes the *Hopf invariant* of f. To bound this we need a bound on the $(L_q\Lambda^2)$-energy of f for $q = \dim V/2$. In fact, if $\|\Lambda^2 \mathcal{D}f\|_{L_q} < A$, then $\|\alpha^*\|_{L_q} \le CA$ and $\|\beta\|_p \le C'A$ for $\frac{1}{p} = \frac{1}{q} - \frac{1}{\dim V}$. Then the L_1-norm of $\beta \wedge \alpha^*$ is bounded by $C''A^2$ provided $\frac{1}{p} + \frac{1}{q} = 1$, i.e. $\frac{1}{q} - \frac{1}{\dim V} + \frac{1}{q} = 1$, which makes $q = \dim V/2$.

Generalization. Let W be an arbitrary $(k-1)$-connected manifold. Then the forms α_i on W have degree $\ge k$ and a simple computation similar to the above shows that every rational homotopy invariant of $f : V \to W$ is bounded by $\|\Lambda^k \mathcal{D}f\|_{L_q}^m$ for $q = \dim V/k$ and a certain positive integer $m \le m_0(\dim V)$. This refines the homotopy finiteness property for maps f with $\|\mathcal{D}f\|_{L_N} \le \mathrm{const}$, $N = \dim V$, (see 2.5) as the norm $\|\mathcal{D}f\|_{L_N}$ bounds the above $\|\Lambda^k \mathcal{D}f\|_{L_q}$.

Remarks

(a) Notice that our $L_q\Lambda^k$-energy for $q = \dim V/k$ is invariant under conformal changes of the Riemannian metric in V. Thus, for a fixed W, the infimum of this energy on the maps in a given homotopy class provides a conformal invariant of V (which measures, in a way, the conformal distance from V to W).

(b) Lower energies also provide non-trivial homotopy information. For example, if a map $f : V \to W$ has a sufficiently small (but yet positive) $L_1\Lambda^k$-energy, then it can be easily homotoped to a map $f' : V \to W$ which sends the k-skeleton of V to the $(k-1)$-skeleton of W. For example, for $k = 2$ we conclude that the induced homomorphism between the fundamental groups factors through a free group,

$$\pi_1(V) \xrightarrow{\quad\quad} \text{Free group} \xrightarrow{\quad\quad} \pi_1(W)$$
$$\underset{f_*}{\underline{\qquad\qquad\qquad\qquad\qquad\qquad}}$$

This raises the problem of a homotopy characterization (of the space) of maps f with some bound on $\|\Lambda^k \mathcal{D}f\|_{L_q}$ for given k and q and suggests the study of discontinuous maps f with $\|\Lambda^k \mathcal{D}f\|_{L_q} < \infty$. (See 3.6.C on some information on these matters.)

Making $q = \frac{3}{2}$ for maps $f : V \to S^2$. If in the above example we choose a sufficiently generic 3-cycle V_i in V we get the $L_q\Lambda^2$-energy of f restricted to V_i of the same order of magnitude as the corresponding energy of f, where the "generic" V_i depending on f is taken from a *compact family* (of 3-cycles or 3-submanifolds in V) *independent* of f. Then the above applies to f on V_i and gives the bound

$$\|\text{Hopf invariant of } f\| \leq \text{const} \left(\|\Lambda^2 \mathcal{D}f\|_{L_{\frac{3}{2}}} \right)^2,$$

for every smooth map $f : V \to S^2$.

This restriction argument works in general and improves the bound on an invariant of f obtained by the evaluation of a product of certain forms on an m-cycle V' in V. Namely, it is more efficient to restrict f to a "generic" V' and apply $(*)_q$ to the induced forms on V' rather than on all of V.

3.6.A. $(\mathbf{L_q \Lambda^k})$-energies in the contact case. Now let V be a contact C-C manifold and see what happens if we evaluate $\Lambda^k \mathcal{D}f$ on the (horizontal) contact subbundle $H \subset T(V)$. We already know that the horizontality condition does not essentially restrict m-cycles for $m < \dim V/2$ and so the above restriction argument works as well with $\mathcal{D}f|H$.

Examples

(a) If $\dim V \geq 2m + 1$, then a smooth map $f : V \to W$, where W is a compact Riemannian manifold, has the norm of the induced cohomology homomorphism $f^* : H^m(W; \mathbb{R}) \to H^m(V; \mathbb{R})$ bounded by const $\|\Lambda^m \mathcal{D}f|H\|_{L_1}$.

(b) If $W = S^2$ as in the previous example, then (the norm of) the Hopf invariant of a smooth map $f : V \to W$ is bounded by const $\|\Lambda^2 \mathcal{D}f|H\|^2_{L_{\frac{3}{2}}}$, provided $\dim V \geq 7$.

Let us indicate a different instance where the restriction of f to horizontal cycles provides a non-trivial homotopy information.

Let V be a compact contact C-C manifold of dimension $2m + 1$ and W be compact Riemannian. Then there exists an $\varepsilon > 0$, such that every smooth map $f : V \to W$ with $\|\Lambda^m \mathcal{D}f|H\|_{L_\infty} \leq \varepsilon$ has trivial homology homomorphism $H_i(V) \to H_i(W)$ for all $i \geq m$.

Idea of the proof. This is obvious for $i = m$ as every m-cycle in V can be made horizontal. Then every i-cycle V' in V for $i = m + m_0$ can be sliced into m_0-dimensional family of horizontal m-cycles. If the f-images of these have small volumes, then $f(V')$ is homologous to zero in V (compare Appendix 1 in [Gro$_{\mathrm{FRM}}$]). For example, if V' is homeomorphic to S^{2m+1}, the slicing of the fundamental cycle can be realized by a degree one map $\alpha : S^m \times S^{m+1} \to S^{2m+1}$ where the slices are the α-images of the spheres $S^m \times s \in S^m \times S^{m+1}$ as s runs over the parameter space, sphere S^{m+1}. One can make this map horizontal on each m-sphere $S^m \times s$ and if $\|\Lambda^m \mathcal{D} f | H\|_{L_\infty}$ is sufficiently small the f-images of these spheres in W have small volumes which makes f (as well as $\alpha \circ f$) homologous to zero.

Question. Does the above remain true for L_∞ replaced by L_q for some $q < \infty$?

The essence of the problem is seen in the following example. Suppose we have a function φ on the unit 2-disk D such that $\|\varphi\|_{L_q}$ is small. We want to slice the disk into curves c_t as in Fig. 10 so that $\|\varphi|c_t\|_{L_1}$ is small for all t.

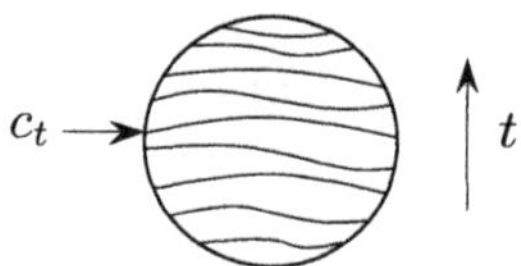

Figure 10

If we try to do it naively with families of parallel straight segments we may run into the "Kakeya" set of Besicovitch: the subset $\{v \in D \mid \varphi(v)| > \lambda\}$ may have small area and yet contain rather long segments in all directions.

To conclude the list of the known homotopy bounds on f in terms of $\mathcal{D} f | H$ we recall that the (C-C conformally invariant) norm $\|\mathcal{D} f\|_{L_N}$, $N = \dim V + 1 = \dim_{\mathrm{Hau}} V$, bounds the number of homotopy classes of maps $f : V \to W$ provided $\dim V \geq 5$ (see 2.5) and we may only repeat the questions we stated earlier for V Riemannian.

Another class of questions arises where V is Riemannian, W is contact C-C of dimension $> 2 \dim V$ and we ask if the horizontality assumption on

$f : V \to W$ reinforces the bound on $\|\Lambda^k \mathcal{D}f\|$ in a homotopically significant way. But the discussion in 3.5.D makes the positive answer unplausible. The same seems to apply to contact maps between contact manifolds with bounds on $\mathcal{D}f|H$. Moreover it may be interesting to translate the above discussion into the intrinsic metric C-C language, where, in fact, $\|\Lambda^k \mathcal{D}f|H\|$ may have several non-equivalent C-C metric counterparts.

3.6.A$'$. Controlled integration of the Rumin complex. Let V be a compact contact manifold of dimension $n = 2m + 1$. We know (see 3.3) that the composition of the exterior d with the restriction to H, i.e. $x \mapsto dx|H$, $\deg x = k - 1$ is an *injection* of the quotient space $\big((k-1)\text{form}\big)/\big(\text{closed }(k-1)\text{-forms}\big)$ into the space of k-form on H (i.e. sections of $\Lambda^k H$), provided $k < m$. So we want to relate the norm on this quotient space coming from the L_p norm on $(k-1)$-forms with L_q of k-forms on H. In other words, given an exact form α on V, we want to integrate it to a form β (which means $d\beta = \alpha$), such that $\|\beta\|_{L_p} \leq \text{const}\,\|\alpha|H\|_{L_q}$. We know how to do this for $\deg \alpha = 1$, but in general it seems hardly possible without modifying the setting (see Remark (a) below). An attractive option is to divide the space of $(k-1)$-forms by a bigger space, namely by $(\text{closed }(k-1)\text{-forms}) + \big((k-1)\text{-forms vanishing on }H\big)$. Then the problem reduces to inverting the operator $d_H : \Lambda^{k-1}(V)/I^{k-1} \to \Lambda^k(V)/I^k$, where I^* is the differential ideal generated by the contact form and I^k stands for $I^* \cap \Lambda^k(V)$ (see 3.3). Now, Rumin shows (see [Rum$_{1;2}$]) that for $k < m$ the operator d_H is *hypoelliptic* in the following sense: the Rumin-Laplace operator $\Delta = (m - k)d_H d_H^* + (m - k + 1)d_H^* d_H$ for a suitable adjoint operator d_H^* satisfies the following L_2-estimate

$$\|L_X L_Y f\|_{L_2} \leq \text{const}(\|\Delta f\|_{L_2} + \|f\|_{L_2}),$$

where X and Y are arbitrary H-horizontal vector fields and the constant depends on X, Y, as well as on the metric on V used in the definition of d_H^*. Then, as in the Riemannian case, we have the (L_2-optimal) solution β of the equation $d\beta = \alpha$ which has $d_H^* \beta = 0$ and one can show that the operator $\alpha \mapsto \beta$ is given by a kernel $K(v, v')$ with $\|K(v, v')\| \leq \text{const}\big(\text{dist}(v, v')\big)^{-(N-1)}$ for $N = \dim_{\text{Hau}} V$ (see [Fol]) and therefore (see 2.4)

$$\|\beta\|_{L_p} \leq C\|\alpha\|_{L_q} \qquad\qquad (*)_q$$

for all q in the interval $1 < q < N$ and $\frac{1}{p} = \frac{1}{q} - \frac{1}{N}$. This $(*)_q$ can be used to bound the homotopy of a map f by observing that d_H-closed forms $\omega \bmod I^k$ on V of degree k for $k < m$ represent the cohomology classes in $H^k(V; \mathbb{R})$ by integrating over *horizontal* k-cycles, or by pairing ω with closed ℓ-forms ω' for $\ell = \dim V - k$ from the annihilator J^ℓ of I^k, where the pairing is the usual one, $(\omega, \omega') \mapsto \int_V \omega \wedge \omega'$. The actual bound on the homotopy invariants obtained this way are weaker than those using restriction of f to horizontal cycles but the above $(*)_q$ brings along a somewhat different perspective.

Question. What is the geometric significance of the controlled integration in the Rumin complex above the middle dimension?

Remarks

(a) Let us comment on the original problem where we to not mod away the horizontal part of β. According to $(*)_q$ every form α on V below the middle dimension integrates to $\bar\beta = \beta + \eta \wedge \beta'$, where $d\bar\beta = \alpha$, and $\|\beta\|_{L_p} \leq C\|\alpha|H\|_{L_q}$. On the other hand, $d(\eta \wedge \beta')|H = d\eta \wedge \beta'|H$, where, recall, η is the 1-form defining H and so $d\eta$ is non-singular on H. Thus, by Lefschetz lemma, $\|\eta \wedge \beta'\| \leq C'\|d(\eta \wedge \beta')|H\|$ everywhere on V. For example, if $\beta = 0$ we have the bounds $\|\eta \wedge \beta'\|_{L_p} \leq \|d(\eta \wedge \beta') \mid H\|_{L_p}$ for all p but for a *generic* L_p-form β' we do not expect anything better, i.e. we cannot significantly change $\eta \wedge \beta'$ by adding to it a *closed* form β'' which would make $\|\eta \wedge \beta' + \beta''\|_{L_p} \leq C\|d(\eta \wedge \beta')|H\|_q$ for $p > q$. (We suggest the reader would prove the non-existence of such β''.) Finally, even if $\beta \neq 0$, the way β is constructed seems to imply that it is "as smooth as possible", and in particular, $\|d\beta|H\|_{L_q} \leq \text{const} \|\alpha|H\|_{L_q}$. (To prove this one needs a suitable bound on $d_H K$ for the kernel $K = K(v, v')$ which, probably, is well known in the hypoelliptic theory.) Thus we conclude to an integration $\alpha \rightsquigarrow \bar\beta$ with $\|\bar\beta\|_{L_p} \leq \text{const} \|\alpha|H\|_{L_q}$ with $p = q$ which cannot be improved to $p > q$.

(b) Probably, much of the above discussion can be given an intrinsic C-C metric meaning in terms of straight cochains and/or (piecewise) Riemannian spaces V_ε approximating V. We suggest the reader would look into this.

3.6.B. Controlled integration and filling in Riemannian manifolds V. Let us start with several remarks on the relation between the controlled integration of forms and filling-in cycles S in V (compare Remark (b) at the beginning of 3.6). First we notice that the controlled integration inequality $(*)_q$, on a compact Riemannian manifold V, i.e.

$$\|\beta\|_{L_p} \leq C\|\alpha\|_{L_q}, \qquad (*)_q$$

remains valid, for $q = \dim V$ and $p = \infty$ if $\deg \alpha = \dim V$. That is, *for every exact form α on V of the top degree, there exists β with $d\beta = \alpha$, such that* $\sup \|\beta\| \le C\|\alpha\|_{L_n}$, $n = \dim V$.

Proof. The form α defines the values of β on all $(n-1)$-cycles S in V which are boundaries via the Stokes formula

$$\beta(S) \underset{\text{def}}{=} \int_S \beta = \int_D \alpha \quad \text{for} \ \ \partial D = S.$$

Since $(\mathrm{Vol}_n D)^{\frac{n-1}{n}} \le \text{const} \, \mathrm{Vol}_{n-1} S$ for a suitable D filling-in S by the isoperimetric inequality, the values of β on S is bounded by

$$\beta(S) = \int_D \alpha \le (\mathrm{Vol}\, D)^{\frac{n-1}{n}} \cdot \int_D \|\alpha\|^n \le \text{const} \, \|\alpha\|_{L_n} \, \mathrm{Vol}_{n-1} S.$$

Then, by the Hahn-Banach theorem the linear functional β can be extended from $(n-1)$-boundaries to all $(n-1)$-chains T in V such that the extended β satisfies the same inequality $\beta(T) \le \text{const} \, \|\alpha\|_{L_n} \, \mathrm{Vol}_{n-1} T$. This makes the pointwise norm of β, thought of as an n-current on V, bounded by $\text{const} \, \|\alpha\|_{L_n}$ and with a little fuss this current can be smoothed to an actual differential n-form with the same bound on the norm.

Remark. The above argument is dual (and essentially equivalent to) to the usual proof of the Sobolev inequality $\|f\|_{L_{n/n-1}} \le \text{const} \, \|df\|_{L_1}$ via the isoperimetric inequality (where the clarifying role of Hahn-Banach was much emphasized by Denis Sullivan).

3.6.B$'$. Thick families of filling-in codimension > 1 in Riemannian manifolds V. Now let $\deg \alpha = k < n$ and let us estimate $\beta(S)$ on a $(k-1)$-cycle in V homologous to zero in terms of α and chains D's filling-in S. For example, if $\deg \alpha = 1$, then S may consist of a pair of points $(v, v') \in V$ and one uses a suitable family of segments between v and v' for D's. (For example for v and v' being the opposite poles of a round sphere one takes all geodesic segments between the poles.) Thus one estimates the difference $\beta(v) - \beta(v')$ for the function β with $d\beta = \alpha$, by the integrals of α along these segments. In particular, one estimates $|\beta(v) - \beta(v')|$ in terms of the norms $\|d\beta\|_{L_{n+\varepsilon}}$ for $\varepsilon > 0$, which leads to the (rather obvious) Sobolev inequality for the C^{δ}-Hölder constant for $\delta = \varepsilon/n + \varepsilon$ of β by $\|d\beta\|_{L_{n+\varepsilon}}$ (which remains valid with $N = \dim_{\mathrm{Hau}} V$

in place of n for all C-C manifolds, see 2.3.E). The same logic applies to $\dim D$'s ≥ 2, where the suitable families of fillings must be sufficiently *thick* as D's approach S. Namely, at each point $v \in S$ the measure of the tangent spaces to the D's at v must be positive in the full linear "pencil" (which is S^{n-k} in this case) of the k-planes in $T_v(V)$ containing $T(S)$. Then one bounds $\beta(S)$ by

$$|\beta(S)| \leq \text{const} \, \|\alpha\|_{L_q}, \; q = n - k + 1 + \varepsilon, \qquad (+)$$

where the constant depends on S and on $\varepsilon > 0$. (If $n = k$ this is valid with $\varepsilon = 0$ as well.) In order to make $(+)$ truly useful, one should reduce the dependence $\text{const}(S)$ to something like $\text{const}(\text{Vol}\,S)$ or $\text{const}(\text{Vol}\,S, \text{Diam}\,S)$. This, probably, can be done in certain cases by refining the filling inequality of Federer-Fleming. Namely, instead of a single filling D of S one may look for a sufficiently thick family of these or (which is easier) to construct a D, such that a *given* positive function φ on V (playing the role of $\|\alpha\|^s(v)$) has $\int_D \varphi \leq \text{const}(\|\varphi\|_{L_q}, \text{Vol}_{n-1}\,S)$. Another possibility is to bound $\beta(S)$ only for certain "standard" cycles S keeping in mind that only a small portion of α matters in $(+)$ for S of a small diameter ρ. In fact, the relevant family of D's may be taken inside a ball B_ρ of radius ρ containing S and then one can use $\|\alpha|B_\rho\|_{L_q}$ in $(+)$ instead of the full L_q-norm. Such a localized version of $(+)$ may suffice for $(*)_q$, $1 < q < n$. (This geometric approach to $(*)_q$ is motivated by possible generalizations where appropriate analytic techniques are not available.)

Remark on the thickness. Let us make more precise our thickness condition on a family of k-dimensional chains (e.g. submanifolds) D in V. We parametrize our family by a measure space M with a probability measure dm and, in fact, "a family of D's" means for us such a measure on the space of all D's.

Next, given a Borel function φ on V, we first integrate φ over each D in V, call this integral $\varphi(D)$, then integrate $\varphi(D)$ over (M, dm) for a given family D_m and call the result of the integration $\varphi(M)$. Notice that $\inf_M \varphi(D_m) \leq \varphi(M)$.

Definition. We say that a family (M, dm) of D_m has q-thickness $\geq C^{-1}$, iff $\varphi(M) \leq C\|\varphi\|_{L_q}$ for all positive Borel functions φ on V. (One usually uses a word like "modulus" but "thickness" seems less confusing.)

Thus our sentence "a standard S can be filled by a q-thick family" means that for each S from a certain given set of "standard" cycles there exists a family of chains D_m filling in S, such that $\{D_m\}$ has q-thickness $\geq C^{-1}$ for a constant $C > 0$ independent of S (from our set). Here is a simple existence theorem for such fillings which has been already used several times.

Let $q > n - k + 1$ and $\{S\}$ be a compact family of piecewise smooth $(k-1)$-cycles in V. Then every $S \in \{S\}$ can be filled by a q-thick family of D's. In particular, an individual cycle S built of finitely many smooth (i.e. smoothly embedded) simplices in V can be filled in by a q-thick family of D's.

3.6.C. Filling-in curves in Riemannian manifolds and $\|\Lambda^2 \mathcal{D}\|$.

Let us apply the above filling ideas directly to a smooth map $f : V \to W$ with a bound $\|\Lambda^2 \mathcal{D} f\|_{L_q} \leq c$ for a given $q > n - 1$ and some $c > 0$. Then every piecewise smooth contractible curve S in V bounds a disk D with a control of the integral of $\|\Lambda^2 \mathcal{D} f(v)\|$ over D and so the image $\Sigma = f(S) \subset W$ bounds a disk Δ in W of area $\leq c \operatorname{const}(S)$. (Actually one can reduce the dependence to $\operatorname{const}(\text{length } S)$ but this is not relevant at the moment.) Let us express the inequality area $\Delta \leq a$ by writing $\operatorname{Fill area} \Sigma \leq a$ (corresponding to the minimal disk filling-in Σ) and let us ask ourselves what is the homotopy structure of closed curves Σ in W with $\operatorname{Fill area} \leq a$. Intuitively, curves Σ with small Fill area look "narrow" like the boundaries of small neighbourhoods of trees in a surface and one expects them to be *simultaneously* contractible. But the mere "narrowness" of curves does not suffice as, for example, the standard sphere S^2 can be sliced into arbitrary narrow curves, all looking like the one in Fig. 11 below.

Figure 11

However, one of the slices must necessarily have $\operatorname{Fill area} \geq \operatorname{area} S^2 / 2$ and the following proposition shows that this is unavoidable.

3.6.C′. Narrow curves proposition. *Let W be a compact Riemannian manifold. If $\pi_1(W)$ acts trivially on $\pi_i(W)$, $i = 2, 3, \ldots$, then the space of closed curves $\{\Sigma\}_a = \{\Sigma \mid \text{Fill area}\,\Sigma \leq a\}$ has finite homotopy type in the space of all closed curves for every $a \geq 0$. That is, the inclusion of $\{\Sigma\}_a$ to the space $\{\Sigma\}$ of all curves factors through a map into a finite polyhedron with the number of cells bounded in terms of a. Furthermore, if $a \leq a_0$ for some sufficiently small $a_0 > 0$ then the subspace of based curves $\{\Sigma_{w_0}\}_a \subset \{\Sigma\}_a$ is contractible in the space of all based curves (i.e loops) in W (where we do not make any assumption on the action of $\pi_1(W)$ on $\pi_i(W)$.*

Sketch of the proof. We use the fact that the space of the maps $S^2 \to W$ with area $\leq a$ has finite homotopy type in the space of all maps $S^2 \to W$, and if a is small the space of based maps is contractible (see 2.5.C). This implies the corresponding properties of the space of fillings Δ of a fixed curve Σ with area $\Delta \leq a/2$ (where "filling" is a map of the disk D^2 to W which restricts on the boundary ∂D^2 to Σ). Finally, one passes from fillings to curves along the standard homotopy route.

Remarks

(a) The above seems to apply to annuli in W joining different curves and yield the uniform local contractibility of the space of curves where the distance is given by the area of the minimal annuli between curves.

(b) If $\pi_1(W)$ acts trivially on $\pi_2(W), \ldots, \pi_k(W)$ then $\{\Sigma\}_a$ has finite homotopy type up to dimension $k - 1$.

Corollary. *If V is a compact simply connected n-dimensional Riemannian manifold, then for each $q > n - 1$ the space of maps $F_c = \{f : V \to W \mid \|\Lambda^2 \mathcal{D}f\|_{L_q} \leq c\}$ has finite homotopy image in all of F, provided $\pi_1(W)$ trivially acts on $\pi_i(W)$, $i \geq 2$. Furthermore, the space F_{c_0} for a small $c_0 > 0$ contracts in F to (the space of) constant maps.*

Sketch of the proof. Join each point $v \in V$ with v_0 by a standard path π where the non-uniqueness is measured by the loops $\pi \circ (\pi')^{-1}$ in V. These loops form a compact family and so their images in W have filling areas controlled by c. In particular, if $c \leq c_0$, these filling areas are small which makes all these loops simultaneously contractible. Then our maps f are simultaneously contractible as well. Similarly, for large c we obtain a homotopy bound on F_c where the details are left to the reader.

Remark. The above seems to generalize to non-simply connected manifolds V, where F_c for small $c > 0$ should be contractible to the space of maps with 1-dimensional images.

Warning. *A map $f : V \to W$ with arbitrarily small norm $\|\Lambda^2 \mathcal{D}f\|_{L_\infty}$ and also with $\|\mathcal{D}f\| \leq$ const may easily have the image of dimension ≥ 2. In fact, there are Lipschitz maps of balls onto balls, e.g. $f : B^3 \to B^2$ which have rank $\mathcal{D}f \leq 2$ almost everywhere on B^3 (see [Bat$_{1;2}$] and [Da-Se]). But these maps can be uniformly approximated by maps with 1-dimensional images as our argument shows.*

Questions. It seems plausible that the above corollary remains valid for every $q > n/2$ and even for $q = n/2$. In fact this is so for maps into classifying spaces of compact Lie groups as follows from a theorem by Uhlenbeck (see 5.4). Then one may go below $n/2$ by looking on the restriction of f to some k-skeleton $V^k \subset V$ and applying the above corollary to $f|V_k$. Thus one sees, for example, that if $q > k - 1$ and $k \geq 2$, then *the bound $\|\Lambda^2 \mathcal{D}f\|_{L_q} \leq c_0$ for a small $c_0 > 0$ forces the restriction $f|V^k$ to be a contractible map $V^k \to W$*, where we assume as earlier that V is simply connected (compare 2.5.G). The question is what happens for smaller q?

One can define "narrow manifolds" of dimension > 1 but one does not expect the narrowness has an effect on homotopy comparable to that for curves. Yet our corollary may have non-trivial generalizations for Λ^j where $j > 2$ as the norm $\|\Lambda^j \mathcal{D}f\|_{L_q}$ does affect the homotopy for $q = n/j$ (see 3.6) and even stronger effect may be expected for $q > n - j + 1$. Yet we do not even know what happens for maps between spheres. For example, for which j, q, n and m is a map $f : S^n \to S^m$ with a sufficiently small norm $\|\Lambda^j \mathcal{D}f\|_{L_q}$ necessarily null-homotopic? Finally, one may allow certain discontinuous maps with $\|\Lambda^j \mathcal{D}f\|_{L_q} < \infty$ and raise the questions similar to those in 2.5.E - H. For example, one may look at the space of measurable maps f for which $\Lambda^j \mathcal{D}f$ is defined on almost all j-vectors (i.e. vectors in $\Lambda^j T(V)$) and $\|\Lambda^j \mathcal{D}f\|_{L_q} < \infty$. Here one should note that each functional $\|\Lambda^j \mathcal{D}f\|_q$ is semicontinuous in the space $F_1^{T(V)}$ (with the topology defined by the norm $\|\mathcal{D}f\|_{L_1}$), i.e. if $f_i \to f$ in this space then

$$\|\Lambda^j \mathcal{D}f\|_{L_q} \leq \liminf_{i \to \infty} \|\Lambda^j f_i\|_{L_q}.$$

Another encouraging sign comes from isosystolic inequalities (see
[Gro_{FRM}]) which may be applied to (the Riemannian metrics induced
by) maps f with $\text{Vol} f \underset{\deg}{=} \|\Lambda^n \mathcal{D} f\|_{L_1} \leq c$ and also to maps with
$\|\Lambda^k \mathcal{D} f\|_{L_1} \leq c$ restricted to k-dimensional subspaces in V.

3.6.D. Thick filling of horizontal curves in contact manifolds. Let
V be a contact manifold of dimension $n \geq 3$. Then the space of closed
piecewise regular horizontal curves (where "regular" means "smoothly
immersed" and the horizontality refers to the implied subbundle $H \subset
T(V)$ of rank $n - 1$) is homotopy equivalent to the space of all closed
curves in V. Furthermore, if $n \geq 5$ then *each contractible piecewise regular
horizontal curve S in V can be filled in by a q-thick family of horizontal
disks D in V, for a given $q > n + 2$.*

Idea of the proof. First let $n = 5$ and V be the space $J^1(\mathbb{R}^2, \mathbb{R})$ of the
1-jets of functions on the (x, y)-plane $\mathbb{R}^2$. Let S, locally, be given by the
1-jet of the zero function at the line $x = 0$ in $\mathbb{R}^2$ and f, g, h be smooth
functions on $\mathbb{R}^2$ with 1-jets vanishing on the line $x = 0$. Thus we obtain
a 3-parametric family of (xy)-planes in V containing S, say

$$\pi : (x, y, a, b, c) \mapsto \left(x, y, J^1_{(x,y)}(af + bg + ch)\right) \in V = J^1(\mathbb{R}^2, \mathbb{R}).$$

(Here and below we do not specify the measure implicit in our notion of
"family" but this is always clear from the context.) We use the standard
coordinates in $J^1(\mathbb{R}^2, \mathbb{R})$, where $J^1_{(x,y)}(\psi) = \left(\psi(x, y),\ \psi_x(x, y),\ \psi_y(x, y)\right)$
and then the Jacobian of the above map $\pi : \mathbb{R}^5 \to \mathbb{R}^5 = J^1(\mathbb{R}^2, \mathbb{R})$ equals

$$\det \begin{pmatrix} f & g & h \\ f_x & g_x & h_x \\ f_y & g_y & h_y \end{pmatrix}. \tag{$*$}$$

Since the 1-jets of the functions f, g, h vanish at $x = 0$, they are divisible
by x^2 and, hence, this determinant is divisible by x^6. It follows that our
family is q-thin (i.e. has q-thickness zero) for $q \leq 7$. On the other hand, for
generic f, g and h (e.g. for $f = x^2$, $g = x^2 y$ and $h = x^3$) this determinant
decays as x^6 (and not faster than that !) for $x \to 0$ and so $(\det)^{-1}$ is
locally p-summable for $p < 6$. This implies that our family of planes has
positive q-thickness for $q > 7$. Then, obviously, every smooth *regular*
curve in V (which is locally equivalent to $J^1 \mid \{x = 0\} = 0$) can be filled
by a q-thick (i.e. having q-thickness > 0) family of disks for every $q > 7$
which proves our claim for the regular case and $n = 5$.

Next, we reduce the case $n > 5$ to $n = 5$ as follows. Again we argue locally and we take a generic m-parametric family of germs of 5-dimensional submanifolds V_α containing our curve S at a given point $s \in S$, where $m = n - 5$. Each V_α is contact for the structure $H_\alpha = T(V_\alpha) \cap H$ and so $S \subset V_\alpha$ can be filled by a 3-dimensional $(7 + \varepsilon)$-thick family of planes as earlier. Then the resulting $(n - 2)$-dimensional family of planes in V is q-thick for every $q > 7 + (n - 5) = n + 2$ as a straightforward computation shows. This proves our claim for *regular* curves S in V for all dimensions ≥ 5.

Finally, let us look at a piecewise regular curve S at a non-smooth point (vertex) $s_0 \in S$. The basic example is where $V = J^1(\mathbb{R}^2, \mathbb{R})$ and S at s_0 is given by the 1-jet of the zero function on the boundary of the positive quadrant $\{x \geq 0, \, y \geq 0\} \subset \mathbb{R}^2$ where s_0 corresponds to $(0, 0)$. Now, as earlier, we use a 3-parametric family $J^1(af + bg + ch)$ where the 1-jets of the functions f, h and h must vanish on S, i.e. on the lines $\{x = 0\}$ and $\{y = 0\}$. Here we take

$$f = \frac{x^2 y^2}{x^2 + y^2} \, , \quad g = \frac{x^2 y^3}{(x^2 + y^2)^{\frac{3}{2}}} \, , \quad h = \frac{x^3 y^2}{(x^2 + y^2)^{\frac{3}{2}}}$$

and observe that the determinant $(*)$ equals $x^6 y^6 (x^2 + y^2)^{-4}$ for these f, g, h. Thus $(\det)^{-1}$ is locally p-summable for $p < 6$ which implies the desired q-thickness of our family for $q > 7$.

This model case applies to $\dim V = 5$ whenever the two tangent vectors to S at the vertex s_0 span an *isotropic* plane in H_{s_0} (for the (curvature) forms $\omega = d\eta | H$ where $\ker = H$) and the general case easily reduces to this situation. Alternatively, one may use generic "conical" families at s_0 filling S (which are *not* C^1-smooth at s_0) similar to those in 3.5.B. (We leave the details to the reader.)

Corollary. *Let W be compact Riemannian, V be a compact contact simply connected C-C manifold of dimension $n \geq 5$ and $q > n + 2$. Then the space of maps $F_c = \{f : V \to W \mid \|\Lambda^2 Df | H\|_{L_q} < c\}$ has finite homotopy image in the space F of all smooth maps $V \to W$, provided $\pi_1(W)$ acts trivially on $\pi_i(W)$ for $i \geq 2$. Furthermore, if $c_0 > 0$ is sufficiently small then $F_{c_0} \subset F$ contracts to (the space of) constant maps (with no assumption on $\pi_1(W)$, compare 3.6.C').*

Counter-example for dim $= 3$. The Hopf map $f_0 : S^3 \to S^2$ can be easily homotoped to an f with $\Lambda^2 \mathcal{D} f | H = 0$ for the standard (horizontal) contact bundle $H \subset T(S^3)$.

Problems. Much of the questions raised in the Riemannian case extend to contact manifolds V, as we want to know what is the homotopy role of the norms $\|\Lambda^j \mathcal{D} f | H\|_{L_q}$ for given j and q, e.g. for $j = 2$ and $q \le n + 2$.

3.6.E. On the global contact geometry. Our study of (V, H) was local and essentially perpendicular to the global contact explosion of the last decade (see [Ben], [El$_{1\text{-}4}$], [Gir], [Ho]). It is unclear at the present moment if our C-C theory can be non-trivially globalized. Namely, we do not know which global contact invariants of (V, H) survive C-C bi-Lipschitz (or quasi-conformal) homeomorphisms. The simplest invariants where the question is already of interest are the Chern classes of the symplectic bundle $(H, d\eta)$. Then come the fillability and overtwist defined by Eliashberg, Hofer-Floer homology etc. Notice, that the behaviour of these invariants is unclear even if the homeomorphisms in question are smooth away from a *finite* subset in V and the contact geometry at such singular points looks very appealing.

4. Pfaffian geometry in the internal light

There are several simple geometric differential objects associated to a polarization (i.e. Pfaffian system) H on V which one wants to visualize in the (internal) C-C metric terms. Eventually one wishes to make the basic Pfaffian invariants and constructions independent of the differential background. But this goal is far from fulfillment.

4.1. A brief metricly guided Pfaffian tour. The basic characteristic of a polarization $H \subset V$ is its rank but even this still can not be recaptured by a robust (e.g. $C^{1-\varepsilon}$-Hölder) metric invariants of the corresponding C-C structure, unless the (first) commutators of the H-horizontal fields span $T(V)$ (as in contact manifolds, for instance) where rank H is determined by the equation

$$\operatorname{rank}(H) + 2(\dim V - \operatorname{rank} H) = \dim_{\mathrm{Hau}}(V, H)$$

(see 1.3.A). Notice that the Hausdorff dimension is sensibly behaved under Hölder maps and so the above metric characterization of rank H is $C^{1-\varepsilon}$-robust.

4.1.A. The H_i-filtration and the type numbers n_i. We denote, as earlier, by $H_i \subset T(V)$ the span of the commutators of order $\leqslant i$ of the H-horizontal fields,

$$H = H_1 \subset H_2 \subset \cdots \subset H_d = T(V)$$

where we assume the situation equiregular, i.e. $n_i = \operatorname{rank} H_i(v)$ independent of $v \in V$. Then the integer vector with the components n_i, where $n_1 < n_2 < \cdots < n_{d-1} < n_d = n = \dim V$ is called the *type* of H and d is the *depth* of H. We know (see 1.3.A) that the sum $\Sigma_{i=1}^{d}\, i(n_i - n_{i-1})$ is a $C^{1-\varepsilon}$-invariant as it equals the Hausdorff dimension of (V, H) but the metric meaning of individual n_i and of d remains obscure. (Notice that V is locally C^α-Hölder equivalent to $\mathbb{R}^n$ for $\alpha = d^{-1}$ and one might think d^{-1} equals the *maximal* α with this property.)

Remark. It seems to be unknown which sequences of numbers n_i may appear as ranks of H_i but these are easily computable for *generic H* where, for example, the inequality $n_1 + \frac{n_1(n_1-1)}{2} \geqslant n$ implies $n_d = n$ for $d \geqslant 3$. The first example with $n - n_2 > 0$ is that of an *Engel structure*, i.e. a generic 2-field H on a 4-space where $n_1 = 2$, $n_2 = 3$, $n_3 = n = 4$. Notice that all Engel structures are mutually locally isomorphic being similar in this respect to contact structures and there are the only generic polarizations with the local uniqueness property (see [Ger$_{\text{EES}}$] about geometry of Engel structures).

4.1.A'. On local connectedness of smooth submanifolds. Naively, one could identify d as the maximum of the Hausdorff dimensions of *smooth* curves c in V. Unfortunately one lacks an adequate C-C metric characterization of smoothness albeit some aspects of smoothness are internally visible. For example, if a C^1-curve c is transversal to H_{d-1}, then the C-C metric on c is Lipschitz equivalent $\sqrt[d]{\text{Euclidean}}$ and so, in particular, $(c, \operatorname{dist}_{\text{C-C}})$ is C^1-*locally connected*, i.e. every two points within distance ε are contained in a (connected!) segment in c of diameter $\leqslant \operatorname{const} \varepsilon$ but this condition is not strong enough to rule out (non-smooth!) curves of large Hausdorff dimension even in the ordinary Euclidean space.

For more general (non-H_{d-1}-transversal) C^k-smooth curves one has some C^α-*connectedness* where the latter inequality is replaced by $\leqslant$ $\operatorname{const} \varepsilon^\alpha$ with α depending on the smoothness of c and its tangency to H_i (compare 4.9). Furthermore, smooth submanifolds $V' \subset V$ also enjoy some local C^α-connectedness which means *contractibility* of each small ε-ball inside the concentric δ-ball for $\delta \leqslant \mathrm{const}\, \varepsilon^\alpha$. (We suggest the reader would make it more precise and specific.) Unfortunately, no such property can distinguish "smooth".

H_i-filtration on curves. The subbundles H_i filter the space $\mathcal{C}$ of smooth (and even Lipschitz) curves in V by $\mathcal{C}_1 \subset \mathcal{C}_2 \subset \cdots \subset \mathcal{C}_d = \mathcal{C}$ for $\mathcal{C}_i$ equal the space of H_i-horizontal curves. The essential metric property of smooth curves $c \in \mathcal{C}_i$ is having the Hausdorff dimension $\leqslant i$ and/or being C^α-Euclidean for $\alpha \geqslant i^{-1}$. This suggests the filtration of $\mathcal{C}$ by $\mathcal{C}_\nu$, $1 \leqslant \nu < \infty$, where ν refers the Hausdorff dimension of $c \in \mathcal{C}_\nu$ and/or the reciprocal of the Hölder exponent α allowing C^α-Hölder parametrization by $t \in \mathbb{R}$. One expects that $\mathcal{C}_\nu$ undergoes particular jumps as ν passes through the integer values. (One tends to think that C^α-curves for $\alpha > \frac{1}{2}$ are H_1-horizontal in a suitable sense, $C^{\frac{1}{3}+\varepsilon}$-curves are H_2-horizontal etc. Maybe, this can be shown if not for individual curves but for suitable families compare 4.9 below.)

Remarks on curvature. Besides the type numbers which manifest the anisotropic nature of the Carnot-Carathéodory geometry there are less apparent infinitesimal invariants but their metric effects may be sometimes more visible than those of the type numbers. For example, for every i_1 and $i_2 \geqslant i_1$ the commutator pairing defines a bilinear form

$$\Omega : H_{i_1} \otimes H_{i_2} \to H_{i_1+i_2}/H_{i_2}$$

generalizing the curvature form for the contact structure (see 3 and 3.2). We saw in 3.2 for type $H = (n-1, n)$ (i.e. for corank $H = 1$) then this curvature may metricly distinguish some H's via the Hausdorff dimension of submanifolds in V.

4.1.B. Submanifolds V' in V of a given type and Thom horizontal homology. Let $V' \subset V$ be a smooth *equiregular* submanifold which means the constancy of the ranks $n_i' = n_i'(V', H) = \mathrm{rank}(H_i' = H \cap T_{v'}(V'))$, $i = 1, \ldots, d$, on V'. These *type numbers* of V' determine its Hausdorff dimension for the metric $\mathrm{dist}_H \mid V'$ by

$$\dim_{\mathrm{Hau}} V' = \sum_{i=1}^{d} i(n_i' - n_{i-1}')$$

and this is the only relation we know (besides $n'_d = n' = \dim V'$, of course). These numbers take the minimal values for *generic* V' transversal to all H_i, that are $m_i = \max(0, n' - (n - n_i))$ and for every integer vector $\{m'_i > m_i\}$ the inequalities

$$n'_i(V', H) \geqslant m'_i \quad, \quad i = 1, \ldots, d, \tag{$*$}$$

impose a non-trivial system of partial differential equation on V'. An instance of this is the relation

$$\sum_{i=1}^{d} i(n'_i - n'_{i-1}) \leqslant M' \tag{$**$}$$

which has a metric interpretation via the Hausdorff dimension. For example, the relation $(**)$ for $(\dim_{\mathrm{Hau}} V =) M' = n' = \dim V'$ is equivalent to the horizontality of V', i.e. to $n'_i(V', H) = n'$, $i = 1, \ldots, d$, and so the non-smooth n'-dimensional subsets V' (e.g. n'-dimensional topological submanifolds) can be viewed as *generalized integral* (i.e. horizontal) *manifolds* of our Pfaffian system (polarization) H.

Let us count the number of P.D.E. (partial differential equations) encoded by $(*)$ and $(**)$. At every point $v \in V$ $(*)$ defines a certain (Schubert) subvariety Σ in the Grassmannian $\mathrm{Gr}_{n'} \mathbb{R}^n$ (for $\mathbb{R}^n = T_v(V)$) and the number in question equals, by definition, to the codimension of this subvariety. Now, clearly,

$$\dim \Sigma = \sum_{i=1}^{d} n_i(m'_i - m'_{i-1}) - (n')^2$$

and

$$\mathrm{codim}\, \Sigma = nn' - \sum_{i=1}^{d} n_i(m'_i - m'_{i-1}).$$

For example, the horizontality of V' is expressed by $(\dim V')$ (corank H) $= n'(n - n_1)$ equations, which corresponds to $(**)$ with $M' = n' = \dim V'$. Next, $(**)$ with $M' = n' + 1$ is given by $nn' - n_1(n' - 1) - n_2$ equations, for $M' = n' + 2$ we have m-equations where $m = \min(nn' - n_1(n' - 1) - n_3, nn' - n_1(n' - 2) - 2n_2)$, and so on. (Notice that the numbers n_i are not arbitrary, e.g. $n_2 - n_1 \leqslant \frac{n_1(n_1 - 1)}{2}$, $n_{i+1} - n_i \leqslant n_i\, n_1$, etc.)

Now we observe that V' in V is locally given by $n - n'$ functions on V' and so the relation $(*)$ is

(-) *underdetermined* : if $\mathrm{codim}\, \Sigma < n - n'$

(0) *determined* : if $\operatorname{codim}\Sigma = n - n'$

(+) *overdetermined* : if $\operatorname{codim}\Sigma > n - n'$.

Thus, in the case $(-)$, we typically expect plenty of C^∞ solutions and this is corroborated by the results in 4.2. Next, in the determined case, $(*)$ may have a reasonably large space of solutions but the existence of these is sometimes hard to prove. Finally, in the truly generic overdetermined case there should not be any C^∞-solutions and under relaxed genericity conditions the solutions must be very special. This will be explained better in 4.? but we should notice here that there is no satisfactory result limiting non-sufficiently smooth solutions. For example, there is no general principle prohibiting integral C^1-manifold of a generic C^∞-Pfaff system in the overdetermined case. In particular, our generic obstructions for C^∞-solutions of the relation $(**)$ (expressing $\dim_{\mathrm{Hau}} V' \leqslant M'$) do not exclude generalized solutions. Yet some of these can be ruled out under special favourable conditions (see 4.5 and 4.11).

Examples

(a) *Horizontality for* $\operatorname{corank} H = 2$. The horizontality of $V' \subset V$ is expressed here (i.e. for $n_1 = n - 2$) by $2n'$ equations and so this is an overdetermined condition for $n' > n/3$ and underdetermined for $n' < n/3$. So, for a generic H integral manifolds are expected (only) of dimension $n' \leqslant n/3$. Yet some non-generic H may have higher dimensional horizontal submanifolds. For instance, the complex holomorphic contact structure viewed as a real polarization of rank $2m-2$ on a $2m$-dimensional manifold has plenty of $2k$-dimensional horizontal submanifolds for $k = (m - 1)/2$ (which, moreover, are complex holomorphic).

(b) *Horizontal surfaces.* If $\dim V' = 2$ then horizontality is overdetermined for $\operatorname{rank} H < \frac{1}{2}n + 1$, where $n = \dim V$, and underdetermined for $\operatorname{rank} H > \frac{1}{2} n + 1$. So (generic) H's below middle dimension are not expected to have integral manifolds of dimension > 1.

Remark on the numbers n_i and genericity. Let L be the free Lie algebra on n_1 generators and Δ_i, $i = 1, 2, 3, \ldots$ denote the rank of the space spanned by the commutators of degree exactly i (so, e.g. $\Delta_1 = n_1$). One knows that

$$\Delta_i = \frac{1}{i} \sum_{j \mid i} \mu(j)\, n_1^{\frac{i}{j}} \quad , \quad i = 1, 2, \ldots,$$

where μ is the Möbius function. Now, for every H of rank i, we have $n_i - n_{i-1} \leqslant \Delta_i$ and if H is generic then, clearly, $n_i = n_i^{\max} \underset{\text{def}}{=} \min\left(n, \Sigma_{j=1}^i \Delta_j\right)$. If we look at the polarizations H with the type numbers n_i prescribed in advance where $n_i < n_i^{\max}$ for some $i \geqslant 2$, we must be aware that these H are subjects to some system of P.D.E. (expressed by the relation type $H = \{n_i\}$) and the notion of genericity among these H should be treated with a respect due to possible complications arizing from this system.

Example where n_2 is prescribed. Fix $n_1 \geqslant 2$ and look at the polarizations H of rank n_1 with given $n_2 = \operatorname{rank} H_2$ written as $n_2 = n_2^{\max} - \delta$ where we assume $n = \dim V \geqslant \Delta_1 + \Delta_2 = n_1 + \frac{n_1(n_1-1)}{2}$. The relation $\operatorname{rank} H_2 \leqslant n_2^{\max} - \delta = n_1 + \frac{n_1(n_1-1)}{2} - \delta$ may be expressed by a system of $\delta(n - n_1 - \delta)$ P.D.E.'s on H. Since H is given by $n_1(n - n_1)$ functions on V, this is an underdetermined system for $\delta n, \ldots, n$ and one expects a reasonable genericity theory for its solutions H (see p.121 in [Gro_{PDR}]). But for $\delta \gg n_1$ this system becomes overdetermined and the solutions should form a rather small space where the idea of genericity may be applied with a caution (if at all).

4.1.B$'$. On the type of a morphisms. Given two Pfaffian systems, i.e. polarized manifolds (V, H) and (V', H'), the *type* of a *morphisms*, i.e. of a smooth map $f : V' \to V$ is given by the numbers

$$n_{ij}(f, v') = \operatorname{rank}_{v'} \ H'_j \cap (\mathcal{D}^{-1}f)(H_i),$$

which reduce to the above n'_i for the case where $H' = T(V')$ and f is an embedding. In general, it is easy to determine the Hölder exponent of f with respect to the C-C metrics dist_H and $\operatorname{dist}_{H'}$ and this give us some metric extract from n_{ij}; but the full metric meaning of the totality of n_{ij} remains unclear.

Exercise. Count the number of P.D. equations on f expressing the inequalities $n_{ij}(f) \geqslant n_{ij}$ for given numbers n_{ij}. (Notice that there are certain inevitable relations between $n_{ij}(f)$, e.g. $n_{11}(f) = n'_1 \Rightarrow n_{ii}(f) = n'_i$ for $i \geqslant 2$.) Find out when the condition "f is C^α-Hölder" is under/overdetermined.

4.1.C. Pfaffian systems in jet spaces. Take a smooth m-dimensional manifold V_0 and observe that the Grassmann manifold $V^1 = \mathrm{Gr}_k V_0$ of k-planes in $T(V_0)$ carries a natural polarization $H^1 \subset T(V^1)$ of corank $m-k$ which is uniquely characterized by the following condition: *the tangential lift (1-jet) of every C^2-smooth k-dimensional submanifold $V_0' \subset V_0$ to V^1 is H^1-horizontal.* (For $\dim V_0' = \dim V_0 - 1$ this is the standard contact structure but for $\mathrm{codim}\, V_0' \geqslant 2$ this H^1 is not so symmetric and looks less beautiful.) This generalizes to the space $V^r = \mathrm{Gr}_k^r(V_0)$ of r-th jets of germs of k-dimensional submanifolds $V_0' \subset V_0$ by observing that V^r carries a natural polarization H^r which is *the minimal subbundle in $T(V^r)$ containing the tangent vectors of the r-jet lifts to V^r of all C^{r+1}-submanifolds $V_0' \subset V_0$ of dimension k.* It is easy to see that this H^r has corank equal the dimension of the space of $(r-1)$-jets of the maps $\mathbb{R}^k \to \mathbb{R}^{m-k}$ at $0 \in \mathbb{R}^k$, where $m = \dim V_0$. Thus

$$\mathrm{corank}\, H^2 = (m-k)\left(1 + k + \frac{k(k+1)}{2} + \cdots + \frac{(k+r-2)!}{(k-1)!(r-1)!}\right).$$

If an integral manifold $V' \subset V^r$ of H^r *diffeomorphically* projects onto some submanifold $V_0' \subset V_0$ then V' equals the r-jet lift $J^r(V_0') \subset V^r$. But there are certain smooth integral manifolds $V' \subset V^r$ where the projection $V' \to V_0$ has a singularity (i.e. not a C^1-immersions).

Examples

(a) *Foliation by the graphs of polynomials.* The space V^r is locally isomorphic to the space of the r-jets of maps $\mathbb{R}^k \to \mathbb{R}^{m-k}$, called $\mathcal{J}^r$, where the graph (image) of the r-jet of each smooth map $f : \mathbb{R}^k \to \mathbb{R}^{m-k}$, denoted $V_f' = J_f^r(\mathbb{R}^k) \subset \mathcal{J}^r$, can be included as a leaf into a H^r-horizontal foliations, namely the one with the leaves V_{f+p}' where p runs over the space of polynomial maps $\mathbb{R}^k \to \mathbb{R}^{m-k}$ of degree r. Now let f be also a polynomial but of degree $> r$ and consider the one parameter family of foliations corresponding to $\lambda f + p$ for $\lambda \in \mathbb{R}$. The tangent space to this foliation at each point $j \in \mathcal{J}^r$, say $T_j(\lambda f + p) \in T_j(\mathcal{J}^r)$, converges as $\lambda \to \infty$ to some k-dimensional subspace $S_j(f)$ in $T_j(\mathcal{J}^r)$ (this is obvious and valid for every C^∞-map f) and the limit field $S(f)$ of k-dimensional subspaces in $T(\mathcal{J}^r)$ is C^∞-smooth away from a proper algebraic subset in $\mathcal{J}^r$. This $S(f)$ is obviously integrable with H^r-horizontal leaves which do not project to smooth k-dimensional submanifolds in $V_0 = \mathbb{R}^m = \mathbb{R}^k \times \mathbb{R}^{m-k}$ anymore. In fact, if f is a homogeneous polynomial map of degree $r+1$

and rank $\geqslant p$ (as defined below) then, clearly, the (k-dimensional) leaves of $S(f)$ project to $(k - p)$-dimensional subvarieties in $\mathcal{J}^{r+1}$ (and hence the projections to $\mathbb{R}^m$ under $\mathcal{J}^{r-1}$ are at most $(k - p)$-dimensional).

Definition. A homogeneous polynomial map $f : \mathbb{R}^k \to \mathbb{R}^{m-k}$ of degree $r + 1$, viewed as a $\mathbb{R}^{m-k}$-valued symmetric $(r + 1)$-form on $\mathbb{R}^k$, $f(x_0, x_1, \ldots, x_r)$, defines a linear map from $\mathbb{R}^k$ to the linear space of such r-forms, say $f' : x \mapsto f(x, x_1, \ldots, x_r)$, and

$$\operatorname{rank} f \underset{\mathrm{def}}{=} \operatorname{rank} f'.$$

For instance, if $\operatorname{rank} f = k$, then the leaves of $S(f)$ are (affine) subspaces in the fibers of the projection $\mathcal{J}^r \to \mathcal{J}^{r-1}$ and as we vary f the tangents to these $S(f)$ span the tangent spaces to the fibers. It follows that the *fibers of the projection* $\mathcal{J}^r \to \mathcal{J}^{r-1}$, *and hence of* $V^r \to V^{r-1}$, *are* H^r-*horizontal*. (Notice that these fibers have dimension $> k$ unless $r = 1$ and $m - k = 1$.)

(b) There are many singular k-dimensional subvarieties $V_0' \subset V_0$ which become desingularized by lifting their non-singular loci to V^r and then taking the topological closure of these lifts in V^r. In fact this desingularization is conjectured (by J. Nash) to work for all algebraic subvarieties V_0' in $V_0 = \mathbb{R}^m$ and sufficiently large $r = r(V_0')$ but even without establishing this conjecture the non-singular loci of the r-jet lifts of singular algebraic varieties $V_0' \subset V_0 = \mathbb{R}^m$ provide us with a pool of smooth k-dimensional H^r-horizontal submanifolds in V^r having their projections to V_0 singular.

(c) The examples (a) and (b) can be brought to an equal footing by considering families of algebraic varieties $V_\lambda' \subset V_0$ where the dimension of V_λ' may jump down at certain values of λ (e.g. V_λ' may degenerate to a single point at $\lambda = 0$) but where the dimension of the r-jet lift of V_λ' to V^r remains constant at these λ.

One gets a better view on integral manifolds in V^r by looking at the natural embedding

$$V^r \subset \underbrace{\operatorname{Gr}_k(\operatorname{Gr}_k(\cdots \operatorname{Gr}_k(V_0) \cdots)}_{r}$$

and by taking the closure $\overline{V}^r$ in this iterated Grassmannian. (Notice that $V^1 = \overline{V}^1$.) Then the s-jets of (potential) integral manifolds $V' \subset V^r$ of H^r are represented by points in $\overline{V}^{r+s}$ where some complication may arise from the possible singularity of $\overline{V}^{r+s}$ at $\Sigma = \overline{V}^{r+s} - V^{r+s}$ (which is, probably, well understood by formal P.D.E. people, e.g. A. Vinogradov & co.).

On C-C regularity of continuous jets. If V_0' is a C^r-submanifold of V_0 its r-jet $V' \subset V^r$ is just continuous but the C-C geometry of (V^r, H^r) suggests certain ways of measuring the tangency of V' to H^r which lead to the following.

Questions. What is the relation between the ordinary $C^{r+\alpha}$-Hölder features of $V_0' \subset V_0$ and the Hölder exponent of the r-jet (lift) map $V_0' \to V' \subset V^r$ with respect to the C-C metric dist_{H^r} on V^r? What is the meaning of the C-C Hausdorff dimension of V'?

4.1.D. Horizontal chains and cycles. Consider singular chains $\Sigma_i c\sigma_i$ in a manifold V where the singular simplices σ_i are *smoothly immersed* into V. If V is polarized by some $H \subset T(V)$ then one filters (the space of) these chains by the types of σ_i and studies the arizing homology theories. The first instance of this was treated by Thom in 1959 (see [Th] who was interested in *horizontal chains* in the above jet spaces (V^r, H^r). Thom viewed such chains, and especially k-cycles, in open subsets $\mathcal{R} \subset V^r$ (where k refers to the dimension of submanifolds $V_0' \subset V_0$ whose r-jets form V^r) as generalized solution of the partial differential relation imposed on k-dimensional submanifolds $V_0' \subset V_0$ by the requirement $J^r(V_0) \subset \mathcal{R}$ (compare [$\mathrm{Gro_{PDR}}$]) and he indicated the idea of the proof of the following statements

(A) *every i-cycle in $\mathcal{R}$ for $i \leqslant k$ is homologous to a horizontal one*

(B) *every horizontal i-cycle for $i < k$ homologous to zero bounds a horizontal $(i+1)$-chain.*

Some aspects of Thom's idea (maps "à fort gradient" and singularities "dents de scie") have been already presented in our contact §3 in the "pleated" disguise of Poenaru. This idea suffices to complete the proof of (A) (see 4.4.A) but (B) appears more difficult. In fact, the (B)-part of the singular homology theory based on smoothly immersed or embedded simplices (without any H in the picture) is a subtle matter only recently settled by F. Lalonde in a satisfactory fashion (see [Lal]. I am not certain Thom insists on *immersed* simplices but this point of view is taken by Briant and Griffiths in [Br-Gr] where the authors indicate a generalization of Thom's theorem, also see [Ge$_{\text{BNCC}}$] on this matter).

4.1.E. Horizontal forms and cohomology. Every finite measure μ on (a compact family of) H-horizontal oriented submanifolds V' of dimension k in V defines a *k-current* which, for a sufficiently "smooth and ample" measure, is represented by a (unique) differential $(n - k)$-form ω' on V which is obviously *H-horizontal* in the following sense

ω' *annihilates every 1-form* η *vanishing on* H, i.e.

$$\eta \mid H = 0 \Rightarrow \eta \wedge \omega' = 0.$$

In other words, ω' vanishes on each hyperplane in $T_v(V)$ containing H_v, $v \in V$, or equivalently, on every $(n - k)$-plane *non-transversal* to H. To see what it means, we represent H locally as the common kernel of $n - n_1$ linear forms, say $\eta_1, \ldots, \eta_{n-n_1}$, for $n_1 = \operatorname{rank} H$ and observe that the above condition is equivalent to divisibility of ω' by $\zeta = \eta_1 \wedge \eta_2 \wedge \cdots \wedge \eta_{n-n_1}$. Thus the H-horizontality condition distinguishes certain subbundle $H\Lambda^{n-k} \subset \Lambda(V) = \Lambda^{n-k}(T(V))$ of rank $n_1!/k!(n_1 - k)!$ (where k should be $\leqslant n_1$ and $H\Lambda^{n-k} \underset{\text{def}}{=} 0$ otherwise).

Next we observe that if the boundaries of the submanifolds V' in the support of μ miss an open subset $U \subset V$, then the resulting form ω is *closed* on U. Thus *de Rham (co)homology of the horizontal forms "corresponds" to Thom's homology of the horizontal cycles* (compare 4.11).

Notice that closed H-horizontal forms annihilate the differential ideal $I^* = I^*(H)$ generated by the 1-forms vanishing on H and so the horizontal cohomology and homology are dual to those of the quotient complex Λ^*/I which are extensively studied in [Ge$_{\text{BNCC}}$] and [Br-Gr] (where the authors point out the duality between their $H^*(\Lambda^*/I)$ and Thom's homology).

Example. *Pure forms.* Let ω be induced from a volume form ω_0 on our $(n - k)$-dimensional manifold V_0 by a smooth map $\varphi : V \to V_0$ of rank $n - k$. Then the horizontality of ω amounts to horizontality of the fibers $\varphi^{-1}(v_0)$, $v_0 \in V_0$. Notice that closed 1-forms and $(n - 1)$-forms are locally generically pure in the above sense and so the passage from (closed) horizontal submanifolds (or cycles) to closed horizontal forms does not essentially enlarges the picture.

Parameter count for closed horizontal forms. Let us evaluate the expected "functional dimension" (of the space) of closed horizontal forms. First, we do that without horizontality by observing that, by definition, the "functional dimension" of the space of k-forms on V equals $\lambda_i = \operatorname{rank}\Lambda^i(T(V))$, i.e. $n!/i!(n-i)!$, while the de Rham cohomology groups have "functional dimension" zero. Therefore, closed i-forms have

$$"f\dim"(\ker d^i) = \lambda_{i-1} - \lambda_{i-2} + \lambda_{i-3} - \cdots (-1)^{i-1}\,\lambda_0 =$$
$$\lambda_i - \lambda_{i+1} + \lambda_{i+2} - \cdots (-1)^{n-i}\,\lambda_n.$$

Then we come to the space of *closed horizontal* forms written as the intersection $H\Lambda^{n-k}\cap\ker d^{n-k}$ where the expected "functional dimension" in the generic case of "transversal" intersection is

$$"f\dim"(H\Lambda^{n-k}\cap\ker d^{n-k}) =$$
$$\operatorname{rank}H\Lambda^{n-k} + "f\dim"\ker d^{n-k} - \lambda_{n-k} =$$
$$n_1!/k!(n_1-k)! + \sum_{j=0}^{n-k}(-1)^{j+n-k-1}\,\lambda_j =$$
$$n_1!/k!(n_1-k)! + \sum_{j=n-k+1}^{n}(-1)^{j-n+k}\,\lambda_j.$$

Notice that this "$f\dim$" is (significantly) greater than the one of the space of H-horizontal k-dimensional manifold (which equals $n-k-k(n-n_1)$) for $2 \leqslant k \leqslant n-2$. For example, if $k=2$ the "functional dimension" of closed horizontal $(n-2)$-forms equals $\frac{n_1(n_1-1)}{2} - n + 1$ and so one generically expects plenty of closed horizontal $(n-2)$-forms on V if $n_1(n_1-1) > 2n-2$ while the horizontal surfaces generically need $2n_1 \geqslant n+2$ for their existence. *Thus, closed horizontal $(n-k)$-forms may be present in abundance without the existence of a single horizontal k-submanifold for* $2 \leqslant k \leqslant n-2$. (But I have no convincing examples actually exhibiting such abundance, compare 4.11.)

Remarks

(a) *Closed forms of a given type.* One could start with submanifolds of given (horizontality) type $(n_1, n_2, \ldots)$ and arrive at similar filtration on de Rham cohomology (compare 4.1.C$'$ and 4.11).

(b) *Hodge theory.* The relation between the notions of horizontality (and type, in general) for submanifolds and forms is quite similar to what happens in the classical Hodge theory where, for example, closed (k, k)-forms correspond to complex submanifolds of real dimension $2k$. In both cases the linearization of the space of submanifolds is achieved via the Plücker embedding of the Grassmann manifold into exterior forms.

(c) *Linearization of* P.D.E. *and Bochner formulae.* A system of s partial differential equations of order r imposed on k-dimensional submanifolds $V_0' \subset V_0$, e.g. on (graphs of) maps $V_0' \to \mathbb{R}^{m-k}$, can be represented by a subvariety $V \subset V^r = \mathrm{Gr}_k^r(V_0)$ of codimension s. Then the solution of the system are represented by H-horizontal k-dimensional submanifolds $V' \subset V$ for $H = T(V) \cap H^r \subset T(V)$ where H^r is the canonical polarization of the jet space V^r. Thus an arbitrary P.D.E. system reduces to a Pfaffian one. (This point of view was emphasized by Thom.) Notice that in many cases one should differentiate our equations several times which corresponds to passing to $V^{(i)} \subset V^{r+i}$ consisting of $(r + i)$-jets of formal solutions of our P.D.E.'s. Only then the Pfaffian system expressed by $H^{r+i} \cap T(V^{(i)})$ becomes truly representative of the original P.D.E.'s (which is essential when the canonical projection $V^{(i)} \to V$ is not surjective). Next we linearize our equations by considering closed H-horizontal $(n - k)$-forms on V, for $n = \dim V$, instead of horizontal submanifolds. (This idea is presented in [Br-Gr] in the dual language of the characteristic homology.) This may greatly enlarge the space of solutions globally as well as locally. The undue enlargement of the space of solutions can be somewhat contained by passing to $V^{(i)}$ over V and even better by replacing the linear span by the *convex hull* of the (space of) actual solutions in the space of (closed) differential forms or currents.

(Not quite Pfaffian) example. Let V be a $2m$-dimensional manifold endowed with an almost complex structure $J : T(V) \to T(V)$. Then *J-holomorphic* submanifolds $V' \subset V$ of real dimension $2k$ are defined by the condition $J(T(V')) = T(V')$ which can be represented by $2k(2m-2k)-2k(m-k) = 2k(m-k)$ partial differential equations imposed on $2m - 2k$ functions (on V'). This P.D.E. system is overdetermined for $2 \leqslant k \leqslant m-1$ and, generically, there is no J-holomorphic submanifolds in

V of dimension $2k \neq 0, 2, 2m$. On the other hand the corresponding bundle of $(2m - 2k)$-forms, which is spanned by the monomials $\Lambda_{i=1}^{k} \eta_i \wedge J\eta_i$ for arbitrary 1-forms η_i on V, has rank $\left(\frac{m!}{k!(m-k)!}\right)^2$, since after the complexifications these monomials span the bundle of $(m - k, m - k)$-forms. For example, J-holomorphic *curves* (i.e. $k = 1$) which themselves has "functional dimension" zero ($2m - 2$ equations against $2m - 2$ functions) give rise to (the space of) closed $(m - 1, m - 1)$-forms having "functional dimension" $m^2 - 2m + 1$ which is strictly positive for $m \geqslant 2$. Next, for $k = 2$ this dimension becomes

$$\left(\frac{m(m - 1)}{2}\right)^2 - \frac{2m(2m - 1)(2m - 2)}{6} + \frac{2m(2m - 1)}{2} - 2m + 1$$

which is > 0 for $m \geqslant 4$, while the "functional dimension" of J-holomorphic surfaces is negative.

Now, if we pass to higher order jets, the corresponding variety $V^{(i)}$ with large i may become empty for $k \geqslant 2$ and, generically, no enlargement of the original (empty) space of J-holomorphic submanifolds takes place. More interestingly, the (formal) *convex hull* of the J-holomorphicity condition is strictly smaller than the linear span. Namely the $(n - k, n - k)$-forms ω in this hull are *positive* in the sense that $\omega(x_1, \ldots, x_{n-k}, Jx_1, \ldots, Jx_{n-k}) \geqslant 0$ for arbitrary tangent vectors x_i, $i = 1, \ldots, n - k$, in V.

Example. The space $\mathbb{C}^m$ contains no compact complex submanifolds of positive dimension k but has plenty of closed (and exact) $(n - k, n - k)$-forms with compact supports. Yet every *positive* closed $(n - k, n - k)$-forms ω with compact support is necessarily zero as on one hand, $\omega \wedge \omega_0 \geqslant 0$ for every positive (k, k)-forms ω_0 with constant coefficients and on the other hand, $\int \omega \wedge \omega_0 = 0$ since ω_0 is exact.

Similar positivity (or convex hull) phenomenon appears in the presence of non-trivial Bochner-Weitzenböck formulae given by certain "positive" forms in the jet spaces of submanifolds (or maps), but an appropriate (non-Pfaffian?) formalism is yet to be developed.

Example. *Holomorphic Pfaffian systems and $\mathbb{C}$-polarizations.* We want to think of complex k-dimensional integral submanifolds as real $2k$-dimensional submanifolds satisfying the Cauchy-Riemann equations as well as the horizontality equations. Then we have the notion of horizontality on form together with the Hodge (p, q)-decomposition. The complex structure provides us with the notion of positivity and closed positive horizontal (k, k)-forms constitute a satisfactory "convex hull" of complex horizontal submanifolds. A particular interesting case is where H is a holomorphic contact structure, possibly with singularities, on a complex projective variety.

Finally we observe that one can formally bring horizontality and holomorphicity on an equal footing with the notion of *complex polarization* which is a $\mathbb{C}$-subbundle H in the complexified tangent bundle $\mathbb{C}T(V)$. But the geometric significance of complex polarizations remains obscure in general.

4.1.E$'$. Intrinsic metric evaluation of horizontality of forms. Let H be an equiregular polarization on V of rank n_1 and $H = H_1 \subset H_2 \subset \cdots \subset H_d$ be the commutator filtration as earlier. Then each (type) integer vector $m_1 \leqslant m_2 \leqslant \cdots m_d = n - k$ defines a subbundle in $\Lambda^{n-k}(V)$ of $(n - k)$-forms ω of cotype $\{m_i\}$, $i = 1, \ldots d - 1$ which vanish on a $(n - k)$-vector $x_1 \wedge \cdots \wedge x_{n-d}$ whenever $x_1, \ldots, x_{m_i+1} \in H_i$ for every $i = 1, \ldots, d - 1$. For example, horizontal forms ω have cotype $\{m_i\}$ for $m_1 = m_2 = \cdots = m_{d-1} = n_1 - k$ (where we assume $n_1 \geqslant k$ as horizontal forms are zero otherwise). Then we set $M = m_1 + 2(m_2 - m_1) + \cdots + d(m_d - m_{d-1})$, with the convention $m_d = n - k$, and let $M^*(\omega)$ be the minimum of $M = M\{m_i\}$ over all $\{m_i\}$ serving as cotype for ω. (If $\{m_i\}$ is a cotype for ω then so is $\{m_i'\}$ with $m_i' < m_i$ and the minimum for M is achieved for the maximal m_i for $i < d$, as m_d is fixed and equals $n - k$.) Finally, for every $(n - k)$-dimensional de Rham cohomology class h define $M^*(h)$ as the maximum of $M^*(\omega)$ over all closed form ω representing h.

Proposition. (Compare 3.3.B.) *Every cohomology class $h \in H^{n-k}(V; \mathbb{R})$ can be represented by a straight (Alexander-Spanier) cocycle c, such that $\|c\|_\varepsilon \lesssim \varepsilon^M$, for $M = M^*(h)$, which means $c(v_0, \ldots, v_{n-k}) \leqslant \mathrm{const}(\mathrm{diam}\{v_0, \ldots, v_{n-k}\})^M$ for all $(n - k + 1)$-tuples of points of diameter $\varepsilon \to 0$, where "diameter" refers to the C-C metric dist_H on M.*

Proof. Let us first visualize the effect of $M^*(\omega)$ on the norm $\|\omega\|_{g_\varepsilon^*}$ for the anisotropically blown up metric g_ε^* (see 1.4.D). This blow up depends on a choice of a Riemannian metric g on V and g_ε^* is characterized by the equality $\|X\|_{g_\varepsilon^*} = \varepsilon^{-i}$ for each unit vector $X \in H_i \ominus_g H_{i-1}$. Thus we see that $\|\omega\|_{g_\varepsilon^*} = O(\varepsilon^{M^*(\omega)})$ for all ω and this property uniquely defines $M^*(\omega)$.

Now we construct c by integrating ω representing h over some "standard" (straight) ε-simplex Δ_ε in V (compare 3.3.B). Notice that, according to 1.4.D-D'' simplices of diameter ε in the C-C metric become roughly of the unit size with respect to $\mathrm{dist}_\varepsilon^*$ corresponding to g_ε^*. Thus, in order to have $\int_{\Delta_\varepsilon} \omega \leqslant \mathrm{const}$, our Δ_ε must be represented by a map σ_ε of the unit simplex Δ to V, such that the Lipschitz constant of σ_ε with respect to $\mathrm{dist}_\varepsilon^*$ is bounded by a constant independent of ε. This is immediate if V is a nilpotent Lie group with a self-similarity $A_\varepsilon : V \to V$, where one may take $\sigma_\varepsilon = A_\varepsilon\, \sigma_1$ for some standard smooth map σ_1 (representing $\Delta_1 \subset V$ with given vertices $v_0, \ldots, v_{n-k}$). Then in the general case, as we know, V can be identified near each point $v_0 \in V$ with a nilpotent Lie group N_{v_0} such that the Riemannian metric of N_{v_0} approximates that of V such that the distance between the two metrics goes to zero in the unit ball around v_0 after the anisotropic ε-blow up of V and N_{v_0} (see 1.4.A, 1.4.D, 1.4.D'). It follows that the above simplex $\Delta_\varepsilon = \sigma_\varepsilon(\Delta)$ is good enough for V as well as for N_{v_0}. $\blacksquare$

Exercises

(a) Evaluate the "functional dimension" of the space of closed form ω of given cotype $\{m_i\}$ and, in particular, with $M^*(\omega)$ equal to a given number M_0.

(b) Determine the degree of the exterior differential relative to cotype, i.e. determine the precise rule $\{m_i\} \mapsto \{m_i'\}$ so that the differential of a form of cotype $\{m_i\}$ has cotype $\{m_i'\}$. Then find the largest M_0 (in terms of n, k and n_i), such that V admits a *non-zero exact form* ω of degree $n-k$ with $M^*(\omega) \geqslant M_0$. (Of course, the truly interesting problem is finding *non-exact closed* forms ω on V and on open subsets $U \subset V$ with $M^*(\omega) \geqslant M_0$ for large M_0. Such forms are expected to exist for a *generic* polarization H whenever the "functional dimension" of the space of closed forms with $M^*(\omega) \geqslant M_0$ is positive).

(c) Extend (a) to the P.D.E. system defined by the relations

$$\mathrm{cotype}(\omega) = \{m_i\} \quad , \quad \mathrm{cotype}(d\omega) = \{m_i'\}$$

for given m_i and m_i'. Do this, in particular, to the relations

$$M^*(\omega) = M_0 \quad , \quad M^*(d\omega) = M_0'.$$

(d) Filter the de Rham complex on V by $M = M^*(\omega)$, define the corresponding graded complex and relate this to the filtration on the (straight) Alexander-Spanier cochains c with $\|c\|_\varepsilon \lesssim \varepsilon^M$ according to the exponent M.

4.2. Analytic techniques for local construction of integral (H-horizontal) submanifolds.

We represent our $H \subset T(V)$ (locally) as the common zero of $n - n_1$ 1-forms $\eta_1, \ldots, \eta_{n-n_1}$ on V for $n_1 = \operatorname{rank} H$, which represent the tautological $H^\perp$-valued form $\eta : T(V) \to H^\perp \underset{\text{def}}{=} T(V)/H$, i.e. the quotient homomorphism, and let $\Omega : \Lambda^2 H \to H^\perp$ be the curvature form of H represented by the differentials $\omega_i = d\eta_i$ on H, $i = 1, \ldots, n - n_1$ (compare §3). We express the horizontality (integrality) condition for an immersion $f : W \to V$ by the system

$$f^*(\eta_i) = 0 \quad , \quad i = 1, \ldots, n - n_1 \,, \qquad (*)$$

which, in fact, contains $k(n - n_1)$ partial differential equations for $k = \dim W$, as every (induced) 1-form on W has k components. (Notice in passing that $(*)$ obviously implies the vanishing of the induced curvature forms, i.e.

$$f^*(\omega_i) = 0,$$

for C^2-maps f and then this conclusion extends to all C^1-maps f satisfying $(*)$ by a straightforward approximation argument. This rules out k-dimensional integral C^1-manifolds through a given point $v \in V$ whenever the forms ω_i on H_v have no common isotropic k-dimensional subspace. If $\omega = \{\omega_i\}$ is generic and $k \leqslant \frac{n_1+1}{2}$, then k-dimensional isotropic subspaces of ω forms a subvariety of codimension $\frac{k(k-1)}{2}$ in the Grassmann manifold $\operatorname{Gr}_k H_v$ having $\dim \operatorname{Gr}_k H_v = k(n_1 - k)$ and so for $n - n_1$ generic ω_i's appearing for a generic H there is no common isotropic subspace in H_v for $(k - 1)(n - n_1) > n_1 - k$, and hence, no k-dimensional integral C^1-submanifold through v).

Linearization of $(*)$. If we deform an immersion f along a field ∂ on $f(W) \subset V$ then the derivative (variations) of the induced form $f^*(\eta_i)$ clearly equals $\omega_i(\partial, \cdot) + d\eta_i(\partial)(\cdot)$ (compare 3.4.1 in [$\mathrm{Gro_{PDR}}$] and [$\mathrm{DA_{ISB}}$])

and, in particular, if ∂ is horizontal the linearized equations $(*)$ become linear *algebraic* in the unknown ∂,

$$\omega_i(\partial, X_j) = \sigma_{ij} \quad , \quad i = 1, \ldots, n - n_1 \ , \ j = 1, \ldots, k \qquad (*)'$$

where X_j is a full frame of vector fields (locally) on W.

4.2.A. Ω-regularity and infinitesimal invertibility. A k-dimensional linear subspace $S \subset T_v(V)$, $v \in V$, is called Ω-*regular* if the above linear system $(*)'$ with $X_j \in S$ is non-singular and hence, admits a solution $\partial \in H_v$ for arbitrary σ_{ij}. If S is H-horizontal, i.e. contained in H_v, then this definition is truly correct being independent of the choice of η_i, and hence of the 2-forms ω_i on $T_v(V)$ representing Ω on H_v. Namely, this regularity amounts to surjectivity of the linear map $\Omega_\bullet : H_v \to \mathrm{Hom}(S, H_v^\perp)$ naturally associated to Ω. In general, the notion of Ω-regularity depends on a choice of η_i but this will cause no difficulty as we shall only need horizontal S at the crucial moment.

Denote by $\mathcal{I}$ the differential operator which assigns to each smooth map $f : W \to V$ the induced forms $\{f^*(\eta_i)\}_{i=1,\ldots,n_1-k}$. If f is Ω-*regular*, i.e. is an immersion with Ω-regular tangent spaces in $T(V)$, then the linearized equations $(*)'$ are algebraically solvable on W and so $\mathcal{I}$ is *infinitesimally invertible* (in the sense of 2.3.1 in $[\mathrm{Gro_{PDR}}]$) on Ω-regular immersions (compare $[\mathrm{DA_{ISB}}]$). Therefore main theorem 2.3.2 of $[\mathrm{Gro_{PDR}}]$ yields the following.

4.2.A$'$. Local h-principle. (see $[\mathrm{DA_{ISB}}]$) *If H is C^∞-smooth then the sheaf of H-horizontal Ω-regular C^∞-immersions $W \to V$ is microflexible and satisfies the local h-principle. In particular, for every Ω-isotropic and Ω-regular linear subspace $S \subset H_v$, $v \in V$, there exists a germ of integral C^∞-submanifold $W \subset V$ at v with $T_v(W) = S$.*

The needed result from $[\mathrm{Gro_{PDR}}]$ is a version of *Nash implicit function theorem* which also applies to C^r-smooth H with sufficiently large r, say $r \geqslant 10$ (probably, less than 10 but 10 is what I am able to see without much thinking) and then the above proposition is valid for C^{r-6}-immersions. Furthermore, if H is real analytic, then the local h-principle (but not microflexibility) remains valid for Ω-regular horizontal C^{an}-immersions (see 2.3.6 in $[\mathrm{Gro_{PDR}}]$). On the other hand germs of (non-Ω-regular) integral manifolds can be sometimes obtained via the Cauchy-Kovalevskayo theorem which may apply in some cases where the C^∞-techniques fail.

Proposition. (compare [DA$_{\mathrm{IC}}$].) *If the system* $(*)'$ *is* $(k+1)$-*under-determined, i.e.* $n_1 - k(n-n_1) \geqslant k+1$, *then every* Ω-*regular* H-*horizontal germ of a* k-*dimensional* C^{an}-*submanifold in* V *extends to a horizontal* $(k+1)$-*dimensional* C^{an}-*germ.*

In fact, the Ω-regularity allows one to resolve $(*)$ with respect to the derivative $\partial = \frac{\partial}{\partial u_{k+1}}$ and the underdeterminancy condition gives room to a ∂ independent of X_i so that the resulting solution becomes an *immersion* of a $(k+1)$-dimensional manifold into W.

4.2.A″. Dimension count for regular isotropic subspaces. Since the map $\Omega_\bullet : H_v \to \mathrm{Hom}(S, H_v^\perp)$ vanishes on $S \subset H_v$ as S is Ω-isotropic, the inequality

$$n_1 - k = \dim H_v/S \geqslant \dim \mathrm{Hom}(S, H_v^\perp) = k(n - n_1)$$

is *necessary* for (the possibility of) S being Ω-regular. Next, we claim that this inequality is also *sufficient* for the existence of an Ω-regular isotropic k-dimensional subspace S in H_v, for a *generic* 2-form Ω on H_v. This is proven in two steps.

Step 1. The forms Ω for which the required $S \subset H_v$ exists constitute a Zariski *open* subset in the space of all forms $\Omega : \Lambda^2 H_v \to H_v^\perp$.

The proof follows by induction on k as the Ω-regularity of a $(k-1)$-dimensional isotropic subspace $S' \subset H_v$ reduces the Ω-isotropic condition on $S \supset S'$ to a *non-singular* system of *linear* equations, namely to

$$\Omega(X, X_i') = 0,$$

for a basis $X_1', \ldots, X_{k-1}'$ in S' and some fixed $X \in S \ominus S'$.

Step 2. If $n_1 - k \geqslant k(n-n_1)$ then the space of Ω's admitting an Ω-regular isotropic S of dimension k is *non-empty.*

To see that, take some $S \subset H_v$, take a bilinear form on $S \oplus (H_v/S)$ corresponding to a *surjective* linear *map* $H_v \to \mathrm{Hom}(S, H_v^\perp)$ vanishing on S and extend this form to an antisymmetric form on $H_v = S \oplus (H_v/S)$ using the natural embedding $A \otimes B \to \Lambda^2(A \oplus B)$ (for $A = S$ and $B = H_v/S$).

Corollary. *(Compare* [DA$_{\text{ISB}}$] *and* [DA$_{\text{IC}}$].*)* If H is a generic C^∞-polarization of rank n_1 and $k \leqslant n_1/(n - n_1 + 1)$ then H admits germs of Ω-regular horizontal k-dimensional C^∞-submanifolds at all $v \in V$ away from a stratified subset $\Sigma \subset V$ of positive codimension. If, moreover, H is C^{an} and $k \leqslant (n_1 - 1)/(n - n_1 + 1)$, then there is an analytic horizontal $(k + 1)$-dimensional germ at every $v \in V - \Sigma$.

Remark. The inequality $k \leqslant (n_1 - 1)/(n - n_1 + 1)$ is equivalent to $n - (k+1) \geqslant (k+1)(n - n_1)$ which expresses the (under)determinancy of the P.D.E.-system for the horizontality of $(k+1)$-dimensional submanifolds in V. Therefore, this system is *overdetermined* for $k > (n_1 - 1)/(n - n_1 + 1)$ and then there is no (even C^∞-smooth) horizontal $(k + 1)$-dimensional submanifolds in V for *generic* (C^∞ or C^{an}) H. In other words, the condition $k \leqslant (n_1 - 1)/(n - n_1 + 1)$ (for the existence of $(k + 1)$-dimensional C^{an}-germs) is sharp. This can not be said about our bound on k in the C^∞-case. Yet some version of the local h-principle and microflexibility for a suitably "regular" class of k-dimensional horizontal germs is expected (for generic H) whenever $n - k > k(n - n_1)$, i.e. where the corresponding system of P.D.E. is underdetermined (compare (2) in (E) of 2.3.8 in [Gro$_{\text{PDR}}$]). On the other hand the existence of integral submanifolds remains highly problematic in the (generic) determined C^∞-case, i.e. for $n - k = k(n - n_1)$.

Exercises

(a) Evaluate the codimension of those (non-generic) Ω which admit no Ω-regular isotropic S of given dimension k and thus find $\operatorname{codim} \Sigma$ for Σ in the above corollary.

(b) Let A, B and S be linear spaces. Study (the space of) pairs $\Omega : \Lambda^2 A \to B$ and $\varphi : S \to A$, where φ is injective Ω-regular and Ω-isotropic. Decide when for a fixed (generic) Ω the space of the above φ's is i-connected for given i and the dimensions of A, B and S (compare 3.3.1 in [Gro$_{\text{PDR}}$]).

4.2.B. Calculus of variations for regular horizontal submanifolds.

Let us make more precise the above remarks on the general notion of regularity which applies to an arbitrary system of partial differential equations imposed on maps $f : W \to V$. Such a system of m equations of order r is represented by an m-codimensional subvariety in the space of r-jets of maps $W \to V$, denoted $\mathcal{R} \subset \mathcal{J}^r$, and our equations are expressed by the inclusion $J^r(f)(W) \subset \mathcal{R}$. (We prefer to work here with maps rather than submanifolds and so use $\mathcal{J}^r$ rather than $\mathrm{Gr}_k^r(V)$. This makes no essential difference as our discussion is local and submanifolds $W^k \subset V^n$ can be represented by maps $\mathbb{R}^k \to \mathbb{R}^{n-k}$, compare 4.1.C). Actual equations appear if we write $\mathcal{R}$ as the zero set of s functions $\Delta_1, \ldots, \Delta_m$ on $\mathcal{J}^r$ and our system becomes $\Delta_i \circ J^r(f) = 0$, $i = 1, \ldots, m$. Differentiating these equations corresponds to lifting (prolongations) of our $\mathcal{R}$ to higher order jet space, say $\mathcal{R}^j \subset \mathcal{J}^{r+j}$ and it is convenient to stabilize $\mathcal{R}$, i.e. to take $\mathcal{R}^\infty \subset \mathcal{J}^{r+\infty}$ where ∞ stands for a large non-specified j. The notion of regularity will apply first to individual jets $x \in \mathcal{R}^\infty$ and then to solutions f of $\mathcal{R}$ where the regularity of f means that for the jets $J^{r+\infty} f(w) \in \mathcal{R}^\infty$ for all $w \in W$.

So we represent our $\mathcal{R}^\infty$ near a point $x \in \mathcal{R}^\infty$ in question by a (non-linear) differential operator $\Delta = \{\Delta_1, \ldots, \Delta_m\}$ and denote by $L = L_f$ the linearization of Δ at f with $J^\infty f(v) = x$. This is a linear differential operator on W defined near the point w under x acting on n-tuples of functions ($n = \dim V$) with the range in m-tuples. Then we look for the right inversion M of L, i.e. a differential operator of certain order s, sending m-tuples to n-tuples, such that $L \circ M = \mathrm{Id}$. It is shown in 2.3.8 of [$\mathrm{Gro_{PDR}}$] that such an M, of a sufficiently high order s, does exist in the generic underdetermined (i.e. $m < n$) case by reducing the identity $L \circ M = \mathrm{Id}$ to a linear system of algebraic equations on (the coefficients of) M. This system, for a sufficiently large s, has more unknown than equations (as we assume $m < n$) and it is proven in [$\mathrm{Gro_{PDR}}$] that this system is non-singular for generic $L = L_f$ and hence has the desired solution M. Notice that the coefficients of this system are made out of coefficients of L and their derivative and so singularity or non-singularity of this system at a given jet $x \in \mathcal{R}^\infty$ makes perfect sense. Thus we call x *regular* if (for a suitable choice of Δ) the above mentioned system on M is non-singular for some (sufficiently large) s.

Observe that the regularity gives us somewhat more than invertibility of L, namely it allows us to "uniformize" (or parametrize) the solutions of the homogeneous P.D.E. system $L(g) = 0$, by taking (locally) the full system of solutions of the system $L \circ M = 0$, say $M_1, \ldots, M_t$. Then every g satisfying $L(g) = 0$ can be written as $\Sigma_{i=1}^t M_i \, h_i$ for some m-tuples of functions $h_1, \ldots, h_t$.

Example. Let L act on pairs of functions in one variable by

$$L : (g_1, g_2) \mapsto g_1' + \ell g_2'$$

where $\ell = \ell(t)$ and g' stands for $\frac{dg}{dt}$. If $g_1' + \ell g_2' = 0$, then

$$g_2 = h'/\ell' \quad \text{and} \quad g_1 = h - \ell h'/\ell' \quad \text{for} \quad h = g_1 + \ell g_2,$$

and so the uniformization (of g_1 and g_2 by h) is achieved whenever ℓ' does not vanish.

Warning. *There are certain non-regular situations where the uniformization is still present. For example, every closed forms on $\mathbb{R}^k$ is exact but the exterior differential is not invertible by a differential operator.*

Now we turn to the calculus of variations where we extremize the value of certain (energy) functional $E(f)$ on the solutions f of some system $\mathcal{R}$ and we want to write down the Euler-Lagrange equations. In order to write these equations we use infinitesimal variations of f supported near a given point $w \in W$. These variations g are just solutions of the system $L_f\, g = 0$ and, in general, it is hard to generate these with the vanishing conditions away from w. But if these g are uniformized by (unrestricted) strings of functions $h_1, \ldots, h_t$, we apply the usual derivation of the Euler-Lagrange to h_i and thus obtain the desired equations on extremal f (restricted by $\mathcal{R}$).

Example. All smoothly immersed horizontal curves in a contact manifold are regular by a trivial computation (essentially reproduced for $n = 3$ in the previous example). Thus C-C geodesics satisfy the geodesic equation and, in particular, are C^∞-smooth. Similarly, the immersed horizontal submanifolds of dimension k are regular for all k, and if they are extremal (i.e. area minimizing) they satisfy the respective Euler-Lagrange equations. (*Warning.* A smooth horizontal map $W \to V$ extremizing some energy may fail to be an immersion and then neither the regularity nor Euler-Lagrange are automatic. Furthermore, there are certain generic polarizations where some immersed horizontal *curves* are non-regular despite claims to the contrary by several people including the present author in exercise (a) p. 84 in [Gro$_{\text{PDR}}$]. This was pointed out to me by L. Hsu, see [Ge$_{\text{VP}}$], [Hsu$_{\text{CVG}}$], [Mont], [Suss], [Bou] and [Pel-Bou] on this matter).

Fat polarizations. A polarization H is called *fat* if every 1-dimensional subspace in H is Ω-regular and 1-dimensional variational problems has been most extensively studied in the fat case (see [Ge$_{\text{CVP}}$], [Ge$_{\text{HPS}}$] and references therein). Here we only notice that "fat" = "contact" for corank $H = 1$ but for corank $H \geqslant 2$ the fatness is locally non-generic (albeit open) condition which seems highly restrictive especially if the underlying manifold is compact without boundary. Standard examples of fat H are provided by the complex contact structure (corank $H = 2$) and the horizontal bundle of the Hopf fibrations $S^{4n+3} \to \mathbb{H}P^n$ (where corank $= 3$).

Question. What could be a meaningful notion of *k-fatness* allowing many (but not all) Ω-regular isotropic k-plane in H for corank $H \geqslant 2$?

Horizontal submanifolds from the infinite dimensional point of view. The space of (smooth) maps $f : W \to V$ can be viewed as an infinite dimensional manifold, say $\mathcal{F}$, where each tangent space $T_f(\mathcal{F})$, $f \in \mathcal{F}$, equals the space of sections of the induced bundle $f^*(T(V)) \to W$. It may be hard (?) to speak of vector fields on $\mathcal{F}$ but one can define (the space of) jets of such "fields" at each point $f \in \mathcal{F}$ by the following (well known) recipe. Given an infinite dimensional vector bundle $p : \mathcal{S} \to \mathcal{F}$, where both $\mathcal{S}$ and $\mathcal{F}$ are spaces of maps between finite dimensional manifolds, one defines the fiber of the 1-jet bundle $J^1(\mathcal{S})$ over $\mathcal{S}$ at s as the space of homomorphisms $h : T_f(\mathcal{F}) \to T_s(\mathcal{S})$ satisfying $(\mathcal{D}p) \circ h = \text{Id}$. For example, if $\mathcal{S} = T(\mathcal{F})$ for the above $\mathcal{F} = \text{Maps}\,(W, V)$, then every such homomorphism h is essentially the same as a homomorphism $T_f(\mathcal{F}) \to T_f(\mathcal{F})$ which is, in turn, given by a "kernel" $K(w, w')$ which is a section of a certain (finite dimensional) vector bundle over $W \times W$. Thus, the functor $\mathcal{S} \to J^1(\mathcal{S})$ keeps us within the category of infinite dimensional manifolds which are spaces of maps between finite dimensional ones. Then, once we have $\mathcal{S}^1 = J^1(\mathcal{S})$, we define $J^2(\mathcal{S})$ as a suitable subspace in $J^1(\mathcal{S}^1)$, etc.

Next we can speak of the Lie bracket between jets of vector "fields" on $\mathcal{F}$ which is a pairing

$$J^r(T(\mathcal{F})) \otimes J^r(T(\mathcal{F})) \to J^{r-1}(T(\mathcal{F}))$$

and if $\mathcal{H} \subset T(\mathcal{F})$ is a polarization in our category, the expression "$\mathcal{H}$ Lie spans $T(\mathcal{F})$" makes perfect sense. For example, if $\mathcal{H}$ corresponds to $H \subset T(V)$, then $\mathcal{H}$ Lie spans $T(\mathcal{F})$ whenever H spans $T(V)$.

Now, let $\mathcal{G} \subset \mathcal{F}$ be a submanifold, for example the subspace of H-horizontal maps $f : W \to V$. We consider the induced polarization $\mathcal{H}' = \mathcal{H} \cap T(\mathcal{F})$ on $\mathcal{G}$ and ask ourselves when the Chow connectivity theorem holds for $\mathcal{H}'$. For example, if $\mathcal{G}$ consists of H-horizontal maps $f : W \to V$, then $\mathcal{H}'$-horizontal curves $\mathbb{R} \to \mathcal{G}$ correspond to H-horizontal cylinders $W \times \mathbb{R} \to V$ and one would like to interpret the existence results for these (e.g. our implicit function theorem) as connectivity theorems for $\mathcal{H}'$.

4.2.C. Partially horizontal submanifolds in V. We want to extend our analysis to k-dimensional submanifolds $W \subset V$ which have rank $(T(W) \cap H) = m$ for a given $m \leqslant k = \dim W$. The corresponding system of P.D.E. now applies to pairs (f, G) where f is a map $W \to V$ and $G \subset T(W)$ is a subbundle of rank m, and the equations express the tangency of the f-image of G to H by

$$f^*(\eta_i) \mid G = 0 \quad , \quad i = 1, \ldots, n - n_1, \tag{+}$$

for the forms η_i defining H (compare $(*)$ at the beginning of 4.2). We linearize $(+)$ (as we did it with $(*)$) using a frame $X_1, \ldots, X_m$ in G and the complementary bundle $G^\perp = T(W)/G$ so that the tangent vectors to the space of G's can be represented by homomorphisms $G \to G^\perp$ or by m-tuples of sections $\partial_1, \ldots, \partial_m$ of $G^\perp$. These ∂_j, $j = 1, \ldots, m$, serve together with the field ∂ (tangent to V along $f(W) \subset V$) as the unknowns of the linearized system which algebrizes for *horizontal* fields ∂ on W (i.e. sections of $H \mid f(W) \underset{\text{def}}{=} f^*(H)$) and reads

$$\omega_i(\partial, X_j) + \eta_i(\partial_j) = \sigma_{ij}. \tag{$+$}'$$

Now, we want to express non-singularity of $(+)'$ and (partial) isotropy of G in an invariant language and we simplify the notations by identifying W with $f(W) \subset V$. Then the inclusion $(T(W), G) \hookrightarrow (T(V), H)$ induces an embedding $G^\perp \hookrightarrow H^\perp$ and consequently a homomorphism

$$\text{Hom}(G, H^\perp) \to \text{Hom}(G, H^\perp/G^\perp).$$

We compose this with $\Omega_\bullet : H \to \text{Hom}(G, H^\perp)$ and denote by $\Omega'_\bullet$ the composed homomorphism $H \to \text{Hom}(G, H^\perp/G^\perp)$. We easily see that

(1) G is *partially Ω-isotropic* in the sense that $G \subset \text{Ker}\,\Omega'_\bullet$;

(2) *non-singularity* of $(+)'$ is equivalent to *surjectivity* of $\Omega'_\bullet$.

We apply the generalized Nash implicit function theorem as earlier and arrive at the *microflexibility and the local h-principle for immersions* f *which are m-horizontal and partially Ω-regular*, i.e. having rank $(\mathcal{D}f)^{-1}(H) = m$ and $\Omega'_\bullet$ surjective. In particular, *if H is C^∞-generic we have a C^∞-germs of m-horizontal partially regular k-dimensional submanifold $W \subset V$ at each $v \in V - \Sigma$, $\operatorname{codim} \Sigma > 0$, provided* $\operatorname{rank} G \leqslant \operatorname{rank} H$, $\operatorname{rank} G^\perp \leqslant \operatorname{rank} H^\perp$ and

$$\operatorname{rank} H - \operatorname{rank} G \geqslant \operatorname{rank} G \operatorname{rank}(H^\perp/G^\perp),$$

i.e. $n_1 - m \geqslant m(n - n_1 - k + m)$. Furthermore, *if H is C^{an}, $n_1 \geqslant m + 1$ and $n_1 - m \geqslant m(n - n_1 - k + m) + 1$, then there is a $(k+1)$-dimensional $(m+1)$-horizontal C^{an}-germ through every $v \in V - \Sigma$.*

Remark. Recall that the $(m + 1)$-horizontality of a $k + 1$-dimensional submanifold in V is expressed by $(m + 1)(n - n_1 - (k + 1) + (m + 1))$ P.D. equations against $n - k - 1$ unknown functions and note that the (under)determinancy of these equations, expressed by the inequality $n - k - 1 \geqslant (m+1)(n - n_1 - k + m)$, is equivalent to $n_1 - m \geqslant m(n - n_1 - k + m) + 1$.

Exercises

(a) Generalize the above to submanifolds $W \subset V$ of a given horizontality type, i.e. with prescribed ranks of the intersections $T(W) \cap H_i$ (compare 4.1.B).

(b) Generalize the exercises from 4.2.A″ to the present context.

On higher order regularity. One expects that m-horizontal k-dimensional germs in generically polarized manifolds (V, H) becomes amenable to the above techniques in the underdetermined case (i.e. for $n - k > m(n - n_1 - k + m)$) with the general notion of regularity indicated in 4.2.B (which implies the infinitesimal invertibility). In particular, one may think there is an m-horizontal germ at each point $v \in V$, if H is C^∞-generic and $n - k > m(n - n_1 - k + m)$.

4.3. The global h-principle for smooth horizontal submanifolds. The microflexibility and local h-principle do not suffices by themselves to construct (global regular) horizontal immersions of k-dimensional manifolds W into V but the theory of *continuous sheaves* (see 2.2 in [Gro$_{\mathrm{PDR}}$]) provides global constructions in the presence of "sufficiently many" $(k + 1)$-dimensional (regular) horizontal germs.

Overregularity. Call an isotropic linear subspace $S \subset H_v \subset T_v(V)$ *overregular* if it is contained in an Ω-regular Ω-isotropic subspace $S' \subset H_v$ with $\dim S' > \dim S$. Then a fiberwise injective homomorphism $T(W) \to T(V)$ is called *overregular* if it sends each $T_w(W) \subset T(W)$, $w \in W$ onto an overregular subspace in H. Finally, an immersion $f : W \to V$ is called *overregular* if the differential $\mathcal{D}f : T(W) \to T(V)$ is overregular. (Notice that our overregularity implies horizontality).

Approximation theorem. *Smooth overregular immersions $W \to V$ satisfy the C^0-dense h-principle. It follows that for every overregular homomorphisms $T(W) \to T(V)$ the underlying map $W \to V$ admits a fine C^0-approximation (which amounts to the uniform approximation if W is compact) by smooth overregular (and hence, horizontal) immersions $W \to V$ (compare 3.5).*

The proof follows from the local analysis of the previous section and the microextension theorem in 2.2.4 of [Gro$_{\mathrm{PDR}}$].

In order to apply the theorem to a k-dimensional manifold W one needs at least one Ω-regular isotropic subspace $S'_v \subset H_v$ of dimension $> k$. What is even better is a continuous fields of such subspaces, say $S'_v \in H_v$ for all $v \in V$, forming a vector bundle S over V. Then every injective homomorphism $T(V) \to S \subset H \subset T(W)$ is overregular. Notice that every S'_v at a given point $v \in V$ extends to such a field in a small neighbourhood $U \subset V$ of v. Furthermore, every S'_v extends to a field on all V if (V, H) admits a transitive locally free action of a Lie group L, i.e. $V = L/\Gamma$ for a discrete group Γ and the polarization H is left-invariant. In both cases the resulting bundle S is trivial and so it receives an injective homomorphism from $T(W)$ whenever the bundle $T(W)$ is also trivial (i.e. W is parallelizable), e.g. if $W = \mathbb{R}^k$.

Corollary. *Every continuous map $\mathbb{R}^k \to U$ for the above U admits a C^0-fine (in particular uniform) approximation by horizontal immersions. This remains true for $U = V$ in the case of V being a Lie group with a left-invariant H admitting a single Ω-regular Ω-isotropic $S' \subset H$ of dimension $k + 1$, and, more generally, for $V = L/\Gamma$.*

Remark. The flagrant deficiency of the above results is an appeal to $(k+1)$-dimensional integral (i.e. horizontal) manifolds in order to produce k-dimensional ones which unpleasantly reduces the range of dimensions where these results may be applied. There are some cases (e.g. contact manifolds see 3.4.3 in [Gro$_{\mathrm{PDR}}$]) where one needs no extra dimension at all and, in general, one probably need much less than overregularity.

Conjecture. *If $n - (k+1) \geqslant (k+1)(n - n_1)$, then Ω-regular horizontal immersions satisfy the C^0-dense h-principle (where the above inequality expresses (under)determinancy of the horizontality condition for $(k+1)$-dimensional germs).*

Exercise. Extend the above theorem to m-horizontal immersions and then to immersions with a prescribed tangency to H_i, $i = 1, \ldots, d$.

h-principle for horizontal curves. It is known (see [Ge$_{\mathrm{HPS}}$] and [Sar]) that *the space of smooth (possibly non-immersed) horizontal paths between two given points in (V, H) is weakly homotopic to the space of all continuous paths between these points.* This generalizes Chow connectivity theorem. Furthermore, (the orbits of) suitably chosen (sufficiently twisted) horizontal vector fields generate (and regularize) enough horizontal curves (compare 1.2.B) to ensure *the dense h-principle for immersed horizontal paths with given ends.* An appropriate form of such h-principle allows, for example, a fine C^0-approximation of a continuous map of 1-dimensional foliations S into V, say $f_0 : S \to V$, by smooth maps $f : S \to V$ which are *horizontal immersions on the leaves,* provided f_0 lifts to an *injective* homomorphism of the (1-dimensional) tangent bundle of the foliation to H (compare [Hsu$_{\mathrm{GHP}}$]).

Warning. *The commutator generation condition for H is not sufficient for the (even higher order) regularity of immersed horizontal curves. In fact there are rigid, i.e. non-microflexible curves (along which the relevant operator is not infinitesimally invertible).*

Example. (borrowed from [Mont]) Let H on $\mathbb{R}^3$ be defined by the form $\eta = y^2 dx + dz$ and c be the unit segment given by $\{x \in [0, 1], y = 0, z = 0\}$. Then every C^1-small perturbation of c can be presented by (the graphs of) functions $y(x)$ and $z(x)$ where the derivative of z is *negative* as $z'(x) = -y^2(x) \leqslant 0$. Since every non-trivial perturbation of c with $z(0) = 0$ has $y^2(x) > 0$ for some $x \in [0, 1]$, the second end has $z(1) < 0$ and so no non-trivial deformation with fixed ends exists. (Yet one can deform $c = c_0$ by reparametrizing the x-interval with maps $\kappa_t : [0,1] \to [0,1]$ fixing the ends, starting from $\kappa_0 = \mathrm{Id}$ and terminating with a smooth map $\kappa_1(x)$ having many critical points $x_i \in [0,1]$. Then the resulting curve $c_1(x) = c_0(\kappa_1(x)) = (x = \kappa_1(x), y = 0, z = 0)$ can be "bent" at the critical points and C^∞-continuously moved keeping horizontality, away from the original position in $\mathbb{R}^3$. In fact, the proof of the above mentioned h-principle for *non-immersed* horizontal curves is based on reparametrizations creating critical points.) Such rigid curves may appear at the first sight impossible paradoxal monsters (only reluctantly, under the continuous pressure from Lucas Hsu and Richard Montgomery I accepted their generic appearance) but, in fact, they are ubiquitous in the nonholonomic geometry (see [Mont$_{\mathrm{SSC}}$]).

Remarks

(a) Generic horizontal curves for the above H meet the "dangerous" hypersurface $y = 0$ at finitely many points and (since η is contact away from H) are microflexible. Probably, *generic* horizontal curves in every H Lie generating $T(V)$ are microflexible[9].

(b) Construction of horizontal curves as orbits of suitable vector fields is in the spirit of *convex integration* (see [Gro$_{\mathrm{PDR}}$] and [Gro$_{\mathrm{CIDR}}$]) which suggests a similar approach to k-dimensional horizontal manifolds for $k \geqslant 2$ viewed as horizontal curves in an infinite dimensional space (see 4.2.B).

4.3.A. On the h-principle for morphisms of a given type.

Suppose W is given a polarization $G \subset T(W)$ and we look for maps $f : W \to V$ with $n_{ij}(f) \geqslant n_{ij}$ for given n_{ij} where $n_{ij}(f)$ are defined as rank $G_j \cap \mathcal{D}^{-1}f(H_i)$. One can work out a suitable notion of regularity and prove the local h-principle as earlier. Then one takes $W' = W \times \mathbb{R}$ with $G' = G \times \mathbb{R}$ and call (a germ of) a map $f : W \to V$ *overregular* if it decomposes as $f = f' \circ \rho$ where $\rho : W \to W'$ is the graph of a smooth function $W \to \mathbb{R}$, and $f' : W' \to V$ is "regular" with $n_{ij}(f') \geqslant n'_{ij} = n_{ij} + 1$, where the "regularity" matches the n'_{ij}-inequalities such that the local h-principle and microflexibility hold for f'. Then the continuous sheaves deliver the global h-principle for overregular maps $f : W \to W$ satisfying $n_{ij}(f) \geqslant n_{ij}$. (We suggest the reader would look at this more specifically.)

---

[9] In fact, this is proven in [Cor$_1$], also see [Cor$_2$].

4.4. Folded integral submanifolds. For many purposes, e.g. for making horizontal cycles, one may use *piecewise* smooth integral submanifolds which are easier to come by than the smooth ones. In fact, the local h-principle and microflexibility suffice to produce certain *folded* (or branched) integral k-dimensional submanifolds without resorting to a use of $(k+1)$-dimensional horizontal germs. An example of such a "manifold" supporting a 1-dimensional cycle is exhibited on Fig. 12 below (compare Poenaru pleating lemma and Fig. 5 and 6 in 3.5).

Figure 12

One may think of folded submanifolds W in V as finite or countable unions of actual smooth k-dimensional submanifolds, $W = \cup_i W_i$, where the intersection between W_i and W_j, whenever non-empty, is also a k-dimensional manifold with piecewise smooth boundary. Such a W is called integral (or horizontal) if such are the pieces W_i for all i. In fact one can define abstract branched (or folded) spaces W built of manifolds and speak of their smooth (horizontal) immersions into V. In this case microflexibility implies the h-principle which we apply to Ω-regular horizontal folded manifolds. Here is a specific result which follows from the folded h-principle (see (D) and (C) in 2.2.7 of [Gro$_{\mathrm{PDR}}$]).

We start with an arbitrary continuous map of a locally finite (at most) k-dimensional polyhedron into V, say $f_0 : W \to V$. We want to approximate f_0 by horizontal *branched* (or *folded*) immersions of another locally finite k-dimensional polyhedron, say $f' : W' \to V$, where f' is called a *branched* (or *folded*) C^∞-*immersion* if W' admits a locally finite covering by compact subpolyhedra, $W' = \cup_i W_i'$, such that f' is smooth on each simplex in W and homeomorphicly sends each W_i' onto a compact C^∞-smooth submanifold in V with boundary. It makes now perfect sense to attribute the properties of H-horizontality and/or Ω-regularity to f' as these apply to the images $f'(W_i')$. Furthermore, we say that the maps f' in question *approximate* f_0 if for every neighbourhood $U \subset V \times W$ of the graph $\Gamma_{f_0} \subset V \times W$ of f_0 and every neighbourhood $U' \subset W \times W$ of the diagonal there exist proper homotopy equivalences $\varphi : W \to W'$, $\varphi' : W' \to W$ and one of our $f' : W' \to V$, such that the graph of $\varphi \circ \varphi'$ is contained in U' and the graph of $f \circ \varphi$ is contained in U.

Folded approximation theorem. *A map $f_0 : W \to V$ admits an approximation by branched (folded) H-horizontal Ω-regular immersions $f' : W' \to V$ if and only if there is a continuous map $w \mapsto S_w \subset H_v$ for $v = f(w)$ which assigns to each $w \in W$ an Ω-regular Ω-isotropic k-dimensional subspace S_w in $H_{f(w)}$.* (These S_w, $w \in W$ form a k-dimensional vector bundle over W which injects into H by an Ω-isotropic and regular homomorphism.)

Idea of the proof. The "only if" claim is obvious as every folded immersion which is Ω-regular and isotropic comes along with such "tangent bundle" S' over W' which then transports to W by $\varphi : W \to W'$. To prove "if" we first use the local h-principle and associate to each S_w a k-dimensional Ω-regular and Ω-isotropic germ in V at $f_0(w)$ tangent to S_w. We take representatives of such germs, called $W_i(S) \subset V$, at some points $v_i = f_0(w_i) \in V$, such that the points $w_i \in W$, $i = 1, 2, \ldots$ form a sufficiently dense discrete subset in W, and then we slightly perturbe $W_i(S)$ using microflexibility to new germs, say $W_i' \subset V$, which intersect according to certain pattern, so that the polyhedron W' *defined* as the abstract union $\cup_i W_i'$ with the identifications corresponding to our pattern is homotopy equivalent to W and then the tautological map $f' : W' \to \cup_i W_i' \hookrightarrow V$ satisfies our specifications. The passage from W_i to W_i' is similar to what we have already seen in Fig. 5, 6 and 12 and is reproduced below in Fig. 13 with W_0 being a circle embedded to the plane.

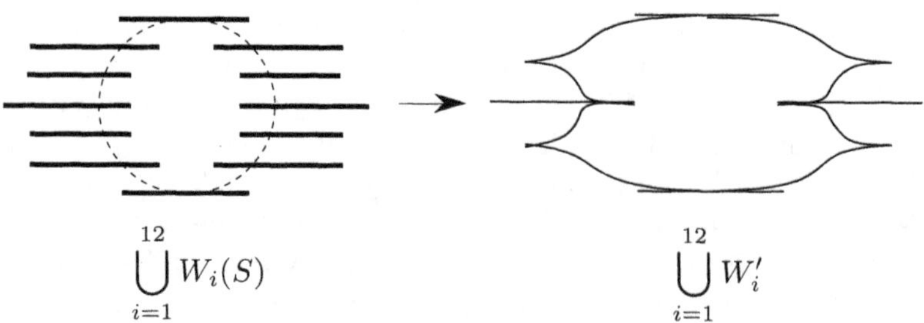

$$\bigcup_{i=1}^{12} W_i(S) \qquad\qquad \bigcup_{i=1}^{12} W_i'$$

Figure 13

We refer to 2.2.7 in [Gro$_{\text{PDR}}$] for details.

Corollary. *Granted "if", the map $f_0 : W \to V$ admits a fine C^0-approximation by piecewise smooth and piecewise horizontal maps $f : W \to V$.*

Proof. Recall $\varphi : W \to W'$ and use $f' \circ \varphi$ for f. Notice that we do not speak here of the Ω-regularity (of f) as this is only needed to oil the microflexibility gear in the machinery of the continuous sheaves but not truly relevant for our applications where we are concerned with the horizontality alone. In fact the notion of regularity can be relaxed in many cases and, probably, brought to a point where it could be (genericly) used for $n - k > k(n - n_1)$ instead of $n_1 - k \geqslant k(n - n_1)$ associated to the Ω-regularity (compare 4.2.A).

The above approximation theorem can be used (in the same way as the one in the previous section) in the presence of a single Ω-regular Ω-horizontal k-dimensional subspace $S \subset H_v$. Namely, *every continuous map of W into a small neighbourhood $U \subset V$ of v admits an approximation by folded horizontal maps $W' \to U$ and this is valid for $U = V$ in the Lie group theoretic case* (compare Corollary in the previous section).

4.4.A. The proof of Thom's theorem (A) in the jet bundle. Let $V = V^r = \mathrm{Gr}_k^r(V_0)$ be the space of r-jets of germs of k-manifolds in V_0 (compare 4.1.B$'$) and $H = H^r$ be the canonical polarization on this V. If we declare *regular* the horizontal submanifolds in V which *regularly* project (i.e. immerse) to V_0 and hence, equal the r-jets of k-dimensional submanifolds in V_0, we arrive at the situation where the folded h-principle may be applied since the microflexibility and the local h-principle are (trivially) valid in this case. It is also obvious that the "if" of the folded approximation theorem is satisfied and so *every $f_0 : W \to V$ can be approximated by folded regular horizontal immersions $f : W' \to V$.* Thus every ℓ-cycle for $\ell \leqslant k$ can be made folded horizontal in every open $\mathcal{R} \subset V$ and (A) of Thom's theorem (see 4.1.B$''$) follows.

4.4.B. Extension of folded and subfolded maps. A map $f : W \to V$ is called *subfolded of rank k* if it is the composition of a folded map $f' : W' \to V$, $\dim W' = k$, and a piecewise linear $\varphi : W \to W'$ (as in the above corollary but now we do not assume $\dim W \leqslant k$). If f' is horizontal and regular, these properties are attributed to f and also the map φ brings the "tangent bundle" of W' to the "if" bundle S over W. We want to decide when f extends to a regular horizontal subfolded map of rank k of a larger polyhedron $W_1 \supset W$.

Extension theorem. *Suppose (compare "if") the bundle S extends to S_1 on W_1 and the tautological homomorphism ("differential" of f) $S \to T(V)$ extends to a fiberwise injective regular horizontal bundle homomorphism of S_1 to $T(V)$. Then f extends to a regular horizontal subfolded map of rank k of W_1 to V where, moreover, this map can be chosen arbitrarily close to the continuous map $W_1 \to V$ underlying the homomorphism $S_1 \to T(V)$.*

The proof is similar to that of the folded approximation theorem.

On folded maps f satisfying $n_{ij}(f) \geqslant n_{ij}$. (Compare 4.3.A.) If W comes with $G \subset T(W)$ and maps $f : W \to V$ take G into account, then the notion of folding becomes problematic unless (W, G) has a sufficient symmetry. A happy case is where $(W, G) = (W_0 \times \mathbb{R}, G_0 \times \mathbb{R})$, as this G is invariant under the diffeomorphisms preserving $\mathbb{R}$-lines in W and one may fold along the graphs $\Gamma \subset W$ of functions $W_0 \to \mathbb{R}$. Then one has the full folded map package (with a suitable notion of regularity) by 2.2.2 and 2.2.7 in $[\text{Gro}_{\text{PDR}}]$.

4.4.C. Deformations of regular horizontal folded maps. An Ω-regular horizontal folded map $f' : W' \to V$ admits "many" horizontal deformations as if they were not restricted by the horizontality condition. This immediately follows from the generalized Nash implicit function theorem in $[\text{Gro}_{\text{PDR}}]$ and the specific feature of the ("many") deformations we need is expressed in the following.

Proposition. *There exists a continuous map $F' : W' \times \mathbb{R}^q \to V$ for some q, (e.g. for $q = 2n$) such that*

(i) $F' \mid W' = W' \times 0 = f'$,

(ii) *the restriction of F' to each "branch" W_i' of W' is smooth, i.e. $F' \mid W_i' \times \mathbb{R}^q$ is smooth for the smooth structure on W_i' induced by the embedding $f' \mid W_i' : W_i' \to V$,*

(iii) *each map $F' \mid W_i' \times \mathbb{R}^q$ is a (smooth) submersion $W_i' \times \mathbb{R}^q \to V$.*

4.5. Lower bounds on the Hausdorff dimension of subsets in (V, H). The folded (and subfolded) horizontal manifolds W mapped into V (whenever they exist) provide us with k-dimensional subsets in V of Hausdorff dimension k. Similarly, m-horizontal submanifolds W (i.e. having rank $H \cap T(W) \geqslant m$) satisfy a certain non-vacuous bound on the Hausdorff dimension. For example, if $H_2 = T(V)$, i.e. the commutators of the horizontal fields span $T(V)$, then these W have $\dim_{\text{Hau}} \leqslant m + 2(k - m)$, and, in general, if $H_d = T(V)$ the inequality becomes $\dim_{\text{Hau}} \leqslant m + d(k - m)$. What is more interesting, however, is a non-trivial *lower* bound on $\dim_{\text{Hau}} V'$ for subsets $V' \subset V$ of a given topological dimension which follows from the global h-principle for folded horizontal (and partially horizontal) submanifolds in V of the complementary dimension (compare 2.3).

Theorem. *If each horizontal space $H_v \subset T_v(V)$, $v \in V$, contains an Ω-regular and Ω-isotropic subspace $S_v \subset H_v$ of dimension k, then every closed $(n - k)$-dimensional subset $V' \subset V$ has $\dim_{\text{Hau}} V' \geqslant \dim_{\text{Hau}} V - k$, where we assume V is equiregular. More generally, if we have m-horizontal and partially Ω-isotropic and Ω-regular $S_v \subset T_v(V)$ at all $v \in V$, then $\dim_{\text{Hau}} V' \geqslant \dim_{\text{Hau}} V - m - d(k - m)$.*

Proof. We use the following characterization of closed subsets V' in V of topological dimension $n - k$.

Alexandroff theorem. ([10]) *There exist $(k - 1)$-dimensional cycles of arbitrary small diameter in the complement $V - V'$ which bound no chains of small diameter in $V - V'$.*

It follows, there exists a k-dimensional polyhedron $W \subset V$ which *stably intersects* V', i.e. every continuous map $f' : W' \to V$ approximating the inclusion $f_0 : W \hookrightarrow V$ in the sense of the definition in 4.4 has *non-empty* pull-back $(f')^{-1}(V')$.

Next, we observe that V can be localized in the present situation to a small neighbourhood $U \subset V$ of a single point $v' \in V' \subset V$ and so the corollary to the folded approximation theorem (see 4.4) provides a branched (folded) H-horizontal (and Ω-regular) immersion $f' : W' \to V$ which stably intersects V'. This f' gives rise to a "large" family of

---

such maps, i.e. $F' : W' \times \mathbb{R}^q \to V$ of 4.4.C where each $F' \mid W_i' \times \mathbb{R}^q$ is a submersion. The stable intersection property of f' implies that for the restriction of f' to some of the branches W_i' of W say to W_0' and so the projection of the pull-back $V_0' \subset W_0' \times \mathbb{R}^q$ of the submersion map $F' \mid W_0' \times \mathbb{R}^q : W_0' \times \mathbb{R}^q \to V$ to $\mathbb{R}^q$ has non-empty interior. Then, by an obvious argument, there is a small ball $B_0 \subset W_0'$ and some $\mathbb{R}^k \subset \mathbb{R}^q$, such that F' is a submersion on $B_0 \times \mathbb{R}^k$ and such that the projection of $V_0'' = V_0' \cap B_0 \times \mathbb{R}^k$ still has non-empty interior. Now $\dim B_0 \times \mathbb{R}^k = n = \dim V$, so our submersion $B_0 \times \mathbb{R}^k \to V$ is an equidimensional immersion and we may assume $V = B_0 \times \mathbb{R}^k$, where $V' = V_0''$ has non-empty interior when projected to $\mathbb{R}^k$ by $\psi : V = B_0 \times \mathbb{R}^k \to \mathbb{R}^k$. Now we argue as in 2.1 and 3.1.A (where the reader should beware of slight discrepancy between the present notations and those in 3.1.A) by using the ball-box theorem and horizontality of the fibers $\psi^{-1}(y)$, $y \in \mathbb{R}^k$ which gives us the desired bound $\dim_{\mathrm{Hau}} V' \geqslant \dim_{\mathrm{Hau}} V - k$, since small ε-balls in V (used in determination of $\dim_{\mathrm{Hau}} V'$) project to approximate boxes in $\mathbb{R}^k$ with the sides

$$\underbrace{\varepsilon \times \cdots \times \varepsilon}_{n_1 - k} \times \underbrace{\varepsilon^2 \times \cdots \times \varepsilon^2}_{n_2 - n_1} \times \cdots \times \underbrace{\varepsilon^d \times \cdots \times \varepsilon^d}_{n_d - n_{d-1}}$$

and thus with volume $\approx \varepsilon^{N-k}$ for $N = \Sigma_{i=1}^d \, i(n_i - n_{i-1}) = \dim_{\mathrm{Hau}} V$. The details of this, as well as the generalization to the m-horizontal case is left to the reader. We also suggest to the reader to try the further generalization to M-*horizontality* for $M = (m_1, \ldots, m_d)$ which refers to submanifolds $W \subset V$ with rank $(T(W) \cap H_i) \geqslant m_i$.

Remarks

(a) It is hard to believe that a Nash type implicit function theorem is indispensable for the lower bound on $\dim_{\mathrm{Hau}} V'$.

(b) It would be interesting to use the above projection argument to bound the Euclidean Hausdorff dimension of V' in terms of the C-C one. One knows in this regard (this was explained to me by J. Bourgain) that the normal projection of a subset $V' \subset \mathbb{R}^n$ with $\dim_{\mathrm{Hau}} V' \geqslant n - k$ onto a *generic* $\mathbb{R}^{n-k} \subset \mathbb{R}^n$ has $\mathrm{mes}_{n-k} > 0$ (see [Fal]) and one wants to extend this result to (generic members of) more general families of smooth maps $\mathbb{R}^n \to \mathbb{R}^{n-k}$. (Actually, even in the topological category one may ask such a question. For example, let $V' \subset \mathbb{C}^n$ have topological dimension $\geqslant 2(n - k)$. Does there exist a *complex* linear map $\mathbb{C}^n \to \mathbb{C}^{n-k}$ so that the image of V' has non-empty interior?) Notice that the argument in [Fal] uses a potential theoretic definition of $\dim_{\mathrm{Hau}}$ in $\mathbb{R}^n$ which seems to nicely fit the C-C framework.

4.6. Horizontal triangulations and a bound on the width of subsets in nilpotent Lie groups.

Let V be a simply connected two step nilpotent group with a left-invariant polarization $H \subset T(V)$ of rank n_1 complementary to the center $V_1 \subset V$ (of dimension $n - n_1$). Recall that such a group is determined up to an isomorphism by the 2-cocycle on $\mathbb{R}^{n_1} = V/V_1$ with values in $\mathbb{R}^{n-n_1} = V_1$ as such cocycles govern $\mathbb{R}^{n-n_1}$-central extensions over $\mathbb{R}^{n_1}$. This cocycle is nothing else but our curvature form $\Omega = \Omega_H : \Lambda^2 H \to H^\perp = T(V)/H$ restricted to the $T_{\mathrm{id}}(V)$ and, in fact, every Ω on $\mathbb{R}^{n_1}$ correspond to some Lie group.

Theorem. *If Ω admits an Ω-regular and Ω-isotropic subspace in $\mathbb{R}^{n_1} = V/V_1 = H_{\mathrm{id}}$ of dimension k then every closed subset $V' \subset V$ satisfies*

$$\text{C-C } wid_{n-k-1} \, V' \leqslant \mathrm{const}_V (\text{C-C } mes_{N-k} \, V')^{\frac{1}{N-k}}, \qquad (*)$$

where the width (as defined in 3.4) and the Hausdorff measure refer to the C-C metric associated to H and some left-invariant Riemannian metric g in V, where $n = \dim_{\mathrm{top}} V$, $N = n_1 + 2(n - n_1) = \dim_{\mathrm{Hau}} V$ and "const_V" signifies "$\mathrm{const}_{H,g}$". Furthermore, the Riemannian width and the Hausdorff measure are related by the inequality

$$\text{Riem-}wid_{n-k-1} \, V' \leqslant \mathrm{const}_g (\text{Riem-}mes_{n-k} \, V')^{\frac{1}{N-k}}. \qquad (**)$$

Proof. First we observe that $(**)$ follows from $(*)$ and notice that $(**)$ is interesting only for *large* mes_{n-k} as for mes_{n-k} small this follows from the purely Riemannian inequality $\text{Riem-}wid_{n-k-1} \, V' \leqslant \mathrm{const}_V (\text{Riem-}mes_{n-k} \, V')^{\frac{1}{n-k}}$ valid for all Riemannian manifolds V of locally bounded geometry, provided $mes_{n-k} \, V' \leqslant \varepsilon$ for some positive $\varepsilon = \varepsilon_V$.

Now, we turn to the proof of $(*)$ and observe, with an obvious generalization of the intersection inequality 3.4.B$'$ to the present context, that $(*)$ would follow from the existence of a suitable "horizontal" triangulation of V (compare 3.4.B). Notice that we do not truly need the Γ-invariance of such triangulation as was required in 3.4.B but we want all simplices of a given dimension to be roughly of unit size (as spelled out below). Then we can make the simplices of any size we want, since V being a *two-step* group, admits a non-trivial self-similarity.

Horizontal triangulation lemma. *There exists a triangulation* Tr *of* V, *such that*

(i) every i-simplex of Tr is the image of the standard simplex Δ^i under a piecewise smooth map $\Delta^i \to V$ with the Lipschitz constant (with respect to the metric g in V) bounded by a constant $C = C_V$,

(ii) for every ball in V the number of simplices in Tr meeting this ball is bounded by a constant depending on the radius (but not on the position) of the ball,

(iii) every k-simplex in Tr is (piecewise) H-horizontal.

Proof. First, notice that (i) and (ii) express the idea of "unit size" indicated above while the guts of the lemma are in (iii) where we shall use our basic assumption of the existence of Ω-regular and Ω-isotropic subspace of dimension k. The presence of such subspace, allows one an approximation of the k-skeleton Tr^k of any triangulation Tr_0 satisfying (i) and (ii) by a subfolded regular horizontal map $\varphi_k : \mathrm{Tr}_0^k \to V$. If $2k \leqslant n - 2$, this map can be made an (horizontal!) embedding by a small perturbation provided by the generalized Nash theorem (compare (E) in 2.3.2 of [Gro$_{\mathrm{PDR}}$]) and the image $\varphi_k(\mathrm{Tr}^k) \subset V$ can be easily extended to the desired triangulation Tr with $\varphi_k(\mathrm{Tr}_0^k)$ serving as the k-skeleton of Tr.

Finally, if $k \geqslant 2n - 2$, one could make φ_k an embedding extendable to Tr by sharpening the folded approximation theorem but this is not necessary as the only cases where $2k > n - 2$ are those of the Heisenberg groups of dimensions 3 and 5 (and k equal 1 and 2 respectively) where our contact discussion in 3.4 applies anyway. (We suggest the reader would check the details of this proof.)

Exercise. Generalize the above theorem to d-step nilpotent groups for all $d \geqslant 2$.

4.7. Lipschitz maps into C-C spaces. We shall extend here the results from 3.5 to general equiregular C-C manifolds (V, H), where we assume that H admits an Ω-regular Ω-isotropic subbundle $S \subset H$ of rank k (e.g. being a nilpotent group of the previous section). Our Lipschitz maps $W \to V$ for $\dim W \leqslant k$ will appear as uniform limits of (Ω-regular) horizontal *subfolded* maps of rank k replacing piecewise horizontal maps of 3.5. As we mentioned earlier, even the *local* extension problem for general (piecewise) horizontal maps into (V,H) appears quite difficult and solved only in the contact case. On the other hand, the extension is easy in the category of subfolded maps of a fixed rank k where all folded maps into V in question have their "tangent" bundles induced from a fixed S on V, as the "if" conditions of the folded approximation (see 4.4) and extension (see 4.4.B) theorem is automatic with such an S. It should be noticed, however, that the contact case is not covered by the present discussion as we did not assume there the existence of S which, in fact, does not necessarily exist even if there is an Ω-regular isotropic $S_v \subset H_v \subset T_v(V)$ for every $v \in V$. (Probably, our present Lipschitz results generalize to the case of a discontinuous field of Ω-regular isotropic subspaces $v \mapsto S_v \subset H_v$.)

The notion of complexity for piecewise smooth maps in 3.5.B obviously extends to the subfolded category and everything remains intact. When this notion applies to the construction and extension of Lipschitz maps one crucially uses rescaling of V and so if (V, H) is a nilpotent group with a self-similarity (or is locally isomorphic to such a group) everything from 3.5.B immediately generalizes to such (V, H). In the general case, as one needs maps of bounded complexity produced on an arbitrary small scale, one needs the following additional argument using the nilpotent Lie group N_v approximating (V, H) at a given point $v \in V$ (see 1.4). The Lie algebra L_v of N_v is, in the present case, a two step nilpotent algebra, $L_v = H_v \oplus H_v^\perp$, where $H_v^\perp = T_v(V)/H_v$, such that $H_v^\perp$ lies in the center of L_v and the commutator rule $H_v \otimes H_v \to H_v^\perp$ is given by the form Ω_v. It follows that N_v has Ω_v-regular isotropic subspaces at all points and so we do have folded maps of bounded complexity in N_v at each scale as N_v comes with a self-similarity. But since N_v approximates (V, v) at the small scale these maps can be moved by small perturbation to corresponding maps into V. (It is useful to think of N_v as a deformation of V. Namely, there exists a smooth family of polarizations H_ε on V at v for $\varepsilon \in [0, \rho]$, $\rho > 0$, such that $H_{\varepsilon > 0}$ on the ρ-ball around v is isomorphic to H on the ε-ball and H_0 equals the implied polarization on N_v, where N_v is locally diffeomorphically mapped to V by $E_v \circ E_0^{-1}$ (see 1.4). Then we have folded ε-families of maps $f_\varepsilon : W \to (V, H_\varepsilon)$, $\varepsilon \in [0, \rho]$, which can be treated as individual maps and which specialize for every ε to the required maps of bounded complexity on the ε-scale.)

Now, everything is ready for the proof of

Lipschitz extension theorem. (Compare 3.5.D.) *Let W be a compact simplicial polyhedron of dimension $\leqslant k$, and $W_0 \subset W$ be a subpolyhedron. Then a Lipschitz map $f_0 : W_0 \to (V, H)$ Lipschitz extends to W if and only if it extends continuously.*

Exercise. State and prove all results from 3.5 in the present situation.

4.7.A. Construction and extension of Hölder maps. If we apply our (folded H-horizontal) extension procedure to a C^α-Hölder map $f_0 : W_0 \to V$ then the resulting extension $f : W \to V$ is also C^α-Hölder. This works for all $\alpha > 0$ if $\dim W \leqslant \operatorname{rank} S$ but for smaller α we can do better using suitable S_i in the i-th commutator bundle $H_i \supset H = H_1$ for $i \geqslant 2$. For example, every H_2-horizontal map is $C^{\frac{1}{2}}$-Hölder and if H_2 admits a subbundle $S_2 \subset H_2$ which is regular isotropic for the curvature form $\Omega_2 : \Lambda^2 H_2 \to H_2^\perp$, then $C^{\frac{1}{2}}$-Hölder maps $W \to V$ becomes available for $\dim W \leqslant k_2 = \operatorname{rank} S_2$ which is an improvement if $k_2 > k = k_1$. Similarly, an S_i of rank k_i gives us $C^{\frac{1}{i}}$-Hölder maps $W \to V$ for $\dim W \leqslant k_i$. Finally, one may wonder what is the role of m-horizontal (and M-horizontal for $M = (m_1, \ldots, m_d)$) maps into (V, H) and some aspect of the problem will be adressed in the next section.

4.8. Dehn isoperimetry in nilpotent Lie groups. (Compare 5.A in [Gro$_{\mathrm{AI}}$].) Let V be a simply connected nilpotent Lie group and $H \subset T(V)$ be a left-invariant polarization corresponding, on the Lie algebra level, to a subspace $H(\mathrm{id}) \subset L$ complementary to $[L, L] \subset L$ (and the curvature form Ω on $H(\mathrm{id}) \subset T_{\mathrm{id}}(V)$ can be identified with the form $\Omega_0 : \Lambda^2(L/[L, L]) \to [L, L]/[L, [L, L]])$ corresponding to the Lie bracketing in the Lie algebra $L = L(V)$. We fix some left-invariant Riemannian metric g on V and we observe that the Lipschitz extension theorem of the previous section specializes to the following.

Quadratic isoperimetric inequality. (C-C *version.*) *Let Ω admits a two dimensional Ω-regular and Ω-isotropic linear subspace in $H(\mathrm{id})$. Then every closed curve in V of finite C-C length bounds a disk of C-C area (i.e. C-C Hausdorff measure mes_2) satisfying*

$$\operatorname{area}(\text{disk}) \leqslant \operatorname{const}_V (\operatorname{length}(\text{curve}))^2. \tag{$*$}$$

In fact there exists a C-C Lipschitz map of the unit disk D to V such that the boundary circle of D parametrizes our curve in V and the implied Lipschitz constant of the map satisfies

$$Lip \leqslant \mathrm{const}'_V \ length(curve).$$

Now we recall that the Riemannian geometry of (V, g) on the large scale is equivalent to the Carnot-Carathéodory one and conclude to the following.

Riemannian isoperimetric inequality. *Every closed curve in V of finite g-length bounds a disk of finite g-area such that*

$$g\text{-area}(disk) \leqslant \mathrm{const}_g (g\text{-length}(curve))^2. \qquad (**)$$

Furthermore, our curve can be parametrized by the boundary of a g-Lipschitz map $D \to V$ with

$$Lip \leqslant \mathrm{const}'_g \ length.$$

Discrete corollary. *Every discrete nilpotent group Γ for which the corresponding Lie algebra admits an Ω-regular isotropic plane (in $L/[L, L]$) satisfies the quadratic isoperimetric inequality.*

Notice that the existence of such a plane is a generic phenomenon for *two* step nilpotent groups with $3n_1 \geqslant 2n + 2$, where n_1 is the rank of the group and n_1 the rank of its abelianization. But d-step groups for $d \geqslant 3$ admit no Ω-regular planes at all and one is faced with a more difficult problem of finding (high order) regular jets in the sense of 4.2. This problem remains open and one does not know if there are any 3-step nilpotent groups with quadratic isoperimetry.

4.8.A. Hölder maps $D \to V$ and isoperimetric inequalities of degree $2i$. We note that H_i-horizontal maps to V are $C^{\frac{1}{i}}$-Hölder (where $H_i \subset T(V)$ is the i-th commutator bundle) and have finite C-C Hausdorff measure mes_{2i} for dist_H in (V, H). We denote by Ω_i he curvature form on the bundle H_i defined by the Lie bracketing as earlier,

$$\Omega_i : \Lambda^2 H_i \to H_i^{\perp} = T(V)/H_i$$

and observe the following.

Theorem. *If H_i contains an Ω_i-regular and Ω_i-isotropic plane then V (and Γ) satisfies the isoperimetric inequality of degree $2i$,*

$$\text{area}_g(\text{filling disk}) \leqslant \text{const}_g(\text{length}_g(\text{curve}))^{2i}.$$

Proof. Lipschitz curves in (V, H) extends to $C^{\frac{1}{i}}$-Hölder disks by the discussion in 4.7.A.

4.8.B. Regular (i,j)-surfaces in V and isoperimetric inequalities of degree $i + j$. A smooth (or folded) map f of a disk D to V is called (i,j), where $j \geqslant i$, if it is H_j-horizontal and also has non-trivial tangency to H_i at all points in D. In other words,

$$\text{rank}(\mathcal{D}f)^{-1}H_j \geqslant 2 \quad \text{and} \quad \text{rank}(\mathcal{D}f)^{-1}H_i \geqslant 1$$

everywhere on D. Images of such maps have C-C $\text{mes}_{i+j} < \infty$ and if these are sufficiently abundant they may be used to fill in closed horizontal curves in V. In order to formulate a sufficient condition for such abundance we use the abstract language of 4.2.B and denote by $\mathcal{R}_{i,j}^{\infty}$ the stabilized differential condition expressing the (i,j)-property of smooth maps $f : D \to V$.

$(i + j)$-isoperimetric inequality. ([11]) *If $\mathcal{R}_{i,j}^{\infty}$ contains a regular jet (in the sense of 4.2.B) then (V, g) satisfies the isoperimetric inequality of degree $i + j$.*

Proof. We proceed as earlier with folded (i,j)-surfaces in (V, H) constructed with the scaling pattern expressed by Fig. 8 and Fig. 9 in 3.5.D. It is easy to see that the C-C Hausdorff measure mes_{i+j} of such a surface filling in our closed curve is finite and, moreover

$$\text{mes}_{i+j} \leqslant \text{const}_V(\text{length})^{i+j}.$$

This C-C isoperimetric $(i + j)$-inequality yields the Riemannian one as earlier and the proof is concluded.

In order to use the $(i + j)$-inequality one needs specific criteria for (the existence of) regular jets in $\mathcal{R}_{i,j}^{\infty}$ some of which are indicated in 5.A$_2$-A$_5$ of [Gro$_{\text{AI}}$]. Here is another such criterion in the spirit

---

[11] Compare 5.A$_4'''$ in [Gro$_{\text{AI}}$].

of Ω-regularity. Let $\eta_1, \ldots, \eta_{n-n_j}$, be 1-forms on V defining $H_j \subset T(V)$ and $\eta_1, \ldots, \eta_{n-n_j}, \ldots, \eta_{n-n_i}$ be the form defining $H_i \subset H_j$. Let $\omega_1, \ldots, \omega_{n-n_j}, \ldots, \omega_{n-n_i}$ be the differentials of these forms and X_1, X_2 two independent tangent vectors at $\mathrm{id} \in V$, where $X_1 \in H_i$ and $X_2 \in H_j$. We associate to X_1 and X_2 the following equations

$$\left. \begin{array}{l} \omega_\nu(\partial, X_1) = \sigma_{\nu,1} \ , \ \nu = 1, \ldots, n - n_i, \\[2ex] \omega_\nu(\partial, X_2) = \sigma_{\nu,2} \ , \ \nu = 1, \ldots, n - n_j, \end{array} \right\} \qquad (*)$$

and call the plane spanned by X_1 and X_2 *regular* if this system in non-singular, for the unknown vector ∂ from $(H_i)_{\mathrm{id}}$. In other words the homomorphism $\Omega^0_{i,j} : H_i \to H_i^\perp \oplus H_j^\perp$ defined at $\mathrm{id} \in V$ by

$$\partial \mapsto \{\omega_1(\partial, X_1), \ldots, \omega_{n-n_i}(\partial, X_1), \omega_1(\partial, X_2), \ldots, \omega_{n-n_j}(\partial, X_2)\}$$

is surjective. The above (X_1, X_2)-plane is called *isotropic* if X_1 and X_2 extend to *commuting* fields, the first in H_i and the second in H_j, which is equivalent to $\omega_\nu(X_1, X_2) = 0$, $\nu = 1, \ldots, n - n_j$. Then one easily sees as earlier that an isotropic regular plane gives rise to a regular jet in $\mathcal{R}^\infty_{i,j}$, since the linearization of the form inducing system (imposed on a map $f : D \to V$ by $f^*(\eta_\nu) = \theta_\nu$) reduces to $(*)$.

It is worthwhile to work out specific examples where this regularity takes place and to find other (practical) criteria for the existence of regular jets.

Remark. Regularity of an H-horizontal map $f : D \to V$ is essentially equivalent to algebraic solvability of the system $L_f \, \partial = \sigma$ where L_f is the linearization of the differential operator D which assigns to each $f : D \to V$ the $(n - n_1)$-tuple of forms on D by $D(f) = \{f^*(\eta_i)\}$, where η_i, $i = 1, \ldots, n - n_1$ are some forms on V defining H. In fact one needs algebraic solvability of $L_f \, \partial = \sigma$ not only for a fixed f but for all nearby f's which do not have to be horizontal. It may happen that $L_f \, \partial = \sigma$ is algebraically solvable for a horizontal f, but not (at least not apparently so) for nearby f's. For example let V be a 3-step group, so that $H_3 = T(V)$ and let L° be the associated graded Lie algebra, (isomorphic to the tangent Lie algebra to (V, H) at $\mathrm{id} \in V$, see 1.4), written as $L^\circ = L_1 \oplus L_2 \oplus L_3$, so that L_1 corresponds to H_1, and $L_1 \oplus L_2$ to H_2. Take two independent vectors X_1 and X_2 in L_1 and define a linear map $\Omega^0 : L_1 \oplus L_2 \to (L_2 \oplus L_3) \oplus (L_2 \oplus L_3) = \mathrm{Hom}(\mathbb{R}^2, L_2 \oplus L_3)$ by

$\partial \mapsto ([\partial, X_1], [\partial, X_2])$. It is easy to see that commuting vectors X_1 and X_2 (should) correspond to horizontal surfaces (as the tangent frame on such surface can be made of commuting fields) and that the surjectivity of Ω^0 is sufficient for algebraic solvability of $L_f\,\partial = \sigma$ for a horizontal f. Yet it is unclear if the existence of commuting X_1 and X_2 with surjective Ω^0 yields (the local h-principle and microflexibility for) horizontal surfaces and/or the quadratic isoperimetric inequality in V. I also must admit I have not worked out any example where such X_1 and X_2 are actually present.

4.8.C. On filling in dimension ≥ 3. The general question is as follows. Let S be a $(k-1)$-dimensional cycle in (V, H) of finite Hausdorff measure mes_ℓ for some $\ell \geq k - 1$. When can it be filled in by a k-dimensional chain D with finite Hausdorff measure mes_m for a given $m \geq \max(\ell, k)$? Furthermore, we want some specific bounds on mes_m in terms of mes_ℓ, e.g. an inequality

$$\mathrm{mes}_m D \leq \mathrm{const}(\mathrm{mes}_\ell S)^\alpha. \tag{?}$$

More geometrically, we may start with a piecewise smooth *horizontal* cycle S and look for a *horizontal* filling D of a controlled k-dimensional volume. This could give us the above (?) for $\ell = k$, $m = k + 1$ and $\alpha = k + 1/k = m/\ell$. Similarly, one may start with a *partially horizontal* cycle having prescribed ranks of the intersections of $T(S)$ with H_i and try to fill it in by a partially horizontal chain D (possibly, with weaker tangencies to H_i than S) with controlled k-volume. Of course, in order to do that, we need "many" horizontal (or partially horizontal) k-dimensional submanifolds in V. But even when such manifolds are abundant, e.g. in contact manifolds V of dimension $n \geq 2k + 1$, we do not know how to fill in horizontal $(k-1)$-dimensional cycles by horizontal k-chains with controlled k-volume. Moreover, our telescoping filling used for $k = 2$ (see Fig. 9 in 3.5.D) may work for $k = 3$ and S being a surface with genus $(S) \leq \mathrm{const}$, e.g. for S homeomorphic to S^2. Here is the idea. First, the general problem can be reduced to the case where $S \subset V$ is *sufficiently regular* in the following sense: every domain $S_0 \subset S$ with area $S_0 \leq \frac{1}{2}$ area S has length $\partial S_0 \geq \mathrm{const}_V(\mathrm{area}\ S_0)^{\frac{1}{2}}$. In particular, the metric balls in S of radii $(,\dots,(\ S)^{\frac{1}{2}}$ have at least quadratic area growth and, consequently, $\mathrm{Diam}\ S \leq \mathrm{const}(\mathrm{area}\ S)^{\frac{1}{2}}$. This is done with the usual cut-and-paste (inductive) regularization techniques which are presented in a sufficient generality in [Gro$_{\mathrm{FRM}}$] (compare 2.5 of the present paper). These techniques work in all dimension (and reduce the problem to S with

$\mathrm{vol}_{k-2}\, \partial S_0 \geqslant \mathrm{const}_V \, (\mathrm{vol}_k\, S_0)^{\frac{k-2}{k-1}}$ for all $S_0 \subset S$ with $\mathrm{vol}\, S_0 \leqslant \frac{1}{2}\,\mathrm{vol}\, S$) and if $k-1 = \dim S = 2$ the cut-and-paste procedure does not increase the genus of S and we assume below S is homeomorphic to S^2. Now we want to fill such H-horizontal S in by an H-horizontal 3-ball (or something topologically similar) of bounded volume. Notice that it is relatively easy to construct a Hölder map of the ball into V extending S on the boundary for a suitable *conformal* parametrization $f : S^2 \to S$ since the regularity of S allows a *Hölder* parametrization of S with the implied Hölder exponent depending on const_V in the isoperimetric inequality for length ∂S_0 (where "Hölder" would turn into Lipschitz if that inequality was sharp with const_V asymptotic to $2\sqrt{\pi}$ for domains $S_0 \subset S$ with area $S_0 \to 0$, compare 2.5.H$'$). But Hölder (unlike Lipschitz) is not sufficiently good for us and we shall indicate below a possible construction of the filling using intrinsic geometry of S. We start by observing that our disk filling S^1 exhibited in Fig. 9 of 3.5.D arises from a sequence of diadic partitions of S^1, where S^1 is divided first into four segments of equal length and then we keep dividing in two as usual. Now we want to produce a similar sequence of partitions of the sphere S (with the intrinsic metric induced from V) and the key property of the partitions we need is expressed in the following.

Tentative proposition. *Let (V, H) be a nilpotent Lie group admitting a regular jet of a 3-dimensional horizontal submanifold (e.g. an Ω-regular Ω-isotropic 3-plane in H) and $S \subset V$ be a (piecewise) smooth horizontal surface. Suppose S admits a sequence of finite partitions denoted P_i with the parts $S_{i,j} \subset S$ which intersect (only) across their boundaries and where P_{i+1} refines P_i for all $i = 1, \ldots$. Assume that*

(i) $\mathrm{diam}\, S_{i,j} \leqslant \varepsilon_i \xrightarrow[i\to\infty]{} 0$,

(ii) every $S_{i,j}$ is divided into at most q parts of P_{i+1} for a fixed q independent of i. Furthermore, this division (partition) of each $S_{i,j}$ into parts of P_{i+1} can be refined to a cell division of S_{ij} with the total number of cells bounded by some constant q',

(iii) the totality of the diameters of $S_{i,j}$ is bounded by

$$\sum_{i=1}^{\infty}\sum_{j} (\mathrm{diam}\, S_{i,j})^3 \leqslant \mathrm{const}(\mathrm{area}\, S)^{\frac{3}{2}}. \qquad (+)$$

Then S can be filled in by a horizontal ball D in (V, H) with

$$\mathrm{Vol}_3\, D \leqslant \mathrm{const}'(\mathrm{area}\, S)^{\frac{3}{2}} \qquad (*+)$$

for $\mathrm{const}' = \mathrm{const}'(V, q')$.

The proof is easy, guided by Fig. 9 in 3.5.D.

Now we need good partitions of S where we are allowed to assume S is sufficiently regular in the above sense. Then one obtains sensible partitions of S, by using, for example, maximal δ-separated nets in S and dividing S into the corresponding (nearest point) Dirichlet domains. The trouble is that the local complexity of such partitions may be unbounded as a small δ-ball in V intersects too many other (disjoint) δ-balls. On the other hand, as the genus of S is assumed bounded (e.g. $S \approx S^2$), this concentration of balls is not significant on the average due to the Gauss-Bonnet formula, and with a suitably generalized averaged version of (ii) one expects some filling D of S exists (where one should not insist on D being the topological ball anymore).

Remark. The above tentative proposition admits a variety of generalizations and refinements. Its role consists in reducing the general filling problem to two subproblems.

(1) Filling (horizontal or partially horizontal) cycles of *bounded complexity* in V.

(2) Abstract combinatorial filling of S (constructed with partitions of S).

Here S may be a general $(k-1)$-cycle with a metric, e.g. a closed (oriented) Riemannian manifold of dimension $k-1$ and an "abstract combinatorial filling" refers to a k-dimensional metric space D isometrically containing S as a closed nowhere dense subset, such that S is homologous to zero in D and such that $D - S$ admits an infinite triangulation into simplices Δ_i, $i = 1, \ldots$, such that

$$\sum_{i=1}^{\infty}(\mathrm{diam}\,\Delta_i)^k \leqslant \mathrm{const}_*(\mathrm{Vol}_{k-1}\,S)^{\frac{k}{k-1}}, \qquad (\star)$$

and where simplices in $D - S$ approaching S have diam $\to 0$. (Notice we do not assume D compact and so the metric completion of $D - S$ may contain apart from S other pieces but of dimension $< k - 1 = \dim S$.) Our, conjecture is that such a D exists for "sufficiently regular" S homeomorphic to S^2 (with the implied const_* depending on this regularity and nothing else). Similar result may hold for higher dimensional S *conformal* to S_0 with a bound on geometry. Finally, we observe that the infimum of

$\Sigma_i(\Delta_i)^\ell$ over all triangulated metric fillings D of S appears an interesting geometric invariant of S. (Compare Fill Vol and Fill Rad in [Gro$_{\mathrm{FRM}}$] and filling on the large scale in [Gro$_{\mathrm{AI}}$].)

Remarks. The "inductive" argument in [Gro$_{\mathrm{AI}}$] for filling circles by disks suggests a slightly different (?) approach to filling 2-spheres by balls.

4.9. Metric properties of submanifolds partitions and maps. The basic invariant of a subvariety $V' \subset V = (V, H)$ is the Hausdorff dimension $\dim_{\mathrm{Hau}} V'$ and $\inf \dim_{\mathrm{Hau}} V'$ over all V' with $\dim_{\mathrm{top}} V' = k$ is a basic metric Hölder-robust invariant of $V = (V, H)$. The major problem in evaluating this invariant, denoted $d_{H|t}(V, k)$, is finding a lower bound on $\dim_{\mathrm{Hau}} V'$ without assuming any smoothness of V' in V. This problem becomes somewhat easier if we replace the Hausdorff dimension by the *Minkowski* (or *entropy*) *dimension* which is defined similar to $\dim_{\mathrm{Hau}}$ with covering by balls but in the definition of $\dim_{\mathrm{Min}}$ all ball covering V' must be of the same radius ε (where eventually $\varepsilon \to 0$ and the number of balls is roughly asymptotic to $(\varepsilon^{\dim_{\mathrm{Min}}})^{-1}$ for $\varepsilon \to 0$ by the definition of $\dim_{\mathrm{Min}}$. Notice that smooth equiregular submanifolds V' have $\dim_{\mathrm{Min}} V' = \dim_{\mathrm{Hau}} V'$ and in all cases $\dim_{\mathrm{Min}} \geqslant \dim_{\mathrm{Hau}}$. Furthermore, if V is equiregular, then the Minkowski dimension can be defined as $\dim_{\mathrm{Min}} = \dim_{\mathrm{Hau}} V - \mathrm{codim}_{\mathrm{Min}} V'$, where the Minkowski codimension is defined with the Hausdorff (which is equivalent to Lebesgue) measure on $V = (V, H)$ as the *critical exponent* α for $\varepsilon^\alpha \, \mathrm{mes}(V' + \varepsilon)$, where $V' + \varepsilon$ refers to the ε-neighbourhood of V' in V with respect to dist_H and "critical α" is the infimum of those α' for which

$$\varepsilon^{\alpha'} \, \mathrm{mes}(V' + \varepsilon) \to 0 \quad , \quad \text{for } \varepsilon \to 0.$$

One can construct further invariants of V by applying $d_{H|t}$ and $d_{M|t}$ to $V' \subset V$ (where "M" stands for "Minkowski") by the following inductive scheme. Let inv be an invariant of metric spaces with values in some set R. For example inv $= \dim_{\mathrm{Hau}}$ takes value in $\mathbb{R}_+$ while $d_{H|t}$ values in $\mathbb{R}_+^{\mathbb{Z}_+} = $ maps $(\mathbb{Z}_+, \mathbb{R}_+)$ where $\mathbb{Z}_+$ corresponds to the topological dimension $k \in \mathbb{Z}_+$. Now, for a general "inv" we consider *all* topological submanifolds $V' \subset V$ of a given dimension k and take the set $\mathrm{in}_k \subset R$ of the values inv $V' \in R$. Thus, for variable k, we obtain a new invariant $\mathrm{inv}_{\mathrm{new}}(V)$ with values in $\mathbb{Z}_+^{2^R}$, namely the one which assigns to each

$k \in \mathbb{Z}_+$ the set $\mathrm{in}_k \in 2^R$. (Of course one could consider more general subsets V' in V and replace $\mathbb{Z}_+$ by the set of $\mathrm{topinv}(V' \subset V)$.) Notice that the smooth counterparts of such invariants with "C^∞-submanifolds" instead of "topological submanifolds" are relatively easy to evaluate and the basic unsolved problem is computing the "topological" metric invariants of (V, H) in terms of the "smooth" ones.

Example (where this does not work). Let $V' \subset V$ be a compact connected smooth submanifold of codimension $n_1 = \mathrm{rank}\, H$ which is transversal to H. Then the metric dist_H on V' is greater than $\sqrt{\overline{\text{Euclidean}}}$ which can be expressed in one of the following six ways (where the properties (1)-(6) below are immediate with the ball-box theorem).

(1) There is a Riemannian metric dist' on V' such that $\mathrm{dist}_H \geqslant \sqrt{\mathrm{dist}'}$ on V'.

(2) There is a Riemannian metric dist' on V' and a proper continuous map $f : V' \to V'$ of degree one, such that every dist_H-ball of radius ε goes to an ε^2-ball for dist' for all $\varepsilon \geqslant 0$.

(3) An ε-*chain* in a metric space is a sequence of points

$$v_0, \ldots, v_i, v_{i+1}, \ldots, v_k$$

where $\mathrm{dist}(v_i, v_{i+1}) \leqslant \varepsilon$ for $i = 0, 1, \ldots, k - 1$, and the length of an ε-chain is defined as $k\varepsilon$. Then we define $\mathrm{dist}'_\varepsilon$ on V' as the infimum of the lengths of the ε-chains in V' between pairs of points for the metric $\mathrm{dist}_H \mid V'$. Then the transversality of V' to H implies that

$$\mathrm{dist}'_\varepsilon(v_1, v_2) \geqslant \mathrm{const}\, \max(\mathrm{dist}_H(v_1, v_2), \varepsilon^{-1} \mathrm{dist}_H^2(v_1, v_2)). \qquad (*)$$

(4) Let $\mathrm{dist}_\varepsilon$ be the Riemannian (or piecewise Riemannian) ε-approximation to dist_H (see 1.4.D). Then $\mathrm{dist}_\varepsilon \mid V'$ is bounded from below by the right hand side of $(*)$.

(5) Given a subset $c \in V$, let

$$\mathrm{length}_\varepsilon\, c = \varepsilon(\text{minimal number of } \varepsilon\text{-balls needed to cover } c).$$

Then define $\mathrm{dist}''_\varepsilon$ as the infimum of the ε-length of curves c joining pairs of points in V'. Then this $\mathrm{dist}''_\varepsilon$ satisfies $(*)$.

(6) Every subset $c \in V'$ of topological dimension $\geqslant 1$ has $\dim_{\mathrm{Hau}} c \geqslant 2$.

We suggest the reader at this point would ponder an inter-relation between these six metric properties of V' and, in particular, show that $(3) \Rightarrow (2)$ by adopting the proof of the "cubical" Besikovič' lemma in §7 of $[\mathrm{Gro_{FRM}}]$.

Next we observe, that if V is a *two* step space, i.e. $H_2 = T(V)$, which means the commutators of H-horizontal fields span $T(V)$, then, in fact, $\mathrm{dist}_H \mid V' \approx \sqrt{\mathrm{Euclidean}}$, and in particular, $\dim_{\mathrm{Hau}} V' = 2\dim_{\mathrm{top}} V'$. On the other hand, if V is d-step for $d \geqslant 3$, then our V' necessarily has $\dim_{\mathrm{Hau}} V' > 2\dim_{\mathrm{top}} V'$.

Now, we want a similar result *without* assuming V' is smooth but just satisfying (some of) (1)-(6). Unfortunately, this does not work in general. Namely, if H admits an integral (i.e. H-horizontal) submanifold $W \subset V$ (on which the metric dist_H is Riemannian) of $\dim W \gg n - n_1$, then one can find a (highly non-smooth) submanifold $V' \subset W$ of dimension $n - n_1$ with $\mathrm{dist}_H \mid V' \approx \sqrt{\mathrm{Euclidean}}$, (as was pointed out to me by S. Semmes).

Question. Are there cases where (1)-(6) imply $\mathrm{dist}_{\mathrm{Hau}} V' > 2\dim_{\mathrm{top}} V'$ or every V admits a topological submanifold V' of codimension n_1 satisfying (1)-(6) and having $\dim_{\mathrm{Hau}} V' = 2\dim_{\mathrm{top}} V'$ (and moreover, having $\mathrm{dist}_H V' \approx \sqrt{\mathrm{Euclidean}}$)?

Local contractibility of V'. We start by observing (with the ball-box theorem) that every smooth *equiregular* $V' \subset V$ (i.e. having rank $(T(V') \cap H_i)$ constant on V' for all $i = 1, \ldots, d$) is *locally contractible* in the sense that every ε-ball in V' (for $\mathrm{dist}_H V'$) is contractible within C_ε-ball for some constant $C = C(V')$ (compare 1.4.B). This remains valid for C^2-smooth not necessarily equiregular V' if V is 2-step, (as again follows from the ball-box theorem and the Taylor remainder formula) but C^∞-submanifolds in d-step spaces for $d \geqslant 3$ do not have, in general, this property. For example, let V be 3-step nilpotent group with the graded Lie algebra, $L = L_1 \oplus L_2 \oplus L_3$ and $V_0' \subset V$ be a 2-dimensional Abelian subgroup corresponding to a subalgebra spanned by some $x \in L_1$ and $y \in L_3$. The metric dist_H on V_0' (isomorphic to $\mathbb{R}^2$) has the ε-balls at the origin equivalent to the boxes $\{|x| \leqslant \varepsilon \, , \, |y| \leqslant \varepsilon^3\}$ and the parabola $V' = \{y = x^2\}$ in $V_0' \subset V$ has highly disconnected balls at the points close to $(x = 0 \, , \, y = 0)$ as a simple consideration shows. On the other hand one can easily show that every C^∞-smooth V' in a d-step manifold V has the local contractibility with the exponent $\alpha = (d-1)^{-1}$, i.e. every ε-ball is contractible within a concentric ball of radius $C\varepsilon^\alpha$.

Including V' into a family. The transversality of V' to H seems hard to characterize in terms of $\mathrm{dist}_H V'$ (as we saw above) and one may try to use more (metric) information on the relative position of such a V' in V. We observe that every smooth V' of codimension n_1 transversal to H, locally, appears as a fiber of a smooth map $\Pi : V \to \mathbb{R}^{n_1}$, say $V' = \Pi^{-1}(0)$ and we have "parallel" submanifolds $V' = V'(x) = \Pi^{-1}(x) \in V$ for all $x \in \mathbb{R}^{n_1}$. Let us enumerate the metric properties of the family $V'(x)$ (which easily follow from the ball-box theorem). To simplify the matter, we assume that our p topologically is a trivial fibration with compact fibers V' over the unit ball $B_1 \subset \mathbb{R}^{n_1}$ around the origin.

(1) The map Π is Lipschitz.

(2) The map Π is *co-Lipschitz*, i.e. the Π-image of every ε-ball around $v \in V$ contains a ball of radius $\mathrm{const}\,\varepsilon$ around $\Pi(v) \in B$.

(2') The adjoint map $\Pi^{\#} : 2^B \to 2^V$ for $\Pi^{\#} : B' \mapsto p^{-1}(B') \subset V$, for $B' \subset B$, is Lipschitz where the spaces of subsets 2^B and 2^V are given their respective Hausdorff metrics.

(3) The partition (foliation) of V into $V'(x)$ is *transversally Lipschitz* which may be expressed in two slightly different ways.

(3a) $\mathrm{dist}_{\mathrm{Hau}}(V'(x), V'(y)) \leqslant \mathrm{const}\,\mathrm{dist}(V'(x), V'(y))$, for all $V'(x)$, where dist on the right hand side refers to the infimum of the distances $\mathrm{dist}(v_1, v_2)$, $v_1 \in V'(x)$, $v_2 \in V'(y)$.

(3b) Let $W \subset V$ be the graph of a continuous section $q : B \to V$. Such a V defines a metric on B (and hence on $W = q(B)$), namely the maximal (or supremal) among metrics dist such that the Π-image of every ε-ball in V with the center in W is contained in the concentric ε-ball for the maximal dist. Then the transversally Lipschitz property claims the Lipschitz equivalence of such metrics for different sections $B \to V$.

(4) There is a (multiply) transitive group of homeomorphisms acting on V which sends fiber of Π to fibers and such that

   (i) the induced homeomorphisms on B are bi-Lipschitz,

   (ii) if V is 2-step that the homeomorphisms are bi-Lipschitz on the fibers,

(iii) if V is d-step, the homeomorphisms are C^α-Hölder on the fibers for $\alpha = (d-1)^{-1}$.

These homeomorphisms of V are constructed with H-horizontal lifts of vector fields on B.

Unfortunately, the above discussion, metrically speaking, hangs in the air as we do not know what kind of partitions (fibrations) may arise without the assumption of smoothness. Here is a specific

Question. When can a given (V, H) be fibered (foliated) by (topological!) submanifolds V_x of a prescribed dimension k, such that one (or all) of the following three conditions is satisfied?

(∗) The foliation of V into V' is transversally C^α-Hölder, for a given α, e.g. for $\alpha = 1$;

(∗∗) every $V'(x)$ has Hausdorff dimension in a given interval, for example the metric dist_H on each fiber $V'(x)$ is C^β-Hölder equivalent to (Euclidean)$^\gamma$ for given β and γ in the interval $0 < \beta, \gamma \leqslant 1$;

(∗ ∗ ∗) the space of V'-fibers (with the Hausdorff distance) has the Hausdorff dimension in a given interval, e.g. being C^δ-Hölder equivalent to a Riemannian space for a given positive $\delta \leqslant 1$.

Now we return from topological dreams to C^∞-reality and indicate some extensions of the above (1)-(4) to more general families of equiregular submanifolds V'_x arising as fibers of a smooth fibrations $\Pi : V \to B$. Such a Π remains Lipschitz but "co-Lipschitz" must be replaced by "co-Hölder" with a suitable exponent $\alpha \geqslant d^{-1}$ (which precise evaluation we leave to the reader). The space of the leaves (fibers) with the Hausdorff metric now may be a more complicated space than Carnot-Carathéodory albeit it has many C-C features (e.g. the finite Hausdorff dimensions; we leave to the reader to decide when it has finite Hausdorff measure). The partition into the leaves is transversally Hölder (find the exponent!) and the geometry of the maximal metric dist (depending on q) also seems less regular (?) than C-C. On the other hand (ii) in the above (4) may be extended to d-step spaces which we explain only for $d = 3$ as follows. Locally, there are two smooth partitions (foliations) of V, one into fibers V' of codimension n_1 transversal to $H = H_1$ as earlier and the second partition with fibers (leaves) V'' of codimension $n_2 (= \mathrm{rank}\, H_2)$ associated to a smooth map $\Pi_2 : V \to B_2 \subset \mathbb{R}^{n_2}$ (where our old Π must be now christened $\Pi_1 : V \to B_1 \subset \mathbb{R}^{n_1}$) such that the fibers V'' are contained in the fibers V' (and, in fact, Π_1 is obtained from Π_2 by composing with a linear projection $\mathbb{R}^{n_2} \to \mathbb{R}^{n_1}$). Now we observe that

(a) each fiber V'' has $\sqrt[3]{\text{Euclidean}}$ geometry;

(b) for each fiber V' the space of fibers V'' in V' with the Hausdorff metric has $\sqrt[2]{\text{Euclidean}}$ geometry and the partition of V' into the fibers V'' is transversally Lipschitz. Furthermore, the implied Lipschitz constant is uniform for all V' in V;

(c) the fibers V' form a transversally Lipschitz foliation with (essentially) Euclidean quotient space (i.e. our old $B \subset \mathbb{R}^{n_1}$).

Now we observe that every diffeomorphism φ of V preserving both partitions is *para-Lipschitz* in the sense that

(a$_1$) φ maps every fiber V'' onto another such fiber by a *bi-Lipschitz* map;

(b$_1$) as $V'(x)$ goes onto another fiber, say $V'(x_1)$ the space of V''-fibers in $V'(x)$ is send to the corresponding space in $V'(x_1)$ by a *bi-Lipschitz* map;

(c$_1$) the map induced on the space of V'-fibers, (i.e. on our $B_1 \subset \mathbb{R}^{n_1}$) is *bi-Lipschitz*.

4.9.A. Parabolic metric spaces. Let us axiomatize the above situation and introduce a class of metric spaces generalizing Carnot-Carathéodory ones. For this we need an auxiliary *parabolic structure* P on a space V, namely a *flag* (i.e. a sequence) of partitions $P_1 > P_2 > \cdots > P_{d-1}$ where "$P_i > P_{i+1}$" means "P_{i+1} refines P_i". The "parts" of P_i are called *fibers* or *leaves* and denoted by V^i, where V^i_v means the fiber passing through a point $v \in V$.

A standard example of P is the *affine* flag of *type* $(n_1 < n_2 < \cdots < n_{d-1})$ on $\mathbb{R}^n$ where V^i's are the affine subspaces of codimension n_i parallel to the linear space $\{x_j = 0, j = 1, \ldots, n_i\} \subset \mathbb{R}^n$. Parabolic structures *homeomorphic* to the affine ones are called in sequel *flat* and only flat structures are relevant for the C-C geometry. Observe that the group of homeomorphisms preserving a flat flag (parabolic structure) is quite large, e.g. transitive on $V \approx \mathbb{R}^n$ but yet is somehow bounded in complexity by the numbers $n_{i+1} - n_i$ as this group is built in a certain way out of Homeo $\mathbb{R}^{n_{i+1} - n_i}$. This limitation of Homeo (V, P) is especially clear for a *full* flat flag P where $n_i = i = 1, \ldots, n-1$, and Homeo (V, P) is "built of" Homeo $\mathbb{R}$.

Finally, we use the notation V/P_i for the space of V^i-leaves (homeomorphic to $\mathbb{R}^{n_i}$ in the flat case) and V^i/P_j, $j > i$, for the space of V^j-leaves inside some $V^i = V_v^i \subset V$. Observe that these V^i/P_j are the fibers of the natural projection $\Pi_i^j : V/P_j \to V/P_i$.

Now, let V, besides a flag P, is given a metric. The idea of *parabolicity* of this metric with respect to P must incorporate (at least) the following three features.

(1) The fibers V^{i+1} inside each V_v^i must be *metrically parallel* in Lipschitz (or at least some Hölder) sense which can be expressed, for example, by requiring the maps $\Pi_i^{i+1} \mid V_v^i \to V_v^i/P_{i+1}$ to be Lipschitz (or Hölder) with the implied Lipschitz (Hölder) constant uniform on (compact subset in) V, where the space V_v^i/P_j is given the Hausdorff distance (between the fibers $V^{i+1} \subset V_v^i$).

(2) The geometry of each quotient space V_v^i/P_{i+1} must be *standard* in a suitable sense. For example we may require each V_v^i/P_{i+1} to be bi-Lipschitz (or bi-Hölder) to $\mathbb{R}^{n_{i+1}-n_i}$ with the metric (Euclidean)$^{\alpha_i}$, (where $\alpha = (i+1)^{-1}$ in the C-C case) and with implied Lipschitz constant uniform on V.

(3) There must exist a sufficiently large (in particular transitive) group of homeomorphisms f of (V, P) which are *para-Lipschitz* for our metric, i.e. bi-Lipschitz $V_v^i/P_{i+1} \leftrightarrow V_{v'}^i/P_{i+1}$, with $v' = f(v)$, for all $i = 1, \ldots, d-1$ and $v \in V$. For example, one may ask for the existence of many refinements of P by *full* flat flags P', such that our metric has all of the above properties with respect to P' and there are many homeomorphisms preserving P' and being para-Lipschitz with respect to P' and P. Furthermore, these refinements must be all mutually para-Lipschitz equivalent. (In the C-C case these "many" P' are taken from the pool of the smooth ones as "para-Lipschitz" is automatic for C^1-smooth homeomorphisms.)

Then a metric space V may be called parabolic if it (locally) admits a compatible flag P with some specific (Hölder) exponents as indicated in (1) and (2). The basic problem is the invariance of these exponents under changes of P. The only invariant one can easily reconstruct with a P is the Hausdorff dimension of V and everything else remains problematic.

Let us indicate further properties of (possible) parabolic structures associated to C-C manifolds which reflect the idea of H_i-horizontality. We observe that the projection $\Pi_v^i : V_v^i \to V_v^i/P_{i+1}$ is bi-Lipschitz on H_{i+1}-horizontal submanifolds in V_v^i and that V_v^i (when it is smooth) contains "many" horizontal curves, so that for each (smooth) curve $c \in V_v^i/P_{i+1}$ the pull-back $\left(\Pi_v^i\right)^{-1}(c) \subset V_v^i$ splits by the horizontal lifts of c to V_v^i.

This suggests the following notion of a *polarization* of a full flat flag P which is a system of n (local) free one parameter groups $X_i(t)$ of homeomorphisms of (V, P), such that $X_{i+1}(t)$ maps every V^i into itself for $i = 1, \ldots, n$. Such polarization is called horizontal for a given metric on V if the actions are para-Lipschitz and the induced metric on each orbit is (uniformly) bi-Lipschitz to $|t_1 - t_2|^{\alpha_i}$ and also if we take the induced action of $X_i(t)$ on V/P_i, $i = 1, \ldots, n - 1$, then again the induced metrics on the orbits are $\approx |t_1 - t_2|^{\alpha_i}$ for some positive exponents $\alpha_i \leqslant 1$. We leave to the reader to think over the meaning (and the existence) of these P and X_i for C-C manifolds.

We conclude by indicating the Riemannian counterpart of metric parabolicity. Such Riemannian manifolds are defined inductively so that if V_1 and V_2 are parabolic and V is a Riemannian fibration over V_1 with the fiber V_2 and an Iso V_2-connection, then V is parabolic (where one may additionally assume that the curvature of the connection is bounded on V_1). Then one obtains a class of (parabolic) manifolds by specifying the building blocks. One possibility is to use for this purpose the Euclidean spaces. A more narrow class appears if we start with $\mathbb{R}$ and allow only oriented fibrations $V \to V_1$ with $\mathbb{R}$-fibers. (Here the curvature is an ordinary q-form on V_1 which is necessarily exact, $\omega = d\eta$, and much depends on the possible rate of growth of η on V_1, where, for example, ω may be assumed bounded.) Parabolic Riemannian manifolds have a variety of distinctive asymptotic metric features (e.g. concerning the isoperimetric profile and the spectrum of Δ) which we shall not discussed here. (It is tempting to think of such Riemannian manifolds as approximations to parabolic metric spaces where the latter may appear as asymptotic tangent cones of the former.)

The above definition of parabolicity parallels nilpotent Lie groups. One may generalize from "nilpotent" to "solvable" which leads to another inductively defined class of spaces, where V belongs to the class if $V_1 = V/\mathrm{Iso}\, V$ does, and where the bottom space reduces to a single point. (Alternatively, one may retain the previous definition of parabolicity with *all* homogeneous spaces as building blocks.)

Example. Let V be built of n copies of $\mathbb{R}$,

$$V = V_n \xrightarrow{\mathbb{R}} V_{n-1} \xrightarrow{\mathbb{R}} V_{n-2} \xrightarrow{\mathbb{R}} \ldots \xrightarrow{\mathbb{R}} V_1 = \mathbb{R},$$

where the corresponding (curvature) q-forms ω_i on V_i, $i = 1, \ldots, n-1$ are bounded. Then an easy induction shows that there are diffeomorphisms $\varphi_i : V_i \leftrightarrow \mathbb{R}^i$, $i = 1, \ldots, n$ where $\|\mathcal{D}\varphi_i\|$ and $\|\mathcal{D}\varphi_i^{-1}\|$ have (at most) polynomial growth and $\omega_i = d\eta_i$ for $\|\eta_i\|$ of polynomial growth. Consequently, V has (at most) polynomial volume growth. (Probably, this remains true for V's built of $\mathbb{R}^{n_i}$'s.)

4.10. Anosov endomorphisms. Let A be an Anosov self-mapping of an infra-nil-manifold V. Such an A comes from an automorphism of a nilpotent Lie group, say $\widetilde{A} : \widetilde{V} \to \widetilde{V}$. We look at the corresponding automorphism a acting on the Lie algebra $L = L(\widetilde{V})$ and we measure the dilation of a on linear subspaces $L' \subset L$ as follows. First, for an individual vector $\ell \in L$ we set

$$\lambda(\ell) = \lambda_a(\ell) = \limsup_{i \to \infty} \ \|a^i \ell\|^{\frac{1}{i}}$$

for a fixed norm $\| \ \|$ in L. Clearly the resulting $\lambda(\ell)$ does not depend on a choice of the norm and for a *generic* $\ell \in L$

$$\lambda(\ell) = \text{spec rad } a \underset{\text{def}}{=} \max |\text{eigenvalues of } a|.$$

Next, we consider the natural action of a on the exterior powers $\Lambda^k L$ and apply the above definition to k-vectors acted upon by $\Lambda^k a$. In particular, we take $\lambda = \lambda_{\Lambda^k a}$ of the k-vectors defined by k-dimensional subspaces $L' \subset L$ and thus define our *dilation* $\lambda(L')$.

Example. If L' is a-invariant, (e.g. $L' = L$) then

$$\lambda(L') = |\text{Det}(a \mid L')|.$$

Finally, we define
$$\text{ent } a = \sup_{L' \subset L} \ \log \lambda(L')$$

and, more generally,

$$\text{ent } a \mid L' = \sup_{L'' \subset L'} \ \log \lambda(L'').$$

This entropy can be computed in terms of the eigenvalues of a as follows. Denote the eigenvalues of a by $\lambda_1, \ldots, \lambda_n$, $n = \dim L$, where each eigenvalue appears with the multiplicity equal to the dimension of the corresponding invariant subspace. Then, obviously,

$$\operatorname{ent} a = \sum_{|\lambda_i| \geqslant 1} \log |\lambda_i| = \sum_{i=1}^{n} \max(0, \log |\lambda_i|) = \sum_{i=n_-+1}^{n} \log |\lambda_i|$$

where $\lambda_{n_-+1}, \ldots, \lambda_n$ are the eigenvalues with $|\lambda_i| \geqslant 1$. Next, let $0 < \mu_1 < \mu_2 < \cdots < \mu_p$, $p \leqslant n$, be the absolute values of the λ_i's, $i = 1, \ldots, n$, and let $L_1 \subset L_2 \subset \cdots \subset L_p = L$ be the corresponding a-invariant subspaces, i.e. L_j is generated by the vectors "belonging" to λ_i with $|\lambda_i| \leqslant \mu_j$. Notice that

$$\lambda(L_j) = |\mu_1|^{d_1} \, |\mu_2|^{d_2} \cdots |\mu_j|^{d_j}$$

for $d_k = \dim(L_k/L_{k-1})$. Finally, for an arbitrary $L' \subset L$ we clearly have

$$\operatorname{ent} a \mid L' = \sum_{\mu_j > 1} d'_j \log \mu_j$$

for $d'_j = \dim(L' \cap L_j / L' \cap L_{j-1})$ under a.

4.10.A. Entropy in codimension 1. It is well known that the topological entropy of A can be computed by the formula

$$\operatorname{ent} A = \operatorname{ent} a = \sum_{i=n_-+1}^{n} \log |\lambda_i|.$$

(This is obvious if you know the definition of ent). We want to evaluate the entropy of A on a compact (possibly non-invariant) subset $V' \subset V$ of codimension one. To do this we take the minimal integer $m \leqslant n_+$, such that the Lie span of the subspace $L_m \subset L$ corresponding to $\lambda_1, \ldots, \lambda_m$ equals L. Now we claim

$$\operatorname{ent}(A \mid V') \geqslant \left(\sum_{i=n_-+1}^{n} \log |\lambda_i| \right) - \log |\lambda_m|. \tag{+}$$

As in the case of $(*)$ in 2.1 this $(+)$ is obvious for smooth hypersurfaces V' since these must be a.e. transversal to the subbundle $H = H_m \subset T(V)$ corresponding to L_m. In the general case, one may just repeat the proof of $(*)$ in 2.1, which says, in effect, that the above transversality property generalizes in a suitable sense to an arbitrary compact subset

V' of dimension $n - 1$. A more suggestive approach reducing the entropy to some generalized Hausdorff dimension is due to Ya. Pesin (see [Pes]) and it runs as follows. Fix a metric in V and $\delta > 0$ and then define ε_k-balls (secretly for $\varepsilon_k = \delta \exp -k$) with respect to A at a point $v \in V$ by

$$B(v, \varepsilon_k, A) = \bigcap_{i=1}^{k} A^{-i}(B(A^i(v), \delta))$$

where $B(A^i(v), \delta)$ is the δ-ball for the given metric. Then for every $V' \subset V$ define the Hausdorff measure of dimension d with coverings of V' by the ε_k-balls and with sums $\Sigma_k \, \varepsilon_k^{\log d}$ (where k numerates balls rather than numbers). Then the Hausdorff dimension defined via the measure as usual. (A more geometric viewpoint on this definition is suggested by the hyperbolic discussion in 0.9 and 1.4.D'.)

Notice that in our infranil case these balls look like boxes of size

$$\delta \left(\underbrace{1 \times 1 \times \quad \times 1}_{n_-} \times \underbrace{\lambda_{n_-+1}^{-k} \times \cdots \times \lambda_n^{-k}}_{n_+} \right)$$

for $n_+ = n - n_-$ and so the resulting Hausdorff dimension does not depend on δ. (For general dynamical systems one eliminates the dependence of δ by letting $\delta \to 0$, see [Pes] for details.)

Remark. The inequality $(+)$ is sharp, as there exists a hypersurface $V' \subset V$ having $\mathrm{ent}(A \mid V') = \left(\Sigma_{i=n_-+1}^{n} \log \lambda_i \right) - \log |\lambda_m|$. Namely, take a smooth V' in V which is everywhere tangent to the subbundle corresponding to $L_{m-1} \subset L$. Thus we have computed $\mathrm{ent}_{n-1}(A)$ that is the infimum of the entropies of A on all compact subsets of codimension one in V.

Notice that from the dynamics point of view the basic reason for $(+)$ is the existence of sufficiently many curves c in V for which $\mathrm{ent}(A \mid c) \leqslant \log \lambda_m$.

4.10.B. Entropy in high codimension. One defines the entropy spectrum of an endomorphism A of a metric space V by setting

$$e_k = \inf_{V'} \; \mathrm{ent}\,(A \mid V'),$$

where the infimum is taken over all compact subsets $V' \subset V$ of topological dimension k. Notice that if $\widetilde{A}$ covers A as earlier then A and $\widetilde{A}$ have equal spectra. Now, we concentrate on the case when $\widetilde{A} : \widetilde{V} \to \widetilde{V}$ is an *expanding* automorphism of the Lie group $\widetilde{V}$ (covering V) which necessarily serves as a self-similarity of some C-C metric in $\widetilde{V}$ (which is unique up to bi-Lipschitz equivalence). Let $\widetilde{\lambda}$ be the scaling done by $\widetilde{A}$, i.e.

$$\operatorname{dist}(\widetilde{A}(v_1), \widetilde{A}(v_2)) = \widetilde{\lambda}\operatorname{dist}(v_1, v_2),$$

for $\operatorname{dist} = \operatorname{dist}_{\text{C-C}}$ and observe that

$$\operatorname{ent} A = N \log \widetilde{\lambda}$$

for $N = \dim_{\text{Hau}}(\widetilde{V}, \operatorname{dist}_{\text{C-C}})$. Similarly, for every compact subset $\widetilde{V}' \subset \widetilde{V}$,

$$\operatorname{ent}(\widetilde{A} \mid \widetilde{V}') = \dim_{\text{Min}} \widetilde{V}'$$

and for $V' \subset V$,

$$\operatorname{ent}(A \mid V') = \dim_{\text{Min}} V'$$

as well. Thus the entropy spectrum reduces to measuring the Minkowski dimension of k-dimensional subsets in $\widetilde{V}$ and our results from 3.1.A and 4.5 apply. For example, *if $\widetilde{V}$ is the Heisenberg group of dimension $n = 2m + 1$, then $e_n = \operatorname{ent} A$ and*

$$e_k/e_n = k/2m + 2 \quad \text{for} \quad k \leqslant m$$

and

$$e_k/e_n = (k + 1)/2m + 2 \quad \text{for} \quad k > m.$$

It would be appealing to make such a computation for every Anosov-Bowen hyperbolic systems in terms of a Markov partition. In fact, it would be equally interesting to extend in full our C-C metric discussion to combinatorially defined (metric and quasi-conformal) spaces, e.g. semi-Markov of [Gro$_{\text{HG}}$], or those underlying *finitely presented* dynamical systems called "hyperbolic" in [Gro$_{\text{HMGA}}$].

Here is a typical question arising in the combinatorial framework.

What is the arithmetic structure of the numbers $e(k)$ and $e(k)/e(\ell)$?

4.11. Horizontal forms on polarized manifolds. We return to the discussion in 4.1.E on horizontal forms and cohomology of (V, H). We write locally H as $\mathrm{Ker}\{\eta_i\}$ for some 1-forms η_i on V, $i = 1, \ldots, m = n - n_1$, $n_1 = \mathrm{rank}\, H$, and denote by $H_j\, \Lambda^{n-k} \subset \Lambda^{n-k}(V) = \Lambda^{n-k}\, T(V)$ the bundle of j-horizontal $(n-k)$-forms α on V, which have, by definition, degree $\geqslant j$ in η_i, i.e. $\alpha = \Sigma_\mu\, \beta_\mu \wedge \zeta_\mu$, where each ζ_μ is the product of j forms among η_i, say $\eta_{i_1} \wedge \eta_{i_2} \wedge \cdots \wedge \eta_{i_j}$. In particular, we are interested in the bundle $H\Lambda^{n-k} = H_m\, \Lambda^{n-h}$ of *horizontal* forms which are divisible by the m-form $\zeta = \eta_1 \wedge \eta_2 \wedge \cdots \wedge \eta_m$ on V. (Notice that ζ is globally defined up to a scalar multiple on V. More invariantly, this is a form with values in the line bundle $\Lambda^m H^\perp$ for $H^\perp = T(V)/H$.)

Now we want to figure out when V supports "many" *closed* (and exact) horizontal forms α. In particular, we wish to realize a given (absolute or relative) de Rham cohomology class by such a form (compare linear lemma in 2.2., contact discussion in 3.3 and 4.1.E). We already know that if V contains "many" regular horizontal jets of k-dimensional submanifolds, e.g. Ω-regular isotropic k-planes in $H \subset T(V)$, then there are "many" (possibly folded) horizontal k-submanifolds in V which can be smoothed (as currents in families) to closed horizontal $(n-k)$-forms. This corresponds to the "non-linear proof" of linear lemma while the "linear proof" should proceed as follows (compare algebraic inversion lemma in 3.3). Given a closed $(n-k)$-form a on V (representing a de Rham class) we want to make it horizontal by adding an exact form. Thus we try to solve the equation $dx = a \bmod \zeta$ on V where $a = b \bmod \zeta$ signify that $a - b$ is divisible by ζ. In other words we denote by $\bar{d}$ the composition of $d : \Lambda^{n-k-1} \to \Lambda^{n-k}$ with the quotient map $\Lambda^{n-k} \to I_\zeta^\perp \underset{\text{def}}{=} \Lambda^{n-k}/H\Lambda^{n-k}$ (where the notation Λ^i sometimes applies to the sheaf of forms as well as the corresponding vector bundle), and want to solve the equation $\bar{d}x = \bar{a}$ for $\bar{d} : \Lambda^{n-k-1} \to I_\zeta^\perp$ and $\bar{a}$ being a section of $I_\zeta^\perp$. Ideally, we want a *differential* operator $\bar{\delta} : I_\zeta^\perp \to \Lambda^{n-k-1}$ inverting $\bar{d}$, i.e. satisfying $\bar{d}\,\bar{\delta} = \mathrm{Id}$. We know such a $\bar{\delta}$ exists for all H (Lie generating $T(V)$) for $k = 1$ (i.e. for $\deg a = n - 1$, (see "linear proof" of linear lemma) and also for *contact* structures H above the middle dimension, i.e. $k < \frac{1}{2}n$, $n = \dim V$ (see 3.3). On the other hand such a $\bar{\delta}$ may be expected for generic H if

$$\text{``}f\dim\text{''}(H\Lambda^{n-k} \cap \mathrm{Im}d) \underset{\text{def}}{=} n_1!/k!\,(n_1 - k)! + \sum_{j=n-k+1}^{n} (-1)^{j-n+k}\, \lambda_j > 0$$

for $n_1 = \mathrm{rank}\, H$ and $\lambda_j = n!/i!\,(n - i)!$ (see 4.1.E).

Questions. Is it actually true that $\bar{d}$ is differentially invertible for generic H when "$f\,\dim$" > 0? If not, what is the correct relation between n, $m = n - n_1$ and k generically needed for the existence of $\bar{\delta}$?

The passage from horizontal k-submanifolds to closed $(n - k)$-forms suggests that the existence of a regular jet (of a horizontal k-germ) at every point $v \in V$ should imply the existence of $\bar{\delta}$. This is confirmed by the linear proof of linear lemma in 2.2 and algebraic inversion lemma in 3.3. But one should not neglect the essential appearance of non-purity in differential forms in the degrees strictly between 1 and $n - 1$ and keep in mind that the formal linearization of the horizontality for submanifolds may lead to something more restrictive than the horizontality of forms, since the ideals in the Grassmann algebra do not always satisfy the Hilbert zero theorem. Namely, let I be the ideal in $\Lambda^*(\mathbb{R}^{n_1})$ generated by some q-forms $\omega_1, \ldots, \omega_m$ (corresponding to the differentials $d\eta_1, \ldots, d\eta_m$, $m = n - n_1$, restricted to $\mathbb{R}^{n_1} = H_v$, where $\eta_1, \ldots, \eta_m$ are the I-forms defining H). This I defines, for each $k = 2, 3, \ldots, n_1$, a subvariety $Z_k = Z_k(I) \subset \mathrm{Gr}_k(\mathbb{R}^{n_1})$ consisting of those k-dimensional subspaces in $\mathbb{R}^{n_1}$ on which all ω_i, $i = 1, \ldots, m$, vanish. But (unlike the polynomial case) these $Z_k(I)$ do not, in general, uniquely determine I, and it may happen that the ("radical") ideal I_V consisting in each degree k of the form vanishing on Z_k is significantly greater than I_V and then the annihilator J_V of I_V may be viewed as the true linearization of Z_k. (Recall that $\omega_i = d\eta_i$ vanish on horizontal submanifold $V' \subset V$ and so the above Z_k contain the possible tangent spaces to such V' at a given point $v \in V$. Then one defines the differential ideal I_V consisting of forms vanishing together with their differentials on the tangents to the horizontal submanifolds in V and the annihilator of I_V may play the role of horizontal forms.) This discussion suggests the following.

Algebraic question. Let $G_k \subset \Lambda^k(\mathbb{R}^{n_1})$ be the subvariety of the *pure k-forms*, i.e. the Plücker image of $\mathrm{Gr}_k(\mathbb{R}^{n_1})$ in $\Lambda^k(\mathbb{R}^{n_1})$ and $L \subset \Lambda^k(\mathbb{R}^{n_1})$ is a linear subspace of dimension ℓ. When does the intersection $L \cap G_k$ linearly span all of $\Lambda^k(\mathbb{R}^{n_1})$? Is this true, for instance, for generic $L \subset \Lambda^2(\mathbb{R}^{n_1})$ where n_1 is large and $\mathrm{codim}\, L$ is small?

Künneth type question. Let (V_i, H_i), $i = 1, 2$ be contact manifolds of equal dimension. Does the above mentioned $\bar{\delta}$ exist for $(V, H) = (V_1 \times V_2, H_1 \times H_2)$ and $\dim V - k \geqslant (\dim V_1 + 1)/2$? (If $m = \dim V - k = 2$ and one tries to solve $dx = a \bmod \zeta$ for $\zeta = \eta_1 \wedge \eta_2$ with $x = \zeta \wedge y$ one does get an algebraic equation, namely $d\zeta \wedge y = a \bmod \zeta$, where, moreover, one may assume $a = a_1 \wedge \eta_1 + a_2 \wedge \eta_2$. But as $d\zeta \wedge d\zeta = 0$, this equation is not solvable unless $d\zeta \wedge a = 0 \bmod \zeta$ which is a non-vacuous relation on a one should reckon with. Notice that according to Z. Ge (see [Ge$_{\text{BNCC}}$]) the vanishing of $H^i(I(H), d)$ for the differential ideal $I(H)$ generated by η_i and $d\eta_i$ follows from the injectivity of the linear map $\underbrace{\Lambda^i H \oplus \cdots \oplus \Lambda^i H}_{m} \to$

$\Lambda^{i+2} H$ given by $(a_1, \ldots, a_m) \mapsto (\omega_1 \wedge a_1, \ldots, \omega_m \wedge a_m)$. Ge shows that for $i = 1$ this injectivity condition is satisfied by fat H (see 4.2.B) and possibly (?) there are meaningful examples for $i \geqslant 2$ and $m \geqslant 2$. Also the Poincaré duality suggests something similar for small $n - i$.)

Stabilization and characteristic cohomology. If we want to linearize not only tangents to horizontal submanifolds by higher order jets of these, we should pass to the manifold V^r of r-jets of germs of k-dimensional submanifolds in V which comes along with a polarization $H^r \subset T(V^r)$. Now we take the (open) subset $V^r_{\text{reg}} \subset V^r$ consisting of regular jets and ask if, for sufficiently large (stable) r, the local cohomology of the complex $(\Lambda^i(V^r_{\text{reg}})/I(H^r), d)$ for $i < k$, where $I(H^r)$ is the differential ideal generated by the 1-forms in V^r_r vanishing on H^r and by their differentials. (If "yes", the proof must come by a gentle algebraic hand waving but finding the minimal "stable" r may require a specific calculation, compare [Vin$_{1;2;3}$] and [Br-Gr].)

Remark. We shall return to this question for $V^r = V$ and $i = 2$ in the framework of *H-connections* over V with possibly non-Abelian structure groups (see §5).

4.11.A. Horizontality via the anisotropic blow-up. We recall the (cotype) horizontality degree $M = M^*(\omega)$ of an $(n - k)$-form ω on (V, H) measuring an averaged horizontality of ω with respect to $H = H_1 \subset H_2 \subset \cdots \subset H_d = T(V)$ (see 4.1.E'). For example, if $d = 2$, and ω is m_1-horizontal but not $(m_1 + 1)$-horizontal, then $M^*(\omega) = m_1 + 2(n - k - m_1)$ and vice versa. Also recall that forms ω of horizontality M integrate to straight cochains c with $\|c\|_\varepsilon \lesssim \varepsilon^M$ (see 4.1.E' and 3.3.B) and the pointwise norm of ω decays as $\mathcal{O}(\varepsilon^M)$ under the anisotropic blow-up g_ε^* of a background metric g on V (which expands the vectors in $H_i \ominus H_{i-1}$ with the rate ε^{-i} for $\varepsilon \to 0$, see 4.1.E′). The latter (characteristic) property of M suggests the following approach to defining interesting horizontalities of tensorial (and non-tensorial) fields on (V, H) as well as constructing such horizontal fields. We consider a distinguished class of fields $\{\omega_\varepsilon\}$ on (V, g_ε^*) (e.g. harmonic forms or spinors, or these belonging to eigenvalues of certain level, or something non-linear like a Yang-Mills field) and try to select among them those which have certain rate of decay (or growth) for $\varepsilon \to 0$ with respect to a suitable norm in the corresponding function space. Such a selection process of ω_ε can be often expressed in terms of H as a horizontality type condition for (possibly non-existant) limit field ω_∞ on V. Then one may ask if (or when) a (suitably renormalized) ω_ε actually converges for $\varepsilon \to 0$ to some field on V and whether the limit field is horizontal in a desirable sense.

Example. If we start with g_ε^*-harmonic *functions* (scalars) on V we obtain in the limit for $\varepsilon \to 0$ Hörmander harmonic functions (i.e. satisfying $\Delta_H \omega = 0$ for the scalar Hörmander-Laplace operator Δ_H (compare [Fuk], [Ge$_{\mathrm{CRM}}$]).

The convergence $\omega_\varepsilon \to \omega_\infty$ looks most promising (especially for linear fields) if (V, H, g) is homogeneous under an action of a Lie group, preferably a compact one, where everything reduces to the representation theory. But we do not indulge ourselves in the luxury of linear algebra but rather indicate a couple of direct analytic consequences of the symmetry allowing a simple comparison between the L_∞ and L_2-norms of the fields in question. Our objective is constructing L_∞-slow growing (or rather L_∞-fast decaying) g_ε^*-harmonic forms ω_ε on V eventually giving rise to "small" straight cocycles on V.

4.11.B. Horizontal cohomology on nilpotent Lie groups and algebras. Let $L = \oplus_{i=1}^d L_i$ be a graded nilpotent Lie algebra and (V, H) the corresponding Lie group with the left-invariant polarization H arizing from $L_1 = H_{\mathrm{id}} \subset T_{\mathrm{id}}(V)$ (where we assume L_1 Lie-generates L as usual). Let $A^t : L \to L$ act on L_i by $\ell \mapsto t^i \ell$, $0 < t < \infty$ and observe that the induced action of A^t on the cohomology $H^*(L)$ also decomposes into eigenspaces, $H^* = \oplus_{M=0}^N H_M^*$, where A^t acts on the M-th component by t^M, where $N = \Sigma_{i=1}^d i \operatorname{rank} L_i = \dim_{\mathrm{Hau}}(V, H)$ and where $H_N^* = H^n(L) = \mathbb{R}$ (and where H of the polarization should not be confused with H^* of the cohomology). We observe that the cohomology H_M^*

is represented by (closed left-invariant) M-horizontal forms on V and we are interested in finding such forms ω with possibly large $M = M^*(\omega)$. To do this we observe that the action of A^t on the cohomology commutes with the cup-product and that the cup-product pairing between H^k and H^{n-k} is faithful by the Poincaré duality (proven on L with an obvious use of the Hodge $*$-operator for some positive quadratic form on L). It follows that there are dual bases h_μ in H^k and h'_ν in H^{n-k}, i.e. with $h_\mu \cup h'_\nu = \delta_{\mu\nu}$, where h_μ and h'_ν are eigenvectors, and consequently

$$M^*(h_\mu) + M^*(h'_\mu) = N, \qquad\qquad (*)$$

for all $\mu = 1, \ldots, \operatorname{rank} H^k = \operatorname{rank} H^{n-k}$, (with $M = M^*(h)$ defined by $h \in H^*_M$).

Corollary. *If n is even and the middle cohomology $H^{n/2}(L)$ does not vanish, then there is a closed left-invariant $n/2$-form ω on V non-homologous to zero in $H^*(L)$ with $M^*(\omega) = N/2$.*

4.11.B′. Horizontal cycles and cocycles on nil-manifolds V/Γ. The above becomes (geometrically) more interesting if V admits a co-compact lattice Γ and we can pass to the (compact nil-manifold) quotient V/Γ. The cohomology $H^*(L)$ obviously (by Poincaré duality in $H^*(L)$) *injects* to $H^*(V/\Gamma; \mathbb{R})$ and thus the relation $(*)$ remains valid in V/Γ, where $M^*(h)$ may be now understood as the maximal M, so that $h \in H^*(V/\Gamma; \mathbb{R})$ is representable by a straight cocycle c with $\|c\|_\varepsilon \leqslant \varepsilon^M$ (compare 4.1.E). In fact, one knows that the homomorphism $H^*(L) \to H^*(V/\Gamma)$ is a bijection and so the eigen splitting $H^*(L) = \oplus_0^N H^*_M$ applies to $H^*(V/\Gamma)$ and carries full information on the M-horizontality of (forms representing) the cohomology of V/Γ. This also gives as a nontrivial lower bound on the *Minkowski dimension* (see 4.9.) of the homology classes $c \in H_*$ which is defined as the infimum of these dimensions of the cycles representing c. To formulate this, define the function M_* on H_* as the dual of M^* on H^*, i.e. let $M_*(c) = \max M^*(h)$ over all $h \in H^*$ for which $\langle h, c \rangle \neq 0$.

Proposition. *The Minkowski dimension of every $c \in H_*(V/\Gamma; \mathbb{R})$ is bounded from below by*

$$\dim_{\mathrm{Min}} c \geqslant M_*(c). \qquad\qquad (+)$$

Proof. If a cycle $c_\bullet$ can be covered by j balls of C-C radius ε, and a straight cocycle $c^\bullet$ has $\|c^\bullet\|_\varepsilon \leqslant \varepsilon^M$ then $c^\bullet(c_\bullet) \lesssim j \, \varepsilon^M$ and the proof follows.

Remarks

(a) The above proposition applies to all compact C-C manifolds, where M^* (and the issuing M_*) can be understood either as the (best) degree of horizontality of forms representing $h \in H^*$ or as the exponent M in the (best) bound on the ε-norms of the straight cochains representing h.

(b) Probably, one can replace $\dim_{\mathrm{Min}}$ in $(+)$ by $\dim_{\mathrm{Hau}}$.

Next, we combine $(+)$ and $(*)$ and conclude to the following.

Corollary. *If two cycles (or homology classes) c and c' in V/Γ have non-zero index of intersection, then*

$$\dim_{\mathrm{Min}} c + \dim_{\mathrm{Min}} c' \geqslant N = \dim_{\mathrm{Hau}} V. \tag{$*+$}$$

This shows, in particular, that if $H_k(V/\Gamma; \mathbb{R}) \neq 0$ then the maxima $M_+(k) = \max\limits_{c \in H_k} \dim_{\mathrm{Min}} c$ and $M_+(n-k) = \max\limits_{c \in H_{n-k}} \dim_{\mathrm{Min}} c$ satisfy

$$M_+(k) + M_+(n-k) \geqslant N. \tag{$++$}$$

Remarks

(a) It is likely that $(*+)$ and $(++)$ remain true for all compact equiregular (or generic) C-C manifold and we shall prove it below for compact homogeneous spaces.

(b) The Minkowski dimension $\dim_{\mathrm{Min}} c$ is not, a priori, an integer. Yet the *proof* of $(*+)$ implies the existence of positive *integers*, say $d(c_\mu)$, satisfying
$$d(c_\mu) + d(c'_\mu) \geqslant N, \tag{$*+)'$}$$
and such that

$$\dim_{\mathrm{Min}} c_\mu \geqslant d(c_\mu) \quad \text{and} \quad d_{\mathrm{Min}}(c'_\nu) \geqslant d(c'_\nu).$$

Consequently, $(++)$ improves to

$$M'_+(k) + M'_+(n-k) \geqslant N, \tag{$++)'$}$$

where M' signifies the greatest integer below M.

(c) The number $M_+(k)$ of a C-C manifold V bounds the Hölder exponent α of a possible "homotopically surjective" (e.g. homeomorphic onto) C^α-map $(V, \mathrm{Riem}) \to (V, \text{C-C})$ by $\alpha \leqslant k/M_+(k)$ which may happen to be better than the general bound $\alpha \leqslant n/N$, for $n = \dim V$ and $N = \dim_{\mathrm{Hau}} V$, or even the slightly improved bound $\alpha \leqslant n-1/N-1$ (which is always valid since all hypersurfaces in V have $\dim_{\mathrm{Hau}} \geqslant N - 1$, see 2.1). For example, if n is even, say $n = 2m$ and N is odd, $N = 2M+1$, then $(++)'$ gives us $M_+(m) \geqslant M + 1$ (provided $H_m(V; \mathbb{R}) \neq 0$) and so $\alpha \leqslant m/M+1$ which is better than $\alpha \leqslant 2m-1/2M$ if $M < 2m-1$.

One can get extra milage from $(++)'$ by deriving a lower bound on $M_+(n-k)$ from an upper bound on $M_+(k)$ in the case where V has sufficiently many horizontal k-submanifolds. For example, such submanifolds are there for generic $H \subset T(V)$ and $k \leqslant n_1/n - n_1 + 1$, where $n_1 = \mathrm{rank}\, H$ (see 4.2.A''). This makes with $(++)'$

$$M'_+(n - k) \geqslant N - k$$

for $N = n_1 + 2(n - n_1)$ and all $k \leqslant n_1/n - n_1 + 1$ for which the k-th homology of V does not vanish, and so the (best) estimate for α one may obtain this way is, roughly,

$$\alpha \leqslant 1 - \frac{N - n}{N - (n_1/n - n_1)}.$$

Example. (inspired by [Pit]) Let $V' \subset V$ be a connected subgroup, such that the intersection $V' \cap \Gamma$ is cocompact in V' and the submanifold $V'/V' \cap \Gamma \subset V/\Gamma$ is not homologous to zero which is equivalent to the existence of a closed left-invariant form ω on V extending the oriented volume form on V'. Then the Minkowski dimension M' of the homology class $[V'/V' \cap \Gamma]$ equals the exponent of the volume growth of V' in V. This means that the R-balls in V' for the distance induced from V have the Riemannian k-dimensional volumes $\approx R^{M'}$ where $k = \dim V'$. This is especially clear if V' is invariant under A^t but, in fact, it is true (and follows by our arguments) for arbitrary $V' \subset V$ without even assuming that the (nilpotent) group V admits any dilation. Furthermore, one can exclude Γ from this discussion and show that if the volume form of V' extends to a closed equivariant form on V, then every infinite k-cycle V'' lying within finite distance from V' and homologous to V' inside some

ρ-neighbourhood of V' has $\dim_{\mathrm{Min}} V'' \geqslant M'$, where $\dim_{\mathrm{Min}}$ refers to the supremum of Minkowski dimensions of the compact parts of V'' (with respect to dist_H) and M' denotes the exponent of the volume growth of V'. Further generalizations concern a submanifold $V' \subset V$ which is not a subgroup but, preferably, of polynomial growth in V and which should be asymptotically non-homologous to zero. The latter may be ensured by the existence of a closed k-form ω on V, $k = \dim V'$, such that ω is positive and does not decay fast on V' (e.g. being equal on V' to the volume forms of V') and on the other hand, ω does not grow too fast (e.g. being bounded) on V. Then one can show in some cases that the volume growth of V' can not be significantly decreased by "small" moves of V' and the Minkowski dimension of the "homology class" $[V']$ can be estimated from below in terms of the rate of the volume growth of V'.

Applications to the lower bound on the filling exponent. Suppose every closed curve in V of length ℓ can be filled in by a disk of area $\lesssim \ell^p$. Then every 2-dimensional homology class in V/Γ grows, under the anisotropic blow-up g_ε^* of g, with the rate $\lesssim \varepsilon^{-p}$ (which is essentially equivalent to $M_+(2) \leqslant p$), as is seen with triangulated cycles where the edges are (short) horizontal curves and the triangles are filled g_ε^*-minimally. It follows that the 2-dimensional filling exponent p of V is bounded from *below* by $M_+(2)$ which implies the following.

Corollary. (^{12}) *Let a simply connected nilpotent group V contain a 2-dimensional subgroup $V' \subset V$ which is not homologous to zero in the sense that the area form of V' extends to a closed left-invariant 2-form ω on V. Then the 2-dimensional filling exponent p of V is bounded from below by the exponent of the area growth of R-balls in V', (i.e. $\mathrm{Area}B(R) \lesssim R^p$ for the balls in V' for the metric $\mathrm{dist}_V \mid V'$).*

Remarks

(a) We do not assume in this Corollary the presence of Γ as we know how to dispose of it. Furthermore, one may relax the invariance assumption on the extended form ω. Namely, it is enough to require ω is *bounded* or even growing on V but rather slowly, say $|\omega(v)| \lesssim (\mathrm{dist}(v_0, v))^\varepsilon$, so that the integral of $|\omega|$ over the boundaries of balls $B(R)$ in V' and near lying manifolds V'' is dominated by area $B(R)$ for $R \to \infty$.

(b) Notice that the fillings involved in the proof of the inequality $M_+(2) \leqslant p$ apply only to some "standard" curves in V. For example if V admits

---

^{12} See [Pit].

a dilation A, then we only need to fill the curves $A^t(c)$ $t \to \infty$ where c runs over a finite set of closed curves. This allows an effective generalization to the k-dimensional fillings for $k \geqslant 3$ via triangulations of cycles and induction by skeletons. For example, if V contains $(k-1)$-regular jets we can make our cycle to be controllably small on the $(k-1)$-skeleton and then the exponent of the k-th filling (inequality) can be estimated from below by $M_+(k)$. (There is much here to be made more precise and specific and this we leave to the reader.)

4.11.B″. Harmonic forms and anisotropic blow-up of V/Γ. Since the (self-similarity) operators $A^t : L \to L$, being automorphisms of the Lie algebra $L = L(V)$, commute with the exterior differential d (on the Grassmann algebra $\Lambda^* L$ = the algebra of left-invariant forms on the Lie group V) so do the projectors on the eigenspaces of A^t acting on $\Lambda^* L$. Furthermore, for every metric g on L with respect to which different eigenspaces of A on L are mutually orthogonal, the eigenspaces of A on $\Lambda^* L$ are also mutually orthogonal and therefore the space of g-harmonic forms, i.e. $(\ker d) \ominus_g (\mathrm{im} d) \in \Lambda^* L$, is A-invariant. Finally we observe that the anisotropic blow-up g_ε^* of the metric g^* left-invariantly extended from $L = T_{\mathrm{id}} V$ to V can be realized by the (expanding) automorphisms $V \to V$ corresponding to A^t, since A^t for $t = \varepsilon^{-1}$ pulls-*back* g to g_ε^*. It follows that *every g-harmonic form on V (and hence on V/Γ) is also g_ε^*-harmonic for all $\varepsilon > 0$.*

Now let us take a metric g on L which does not necessarily agree with A and look at g_ε^*-harmonic forms for $\varepsilon \to 0$. It is easy to see that the quadratic forms $g_\varepsilon' = (A^\varepsilon)^* \, g_\varepsilon^*$ on L converge for $\varepsilon \to 0$ to some positive definite quadratic form g^1 which does agree with A (i.e. the eigenspaces of A are g^1-orthogonal) and consequently the space $\mathcal{H}_{g_\varepsilon^*}$ of g_ε^*-harmonic form converge to $\mathcal{H}_{g^1}$ for $\varepsilon \to 0$.

Example. Let L be the Heisenberg algebra of rank $n = 2m + 1$ where the de Rham complex of L (i.e. of left-invariant forms on the Heisenberg group) is generated by x_i, y_i and z, $i = 1, \ldots, m$ with the relations $dx_i = dy_i = 0$, $dz = \Sigma_{i=1}^m x_i \wedge y_i$. Then one easily sees that every cohomology class $h \in H^k(L)$, $k \leqslant m$ can be represented by an (exterior) polynomial in x_i and y_i, i.e. by some $\lambda \in \Lambda^k(\mathbb{R}^{2m})$ and $h = 0 \Leftrightarrow \lambda$ is divisible by $\omega = \sum_{i=1}^m x_i \wedge y_i$. In other words $H^{\leqslant m}(L) = \Lambda^{\leqslant m}(\mathbb{R}^{2m})/I_\omega$, where I_ω denotes the ideal in the Grassmann algebra $\Lambda^*(\mathbb{R}^{2m})$ spanned by ω.

On the other hand, every $h \in H^{>m}(L)$ can be represented by $z \wedge \gamma$, $\gamma \in \Lambda^*(\mathbb{R}^{2m})$, where $d(z \wedge \gamma) = \omega \wedge \gamma = 0$. Thus $H^{>m}(L)$ identifies with the kernel of the multiplication operator $\wedge \omega : \Lambda^*(\mathbb{R}^{2m}) \to \Lambda^*(\mathbb{R}^{2m})$, i.e. the annihilator $J_\omega \subset \Lambda^*(\mathbb{R}^{2m})$ of I_ω. It follows that $H^*(L)$ equals the space of equivariant Rimin's forms on the Heisenberg group (see 3.3.A).

Exercise. Prove the above using Gysin exact sequence,

$$\cdots \to H^{k-1}(L) \to \Lambda^k(\mathbb{R}^{2m}) \overset{\wedge\omega}{\to} \Lambda^{k+2}(\mathbb{R}^{2m}) \to H^{k+2}(L) \to \cdots$$

and Lefschetz lemma from 3.3.

We conclude by observing that A^t acts on L by $\{x_i, y_i, z\} \mapsto \{tx_i, ty_i, t^2z\}$ and $M^*(h) = \deg h$ for $\deg h = k \leqslant m$ and $M^*(h) = k+1$ for $\deg h = k > m$.

Questions

(a) It remains unclear what happens to harmonic spaces $\mathcal{H}_{g_\varepsilon^*}(V/\Gamma)$ where g on V/Γ *does not* descend from a *left-invariant* metric on V^{13}.

(b) What is the behaviour of (harmonic) spinors on L under the action of A and the blow-up of g? (Beware that the spinors themselves, not only Dirac, depend on g.)

4.11.C. Lower bound on the Minkowski dimension of cycles in compact homogeneous spaces (V, H). We want to show that *every two cycles c and c' in V have*

$$\dim_{\mathrm{Min}} c + \dim_{\mathrm{Min}} c' \geqslant N \tag{$*+$}$$

whenever $c \cap c' \neq 0$ and thus

$$M_+(k) + M_+(n-k) \geqslant N \tag{$++$}$$

if $H_k(V; \mathbb{R}) \neq 0$ (compare 4.10.B'). To prove this we invoke our anisotropic blow-up g_ε^* of a back-ground metric g on V where we observe that the k-dimensional volume of every k-dimensional homology class c in V satisfies

$$\dim_{\mathrm{Min}} c < M \Rightarrow \mathrm{Vol}_{g_\varepsilon} c = o(\varepsilon^{-M}).$$

Now $(*+)$ follows from the following.

---

[13] This was clarified in [Ge$_{\mathrm{ALRC}}$].

Homogeneous antisystolic inequality. *Let V be a compact homogeneous Riemannian manifold and c and c' be two cycles of complementary dimensions in V. Then their intersection index is bounded by*

$$c \cap c' = \mathrm{const}_n (\mathrm{Vol}\, c)(\mathrm{Vol}\, c')/\mathrm{Vol}\, V. \qquad (\star)$$

Proof. Let us translate c and c' by the isometry group G of V and average them as currents over the Haar measure of G. The resulting currents are G-invariants, hence smooth, and are given by closed differential forms, say ω and ω' representing dual cohomology classes of c and c'. In particular, $[\omega \wedge \omega'](V) = c \cap c'$. On the other hand the pointwise norm of ω is (obviously) bounded by

$$\|\omega\| \leqslant \mathrm{const}'_n \, \mathrm{Vol}\, c / \mathrm{Vol}\, V$$

and similarly,

$$\|\omega'\| \leqslant \mathrm{const}'_n \, \mathrm{Vol}\, c' / \mathrm{Vol}\, V.$$

This implies the desired bound on $c \cap c' = [\omega \wedge \omega'](V)$.

4.11.C$'$. Small harmonic forms and a lower bound on $\dim_{\mathrm{Min}}$ for non-compact homogeneous spaces. We want to make the above argument more constructive by exhibiting invariant forms ω with large $M^*(\omega)$, (or at least with small pointwise norms with respect to g_ε^*) without any "ad absurdum" assumptions on cycles in V. Recall that the closed forms ω with minimal L_2-norms in their respective cohomology classes are the harmonic ones and since V is homogeneous the harmonic ω are invariant and so their pointwise norms are related to the L_2-norms by

$$\|\omega\|_{L_2} = \|\omega\|(\mathrm{Vol}\, V)^{\frac{1}{2}}.$$

Furthermore, the L_2-norms on the spaces harmonic forms of complementary dimensions are *Poincaré dual* which means the existence of L_2-orthonormal bases $h_\mu \in \mathcal{H}^k_{g_\varepsilon^*}$ and $h'_\nu \in \mathcal{H}^{n-k}_{g_\varepsilon^*}$, such that $h_\mu \cup h'_\nu = \delta_{\mu\nu}$. This applies to all metric g_ε^* for $\varepsilon \to 0$ and gives us sufficiently many "small" forms ω to prove $(\star)$. This also gives us a bound on $\|c\|_\varepsilon$ for straight cocycles c's representing h's, as every ω on V with $\|\omega\| \leqslant \varepsilon^{-M}$ integrates to a straight cochain (cocycle if $d\omega = 0$) c with $\|c\|_\varepsilon \leqslant \mathrm{const}\, \varepsilon^{-M}$ for "const" independent of ε. Unfortunately, the harmonic forms $h \in \mathcal{H}^*_{g_\varepsilon^*} = H^*(V)$ where we have good bound on $\|h\|$ may (?) a priori strongly vary with ε and so we can not get a (interesting) bound $\|c\|_\varepsilon \lesssim \varepsilon^{-M}$ for a fixed c and all $\varepsilon \to 0$. But yet we have the following weak form of $(*)$ from 4.10.B.

There exist Poincaré dual bases $h_\mu \in H^k$ and $h'_\nu \in H^{n-k}$, a sequence $\varepsilon_i \to 0$ and sequences of straight cocycles $c_\mu(i)$ and $c'_\nu(i)$, $i = 1, 2, \ldots$, such that

$$(i) \quad [c_\mu(i)] \to h_\mu \quad and \quad [c'_\nu(i)] \to h'_\nu \quad for \quad i \to \infty,$$

$$(ii) \quad \|c_\mu(i)\|_\varepsilon \leqslant \mathrm{const}\, \varepsilon^{-M_\mu(i)} \quad and \quad \|c'_\nu(i)\| \leqslant \mathrm{const}\, \varepsilon^{-M'_\nu(i)}$$

for some $M_\mu(i)$ and $M'_\nu(i)$ satisfying

$$M_\mu(i) + M'_\mu(i) \leqslant N \ , \ i = 1, 2, \ldots \tag{$*-$}$$

and converging to some constants, say $M_\mu(i) \to M^-(h_\mu)$ and $M'_\nu(i) \to M^-(h_\nu)$ for $i \to \infty$.

Now we observe that the above applies to equivariant forms on arbitrary (not necessarily compact) homogeneous space (V, H, g) in-so-far as the equivariant cohomology satisfies the Poincaré duality. For example, this is the case if V is a *unimodular* Lie group, i.e. Trace $ad_x = 0$ for every $x \in L = L(V)$. Then "small" forms and (straight cycles) descend to the quotients V/Γ for all discrete subgroups Γ in V and lead to non-trivial lower bound on the Minkowski dimension on the homology $H_*(V/\Gamma)$. For example, if $H^k(L) \neq 0$ (where L is the Lie algebra of V), *then the maximal Minkowski dimensions of the k-cycles in $H^k(V/\Gamma)$ and $H_{n-k}(V/\Gamma)$ satisfy, as earlier, the "anti-systolic" inequality*

$$M_+(k) + M_+(n-k) \geqslant N(= \dim_{\mathrm{Hau}}(V, H)).$$

In particular, if $k = n/2$, then there exists a class $h \in H_k(V/\Gamma)$ with $\dim_{\mathrm{Min}} h \geqslant N/2$.

Remark. It is unclear with this argument if $M_+(k)$ is an integer and/or if the numbers $M_+(k)$ and $M_+(n-k)$ can be minorized by integers $M'_+(k)$ and $M'_+(n-k)$ satisfying $M'_+(k) + M'_+(n-k) \geqslant N$ (compare $(++)'$ in 4.10.B$'$).

On the upper systolic bound. Let $M_-(k)$ denotes the infimum of the Minkowski dimensions of the non-zero homology classes in $H_k(V; \mathbb{R})$. If $k = 1$ and $H_1(V; \mathbb{R}) \neq 0$, then, obviously,

$$M_-(1) + M_-(n-1) \leqslant N$$

as $M_-(1) = 1$ and $M_-(n-1) = N-1$, but this inequality seems unlikely to be true in general (even with $\dim_{\mathrm{Hau}}$ instead of $\dim_{\mathrm{Min}}$) for $2 \leqslant k \leqslant n-2$.

In fact, one expects a strong failure of intersystolic inequalities for the (anisotropically blown-up) manifolds $V_\varepsilon^* = (V, g_\varepsilon^*)$, where, conjecturally,

$$\mathrm{syst}_k(V_\varepsilon^*)\, \mathrm{syst}_{n-k}(V_\varepsilon^*)/ \mathrm{Vol}(V_\varepsilon^*) \to_{\varepsilon \to 0} \infty,$$

for most (V, H) and $2 \leqslant k \leqslant n - 2$, (contrary to what happens to the *stable systoles* where the volume of the cycles is replaced by the $\mathbb{R}$-*mass*, compare [Gro$_{\mathrm{SISI}}$]). Unfortunately, we have no way of evaluating the volumes of the homology classes in $H_k(V_\varepsilon^*)$ for $\varepsilon \to 0$ even for homogeneous (V, H) (while the $\mathbb{R}$-mass is computable in the homogeneous case by averaging over $G = \mathrm{Iso}\, V_\varepsilon^*$).

Remark. The systolic situation in V_ε^* is somewhat similar to that in *almost Kähler manifolds* V where the underlying complex structure is non-integrable (while the structure 2-form is closed). There , one easily sees the failure on the sharp isosystolic inequality for even $k \geqslant 4$ but does not know if non-sharp inequality remains valid (compare [Gro$_{\mathrm{SISI}}$] and [Gro$_{\mathrm{MGKM}}$]).

4.11.C″. On the limit of $\mathcal{H}_{g_\varepsilon^*}$ for $\varepsilon \to 0$. Let L be a Lie algebra with a (horizontal) subspace $H \subset L$ which Lie generates L and $H = H_1 \subset H_2 \subset \cdots \subset H_d = L$ be the filtration by the commutators, i.e. $H_i = [H, H_{i-1}]$. We denote by $L^\bullet$ the corresponding graded space $L^\bullet = \oplus_{i=1}^d H_i/H_{i-1}$ with the natural structure of a graded nilpotent Lie algebra. We give L a metric g, identify $L^\bullet$ with L via $H_i/H_{i-1} = H_i \ominus_g H_{i-1}$, introduce the linear operators $A^t : L^\bullet \to L^\bullet$ acting by t^i on $H_i \ominus H_{i-1}$ and let $g_\varepsilon^* = (A^t)^* g$ for $t = \varepsilon^{-1}$. Then we observe that the space of g_ε^*-harmonic forms of $L = L^\bullet$ say $\mathcal{H}_\varepsilon \subset \Lambda^*(L)$ converges, as $\varepsilon \to 0$, to a space $\mathcal{H}_0 \subset \Lambda^*(L)$ which is contained in the space $\mathcal{H}(L^\bullet) \subset \Lambda^*(L^\bullet = L)$ of the harmonic form of the algebra $L^\bullet$. This leads to the following.

Questions

(1) When (if) does $\mathcal{H}_0$ non-trivially intersect the image of the differential $d = d_L$ on $\Lambda^*(L)$?

(2) In which case does there exist a natural boundary operator on the cohomology of $L^\bullet$ say $d^\bullet$ on $H^*(L^\bullet)$, such that the cohomology of the complex $(H^*(L^\bullet), d)$ equal $H^*(L)$? (We know $d^\bullet$ does exist for rank $L/H = 1$ by a Lie algebraic adaptation of the Rumin complex, compare example in 4.10.B″, but in the general case one may need a spectral sequence rather than a single complex, compare [Br-Gr].)

Next we pass to the anisotropic blow-up of a manifold $(V, H, g) \rightarrow (V, H, g_\varepsilon^*)$ and observe that for each point $v_0 \in V$ there is a (pointed) limit of $(V, v, H, g_\varepsilon^*)$ for $\varepsilon \rightarrow 0$ to a graded nilpotent Lie group $(V_v, v, H(v), g(v))$ where harmonic forms on V subconverge to left-invariant harmonic forms on V_v. This suggests (generalizing the Rumin complex) some operator $d^\bullet$ on sections of the bundle of the cohomologies of the Lie algebras $L_v = L(V_v)$ but I have nothing to say in this regard. (There is Vinogradov's spectral sequence in this picture, see $[\mathrm{Vin}_{1;2;3}]$, $[\mathrm{Br\text{-}Gr}]$, but its role in our game is unclear.)

5. Anisotropic connections

Let V be a smooth manifold with a polarization $H \subset T(V)$ and $p : X \rightarrow V$ be a G-bundle for some Lie group G. Geometrically speaking, an H-connection in X is given by parallel transport in X along H-horizontal curves in V or, equivalently, rectifiable curves for the C-C geometry in V corresponding to H. Infinitesimally, an H-connection ∇ in a principal bundle X is defined by a G-invariant subbundle $\tilde{\nabla} \subset T(X)$ such that the differential $\mathcal{D}_p : T(X) \rightarrow T(V)$ sends each $\tilde{\nabla}_x$ isomorphically onto H_v for $v = p(x) \in V$.

5.1. Curvature Ω_∇. Recall the curvature $\Omega_H : H \wedge H \rightarrow H^\perp = T(V)H$ (which can be represented by k scalar 2-forms $\omega_1, \ldots, \omega_k$ for $k = \mathrm{rank}\, H^\perp$ for $\omega_i = d\eta_i$ where $\eta_1, \ldots, \eta_k$ are 1-forms on V locally defining H, i.e. $H = \underset{i}{\cap}\, \mathrm{Ker}\, \eta_i$) and let $K \subset H \wedge H$ denote the kernel of Ω_H. We denote by $L(X) \rightarrow V$ the adjoint Lie algebra bundle over V and we think of linear forms on K with values in $L(X)$ as of partially defined $L(X)$-valued 2-forms on H. In particular, we have the curvature tensor $\Omega_\nabla : K \rightarrow L(X)$ which is such a form defined as follows. Let $k \in K_v \subset K$ be represented by $k = (\sum_i h_i \wedge h_i')(v)$ for some H-horizontal vector fields h_i near v. Then, by the definition of Ω_H, its vanishing at k means that

$$\sum_i [h_i, h_i'] \in H.$$

Then we lift all h_i and h_i' to $\tilde{\nabla}$-horizontal G-invariant fields $\tilde{h}_i$ and $\tilde{h}_i'$ on X and take the sum of their Lie brackets, $\sum_i \left[\tilde{h}_i, \tilde{h}_i'\right]$. Since the bracketing agrees with the lift, the field $\sum_i \left[\tilde{h}_i, \tilde{h}_i'\right]$ lies in $\tilde{H} = (\mathcal{D}_p)^{-1}(H) \subset T(X)$ and as $\tilde{\nabla}$ also lies in $\tilde{H}$ we may project to the vertical bundle $T_{\mathrm{ver}}(X) = \tilde{H}/\tilde{\nabla}$ consisting of the vectors tangent to the G-fibers of X. This projection clearly is G-invariant and thus defines a vector in $L(X)$ at v which is just the desired value of Ω_∇ at k.

Remarks

(a) Partial connections in this context were introduced by Z. Ge in [Ge$_{\mathrm{BNCC}}$] who equivalently defines Ω_∇ as an $L(X)$-valued 2-form on V modulo the differential ideal generated by H i.e., by the 1-forms vanishing on H.

(b) Intuitively, the value of the curvature on an infinitesimal disk $D \subset V$ tangent to a bivector equals the holonomy around the boundary circle S of D. In our case this only makes good sense for H-horizontal disks D which necessarily have $\Omega_H|D = 0$ and thus the domain of Ω_∇ is restricted to the kernel of Ω_H.

5.1.A. On the equation $\Omega_\nabla = 0$. To grasp the meaning of Ω_∇ let us analyze the equation $\Omega_\nabla = 0$. Unlike the usual case this equation does not always imply local splitting (triviality) of (X, ∇). For example, every connection ∇ over a three-dimensional contact manifold (V, H) has $\Omega_\nabla = 0$, since the form Ω_H on $H \wedge H = \mathbb{R}$ has trivial kernel K. But *if every small smooth generic H-horizontal closed curve S bounds an H-horizontal disk D in V, then the vanishing of the curvature makes the holonomy over every small horizontal curve trivial.* In fact, the connection ∇ splits over D whenever Ω_∇ vanishes on D.

On the other hand, *if ∇ has trivial holonomy over all smooth closed horizontal curves in some domain V_0 in V then ∇ splits over V_0.* To construct a splitting, i.e. a horizontal frame field over V_0, we take such a frame f_0 at a point $v_0 \in V_0$ and then parallelly transport it to other $v \in V_0$ along horizontal paths in V_0 (where we assume V_0 is connected). Since the holonomy is trivial, the result of such a transport is independent of the underlying path and defines, indeed, the required frame field over V_0.

Corollary. *Let H admit a regular H-horizontal jet of a 2-germ at each point $v \in V$ (see 4.2). Then the equation $\Omega_\nabla = 0$ makes ∇ locally split.*

Proof. The existence of regular jets ensures filling curves by disks (see 4.7).

Remark. The above Corollary shows, in particular, that the complex $(\Lambda^*(V)/I(H), d)$ is locally exact in degree one, where $I(H)$ denotes the differential ideal spanned by H. One may try a similar argument in higher degrees, using, for example, *h-horizontal* triangulations Tr of V, where all simplices up to dimension k are horizontal. Then every i-form ω with $dw = 0 \bmod I(H)$ for $i < k$ gives us a closed cochain which is locally a coboundary, say $\omega_\nabla = \delta u$ and then one may try to go back from cochains in Tr to forms in Λ^*/I by either some smoothing process, or by scaling Tr to Tr_ε with simplices of size $\varepsilon \to 0$.

Example. Suppose we want to find a 1-form $\lambda \bmod I(H)$ so that $d\lambda = \omega \bmod I(H)$ where $dw = 0| \bmod I(H)$. If we have in our possession local filling of circles by disks and 2-spheres by 3-balls, the given information on λ is encoded (via $d\lambda$) in the integrals of λ over all (small) closed H-horizontal circles. To have the full $\lambda \bmod I(H)$ we need the integrals of λ on all (not only closed) horizontal curves and so we are faced with an extension problem à la Hahn-Banach and the above triangulation may suggest a suitable functional set-up (compare our abortive discussion in 4.11).

Questions

(a) Are there examples of polarizations H with high dimensional kernels $K \subset H \wedge H$ of Ω_H admitting non locally split connections ∇ with $\Omega_\nabla = 0$ (on K)? (Somewhat non-convincing examples are suggested by *integrable* polarizations H and by Cartesian products of 3-dimensional contact manifolds.)

(b) Recall that the implication

$$\Omega_\nabla = 0 \Rightarrow (\nabla \text{ locally splits}) \qquad (\Rightarrow)$$

for $\mathbb{R}$-bundles is equivalent to the local exactness of the complex $(\Lambda^*(V)/I(H), d)$ in degree one. Does this local exactness yield $(\Rightarrow)$ for all ∇ (with possibly non-Abelian structure groups)?

Notice that ∇ is locally trivial if and only if it has trivial holonomy along all small closed H-horizontal curves and so a trivialization of ∇, if it exists, is essentially unique. This shows that $\Omega_\nabla = 0$ is an integrable system in the sense of Frobenius and so verification of ($\Rightarrow$) reduces to an infinitesimal computation. To see this more explicitly, let us observe that ($\Rightarrow$) is equivalent to a possibility of extending our H-connection ∇ over (V, H) with $\Omega_\nabla = 0$ to an actual connection (define over all of $T(V)$), say ∇^+ with $\Omega_{\nabla^+} = 0$. Such an extension, if it exists, is unique and can be constructed as follows. We recall the filtration $H = H_1 \subset H_2 \subset \cdots \subset H_j \subset \cdots \subset H_d = T(V)$ where H_j is generated by the commutators of H-horizontal fields of order j. Then we extend ∇ from $H_1 = H$ to $\nabla^{(2)}$ on H_2 according to the formula

$$\nabla^{(2)}_{[h_1, h_2]} = \nabla_{h_1} \nabla_{h_2} - \nabla_{h_2} \nabla_{h_1} \qquad (+)$$

for h_1, h_2 in H and then proceed in the same way for H_3, H_4, etc up to $\nabla^+ = \nabla^{(d)}$. Notice that the vanishing of Ω_∇ makes $(+)$ correctly defined, and, in order to proceed further, $\nabla^{(2)}$ must be an actual orthogonal connection (which is by no means automatic) and its H_2-curvature must be zero. Then one can extend $\nabla^{(2)}$ to $\nabla^{(3)}$ on H_3 etc. Thus the local triviality of ∇ amounts to a chain of differential relations incorporating the equations $\Omega_\nabla = 0$, $\Omega_{\nabla^{(2)}} = 0, \ldots$, as well as the "orthogonal connection" properties of $\nabla^{(i)}$ for $i \geq 2$. Notice that in this chain each following relation makes sense due to the preceding ones.

Characteristic forms. Z. Ge points out in [Ge$_{\text{BNCC}}$] that the curvature Ω_∇ can be used to define in the obvious way the Chern-Weil characteristic classes of our bundle X over V in the cohomology of the complex $(\Lambda^*(V)/I(H), d)$. For example, if V is a contact manifold, then this cohomology below the middle dimension is canonically isomorphic to the de Rham cohomology by Rumin theorem and so these classes land in the ordinary cohomology of V where they coincide with the classical characteristic classes. It is unclear what applications these classes may have.

Remark. If one does not insist on the curvature of ∇ but allows higher order infinitesimal formulae one may hope to recapture *ordinary* de Rham representatives of the characteristic classes of X in terms of an H-connection ∇ for an arbitrary H (Lie spanning $T(V)$). This was pointed out to me by A. Šwarc who was interested in quantization of Chern-Simons kind of invariants over 3-dimensional contact manifolds.

5.2. Norms and metric associated to orthogonal connections.
From now on we assume G is a compact group and $X \to V$ is a G-vector bundle. In fact, we make no essential use of G anymore and work with an arbitrary Euclidean bundle with a Euclidean H-connection ∇ on V (i.e. $G = O(r)$ for $r = \operatorname{rank} X$ and the standard representation of $O(r)$ on $\mathbb{R}^r$). Now we may speak of the pointwise norm $\|f(v)\|$ of a section $f : V \to X$ and, if H is endowed with a Riemannian metric, which we always assume, we also have the norm $\|\nabla f(v)\|$ (which reduces to $\|\mathcal{D}f|H\|$ for the trivial (split) bundles (X, ∇)).

The local Sobolev inequalities (see 2.4) for sections with (small) compact support extend to these norms,

$$\|f\|_{L_p} \le \operatorname{const} \|\nabla f\|_{L_q} \tag{$*$}_q$$

for all q in the interval $1 \le q < N = \dim_{\mathrm{Hau}} V$ and $\frac{1}{p} = \frac{1}{q} - \frac{1}{N}$. This follows from 2.4 applied to the function $\|f(v)\|$ on V. An interesting feature here is the independence of const of ∇.

Next we want to study the (Hölder) continuity of f with $\|\nabla f\|_{L_q} < \infty$ for $q \ge N$. For this we need some metric on X compatible with ∇. An obvious candidate is obtained as follows. Let $\tilde{H} = (\mathcal{D}p)^{-1}(H) \subset T(X)$ and observe that $\tilde{H} = T_{\mathrm{ver}}(X) \oplus \tilde{\nabla}$. The subbundle $\tilde{\nabla}$ inherits a metric from H while $T_{\mathrm{ver}}(X)$ has a metric from the Euclidean structure on X. Thus $\tilde{H}$ also has a Riemannian metric and we give X the associated C-C metric.

Warning. *We have not made any regularity assumptions on ∇ and the above C-C metric on X may degenerate for a general Borel measurable ∇. For example, if arbitrarily short loops at some point $v \in V$ have definite holonomy, then this metric may become zero for some pairs of distinct points in the fiber $X_v \subset X$. Yet this does not happen for continuous (or even bounded) connections ∇. (Recall that every H-connection ∇ can be written locally as (trivial connection) $+$ ($L(G)$-valued 1-form on H) and the continuity of ∇ refers to this 1-form viewed as a section of a bundle over V.)*

Observation. *Every section f with $\nabla f \in L_q$ for $q > N$ is C^α with $\alpha = (q - N)/N$ where the implied bound on the Hölder constant $L_\alpha(f) \leq$ const $\|\nabla f\|_{L_q}$ has const independent of ∇.*

This is shown by the same straightforward computation as was used in 2.3.E, 2.4 and 2.5 with "standard pencils" of H-horizontal curves between given points v_1 and v_2 in V. A generic curve in such a pencil, say $\gamma = \gamma(t)$, has well controlled integral $\int_\gamma \|\nabla_t f(\gamma(t))\| dt$ which implies the desired bound on L_α.

The case $q = N$. If our H-connection ∇ is continuous (or at least Borel measurable bounded) then the metric in X is locally equivalent to the product metric on $V \times \mathbb{R}^m$, $m = \operatorname{rank} X$, and so our regularization from 2.5 applies here. Thus every section f with $\|\nabla f\|_{L_N} < \infty$ can be approximated by continuous sections f_ε, $\varepsilon \to 0$, where "approximation" means a.e. convergence as well as $\|\nabla f - \nabla f_\varepsilon\|_{L_N} \to 0$ for $\varepsilon \to 0$. Furthermore, if f is *taut*, in the sense that no deformation with small support diminuishes $\|\nabla f(v)\|$, then the logarithmic modulus of continuity of f is bounded in terms of $\|\nabla f\|_{L_N}$. Furthermore, if f is locally $\|\nabla f\|_{L_N}$-minimizing (or only quasi-minimizing) then it is Hölder.

(codim 1)-*Reminder.* All these properties of $\|\nabla f\|_N$ were established earlier (for split connections) under the (codim 1)-stability assumption on (V, H) and this assumption is needed in the present (non-split) case as well.

Generalization. The above Hölder continuity results for $q > N$ as well as for $q = N$ make sense for an arbitrary (non-vector) G-bundle where the fiber W is a G-manifold with a G-invariant metric (and where we do not have to assume G is compact). Our Hölder observation for $q > N$ obviously extends to this case with const independent of W and ∇; furthermore, the above statements (and their proofs) for $q = N$ also generalize with no difficulty if W has locally bounded geometry. This leads to the following

Theorem. *Let (V, H) be* (codim 1)-*stable. Then*

(i) *the existence of a measurable section $f : V \to X$ with $\|\nabla f\|_{L_N} \leq c < \infty$ implies the existence of a continuous section $f' : V \to X$ with $\|\nabla f'\|_{L_N} \leq c + \varepsilon$ for an arbitrary small $\varepsilon > 0$;*

(ii) *if V and W are compact and $\pi_1(W)$ acts trivially on $\pi_n(W)$ for $n = \dim V$, then there are at most finitely many homotopy classes of sections $f : V \to X$ with $\|\nabla f\|_{L_N} \le c$ for every constant c;*

(iii) *if V and W are compact and $\pi_1(W)$ acts trivially on $\pi_i(W)$ for $i = n, n+1, \ldots$, then the space F_c of continuous sections $f : V \to X$ with $\|\nabla f\|_{L_N} \le c$ has finite homotopy type in the space $F = F_\infty \supset F_c$ of all continuous sections.*

Notice that all this is well known in the Riemannian case (i.e. where $H = T(V)$) and due to K. Uhlenbeck.

Remark on Kac-Feynman formula and Kato inequality. One knows that the heat flow in a vector bundle with a Euclidean connection ∇ over a Riemannian manifold V decays the slowest if the bundle is split, which gives one a lower bound on the spectrum of $\nabla^*\nabla$ in terms of the Laplacian spectrum of V. Apparently, this generalizes to our H-connections. Furthermore one expects similar bounds for the (non-linear) L_p- and L_{pq}-spectra in the sense of [Gro$_{\text{DNLS}}$].

5.3. L_q-distance in the space of connections. Let X_1 and X_2 be Euclidean vector bundles of the same dimension over V and $Y = \mathrm{Iso}(X_1, X_2)$ be the (non-vector) bundle of isometric homomorphisms $X_1 \to X_2$. If X_1 and X_2 come along with H-connections ∇_1 and ∇_2 then these give rise to a natural H-connection ∇ on Y and so one may speak of the norms $\|\nabla f\|_{L_q}$ of our (smooth, continuous or measurable) homomorphisms $f : X_1 \to X_2$ viewed as (smooth, continuous or measurable) sections $f : V \to Y$.

Example. A smooth homomorphism f has $\|\nabla f\|_{L_q} = 0$ if and only if it sends the connection ∇_1 to ∇_2.

To get a clearer picture of $\|\nabla f\|$ we identify X_1 and X_2 by f and thus define the difference $\nabla_1 - \nabla_2$ which is an $L(G)$-valued form on V. Thus we see that $\|\nabla f\| = \|\nabla_1 - \nabla_2\|$ which implies that

(i) $\|\nabla f\|_{L_q} = \|\nabla f^{-1}\|_{L_q}$,

(ii) $\|\nabla(f_1 \circ f_2)\|_{L_q} \le \|\nabla f_1\|_{L_q} + \|\nabla f_2\|_{L_q}$, for $X_1 \xrightarrow{f_1} X_2 \xrightarrow{f_2} X_3$.

It follows that $\inf_f \|\nabla f\|_{L_q}$ over all measurable $f : X_1 \to X_2$ defines a metric (which may be sometimes infinite) on the set of isomorphism classes of bundles with H-connections over V. This metric is called L_q and denoted dist_{L_q}.

Example. If V is (codim 1)-stable and $\mathrm{dist}_{L_N}(X_1, X_2) < \infty$ for $N = \dim_{\mathrm{Hau}} V$, then X_1 and X_2 are isomorphic as vector bundles since the existence of a measurable section $f : V \to Y = \mathrm{Iso}(X_1, X_2)$ with $\|\nabla f\|_{L_N} < \infty$ implies the existence of a continuous one.

5.3.A. On L_q-non-flatness of ∇. Let X be a Euclidean bundle with an H-connection ∇ over (V, H) and let us discuss several invariants of ∇ measuring its L_q-distance from a (locally) split connection.

Example: connections over S^n. Let V be homeomorphic to the n-sphere covered by two balls B_+ and B_- and d_+ and d_- be the L_q-distances from the bundles $(X, \nabla)|B_+$ and $(X, \nabla)|B_-$ to the split bundles over these balls. This means we have orthonormal frames f_+ over B_+ and f_- over B_- with $\|\nabla f_\pm\|_{L_q} \le d_\pm + \varepsilon$. Then we have the usual map of the annulus $A = B_+ \cap B_-$ to the structure group $G = O(r)$, $r = \mathrm{rank}\, X$, call it $\varphi : A \to G$ (defined by the relation $\varphi f_+ = f_-$), which clearly satisfies

$$\|\mathcal{D}\varphi | H\|_{L_q} \le d_+ + d_- + 2\varepsilon.$$

Then the same inequality is satisfied (up to a constant) on some $(n-1)$-sphere of this annulus $A = I \times S^{n-1}$, and so, for example, if $d_+ + d_- \le c_0$ for some small $c_0 > 0$, then this map $\varphi : S^{n-1} \to G$ is *contractible* provided $q > N - 1$ and (V, H) is $(n-1)$-stable (where N is the Hausdorff dimension of V and V is assumed equiregular, compare 2.5). Now, if our frames f_+ and f_- where *continuous*, this would make the bundle $X \to V$ *topologically trivial* but this is not so for general measurable f_+ and f_-. Yet, if $q \ge N$, we can always make f_+ and f_- continuous (see 5.2.) and then the triviality of $X \to V$ is ensured whenever $d_+ + d_- \le c_0$. Furthermore, if $q \ge N$, we also see that the number of mutually non-equivalent bundles over $V = S^n$ which may have $d_+ + d_- \le c$ for some H-connection ∇ on X is finite and bounded by some constant depending on c.

Let us generalize the above to an arbitrary compact V which is covered by finitely many balls $B_i \subset V, i = 1, \ldots, m$. We measure the non-flatness of ∇ by the L_q-distances of $(X, \nabla)|B_i$ from split bundles, call them d_i, and claim that in the (codim 1)-stable case and for $q \geq N$ the sum $\sum_{i=1}^{m} d_i$ gives us a bound on a possible (topological) type of X. Namely

(I) *There are at most finitely many mutually non-equivalent bundles X over V with $\sum_{i=1}^{m} d_i < c$, for every fixed $c > 0$, and their number is bounded by some constant depending on c (as well as on V, B_i, H and rank X).*

(II) *There exists a positive constant $c_0 = c_0(V, B_i, H, \operatorname{rank} X)$, such that the inequality $\sum_{i=1}^{m} d_i \leq c_0$ implies that X admits a locally split (flat) connection. In particular, if V is simply connected then $X \to V$ is topologically trivial.*

Proof of (I). Every frame f_i on B_i is given by $r = \operatorname{rank} X$ sections, say f_i^j, $i = 1, \ldots, m$, $j = 1, \ldots, r$, and each of these sections can be cut off near the boundary of B_i and thus extended to all of V. These sections define a homomorphism of the split bundle of rank mr into X, say $\pi : Y \to X$, which is easily seen to be surjective and having $\|\nabla\pi\|_{L_q} < c'$, where $c' \leq c_1 c$ for some $c_1 = c_1(V, B_i)$. We assign to each $v \in V$ the orthogonal complement of the kernel of π in the fiber $Y_v = \mathbb{R}^M$, $M = mr$, that is $v \mapsto (\ker \pi_v)^\perp \subset \mathbb{R}^M$, and thus map V to the Grassmann manifold, say by $\psi : V \to Gr_r\mathbb{R}^M$. As we can assume the frames f_i continuous (for $q \geq N$) this ψ is continuous and obviously is a classifying map for $X \to V$, i.e. it induces on V an r-bundle isomorphic to X. In fact, our extended sections f_i^j define a fiberwise injective morphism of X to the canonical bundle over $\operatorname{Gr}_r \mathbb{R}^M$, say $\tilde{\psi} : X \to X_{\operatorname{can}}$. (To see this, one should actually use the dual frames of linear functions on X which give us an injective homomorphism $X \to Y$ but this duality may be absorbed by the Euclidean structure in the fibers.) Then it is easy to see that $\tilde{\psi}$ and ψ are L_q-controlled over V by the L_q-norms of ∇f_i^j and for ψ; this means

$$\|\mathcal{D}\psi|H\|_{L_q} \leq c''$$

for some c'' determined by c. This bounds the homotopy class of ψ (see 2.5.B) and thus the equivalence class of X. Q.E.D.

Remark. The above argument appeals to the L_q-non-flatness of (X, ∇) measured by injective homomorphisms $h : X \to Y$, where Y is a split bundle, and the non-flatness is measured by

$$\|\nabla h\|_{L_q} + \|h\|_{L_q} + \|h^{-1}\|_{L_q}, \tag{$*$}$$

where h^{-1} denotes the inversion of h on its image in Y. If rank $Y =$ rank X this non-flatness is essentially the same as the L_q-distance between X and Y.

Question. Let V be compact and X be (known) topologically trivial, e.g. V be contractible. Suppose there is an injective $h : X \to Y$, where rank $Y >$ rank X, with $\|\nabla h\|_{L_q} + \|h\|_{L_q} + \|h^{-1}\|_{L_q} \le c$. Can one bound the L_q-distance of X to the split bundle of rank $r =$ rank X in terms of c?

Exercise. Assume the frames f_i *continuous* and prove (I) for all $q > N - 1$.

Proof of (II). Our frames, which we may assume continuous, define a 1-cocycle φ on the nerve of the covering of V by B_i with values in the sheaf of continuous maps of V into the structure group $G = O(r)$, where each implied map $\varphi_{ij} : B_i \cap B_j \to G$ is defined by the formula $\varphi_{ij} f_i = f_j$ on $B_i \cap B_j$. If f_i and f_j have small norms $\|\nabla\|_{L_q}$, then the norm $\|\mathcal{D}\varphi_{ij}|H\|_{L_q}$ is also small and then, for $q > N$, this φ_{ij} is uniformly ε-close to a constant map, where $\varepsilon \to 0$ for $c_0 \to 0$. Then a simple limit argument (left to the reader) allows us to construct a new cocycle $\bar{\varphi}$, where all $\bar{\varphi}_{ij}$ are *constant* maps $B_i \cap B_j \to G$, which gives us the required flat structure in X. Finally, the case $q = N$ in the (codim 1)-stable case is handled with a (taut) regularization as earlier (where the details are left to the reader).

Remarks

(a) If one applies the proof of (I) to a flat bundle, one concludes to the (well known) finiteness of equivalence classes of such bundles over V (which may be derived from the finiteness of the number of connected components of representations $\pi_1(V) \to G = O(r)$).

(b) The proof of (II) suggests the infimum of L_q-norms of cocycles defining X over V (covered by B_i) as a measure of non-flatness of X where ∇ is not present any more.

Exercise. Prove (I) by the cocycle argument employed in the proof of (II).

5.3.B. Non-flatness measured by horizontal monodromy and curvature. The *monodromy* of an H-connection ∇ at a closed H-horizontal curve S in V is the conjugacy class of the ∇-parallel transport along S, denoted $\nabla(S) \in G/\mathrm{Conj}$ for the structure group G $(=O(r))$. The function $S \mapsto \nabla(S)$ is the basic invariant of ∇ which can be made numerical by composing with some $\sigma : G/\mathrm{Conj} \to \mathbb{R}$ and integrating over a measure on the space of horizontal curves S in V. An important function σ in this regard is the "norm" $\|g\| \underset{\mathrm{def}}{=} \mathrm{dist}(g, \mathrm{id})$ for the standard bi-invariant metric on $G = O(r)$ corresponding to the Killing form. This "norm" is subadditive on closed curves: if S is decomposed into S_i, $i = 1, \ldots, m$, then

$$\|\nabla(S)\| \leq \sum_{i=1}^{m} \|\nabla S_i\|,$$

where a typical decomposition is as in Fig. 14

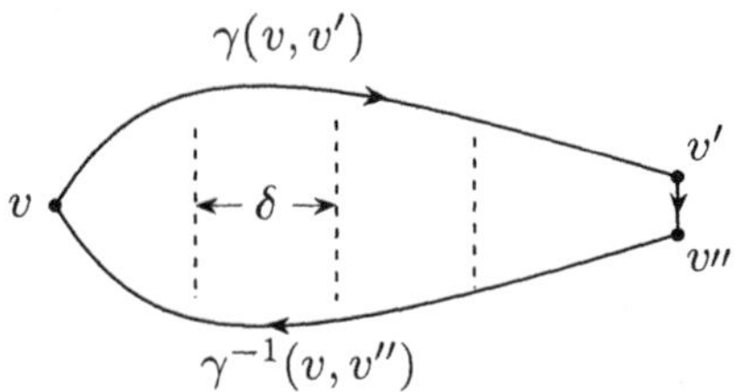

Figure 14

The values of $\nabla(S)$ on infinitesimal curves at a point $v \in V$ can be expressed in terms of suitable curvatures of ∇ at v and their jets. In particular, if S bounds our H-horizontal disk D, then $\|\nabla(S)\| \leq C \int_D \|\Omega_\nabla\|$ for a universal constant C depending on how we normalize (the norm of) the curvature. In what follows, we absorb C into such a normalization, i.e. make it 1.

If the function $\|\nabla(S)\|$ vanishes on all closed horizontal curves, then, obviously, the bundle (X, ∇) splits and, in particular, X is topologically trivial. Furthermore, if $\|\nabla(S)\| = 0$ on all sufficiently short curves, then (X, ∇) locally splits and so X is a flat bundle coming from some representation $\pi_1(V) \to G$. Then an obvious limit argument shows that if V is compact then there exists $\varepsilon > 0$, such that

(I) *if* $\|\nabla(S)\| \leq \varepsilon$ *for all closed horizontal curves* S *in* V *then* X *is topologically trivial;*

(II) *if* $\|\nabla(S)\| \leq \varepsilon$ *for all short horizontal curves, to be specific for all* S *of length* ≤ 1, *then* X *is equivalent to a flat bundle.*

Corollary. *Let* H *admit an* Ω-*regular* Ω-*isotropic jet of 2-germs at each point* $v \in V$. *Then the inequality* $\|\Omega_\nabla\| \leq c_0$ *for a small* $c_0 = c_0(V) > 0$ *makes the conclusion of (II) hold true.*

Proof. One can fill in short curves S by horizontal disks D of small area (see 4.7) and bound $\|\nabla(S)\|$ by $\int_D \|\Omega_\nabla\| \leq \|\Omega_\nabla\|$ area D.

Remark. The conclusions of (I) and (II) remain valid if the bound $\|\nabla(S)\| \leq \varepsilon$ is assumed only for "standard" closed curves, e.g. having their complexity (see 3.5, 4.7.) bounded by a fixed large constant. Thus one does not need in the above Corollary the full geometric force of the isoperimetric inequality but only the analytic part (filling curves of bounded complexity) depending on the implicit function theorem. Probably, the latter can also be removed from the present discussion with some kind of "almost horizontal disks" filling in "very standard infinitesimal curves".

5.3.C. Radial gauge fixing. Let us make the above remarks clear by using "standard short segments" in V between near points. Namely, we associate to each $v \in V$ a horizontal vector field Y_v in a small ball $B_v(\rho) \subset V$, such that the forward orbits of Y_v converge to v and such that the length of an orbit from $v' \in B_v(\rho)$ to v is bounded by const dist (v, v') (compare 2.3.A). Thus we join every two close by points v and v' in V by a horizontal curve, say $\gamma(v, v')$ (where, in general, $\gamma(v', v) \neq \gamma^{-1}(v, v')$), such that

$$\text{length}\, \gamma(v, v') \approx \text{dist}(v, v').$$

A *radial gauge* in $B_v(\rho)$ is a frame f of X over $B_v(\rho)$ obtained by parallel transport of a frame at v along the orbits of some radial field Y_v. The values of such f at two near points v' and v'' can be compared with the parallel transport along some (standard) path $\gamma(v', v'')$ and the result of this transport, (which is an element of the group G) equals (up to conjugation) to the monodromy around the triangle $S = \gamma(v, v')\gamma(v'v'')\gamma^{-1}(v, v'')$, see Fig. 15 below.

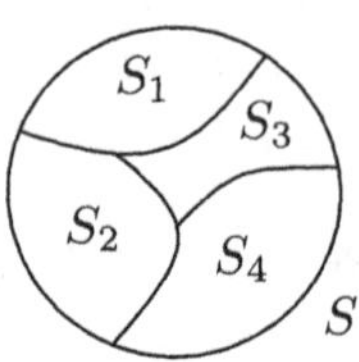

Figure 15

This monodromy can be estimated under the assumptions of the above Corollary by

$$\text{const}\,\|\Omega_\nabla\|(\text{length } S)^2 \leq \text{const}'\,\|\Omega_\nabla\|\,(\text{dist}(v,v') + \text{dist}(v',v''))^2 \,.$$

This can be improved for $v'' \to v'$ since for small $\text{dist}(v'',v')$ the triangle becomes narrow and can be filled-in more efficiently. Namely, the distance between the Y_v-orbits of v' and v'' is bounded by $\text{const}_\rho\,(\text{dist}(v',v''))^{\frac{1}{d}}$ where d is the depth of the filtration $H = H_1 \supset H_2 \supset \cdots H_d = T(V))$, see 4.9) and one can subdivide S into m pieces S_i of size $\delta \approx (\text{dist}(v',v''))^{\frac{1}{d}}$ for $m \approx \rho/\delta$. Then we fill in each S_i by a disk D_i of area $\lesssim \delta^2$ and thus obtain a filling D of S with area $\lesssim \rho\,(\text{dist}(v',v''))^{\frac{1}{d}}$. This shows that the radial frame (gauge) is $C^{\frac{1}{d}}$-Hölder with the implied Hölder constant bounded by $C\rho\|\Omega_\nabla\|$ which yields the Corollary by the argument of 5.3.A and also gives a topological bound on X in terms of $c = \sup\|\Omega_\nabla\|$ for every (not only small) $c > 0$ (compare 5.3.A).

5.4. Geometric and topological effects of the bound $\|\Omega_\nabla\|_{L_q} \leq c$.

We saw in the previous section that the bound $\|\Omega_\nabla\|_{L_\infty} \leq c_0$ for small c_0 makes X geometrically almost flat and topologically flat and we want to prove this for some $q < \infty$. The corresponding result for the ordinary connections (where $H = T(V)$) due to Karen Uhlenbeck (see [Uhl]) reads

(I) *Let V be a compact simply connected Riemannian manifold. Then there exists a constant $c_0 > 0$ (depending on V and $r = \text{rank}\,E$) such that every Euclidean bundle (E, ∇) of rank r over V with $\|\Omega_\nabla\|_{L_{n/2}} \leq c_0$ admits a continuous orthonormal frame f over V satisfying*

$$\|\nabla f\|_{L_n} \leq \text{const}\, c_0 \qquad\qquad (*)$$

for $\text{const} = \text{const}\,(V, r)$.

Remarks

(a) The norm $\|\nabla f\|$ is essentially the same thing as the norm of the $L(G)$-valued 1-form ∇_f representing ∇ in the frame f. The full Uhlenbeck theorem says that the gradient (i.e. the 1-jet) of this form for a suitable f satisfies

$$\|\operatorname{grad} \nabla_f\|_{L_{n/2}} \leq \operatorname{const} c_0 \qquad (**)$$

and then $(*)$ follows from $(**)$ by the Sobolev inequality.

(b) If V is not simply connected, the Uhlenbeck theorem applies to coverings of V by balls and yields a local information on X which can be globalized by the argument employed in 5.2.A. In fact, the proof of the theorem in the general simply connected case is obtained with such a covering.

(c) If the structure group G is Abelian, the Uhlenbeck theorem reduces to the controlled integration of exact 2-forms (with the corresponding Green kernel, see 3.6). Then the general case follows by an implicit function argument.

(d) The Uhlenbeck theorem implies (see 5.2.A) that every bundle (X, ∇) over a compact simply connected Riemannian manifold V with small norm $\|\Omega_\nabla\|_{L_{n/2}}$ is topologically trivial. Consequently every map φ of V into the classifying space $\operatorname{Gr}_r \mathbb{R}^M$ with small $\|\Lambda^2 \mathcal{D}\varphi\|_{L_{n/2}}$ is contractible. (We mentioned earlier that the similar property is unknown for more general simply connected manifolds W in place of $\operatorname{Gr}_r \mathbb{R}^M$.)

(e) Uhlenbeck's theorem has non-trivial implications for the bundles (X, ∇) where $\|\Omega_\nabla\|_{L_{n/2}} \leq c$ for a fixed but not necessarily small $c > 0$. In fact, the norm $\|\Omega_\nabla\|_{L_{n/2}}$ is invariant under scaling the metric in V and so (X, ∇) can be studied over small balls in V rescaled to the unit size. Then almost all of V can be covered by balls B_i with $\|\Omega_\nabla|B_i\|_{L_{n/2}} \leq c_0$ and the remaining "bad part" of V reduces to (arbitrarily small) neighbourhoods of finitely many points in V where (only) Uhlenbeck's bubbling may occur (compare 2.5). The topological corollary of this is *the finiteness of the topological equivalence classes of bundles with* $\|\Omega_\nabla\|_{L_{n/2}} \leq c$. But this corollary can be easily derived by pure topology by observing that the $L_{n/2}$-norm of the curvature bounds the rational characteristic classes of E via the Chern-Weil formula and these classes determine E up to a finite number of possibilities.

(f) Finally we point out (following Uhlenbeck) an (immediate) asymptotic corollary of her theorem.

Let (X, ∇) be a bundle over $\mathbb{R}^n$ with $\|\Omega_\nabla\|_{L_{n/2}} < \infty$. Then X extends to a bundle $X^\bullet$ on the sphere S^n obtained by the 1-point compactification of $\mathbb{R}^n$ and ∇ extends to a measurable connection $\nabla^\bullet$ on $X^\bullet$ which admits a continuous frame f near the infinity $v_\infty \in S^n$ with $\|\nabla^\bullet f\|_{L_n} < \infty$.

Now we can state our basic problem for H-connections ∇ on bundles X over (V, H). We want to know for which q (depending on (V, H)) the bound on $\|\Omega_\nabla\|_{L_q}$ has an effect similar to the conclusion of Uhlenbeck's theorem and issuing corollaries. For example, when does the bound $\|\Omega_\nabla\|_{L_q} < c$ allow at most finitely many topological non-equivalent bundles X and imply triviality of X for small $c = c_0$? What is the integral curvature condition on an H-connection (or an ordinary one) over a simply connected nilpotent Lie group which would allow a good one-point compactification? (This question makes sense for more general open Riemannian and C-C manifolds.)

5.4.A. Gauge theory over contact manifolds. Since Rumin theory provides controlled integration of 2-forms on contact manifolds V of dimension $n \geq 5$ (see 3.6), the full Uhlenbeck package seems to generalize to this case with L_N in place of L_n for $N = n + 1 = \dim_{\text{Hau}} V$. Namely, *every (X, ∇) with small $\|\Omega_\nabla\|_{L_{N/2}}$ must admit continuous frames f with small $\|\nabla f\|_{L_N}$ over simply connected (regions in) V*, etc. (We suggest the reader would check that Uhlenbeck's proof indeed transplants to Rumin's hypoelliptic framework of contact manifolds.)

An especially attractive case is that of $\dim V = 5$, where $\operatorname{rank} H = 4$ and one can define the Yang-Mills equation. In particular, if H is given by a connection in an S^1-bundle over a symplectic 4-manifold V_0, one may lift the Yang-Mills fields from V_0 to V, the total space of the S^1-bundle $V \to V_0$, and try to express Donaldson invariants of V_0 in terms of contact geometry of V. (But this is just a wishful thinking not corroborated by a serious evidence.)

5.4.B. Monodromy control by thick fillings. In order to bound the monodromy over a curve $S \subset V$ it suffices to find a horizontal filling D of S with small integral of $\|\Omega_\nabla\|$ over this D and if we have a bound on $\|\Omega_\nabla\|_{L_q}$ such a D can be chosen as a member of a given q-thick family of fillings of S (see 3.6.D). If we look for such a family consisting of Ω-regular disks, the question becomes essentially (differential) algebraic as jets of such families at S extend to germs by the generalized Nash implicit function theorem but the specific evaluation of the minimal $q = q(H)$ requires a computation in particular cases which was performed so far only for contact manifolds (see 3.6.D). Recall that if our family is given by a smooth map $f : D \times B^{n-2} \to V$ which is regular (i.e. immersive) on the disks and if the Jacobian of f decays as (or no faster than) ρ^j where $\rho = \rho(d, b)$ is the distance of $d \in D$ to the boundary $S = \partial D$, then this family is q-thick for every $q > j + 1$. For example, the obvious $(n - 2)$-dimensional pencil of disks in $\mathbb{R}^n$ filling the standard circle is q-thick for $q > n - 1$. We have constructed in 3.6.D contact pencils f with $j = n + 1$ and hence q-thick for $q > n + 2$ and now we make three extra points.

(1) Start with a generic H of rank $n_1 > (2n + 2)/3$ which ensures the existence of regular Ω-isotropic plane at a generic point in V (see 4.2.A''). In fact, this also gives us such a plane through a generic horizontal vector. Assuming this genericity one can produce families $f : D \times B^{n-2} \to V$ with decay (at most) $a_1 \, \rho^j$ for some $j < \infty$. This gives us q-thick fillings, for $q = j + 1 + \varepsilon < \infty$, if sufficiently many horizontal curves S in order to derive basic topological finiteness results. For example, *there are at most finitely many topologically non-equivalent bundles* $X \to (V, H)$ *with* $\|\Omega_\nabla\|_{L_q} \leq c$. It would be interesting to have a specific evaluation of j and q for general H but we indicate this below only in two cases close to the contact one.

(2) Let $(V, H) = (V_1 \times V_2, H_1 \times H_2)$ for contact (V_i, H_i) of dimensions ≥ 5. Then H_i-horizontal disks $\varphi_i : D \to V_i$, $i = 1, 2$. Define a horizontal bi-disk in V, namely $\varphi = \varphi_1 \times \varphi_2 : D \times D \to V$ and then families of disks in V_i with the Jacobian decays of degrees j_i, $i = 1, 2$, give us a family in V with $j = j_1 + j_2 + 1$. Thus we obtain q-thick fillings of horizontal closed curves S in V for every $q > n + 3$. Furthermore, since the Ω-regularity etc is stable under small perturbation of H, these q-thick fillings also exist for polarizations H on V close to $H_1 \times H_2$ and so the basic topological finiteness results hold true for these H and q.

(3) Let (V, H) be the 1-jet bundle of maps $V_0 \to \mathbb{R}^2$. This H is very much similar to the above and the evaluation of q is left to the reader. We also suggest the reader would look at the thick filling of k-dimensional cycles for $k > 1$.

Final Remarks. The above results are of preliminary nature as we strive for $q = N/2$ which would give the best possible result with the (scale invariant) norm $\|\Omega_\nabla\|_{L_{N/2}}$. One way to do this is offered by an extension of hypoelliptic theory to the complex $(\Lambda^*(V)/I(H), d)$ between degrees 1 and 2 which is needed for an analytic proof à la Uhlenbeck. A step in this direction is made by Z. Ge in [Ge$_{\text{BNCC}}$] but his hypoellipticity criterion (in degree one) seems rather restrictive. Another (less realistic but geometrically attractive) possibility is to combine the thick filling idea with some geometric "controlled integration" (over suitable measures in the space of curves) which would lead us directly to bounds on ∇ in terms of Ω_∇.

Inverse Kato inequalities. The existence of frames and sections f with small integral norms of ∇f is opposite to Kato inequality (see 5.2) and one may expect further upper bounds on the spectrum of the operator $\nabla^*\nabla$ in terms of the Hörmander Laplacian ∇_H on (scalars on) V and integral bounds on $\|\Omega_\nabla\|$. For example, one wishes to estimate the number λ, such that the eigensections of $\nabla^*\nabla$ below λ span each fiber of our bundle $X \to V$. It would be interesting to estimate this λ in terms of $\|\Omega\|_{N/2}$ (and V) by a purely linear argument which would provide an alternative to Uhlenbeck's approach.

References

[Bass] Bass H., The degree of polynomial growth of finitely generated nilpotent group. Proc. Lond. Math. Soc. (3) 25, pp. 603–614, 1972.

[Bell] Bellaïche A., The tangent space in sub-Riemannian geometry. This volume.

[Be-Ve] Berestovskii V.N., Vershik A.M., Manifolds with intrinsic metrics and nonholonomic spaces. Advances in Sov. Math. vol (**7**) AMS 1992.

[Beth] Bethuel F., The approximation problem for Sobolev maps betwen two manifolds. Acta. Math., **167** (1991), pp. 153–206.

[Ben] Bennequin D., Entrelacements et équations de Pfaff. Soc. Math. France, Astérisque **107–108** (1983), 83–161.

[Bou] Valère Bouche L., The geodesics' problem in Sub-Riemannian geometry. About the R. Montgomery's example simplified by I. Kupka. Preprint du L.A.M.A. 92-06 (1992).

[Br-Gr] Bryant R.L. and Griffiths P.A., Characteristic Cohomology of Differential systems (I): General Theory. J.-Amer.-Math.-Soc. **8** (1995), no. 3, 507–596.

[Brock] Brockett R.W., Nonlinear Control Theory and Differential Geometry, Proc. of the Int. Congress of Mathematicians, Warszawa, (1983).

[Bu-Za] Burago Y.D. and Zalgaller V.A.: Geometric inequalities. Springer, Berlin, Heidelberg 1988. [Russian edn.: Nauka, Moscow, 1980].

[Cho] Chow W.L., Über Systeme von linearen partiellen Differentialgleichungen erster Ordnung, Math. Annalen **117** (1939), pp. 98–105.

[Cor] Corlette K., Hausdorff dimensions of limit sets I, preprint (1989).

[D-G] D'Ambra G. and Gromov M., Lectures on transformation groups: geometry and dynamics, Surveys in Differential Geometry (Supplement to the Journal of Differential geometry), **1** (1991), pp. 19–111.

[DA$_{C1}$] D'Ambra G., Nash C^1-embedding theorem for Carnot-Carathéodory metrics. Differential-Geom.-Appl. **5** (1995), no. 2, 105–119.

[DA$_{IC}$] D'Ambra G., Construction of connection inducing maps between principal bundles. J. Diff. Geom. **26**, (1987), pp. 67–79.

[DA$_{IS}$] D'Ambra G., Induced subbundles and Nash's implicit function theorem. Differential-Geom.-Appl. **4** (1994), no. 1, 91–105.

[Da-Se] David G., Semmes S., Lipschitz mappings between dimensions, Unpublished manuscript.

[Dek] Dektjarev I.M., Problems of value distribution in dimension higher than one, Uspeki Mat. Nauk 25 ; **6** (1970), pp. 53-84

[E-L] Eells J. and Lemaire L., Another report on harmonic maps. Bull. London. Math. Soc. **20** (1988) 385–524, pp. 388.

[El$_1$] Eliashberg Ya., Classification of overtwisted contact structures on manifolds, Invent. Math. **98** (1989), pp. 623–637.

[El$_2$] Eliashberg Ya., Contact 3-manifolds, twenty years since J. Matinet's work. Ann. Inst. Fourier **42** (1992), pp. 165–192.

[El$_3$] Eliashberg Ya., New invariants of open symplectic and contact manifolds, J. Amer. Math. Soc. **4** (1991), pp. 513–520.

[El$_4$] Eliashberg Ya., Filling by holomorphic disks and its applications, London Math. Soc. Lect. Notes Ser. **151** (1991), pp. 45–67.

[Fal] Falconer K., Fractal Geometry. Mathematical Foundations and Applications. John Wiley and Sons. 1990.

[Fef-Ph] Fefferman C.L. and Phong D.H., Subelliptic eigenvalue problems, Proceedings of the Conference on Harmonic Analysis in Honor of Antoni Zygmund, Wadsworth Math. Series, pp. 590–606 (1981).

[Fol] Folland G.B., Subelliptic estimates and function spaces on nilpotent Lie groups, Arkiv för Mat. **13**, pp. 161–207 (1975).

[Fuk] Fukawa K., Collapsing of Riemannian manifolds and eigenvalues of Laplacian operators, Invent. Math. **87** (1987), pp. 517–547.

[G-E] Gromov M. and Eliashberg J., Construction of nonsingular isoperimetric films, Trudy Steklov Inst. **116**, (1971), pp. 18–33.

[G-L-P] Gromov M., Lafontaine J., Pansu P., Structures métriques pour les variétés Riemanniennes, Cedic-Fernand Nathan, Paris (1981).

[Ge$_{BNCC}$] Ge Z., Betti numbers, characteristic classes and sub-Riemannian geometry, Illinois J. of Mathematics, **36** (1992), pp. 372–403.

[Ge$_{CRM}$] Ge Z., Collapsing Riemannian metrics to sub-Riemannian metrics and Laplacians to sub-Laplacians, Canadian J. Math., 1993, pp. 537–552.

[Ge$_{GG3}$] Ge Z., On the Global Geometry of Three-dimensional Sub-Riemannian Manifolds. II On Sub-Riemannian Metrics and $\widetilde{SL_2}R$-geometry. The Fields Institute for Research in Mathematical Sciences, Preprint 1994.

[Ge$_{HPS}$] Ge Z., Horizontal paths space and Carnot-Carathéodory metrics, Pacific J. of Mathematics, **161** (1993), pp. 255–286.

[Ge$_{VP}$] Ge Z., On a variational problem and the spaces of horizontal paths, Pacific J. of Mathematics, **149** (1991), pp. 61–93.

[Gir] Giroux E., Topologie de contact en dimension 3, Séminaire Bourbaki, 1992-93, n° 760.

[Good] Goodman R.W., Nilpotent Groups, the Structure, and the Application to Analysis, Lectures Notes in Math., vol. 562, Springer, 1977.

[Gro$_{AI}$] Gromov M., Asymptotic invariants of infinite groups. Geometric group theory, Vol. 2 (Sussex, 1991), 1–295. London Math. Soc. Lecture Note Ser., **182**. Cambridge Univ. Press, Cambridge, 1993.

[Gro$_{AI}$] Gromov M., Asymptotic geometry of homogeneous spaces, Proceedings Rend. Sem. Mat. Univ. Politec. Torino (1983), Conférence on Homogeneous Spaces in Symp. Math., pp. 59–60.

[Gro$_{CIDR}$] Gromov M., Convex integration of differential relations, Izv. Akad. Nauk. S.S.S.R. 33, **2**, (1973), pp. 329–343.

[Gro$_{DNLS}$] Gromov M., Dimension, non-linear spectra and width, Springer-Verlag, Lecture Notes in Mathematics, **1317** (1988), pp. 132-1-85.

[Gro$_{FPP}$] Gromov M., Foliated plateau problem, parts I, II, Geometric and Functional Analysis 1 : **1** (1991), pp. 14–79 ; (1991), pp. 253–320.

[Gro$_{FRM}$] Gromov M., Filling Riemannian manifolds, Journal of Differential Geometry **18** (1983), pp. 1–147.

[Gro$_{GPG}$] Gromov M., Groups of polynomial growth and expanding maps, Publications Mathématiques IHES **53** (1981), pp. 53–73.

[Gro$_{HED}$] Gromov M., Homotopical effects of dilatation. Journal of Differential Geometry **13** (1978) 303–310.

[Gro$_{HG}$] Gromov M., Hyperbolic groups, Essays in Group Theory, S. Gersten editor, MSRI Publications n° **8**, Springer (1987), pp. 75–265.

[Gro$_{HMGA}$] Gromov M., Hyperbolic manifolds, groups and actions, in "Riemannian Surfaces and Related Topics", Ann. Math. Studies **97** (1981), pp. 183–215.

[Gro$_{MIKM}$] Gromov M., Metric invariants of Kähler manifolds, Differential geometry and topology (Alghero, 1992), 90–116. World Sci. Publishing, River Edge, NJ, 1993.

[Gro$_{PDR}$] Gromov M., Partial differential relations, Springer-Verlag (1986).

[Gro$_{PLI}$] Gromov M., Paul Levy's isoperimetric inequality, Preprint IHES (1980).

[Gro$_{SAP}$] Gromov M., Stability and pinching. Sessions on Topology and Geometry of Manifolds (Italian) (Bologna, 1990), 55–97. Univ. Stud. Bologna, Bologna, 1992.

[Gro$_{SISI}$] Gromov M., Systoles and intersystolic inequalities. Preprint IHES (1993).

[Gro$_{WRI}$] Gromov M., Width and related invariants of Riemannian manifolds, Astérisque **163–164** (1988), pp. 93–109.

[Hart] Hart P., Ordinary differential equations. John Wiley and Sons. New York-London-Sydney. 1964.

[Ham] Hamenstädt U., Some regularity theorems for Carnot-Carathéodory metrics, J. Diff. Geom. **32** (1991), pp. 192–201.

[Herm] Hermann R., Differential Geometry and the Calculus of Variations, Math. Sci. Eng. **49**, Academic Press, New-York, 1968.

[Ho] Hofer H., Pseudoholomorphic curves in symplectizations with applications to the Weinstein conjecture in dimension three, Inv. Math. **114** (3), pp. 515–585. 1993.

[Hol] Holopainen I., Positive solutions of quasilinear elliptic equations on riemannian manifolds. Proc. London Math. Soc. (3) **65** (1992), pp. 651–672.

[Hol-Rick] Holopainen I. and Rickman S., Quasiregular mappings of the Heisenberg group. Preprint (1991).

[Hör] Hörmander L., Hypoelliptic second order differential equations, Acta Math. **119**, pp. 147–171 (1967).

[Hsu$_{\text{CVG}}$] Hsu L., Calclus of Variations via the Griffiths formalism, Journal of Differential Geometry **36** (1992), pp. 551–589.

[Hsu$_{\text{GHP}}$] Hsu L., Gromov's h-principle for strongly bracket-generating distributions (IAS preprint: 1993)

[Hsu$_{\text{SRCT}}$] Hsu L., Sub-Riemannian Comparison Theorems for Contact Manifolds (IAS preprint: 1993).

[Je-Sa] Jerison D. and Sánchez-Calle A., Subelliptic, second order differential operators. Lect. Notes in Math. **1277**, pp. 46–77, Springer-Verlag 1987.

[Jo] Jost J., Equilibrium Maps between Metric Spaces. Ruhr-Universität Bochum, Preprint.

[Kar] Karcher H., Riemannian center of mass and mollifier smoothing. Comm. Pure and Appl. Math. **30** (1977), 509–541.

[Kari] Karidi R., Geometry of balls in nilpotent Lie groups. Duke-Math.-J. **74** (1994), no. 2, 301–317.

[Kor] Korevaar N., Upper bounds for eigenvalues of conformal metrics. J. Differential Geometry **37** (1993), pp. 73–93.

[Kor-Rei] Koranyi A. and Reimann H.M., Foundations for the theory of quasiconformal mappings on the Heisenberg group. Preprint 1991.

[La] Lalonde F., Homologie de Shih d'une submersion (homologies non singulières des variétés feuilletées). Supplément au Bulletin de la Société Mathématique de France. 1978, Tome **115**, Fascicule 4.

[Lee] Lee Y.I., The metric properties of Lagrangian surfaces. Dissertation of Stanford University for the degree of doctor of philosophy. 1992.

[Mar] Margulis G., Discrete groups of motions of manifolds of non-positive curvature (Russian), ICM 1974. Translated in A.M.S. Transl. (2) **109** (1977), pp. 33–45.

[Mit$_1$] Mitchell J., A Local Study of Carnot-Carathéodory Metrics. Ph.D. Thesis. Stony Brook 1982.

[Mit$_2$] Mitchell J., On Carnot-Carathéodory metrics, J. Differ. Geom. **21** (1985), pp. 35–45.

[Mont] Montgomery R., Survey of singular geodesics, this volume.

[Most] Mostow G.D., Strong rigidity of symmetric spaces, Ann. Math. Studies **78**, Princeton (1973).

[N-S-W] Nagel A., Stein E.M. and Wainger S., Balls and metrics defined by vector fields I : Basic properties, Acta Math. **155**, pp. 103–147 (1985).

[Nag] Nagata J-I., Modern dimension theory, North-Holland, 1965.

[Pan$_{CBN}$] Pansu P., Croissance des boules et des géodésiques fermées dans les nil-variétés, Ergod. Th. dynam. Syst. **3** (1983), pp. 415–445.

[Pan$_{InIs}$] Pansu P., Une inégalité isopérimétrique sur le groupe d'Heisenberg, C.R. Acad. Sci. Paris **295** (1982), pp. 127–131.

[Pan$_{QIR1}$] Pansu P., Métriques de Carnot-Carathéodory et quasi-isométries des espaces symétriques de rang un, Annals of Maths. **129**: 1 (1989), pp. 1–61.

[Pan$_{QM}$] Pansu P., Quasiconformal mappings and manifolds of negative curvature, in "Curvature and Top of Riemannian Manifolds" (Shiohama et al., eds), Lect. Notes in Math. **1201** (1986), pp. 212–230, Springer-Verlag.

[Pan$_{Th3}$] Pansu P., Géometrie du groupe d'Heisenberg. Thèse Univ. Paris VII, 1982.

[Pel-Bou] Pelletier F., et Valére Bouche L., Le problème des géodésiques en géométrie sous-riemannienne singulière-II, C.-R.-Acad.-Sci.-Paris-Ser.-I-Math. **317** (1993), no. 1, 71–76.

[Pes] Pesin Ya. B, Dimension type characteristics for invariant sets of dynamical systems. Uspekhi Mat. Nauk **43**: 4 (1988), pp. 95–128.

[Pit] Pittet C., Isoperimetric Inequalities in Nilpotent Groups. Preprint 1994.

[Rash] Rashevski P.K., About connecting two points of complete nonholonomic space by admissible curve, (in Russian), Uch. Zapiski ped. inst. Libknexta, no. **2**, pp. 83–94, (1938).

[Ru] Ruh E.A., Almost Lie groups, Proc ICM-1986, Berkeley, pp. 561–564, AMC 1987.

[Rum$_1$] Rumin M., Formes différentielles sur les variétés de contact. Thèse de Doctorat. Orsay 1992.

[Rum$_2$] Rumin M., Un complexe de formes différentielles sur les variétés de contact, C.R. Acad. Sci. Paris, t. **310** (1990), serie I., pp. 401–404.

[Sar] Sarychev A.A.V., On the homotopy type of the spaces of trajectories of nonholonomic dynamic systems, Dokl. Acad. Sci. USSR, v. **314** (6), 1990.

[Sch-Uhl] Schoen R. and Uhlenbeck K., A regularity theory for harmonic maps. J. Differential Geom., **17** (1982), pp. 307–335.

[Sin] Singer I.M., Infinitesimally homogeneous spaces, Comm. Pure Appl. Math. **13**, pp. 685–697, 1960.

[Ste] Stein E.M., Singular integrals and differentiability properties of functions, Princeton Univ. Press (1970).

[Stri] Strichartz R.S., Sub-Riemannian Geometry, J. Differential Geometry **24** (1986), pp. 221–263.

[Suss] Sussmann H.J., Abnormal sub-Riemannian minimizers, this volume.

[Th] Thom R., Remarques sur les problèmes comportant des inégalités différentielles globales, Bull. Soc. Math. France **87** (1959), pp. 455–461.

[Uhl] Uhlenbeck Karen K., Connections with L^p Bounds on Curvature. Commun. Math. Physics. **83**, pp. 31–42 (1982).

[Var] Varopoulos N. Th., Small Time Gaussian Estimates of Heat Diffusion Kernels. II. The Theory of Large Deviations. J. of Functional Analysis, Vol **93**, No 1. (1990).

[Var-Sa-Co] Varopoulos N. Th., Saloff-Coste L., Coulhon Th., Analysis and Geometry on Groups, Cambridge University Press (1993).

[Ver-Ger] Vershik A.M., Gershkovich V. Ya., nonholonomic Manifolds and Nilpotent Analysis, J. Geom. and Phys. vol. **5**, no. 3, (1988).

[Ver-Ger] Vershik A.M., Gershkovich V. Ya., nonholonomic Geometry and Nilpotent Analysis, J. Geom. and Phys. 5 3 (1989), pp. 407–452.

[Vin$_1$] Vinogradov A.M., Geometry of nonlinear differential equations, J. Soviet Math. **17** (1981), pp. 1624–1649.

[Vin$_2$] Vinogradov A.M., The C-spectral sequence, Lagrangian formalism and conservation laws I, II, J. Math. Anal. Appl. **100** (pp. 1–129), 1984.

[Vin$_3$] Vinogradov A.M., Local symmetries and conservation laws, Acta Applicandae Mathematicae **2** (21–78), 1984.

Progress in Mathematics, Vol. 144, © 1996 Birkhäuser Verlag Basel/Switzerland

Survey of singular geodesics

RICHARD MONTGOMERY[*]

The existence of singular minimizers in the calculus of variations has been known at least since the time of Carathéodory. Several authors had claimed that such minimizers do not exist in the context of sub-Riemannian geometry. Recently the author gave an example where the singular extremals are minimizers and are stable under perturbation. This demonstrates that they are of central importance to sub-Riemannian geometry. We describe the example and the current state of knowledge regarding singular geodesics. Several open problems are posed.

1. Introduction

We begin with some relevant definitions and notation. A *sub-Riemannian structure* on a n-dimensional manifold is a smoothly varying distribution of k-planes together with a smoothly varying inner product on these planes. The *dimension* of the sub-Riemannian manifold is the pair (k, n). The manifold is denoted by Q, and the distribution by D, $D \subset TQ$. The inner product will be written $\langle \cdot, \cdot \rangle$. A path will be called *horizontal* if it is absolutely continuous and its derivatives lie in D wherever they exist. We define the length ℓ of such a path in the usual Riemannian manner:

$$\ell(\gamma) = \int \sqrt{\langle \dot{\gamma}(t), \dot{\gamma}(t) \rangle} dt.$$

The sub-Riemannian distance $d(q_0, q_1)$ between two points q_0 and q_1 is also defined as in Riemannian geometry:

$$d(q_0, q_1) = \inf(\ell(\gamma)),$$

where the infimum is taken over all horizontal paths which connect q_0 and q_1. The distance is taken to be infinite if there is no such path. In this manner, every sub-Riemannian manifold is a metric space. Chow's theorem asserts this distance is finite if the manifold is connected and the distribution is bracket generating. We refer the reader to other contributions to this volume for precise statements of Chow.

---

[*]Mathematics Department University of California, Santa Cruz.
Santa Cruz, CA, 95064. U.S.A.

Definition 1. A path which realizes the distance between its endpoints is called a *minimizing geodesic* or simply a *minimizer*

Remark 1. Gromov (in this volume) and a number of other authors call sub-Riemannian geometries Carnot-Carathéodory metrics.

As in Riemannian geometry, the problem of finding sub-Riemannian geodesics can be formulated in several equivalent ways. Instead of minimizing the length we could minimize the time it takes to travel between the two points, subject to the constraint that the speed of travel $\sqrt{\langle \dot\gamma(t), \dot\gamma(t) \rangle}$ is less than or equal to 1. Or we could fix the time interval to be $[0,1]$ (or any other interval) and minimize the integrated kinetic energy $\frac{1}{2}\langle \dot\gamma(t), \dot\gamma(t) \rangle dt$. We will use the last formulation here.

Our problem then becomes to find a path $\gamma : [0,1] \to Q$ which minimizes the integral

$$E(\gamma) = \frac{1}{2} \int_0^1 \langle \dot\gamma, \dot\gamma \rangle dt$$

subject to the constraints:

(a) $\dot\gamma(t) \in D$ whenever this derivative exists,

(b) $\dot\gamma$ is square integrable (and in particular the derivative $\dot\gamma(t)$ exists for almost all t),

(c) $\gamma(0) = q_0$,

(d) $\gamma(1) = q_1$.

Again, a solution to this problem is called a *minimizing sub-Riemannian geodesic* or simply a *minimizer*.

Let Ω_D, $\Omega_D(q_0)$, and $\Omega_D(q_0, q_1)$ be the space of all curves satisfying the constraints (a)-(b), respectively (a)-(c), and (a)-(d). These path spaces do not depend on the choice of inner product $\langle \cdot, \cdot \rangle$ on D for if $\dot\gamma$ is square integrable with respect to one smooth metric on D then it is integrable with respect to any other. Now Ω_D and $\Omega_D(q_0)$ can be given the structure of Hilbert manifolds as is easily seen by reformulating the minimization question in the standard language of optimal control. To do this choose a (local) framing of D say $X_1, X_2, \ldots, X_k$. These are linearly independent vector fields which span D pointwise and which we may suppose to be orthonormal. The sub-Riemannian geodesic problem is equivalent to the optimal control problem whose underlying system is

$$\dot q = \Sigma_{a=1}^k u^a(t) X_a(q) \tag{1}$$

and whose cost functional is

$$E = E[u] = \int_0^T \frac{1}{2}\Sigma(u^a(t))^2 dt. \tag{2}$$

The problem is to find controls $t \to u(t) = (u^1(t), \ldots, u^k(t))$, $0 \le t \le 1$, which steer between the two given points q_0 and q_1 in time 1, and in such a way as to minimize E over all such controls. The control vector $u = (u^1, \ldots, u^k)$ at time t is of course just the coordinates of $\dot{\gamma}(t)$ relative to the framing. Since we want $E < \infty$ the path $u(t)$ should be in the space $L_2 = L_2([0,1], \mathbb{R}^k)$ of square integrable controls. The manifold structure on the path spaces Ω_D, $\Omega_D(q_0)$, is thus defined by putting the L_2 topology on the space of controls $u(t)$. We will call this topology the H_1-topology since it is restriction of the usual H^1 topology on all curves.

Now we introduce the endpoint map.

Definition 2. The **endpoint map** is the map:

$$\text{end} = \text{end}_{q_0} : \Omega_D(q_0) \to Q$$

which assigns to each curve its endpoint: $\text{end}_{q_0}(\gamma) = \gamma(1)$.

Thus $\Omega_D(q_0, q_1) = \text{end}_{q_0}^{-1}(q_1)$ and so our problem can be reformulated as minimizing $E(\gamma)$ subject to the constraint $\text{end}_{q_0}(\gamma) = q_1$.

It is well-known that end is a smooth map and its derivative $d(\text{end})$ has been calculated in a number of places. This derivative can calculated by applying the variation of parameters formula from ODEs to system 1. See for example the text by Pontrjagin et al [27], the first chapter of Bismut's book [3], the top of p. 57 of Sontag's text [29], or numerous articles by Agrachev and Gramkrelidze-Agrachev [1].

Definition 3. A *singular curve* $\gamma \in \Omega_D$ is a singular point of $\text{end}_{\gamma(0)}$. A curve $\gamma \in \Omega_D$ is *regular* if it is not singular, i.e. if end is a submersion at γ. A minimizing geodesic is called regular or singular if it is regular or singular as a curve.

Clearly the property of being singular depends only on the distribution D and not at all on the choice of cost L.

Question 1. Is every minimizing geodesic regular?

Answer. No. A counterexample is provided in the next section.

Our example is actually stronger than this. The regular minimizing geodesics are characterized by a system of ODEs which are quite similar to the Riemannian geodesic equations. We will call this the sub-Riemannian geodesic equations. Now in Riemannian geometry every minimizing geodesic solves the geodesic equations. Our example shows that this is false in sub-Riemannian geometry: there are minimizing geodesics which do not solve the sub-Riemannian geodesic equations. Moreover they are topologically stable under perturbations of the distribution and the metric.

It will help to give the finite-dimensional version of this phenomenon. Recall the method of Lagrange multipliers. Suppose we are trying to minimize a function $f : \mathbb{R}^n \to \mathbb{R}$ subject to a constraint $G = c$ where $G = (G_1, \ldots, G_k) : \mathbb{R}^n \to \mathbb{R}^k$ is a vector valued function, c is a constant vector, and $k < n$. Of course G plays the role of the endpoint map for studying sub-Riemannian geodesics. The method of Lagrange multipliers tells us to consider the system of $n + k$ equations:

$$\lambda_0 df(x) + \Sigma \lambda_i dG_i(x) = 0 \; ; \; G_i(x) = c_i \tag{3}$$

in the variables (x, λ) where we have introduced the *Lagrange multipliers* $\lambda = (\lambda_0, \lambda_1, \ldots, \lambda_k)$ and df, dG_i denote the (total) differentials of the corresponding functions and so are covectors in $\mathbb{R}^n$. ($df = \Sigma_i \frac{\partial f}{\partial x^i} dx^i$.) The multipliers must satisfy $\lambda \neq (0, 0, \ldots, 0)$. The method asserts that any minimizing x must be the first part of a solution (x, λ) to this system.

Often we only teach our calculus students the case where $\lambda_0 = 1$, this being equivalent to the case $\lambda_0 \neq 0$ upon dividing by λ_0. But the solutions (x, λ) with $\lambda_0 = 0$ cannot be ignored. To understand their meaning, note that for such a point $(\lambda_1, \ldots, \lambda_k) \neq (0, \ldots, 0)$ annihilates the image of $dG(x)$. (Here, $dG(x)$ denotes the Jacobian matrix of the transformation G. It is the matrix with ith row $dG_i(x)$.) Now recall that a point x is called *regular* for the mapping G if the differential $dG(x)$ is onto, and *singular* if it is not. Also recall the implicit function theorem which asserts that if x is a regular point with $G(x) = c$ then the surface constraint set $G = c$ is a smooth $n - k$ dimensional manifold in a neighborhood of x. *Thus the solutions (x, λ) for which $\lambda_0 = 0$ correspond to the singular points of G and are possible nonmanifold points for G.* In other words: insisting that $\lambda \neq 0$ is the same as ignoring the nonmanifold points of the constraint set.

Definition 4. A solution (x, λ) to the Lagrange multiplier equations is called a *regular extremal* or *normal extremal* if the multiplier λ satisfies $\lambda_0 \neq 0$. It is called a *singular* or *abnormal extremal* if it satisfies $\lambda_0 = 0$. A minimizer x for the constrained problem is called *strictly singular* if every nonzero multiplier λ for which (x, λ) solves the multiplier system satisfies $\lambda_0 = 0$.

The sub-Riemannian geodesic equations characterize the normal extremals in our situation. There is a slight subtlety here alluded to at the end of the definition. This has to do with the fact that one minimizer x can correspond to a number of different extremals (x, λ). If we call a minimizer x singular or regular according to whether or not the corresponding solution (x, λ) to the multiplier system is regular or singular then a minimizer can be both regular and singular. For it could have one multiplier with $\lambda_0 = 0$ and another with $\lambda_0 \neq 0$.

It is good to have a simple example in mind. Suppose that the set $\{G = c\}$ is the upper half $z \geq 0$ of the standard cone $x^2 + y^2 = z^2$ in $\mathbb{R}^3$ including the vertex $z = 0$. Take f to be any linear function for which the angle between $\mathrm{grad}(f)$ and the z-axis is less than 45 degrees. Then the cone point will be a strictly singular minimizer. If the angle is exactly 45 degrees then the minimizer is not strict and the minimizing plane $f = 0$ is tangent to the cone along a generator. In this case the cone point 0 is simultaneously regular and singular since it sits under extremals $(0, \lambda)$ of both types. This property of being both regular and singular is clearly a lucky accident of the choice of cost function f.

1.1. The differential equations, Pontrjagin's principle and the main result.

The method of Lagrange multipliers as applied to problems in optimal control, and hence to our sub-Riemannian geodesic problem, is the Pontrjagin maximum principle [27]. The regular extremals are called either regular or normal extremals by practitioners of the Pontrjagin principle. They can be characterized as the solutions to a certain Hamiltonian system of ODEs on the cotangent bundle T^*Q of Q. This is the system referred to above as the sub-Riemannian geodesic equations. The singular extremals are called either singular or abnormal extremals by practitioners of the Pontrjagin principle. (For systems with drift, that is where D is an affine subspace not passing through the zero section, the notions of abnormal and singular no longer coincide. For a thorough discussion of this point see the appendix to the paper of Liu and Sussmann).

We will recall the basic elements of the Pontrjagin principle. The multipliers for the problem consist of pairs

$$(\lambda_0, p(t)) \neq (0,0), \lambda_0 \in \mathbb{R}, \lambda_0 \geq 0, p(t) \in T^*_{\gamma(t)}Q, 0 \leq t \leq 1.$$

The curve $t \to (\gamma(t), p(t))$ is to be an absolutely continuous curve.

Let X_a be the orthonormal framing for D as above and introduce the fiber-linear functions $P_a : T^*Q \to \mathbb{R}$ by

$$P_a(q,p) = p(X_a(q)).$$

In other words, think of the X_a as fiber linear functions on the dual vector bundle. (Here $p \in T^*_qQ$.) Form $\mathcal{H} = \lambda_0 \frac{1}{2}\Sigma(u^a)^2 - \Sigma u^a P_a$ According to the maximum principle, we minimize this expression over the controls u, for fixed λ_0 and fixed $(q,p) \in T^*Q$ thus obtaining a Hamiltonian of q and p alone. The resulting solutions $(q(t), p(t))$ to Hamilton's equations are called Pontrjagin extremals and the theorem of Pontrjagin and coworkers asserts that the minimizers are among their projections $q(t)$.

The two cases of regular and singular can be reduced to the cases $\lambda_0 = 1$ and $\lambda_0 = 0$. In the regular case we obtain:

$$H = H_{\text{regular}} = \frac{1}{2}\Sigma P_a^2.$$

This is a rank k fiber quadratic form on the cotangent bundle. Note that if $k = n$ this is exactly the Hamiltonian governing geodesic flow for a Riemannian metric. In the second case the P_a must all be identically zero along the curve, for otherwise the quantity $\mathcal{H}$, being linear, has no minima! The Hamiltonian is identically zero, but there are still some dynamics. The singular extremals are the solutions to Hamilton's equations for the Hamiltonian $\Sigma u^a P_a$ subject to the constraint P_a are identically zero.

There is a geometric way of saying what the singular extremals are which we learned from Lucas Hsu [16]. The constraint $P_a = 0$ says that the curve of covectors $p(t)$ annihilates D at each time t. Let $D^\perp \subset T^*Q$ be the bundle of covectors which annihilate D. Let ω be the restriction of the canonical two-form ($\Sigma dp_i \wedge dq^i$ in canonical coordinates) to $D^\perp$. A curve $\zeta(t) = (q(t), p(t))$ in $D^\perp$ is called characteristic if it is in the kernel of ω, that is, if for all t and all vectors v tangent to $D^\perp$ at $\gamma(t)$ we have $\omega(\dot\zeta, v) = 0$. Hsu's result asserts that the smooth singular extremals are exactly the characteristics of $D^\perp$. His result can be extended without serious difficulty to the case of interest here in which ζ is only absolutely continuous.

In order to continue the discussion it is best to restrict D to be a bracket-generating distribution. This is because if D is not bracket generating then every curve is singular. For in this case Q is foliated by leaves such that every horizontal curve must lie in one such leaf. Any curve of covectors annihilating the tangent to the leaf provides a singular extremal.

Main result. We give we the first example of a bracket-generating distribution which admits a minimizing curve which does not satisfy the geodesic equations. In other words, the curve is an example of a strictly singular minimizing geodesic. Our curve is globally minimizing but may be very short. In contrast, other authors studying this phenomenon have investigate the phenomenon of being locally minimizing within the space of curves (say with our topology). Said differently, our example is a local-in-time but global-in-path-space minima. The example is topologically stable in the sense that it persists under perturbation of both the distribution (control system) and metric (cost).

Our example provides a counterexample to theorems asserting that all minimizers satisfy the geodesic equations (i.e. come from regular extremals) found in papers of Rayner, Strichartz, Taylor, Hamenstädt, and in the diplomarbeit of C. Bär. See [28],[30], [33],[14]. To be fair, Strichartz retracted his proof [31], and Hamenstädt was simply quoting Bär [2].) Gaveau [10] had claimed to have a counterexample to these theorems before us, but Brockett [5] found an irreparable error in it.

A careful reading of Hermann [15] and of Bismut [3] suggests that these authors believed examples such as ours may have existed. Examples similar in spirit to ours are well known for other problems in optimal control.

2. The example and its properties

2.1. The basic example. Let x, y, z be coordinates on a three-dimensional space. Consider the distribution D defined by the Pfaffian system

$$dz - y^2 dx = 0. \tag{4}$$

Alternatively, consider the control system:

$$\dot{x} = u$$

$$\dot{y} = v \tag{5}$$

$$\dot{z} = y^2 u.$$

This distribution is bracket-generating everywhere. It is not regular which means different numbers of Lie brackets are required to generate the entire tangent space $\mathbb{R}^3$ at different points. It is of contact type off of the plane $y = 0$ and so one Lie bracket suffices to generate. To see this set $\theta = dz - y^2 dx$ and recall that the contact condition can be formulated $\theta \wedge d\theta \neq 0$. Now

$$\theta \wedge d\theta = 2y dx \wedge dy \wedge dz \tag{6}$$

which is zero precisely when $y = 0$. On the plane $y = 0$ we need one extra Lie bracket so the growth vector is $(2, 2, 3)$ at these points. The singular curves are the horizontal curves which lie in the plane $y = 0$. They are line segments of the form

$$y = 0, \; z = z_0 = (\text{constant}), \; x_0 \leq x \leq x_1. \tag{7}$$

(See [23], or [20], or for more details.)

The general metric on this distribution can be expressed in the form $ds^2 = E dx^2 + 2F dx dy + G dy^2$ with $EG - F^2 > 0$, $E, G > 0$, and E, G, F functions of x,y,z. Thus the sub-Riemannian geodesic problem for our distribution is to minimize $\int_0^\tau ds^2$ among all paths $q(t) = (x(t), y(t), z(t))$, $0 \leq t \leq \tau$, satisfying the condition $\dot{z} = y^2 \dot{x}$ and the end point conditions $q(0) = (0, 0, 0)$, $q(\tau) = q_1 = (x_1, 0, 0)$, with $x_1 \neq 0$.

Theorem 1. (Montgomery, [20]). *The singular curves (7) are locally minimizing geodesics (solutions to the above problem) for any sub-Riemannian structure whose underlying distribution is defined by equation (4)and whose metric has the form $ds^2 = e(y)^2 dx^2 + g(y) dy^2$. Specifically, there is a positive number $\tau_* = \tau_*(e, g)$ such that for any $\tau < \tau_*$ the arc $(t, 0, z_0), 0 \leq t \leq \tau$ of the singular curve passing through $(0, 0, z_0)$ is the unique minimizing sub-Riemannian geodesic between its endpoints. These curves satisfy the geodesic equations if and only if $de/dy|_{y=0} = 0$.*

Since the author's original proof Liu found an integral inequality which leads to a much simpler proof valid for general metrics.

Theorem 2. (Liu and Sussmann; [18]) *The singular curves (7) are locally minimizing geodesics (in the same sense as the previous theorem) for any sub-Riemannian structure whose underlying distribution is defined by equation (4).*

Ivan Kupka [17] gave another proof of theorem 1 similar in spirit to the authors. The structure of both proofs is the following. The only other candidate minimizers are the regular extremals which solve specific differential equations whose coefficients depend on e and g, namely the sub-Riemannian geodesic equations. These equations are analyzed in an asymptotic limit and it is shown by direct calculation that all solutions satisfying the endpoint conditions are longer. In short, we sort through all competing regular extremals. The arguments involved are quite lengthy. The author's analysis is based on the fact that the differential equations can be interpreted as the equations of a planar charged particle in a magnetic field. This provides enough intuition to do the necessary analysis. Kupka chooses a simple form of the metric for which the geodesic equations can be solved by elliptic functions then analyzes the resulting elliptic integrals describing the endpoint conditions and lengths. At the time of this conference Liane Valère [34] obtained to a simple direct analytic proof based on inequalities of the fact that these singular curves are locally-in-H^1-minimizing. Since the conference Wensheng Liu and Hector Sussmann [18] proved Theorem 2, the case of general metrics stated above. Their proof generalizes to apply to generic singular curves for sub-Riemannian structures of dimension $(2, n)$ instead of just $2, 3$). (See the "note in proof".)

2.2. Genericity. The distribution of eq.(1) is topologically stable, the meaning of which we will momentarily explain. This was proved by Martinet [19]. Consider any rank 2 distribution D on a 3-manifold. Locally D can be defined by the Pfaffian system:

$$\alpha = 0$$

for some nonvanishing one-form α. Write $\alpha \wedge d\alpha = f d^3 x$ where $d^3 x$ is a (local) volume form and f is a function. Suppose that at some point p the function f satisfies

$$f(p) = 0, \quad df(p) \neq 0.$$

Martinet's theorem [19] says that there exist coordinates centered at p such that D is defined by the equation (4). In particular, Martinet's theorem implies that the set of distributions which can be put into the local form (4) forms an open set relative to the Whitney-C^2 topology on the space of distributions. This is the meaning of stable.

The geometric significance of the singular curves (7) is as follows: Consider the surface $\{f = 0\}$. This is the locus of points at which the distribution D fails to be contact. D intersects this surface transversally thus defining a line-field on it. The lines (7) are the lines of this field.

We take a moment in order to translate from the language of forms into the dual language of vector fields. The condition on the one-form α is equivalent to saying that the distribution D admit a frame X, Y of vector fields near p such that their Lie bracket $[X, Y]$ at p depends linearly on $X(p)$ and $Y(p)$ but for which $[X, [X, Y]](p)$ is linearly independent. Moreover if we define the function f by $f = dx \wedge dy \wedge dz(X, Y, [X, Y])$ then $df(p) \neq 0$. Alternatively, f is the triple product $(X \times Y) \cdot [X, Y]$.

2.3. Rigidity. Our singular curves have a surprising property called C^1-rigidity. I believe it is one of the underlying reasons behind our Theorem. The term C^1-rigid was coined by Bryant and Hsu [8].

Definition 5. A C^1 integral curve c of a distribution D is called C^1-rigid if every sufficiently C^1-close curve to c which is an integral curve of D and shares the same endpoints as c is a reparametrization of c.

In other words, a C^1-rigid curve is an isolated point in the space of all unparametrized D-curves with fixed endpoints.

Let $q(t) = (x(t), y(t), z(t))$ be any curve satisfying our control law $dz - y^2 dx = 0$ and having its endpoints $q(0) = (x_0, 0, z_0)$, $q(\tau) = (x_1, 0, z_0)$ in common with one of the lines $\ell(t) = (x_0 + t, 0, z_0)$ of equation (7). Then

$$z_1 - z_0 = 0 = \int dz = \int_c y^2 dx = \int_0^\tau y(t)^2 \frac{dx}{dt} dt.$$

In particular if $\frac{dx}{dt} \geq 0$ then $y(t) \equiv 0$ and consequently q is a reparametrization of the line ℓ! We have just proved that the singular curves of the theorem are C^1-rigid.

Being an isolated point, a C^1-rigid curve is automatically a *local* minimum (and maximum!) *relative to the C^1-topology* for any functional on the space of all D-curves with its endpoints. Here "local" means local in the C^1 topology on the space of D-curves. *The central difficulty in proving Theorem 1 is that the singular curves are not C^0 or even Sobolev H^1 rigid.* In fact, for any bracket generating distribution there are no D-curves which are isolated points relative to these other topologies.

Liu and Sussmann have found an example of a bracket-generating distribution on $\mathbb{R}^3$ with a C^1-rigid curve which is not locally minimizing. However its form is topologically unstable. It seems likely that with some additional stability assumptions every C^1 rigid curves are locally minimizing.

2.4. Higher dimensions. Perhaps the most interesting rank 2 distribution occurs in dimension 4. It is stable in the same sense as described above. It is regular with growth vector $(2, 1, 1)$. It is the only stable regular distribution besides the contact distributions. See Gershkovich and Vershik [35], [36] and also [21] regarding this fact. It admits a local frame X, Y whose brackets define the free 2-step nilpotent Lie algebra on 2 generators X, Y:

$$[X, Y] = Z, [Y, Z] = W.$$

This algebra is sometimes called the Engel algebra and we call such a distribution an Engel distribution, or Engel structure. The C^1-rigid curves of an Engel distribution are the integral curves of the vector field X. (The vector field X is special because it does not generate the whole 4-dimensional tangent space under bracketing with D.) For more discussion and a proof of the stability of the Engel distribution see p. 50 of [6]. For a detailed account of global properties of Engel distributions see Gershkovich's forthcoming monograph [12].

A special case of the following theorem was proved by the author [20].

Theorem 3. (Sussmann) *The C^1-rigid curves of an Engel distribution are locally minimizing geodesics for any sub-Riemannian structure on this distribution.*

Probably the simplest geometric model for the Engel distribution is as the space M^4 of *contact line elements* for contact distribution E on a three-manifold Q. E can be expressed locally as the Pfaffian system:

$$dz - ydx = 0. \tag{8}$$

Let w denote the "slope" of a contact line (a line $\ell \subset E_{(x,y,z)}$) relative to the linear coordinates dy and dx on $E_{(x,y,z)}$. In other words, $w = \frac{dy}{dx}$ is the affine coordinate on the projective line $\mathbb{P}E_{(x,y,z)}$. Then the Engel distribution D is defined by adding the equation:

$$dy - wdx = 0$$

to the Pfaffian equation which defines E. The set of tangents to any regular E-curve (Legendrian curve) defines a curve in M and these are precisely the (projectable) D-curves.

The Engel frame X, Y is $Y = \frac{\partial}{\partial x} + y\frac{\partial}{\partial z} + w\frac{\partial}{\partial y}$ and $X = \frac{\partial}{\partial w}$ relative to these coordinates. Thus the vertical curves (rotate the contact line, keeping its point of contact on the 3-manifold fixed) are the singular curves in this example. We thank Robert Bryant for describing this geometric model to us.

In 5 dimensions there are rank 2 distributions D with the property that tangent to every *direction* $v \in D$ there is a C^1 rigid curve. The author has verified his conjecture for a very special class of sub-Riemannian structures having this type of underlying distribution. Robert Bryant has observed that such distributions are generic in these dimensions. Elie Cartan wrote an inspirational paper [9] on such distributions.

Regarding general rank 2 distributions on n-dimensional spaces, $n \geq 4$ we have:

Theorem 4. (Bryant-Hsu)[8] *Let D be a rank 2 distribution on a space of dimension 4 or greater. Suppose that $D + [D, D]$ is not involutive at a point p. Then there is a C^1-rigid curve passing through p. Moreover, there exist coordinates $w, x, y, z, v_1, v_2, \ldots v_{n-4}$ centered at p and a local framing W, X of D such that $W = \frac{\partial}{\partial w}$, $X = \frac{\partial}{\partial x} + y\frac{\partial}{\partial z} + w\frac{\partial}{\partial y} + \Sigma F_i \frac{\partial}{\partial v_i}$.*

3. Some open questions

1. Is every minimizing geodesic on a smooth sub-Riemannian manifold smooth?

2. Can you hear a singular curve? More specifically, if $\triangle$ is a sub-Laplacian whose underlying distribution admits C^1-rigid curves, do the existence or lengths of these singular curves show up in the spectrum or heat kernel of $\triangle$?

3. How do you find conjugate points along a locally minimizing singular geodesic?

   Kupka and collaborators have begun a numerical study of this question. It is possible that recent results of Agrachev and Sarychev have some bearing on this question.

4. What conditions guarantee that a C^1-rigid curve is locally minimizing for any metric? What about other singular curves?

Acknowledgments. I would like to thank Robert Bryant, Lucas Hsu, and Héctor Sussmann for valuable discussions. I would like to express my deep thanks to André Bellaïche and J.-J. Risler for inviting me to this conference and for their hospitality while in Paris.

4. Note in proof

There have been a number of developments since this paper was submitted two years ago. Liu and Sussmann have proved that all of the rigid curves found by Bryant and Hsu (plus a few extras) are locally minimizing in the sense of Theorem 1 above. In regards to the open questions all but one of them have been answered to a significant extent.

Questions 3 and 4 have been answered fairly completely by Agrachev and Sarychev. They show that for an extremal to be C^1-rigid it must satisfy $p(D+[D,D]) = 0$ and that the quadratic form $p([v,[\dot{\gamma},v]]$, $v \in D$ must be positive along γ. These conditions guarantee local-in-H^1 minimality. The conjugate point or Morse theory of singular curves is formulated in symplectic-geometric terms and is quite elegant.

Question 2 has been answered affirmatively by the author for the case of the original example and more generally for metrics of the form $E(x,y)dx^2+2F(x,y)dxdy+G(x,y)dy^2$. See [24]. The associated subLaplacian in 3-space decomposes under translation in the z-variable, assumed periodic, into the direct sum of a countable collection of 2-dimensional covariant Laplacians. These Laplacians are the Schrödinger operators governing the behaviour of a charged planar quantum particle of spin zero in the magnetic field whose zero locus is the projection in the xy plane of the singular curves. The operator direct sum is indexed by the particle's charge which can be interpreted as the Fourier variable dual to z. The singular geodesics turn out to dominate the spectral asymptotics of the sub-Laplacian in the correct limit which is the charge tending to infinity. The effect of the singular geodesics on the heat kernel has not yet been studied.

Question 1 remains open.

References

[1] A. Agrachev and A. Sarychev, "Abnormal sub-Riemannian geodesics: Morse Index and Rigidity", preprint, 1993.

[2] C. Bär, "Carnot-Carathéodory-Metriken", Diplomarbeit. Bonn, 1988.

[3] J.-M. Bismut, "Large Deviations and the Malliavin Calculus", Birkhäuser, 1984.

[4] G.A. Bliss, "Lectures on Calculus of Variations", *Univ. of Chicago Press*, 1946.

[5] R.W. Brockett, "Nonlinear Control Theory and Differential Geometry", *Proc. of the Int. Congress of Mathematicians*, Warszawa, 1983.

[6] R. Bryant, S. Chern, R. Gardner, H. Goldschmidt, P. Griffiths, "Exterior Differential Systems", M.S.R.I Publications, **18**, Springer-Verlag, 1991.

[7] C. Carathéodory, "Untersuchungen uber die Grundlagne der Thermodynamik", *Math. Ann.*, **67**, 355-386, 1909.

[8] R. Bryant and L. Hsu, "Rigidity of Integral Curves of Rank Two Distributions", *Invent. Math.*, **114**, 435–461, 1993.

[9] E. Cartan, "Les Systèmes de Pfaff à cinq variables et les équations aux dérivées partielles du second ordre", *Ann. École Norm. Sup.*, **27**, 3, 1910; also in his collected works, vol. 2, pp. 927.

[10] B. Gaveau, "Principe de moindre action, propagation de la chaleur, et estimées sous-elliptiques sur certains groupes nilpotents", *Acta Math.*, **139**, 95–153, 1977.

[11] Ge Zhong, "On a constrained variational problem and the space of horizontal paths", *Pac. J. Math.*, **149**, 61–94, 1993.

[12] V. Gershkovich, "Global Properties of Engel Distributions", preprint, Un. of Melbourne, Australia, 1993.

[13] M. Gromov, "Carnot-Carathéodory spaces seen from within", this volume.

[14] U. Hamenstädt, "Some Regularity Theorems for Carnot-Carathéodory Metrics", J.Diff.Geom., **32**, 819–850, 1990.

[15] R. Hermannn, "Some Differential Geometric Aspects of the Lagrange Variational Problem", *Indiana Math. J.*, 634–673, 1962.

[16] L. Hsu, "Calculus of Variations via the Griffiths Formalism", J.Diff.Geom., **36**, 3, 551–591, 1991.

[17] I. Kupka, "Abnormal Extremals", preprint, 1992.

[18] W.-S. Liu and H.J. Sussmann, "Shortest paths for sub-Riemannian metrics on rank two distributions", *Trans. A.M.S.*, 1994.

[19] J. Martinet, "Sur les singularités des formes différentielles", *Ann. Inst. Fourier*, **20**, 1, 95–178, 1970.

[20] R. Montgomery, "Abnormal Minimizers", *SIAM J. Control and Opt.*, **32** (1994), no. 6, 1605–1620.

[21] R. Montgomery, "Generic Distributions and Finite Dimensional Lie Algebras", *Journal of Differential Equations*, **103**, 387–393, 1993.

[22] R. Montgomery, "Singular Extremals on Lie Groups", to appear, MCSS, 1995.

[23] R. Montgomery "Abnormal Optimal Controls and Open Problems in Nonnonholon-holonomic Steering", NOLCOS conference proceedings, Bordeaux, France, 1992.

[24] R. Montgomery, "Hearing the zero locus of a magnetic field", *Comm. Math. Phys.*, 1994.

[25] M. Morse and S. Myers, "The Problems of Lagrange and Mayer with Variable Endpoints", *Proc. of the Am. Acad.*, **66**, 6, 236–253, 1931.

[26] P. Pansu, "Métriques de Carnot-Carathéodory et quasi-isométries des espaces symétriques de rang un", *Ann. of Math.*, **129**, 1–60, 1989.

[27] L. S. Pontrjagin, V. G. Boltyanskii, R. V. Gamkrelidze, and E. F. Mishchenko, "The Mathematical Theory of Optimal Processes", Wiley, Interscience, 1962.

[28] C. B. Rayner, "The Exponential Map for the Lagrange Problem on Differentiable Manifolds", *Phil. Trans. of the Royal Soc. of London*, ser. A, Math. and Phys. Sci., 1127, **262**, 299–344, 1967.

[29] E. D. Sontag, "Mathematical Control Theory", Springer-Verlag, New-York, 1990.

[30] R. Strichartz, "Sub-Riemannian Geometry", *J. Diff. Geom.*, **24**, 221–263, 1983.

[31] R. Strichartz, "Corrections to 'Sub-Riemannian Geometry' ", *J. Diff. Geom.*, **30**, 2, 595–596, 1989.

[32] H. J. Sussmann, "A Cornucopia of sub-Riemannian minimizers", this volume.

[33] T.J.S. Taylor, "Some Aspects of Differential Geometry Associated with Hypoelliptic Second Order Operators", *Pac. J. Math.* **136**, 2, 355–378, 1989.

[34] L. Valère, "The Geodesic Problem in sub-Riemannian Geometry", preprint, 1992.

[35] A.M. Vershik and V. Ya Gerhskovich, "An estimate of the functional dimension of the space of orbits of germs of generic distributions", *Math. USSR, Zametki*, **44:45**, 596–603, 1988.

[36] A.M. Vershik and V. Ya Gerhskovich, "Nonholonomic Dynamical Systems, Geometry of Distributions and Variational Problems", in *Dynamical Systems VII* ed. V.I. Arnol'd and S.P. Novikov, **16** of the Encyclopaedia of Mathematical Sciences series, Springer-Verlag, NY, 1994. (Russian original, 1987.)

Progress in Mathematics, Vol. 144, © 1996 Birkhäuser Verlag Basel/Switzerland

A cornucopia of four-dimensional abnormal sub-Riemannian minimizers[‡]

HÉCTOR J. SUSSMANN[*]

"The skull seems broken as with some big weapon, but there's no weapon at all lying about, and the murderer would have found it awkward to carry it away, unless the weapon was to small to be noticed."

"Perhaps the weapon was too big to be noticed," said the priest, with an odd little giggle.

Gilder looked round at this wild remark, and rather sternly asked Brown what he meant.

"Silly way of putting it, I know," said Father Brown apologetically. "Sounds like a fairy tale. But poor Armstrong was killed with a giant's club, a great green club, too big to be seen, and which we call the earth. He was broken against this green bank we are standing on."

"How do you mean?" asked the detective quickly.

Father Brown turned his moon face up to the narrow façade of the house and blinked hopelessly up. Following his eyes, they saw that right at the top of this otherwise blind back quarter of the building, an attic window stood open.

"Don't you see," he explained, pointing a little awkwardly like a child, "he was thrown down from there?"

> G.K. Chesterton, "The Three Tools Of Death," in *The Innocence of Father Brown*, The Father Brown Omnibus, Dodd, Mead & Co., New York (1983), p. 117.

Abstract. We study in detail the local optimality of abnormal sub-Riemannian extremals for a completely arbitrary sub-Riemannian structure on a four-dimensional manifold, associated to a two-dimensional bracket-generating regular distribution. Using a technique introduced in earlier work with W. Liu, we show that large collections of simple (i.e.

[‡]This work was done while the author was a visitor at Institute for Mathematics and its Applications (I.M.A.), University of Minnesota, Minneapolis, Minnesota 55455, U.S.A. A preprint of this paper, entitled "A cornucopia of abnormal sub-Riemannian minimizers, Part I: the four-dimensional case," is # 1073 of the I.M.A. Preprint series, December 1992.

[*]Department of Mathematics, Rutgers Un., New Brunswick, NJ 08903, U.S.A. Supported in part by the National Science Foundation under Grant DMS92-02554.

without double points) nondegenerate extremals exist, and are always uniquely locally optimal. In particular, we prove that the simple abnormal extremals parametrized by arc-length foliate the space (i.e. through every point there passes exactly one of them) and they are *all* local minimizers. Under an extra nondegeneracy assumption, these abnormal extremals are strictly abnormal (i.e. are not normal). (In the forthcoming paper [6] with W. Liu we show that in higher dimensions there are large families of "nondegenerate abnormal extremals" that are local minimizers as well. In dimension 3, for a regular distribution there are no nontrivial abnormal extremals at all, but if the distribution is not regular then, generically, there are two-dimensional surfaces that are foliated by abnormal extremals, all of which turn out to be local minimizers.) This adds up to a picture which is rather different from the one that appeared to emerge from previous work by R. Montgomery and I. Kupka, in which an example of an abnormal extremal for a nonregular distribution in $\mathbb{R}^3$ was studied and shown to be locally optimal with great effort, by means of a very long and laborious argument, and then this example was used to produce a similar one for a regular distribution in $\mathbb{R}^4$. All this may have given the impression that abnormal extremals are hard to find, and that proving them to be minimizers is an arduous task that can only be accomplished in some very exceptional cases. Our results show that abnormal extremals exist aplenty, that most of them are local minimizers, and that in some widely studied cases, such as regular distributions on $\mathbb{R}^4$, this is in fact true for *all of them*.

1. Introduction

Ever since the early work of Brockett [2] and Strichartz [9], [10] on sub-Riemannian geometry, it has been clear that sub-Riemannian minimizers fall into two not mutually exclusive categories, namely, the "normal" and "abnormal" extremals. Normal extremals are obviously smooth, and satisfy equations that in many ways resemble those of Riemannian geodesics. (In [9], it was stated that sub-Riemannian minimizers are necessarily smooth, and this was derived from an assertion equivalent to the proposition that all minimizers are normal extremals. Subsequently, it was noticed that the proof of this assertion involved an invalid application of the Pontrjagin Maximum Principle, and that a truly correct analysis based on this result from Optimal Control Theory implied the possibility that a minimizer might be "abnormal." In [10], it was pointed out that the result of [9] remained valid for a very restrictive class of sub-Riemannian manifolds, namely, those that obey the "strong bracket-generating condition.")

Until recently, it was not clear whether strictly abnormal extremals that actually are minimizers can exist. ("Strictly abnormal" means "abnormal and not normal," cf. below.) This question was answered in recent work by R. Montgomery [7], who gave one example of a sub-Riemannian structure in $\mathbb{R}^3$, associated to a two-dimensional subbundle E of the tangent bundle, for which there exists a strictly abnormal uniquely optimal extremal. (We call an admissible trajectory *optimal* if it minimizes length among all admissible trajectories with the same initial and terminal points, and *uniquely optimal* if it is the only optimal trajectory joining these two points, up to reparametrization of the time interval.) Montgomery's optimality proof is rather lengthy and involved, making it desirable to find simpler ways of establishing the result. I. Kupka provided in [3] a different proof, also quite lengthy, based on a detailed analysis of the solutions of the differential equation defining the normal extremals. The Montgomery-Kupka examples are for a two-dimensional distribution in $\mathbb{R}^3$ which of necessity cannot be regular (the definition of a "regular distribution" is given below), since regular 2-dimensional distributions in $\mathbb{R}^3$ are strongly bracket generating and hence have no abnormal extremals. But, starting from these examples, one can construct (essentially by adding an extra variable) examples of minimizing strictly abnormal extremals for a regular distribution in $\mathbb{R}^4$. However, due to the extreme complexity of the proofs, these results have failed to yield a true understanding of the real reason why the particular abnormal extremals considered there happen to be optimal, and have created the impression that abnormal extremals are very hard to find, and that proving them to be optimal may only be possible in some very exceptional situations, and may require very hard work and a large amount of luck.

In this note we shall attempt to correct that impression, by showing that abnormal extremals exist in large numbers, that most of them are optimal, and that there is a very simple technique—essentially due to W. Liu—for proving optimality for a very wide broad range of situations. To make our point clear, we will concentrate on the most dramatic case, which also happens to be the situation that is universally recognized as the simplest, namely, that of a *four-dimensional* manifold M with a sub-Riemannian metric arising from a *two-dimensional regular distribution* E. We will show that the following facts are *always*[14] true:

1. There is a line subbundle L of E such that the abnormal extremals are exactly the integrals curves of L.

2. All the abnormal extremals that are parametrized by arc-length and are simple (i.e. contain no loops) are locally optimal.

3. Optimality can be proved by a simple technique involving elementary inequalities.

4. All these abnormal extremals are actually strictly abnormal, provided that a simple generic condition (stated below) is satisfied.

In other words: to find optimal abnormal extremals for the simplest regular case (i.e. a 2-dimensional distribution in $\mathbb{R}^4$), one need not think hard and wonder where to look and how to select an example. The examples are everywhere, they are *all* locally optimal, and the proof of this fact just involves some elementary inequalities.

A crucial difference between our point of view and that of previous authors who have studied the problem is that we make systematic use of a control-theoretic approach, and in particular work with systems of vector fields and use properties of vector fields and Lie brackets to make appropriate choices of coordinate charts, rather than carry out calculations in terms of differential forms. It is our belief that the vector field formulations are more natural and geometric, and in addition are also better for effective calculation. In our view, the present paper provides support for this assertion. We hope the reader will be persuaded that abnormal extremals, which have appeared somewhat mysterious to geometers eager to pursue the analogy with Riemannian geometry, are not at all surprising to a mathematician who operates from an Optimal Control perspective, since a routine application of the Pontrjagin Maximum Principle leads to them immediately. Similarly, when one uses the language of vector fields and Lie brackets to translate the conclusions obtained from the Maximum Principle into useful information, one is led directly to the canonical forms for abnormal extremals derived below, which lend themselves to an easy optimality proof.

---

[14] We emphasize that this is ALWAYS true, not just "generically" or "almost always."

The techniques of this paper can also be easily applied to study non-degenerate abnormal extremals in higher dimensions, and the abnormal extremals that arise in dimension 3 for nonregular generic distributions. This will be done in [6].

2. Sub-Riemannian manifolds and abnormal extremals

A BRIEF REVIEW

If M is a C^∞ manifold, and $p \in M$, we use T_pM, T_p^*M to denote, respectively, the tangent and cotangent spaces of M at p, and TM, T^*M to denote the tangent and cotangent bundles of M. If $\lambda \in T_p^*M$, $v \in T_pM$, we write $\lambda(v)$, $\langle \lambda, v \rangle$ or, simply, λv, to denote the value at v of the linear functional λ. A subbundle E of TM is sometimes called a *distribution* on M. A *nonholonomic subbundle* (also known as a *bracket-generating distribution*) is a subbundle E of TM such that the Lie algebra $L(E)$ of vector fields generated by the global C^∞ sections of E has the *full rank property*, i.e. satisfies $\{X(p) : X \in L(E)\} = T_pM$ for all $p \in M$.

If E is a C^∞ subbundle of TM, we use $\Gamma(E)$ to denote the set of all C^∞ sections of E defined on open subsets of M. For a positive integer k, we let $\Gamma^k(E)$ denote the set of all vector fields X such that the domain of X is an open subset of M, and X is a linear combination of iterated brackets of degree $\leq k$ of members of $\Gamma(E)$. For $p \in M$, we let $E^k(p)$ denote the set $\{X(p) : X \in \Gamma^k(E)\}$. We write $\mu_E^k(p) = \dim E^k(p)$. The subbundle E is called *regular* if for every k the integer $\mu_E^k(p)$ is independent of p.

An *E-admissible arc* is an absolutely continuous curve γ on M, defined on some compact interval $[a, b]$, such that $\dot\gamma(t) \in E(\gamma(t))$ for almost all $t \in [a, b]$. If E is nonholonomic and M is connected, then any two points in M can be joined by an E-admissible arc.

A C^∞ *Riemannian metric* on E is a C^∞ section $p \to G_p$ of the bundle $E^* \otimes E^*$ such that for each $p \in M$ the bilinear form $E(p) \times E(p) \ni (v, w) \to G_p(v, w) \in \mathbb{R}$ is symmetric and strictly positive definite. A *sub-Riemannian structure* on a manifold M is a pair (E, G) where E is a nonholonomic C^∞ subbundle of TM and G is a C^∞ Riemannian metric on E. A *sub-Riemannian manifold* is a triple (M, E, G) such that M is a C^∞ manifold and (E, G) is a sub-Riemannian structure on M. We call a sub-Riemannian manifold (M, E, G) *regular* if the subbundle E is regular. One can always construct a Riemannian metric on any subbundle

E of TM by just taking a Riemannian metric on TM and restricting it to E. If $p \in M$, $v \in E(p)$, then the *length* $\|v\|_G$ of v is the number $G_p(v,v)^{1/2}$. The *length* $\|\gamma\|_G$ of an E-admissible arc $\gamma : [a,b] \to M$ is the integral $\int_a^b \|\dot{\gamma}(t)\|_G dt$. If $p, q \in M$, then the infimum of the lengths of all the E-admissible curves γ that go from p to q is the *distance* from p to q, and is denoted by $d_G(p,q)$. If M is connected and E is nonholonomic, then $d_G(p,q) < \infty$ for all p, q, and $d_G : M \times M \to \mathbb{R}$ is a metric whose associated topology is the one of M. An E-admissible curve $\gamma : [a,b] \to M$ such that $d_G(\gamma(a), \gamma(b)) = \|\gamma\|_G$ is called a *minimizer*.

An E-admissible curve γ is *parametrized by arc length* if $\|\dot{\gamma}(t)\|_G = 1$ for almost all t in the domain of γ. If $\gamma : [a,b] \to M$ is E-admissible, then we can define $\tau(t) = \int_a^t \|\dot{\gamma}(s)\|_G ds$, so τ is a monotonically nondecreasing function on $[a,b]$ with range $[0, \|\gamma\|_G]$. Moreover, if $t_1 < t_2$ but $\tau(t_1) = \tau(t_2)$, then $\gamma(t_2) = \gamma(t_1)$. So we can define $\tilde{\gamma} : [0, \|\gamma\|_G] \to M$ by letting $\tilde{\gamma}(s) = \gamma(t)$ if $\tau(t) = s$. Then, if $s_1 < s_2$, and $s_i = \tau(t_i)$ for $i = 1, 2$, the points $\tilde{\gamma}(s_1)$ and $\tilde{\gamma}(s_2)$ can be joined by the restriction of γ to the interval $[t_1, t_2]$, whose G-length is $s_2 - s_1$. So $d_G(\tilde{\gamma}(s_1), \tilde{\gamma}(s_2)) \leq s_2 - s_1$. If $\hat{G}$ is a Riemannian metric on M (i.e. a metric defined on the whole tangent bundle TM) that extends G, then the $\hat{G}$-distance $d_{\hat{G}}(\tilde{\gamma}(s_1), \tilde{\gamma}(s_2))$ is a fortiori $\leq s_2 - s_1$. So $\tilde{\gamma}$ is Lipschitz as a map into $(M, d_{\hat{G}})$. Clearly, $\gamma = \tilde{\gamma} \circ \tau$. Since $\tilde{\gamma}$ is Lipschitz and τ is integrable, we have $\int_0^s \|\dot{\tilde{\gamma}}(\sigma)\|_G d\sigma = \int_0^t \|\dot{\gamma}(\theta)\|_G d\theta = s$, if $s = \tau(t)$. So $\|\dot{\tilde{\gamma}}(s)\| = 1$ for almost all s. Therefore $\tilde{\gamma}$ is parametrized by arc length.

In particular, every minimizer γ is equivalent modulo reparametrization to an arc γ^* which is parametrized by arc length and is *time-optimal* for the control problem Σ (i.e. goes from its initial point p to its terminal point q in time not greater than that of any other trajectory of Σ that goes from p to q), where Σ is the class of all E-admissible arcs δ that satisfy $\|\dot{\delta}(t)\| \leq 1$ for almost all t. (It is clear that, if γ is a minimizer, then the arc $\tilde{\gamma}$ constructed above is a solution of the minimum time problem.) Conversely, it is easy to see that, if γ is a solution of the minimum time problem, then γ is a minimizer parametrized by arc length. So the class of solutions of the minimum time control problem for Σ coincides with the class of minimizers that are parametrized by arc length.

The solutions of the minimum time problem satisfy a necessary condition for optimality given by the Pontrjagin Maximum Principle (cf. [1], [4], [8]). A trajectory that satisfies this condition is called a *Pontrjagin extremal*. To state the condition, we need to define, for an arbitrary *E*-admissible curve, what is meant by an *H-minimizing adjoint vector* along γ. We would like to say that an *H*-minimizing adjoint vector is an adjoint vector that is *H*-minimizing. Unfortunately, the concept of an adjoint vector, by itself, is not intrinsic, since it depends on choosing an orthonormal basis of sections of *E*. So we will first define the concept of an adjoint vector relative to a basis $\mathbf{f}$ (or $\mathbf{f}$-*adjoint vector*), and then the concept of an $\mathbf{f}$-*H*-minimizing adjoint vector. The latter turns out to be intrinsic, and this will give us the desired definition.

We first define what is meant by an $\mathbf{f}$-*adjoint vector*, assuming that γ is such that the set $\gamma([a,b])$ is entirely contained in an open set Ω and that $\mathbf{f} = (f_1, \ldots, f_m)$ is a basis of smooth sections of *E* on Ω. Under this assumption we can express γ as a trajectory of the control system $\dot{x} = u_1 f_1(x) + \ldots + u_m f_m(x)$, that is, as a solution of the differential equation $\dot{x}(t) = u_1(t)f_1(x(t)) + \ldots + u_m(t)f_m(x(t))$ for some *m*-tuple $(u_1, \ldots, u_m)$ of real-valued integrable functions on $[a,b]$. (Notice that the control functions u_i are uniquely determined by γ and $\mathbf{f}$, since the f_i are linearly independent at each point.) If in addition Ω is also the domain of a coordinate chart $\kappa = (\kappa_1, \ldots, \kappa_n)$ of *M*, then we define an $\mathbf{f}$-*adjoint vector along* γ to be a vector-valued absolutely continuous function $\lambda^\kappa : [a, b] \to \mathbb{R}_n$ that satisfies the adjoint equations of the Pontrjagin Maximum Principle:

$$\dot{\lambda}^\kappa(s) = -\sum_{i=1}^{m} u_i(s)\left(\lambda^\kappa(s)\frac{\partial f_i}{\partial \kappa}(\gamma(s))\right) .$$

(We use $\mathbb{R}^n$, $\mathbb{R}_n$ to denote, respectively, the spaces of *n*-dimensional column and row vectors. We also use the familiar convention of thinking of *n*-tuples of coordinates of points or of components of tangent vectors as columns, so the f_i are columns of functions and the $\frac{\partial f_i}{\partial \kappa}(x)$ are square matrices. The columns of $\frac{\partial f_i}{\partial \kappa}$ are the partial derivatives $\frac{\partial f_i}{\partial \kappa_\ell}$, for $\ell = 1, \ldots, n$. Then λ^κ is a row vector, and $\lambda^\kappa \frac{\partial f_i}{\partial \kappa}$ is therefore a row vector as well.)

It is well known—and easy to prove— that, if we think of $\lambda^\kappa(t)$ as the *n*-tuple of components with respect to κ of a covector $\lambda(t)$ at $\gamma(t)$ then, if the adjoint equation holds for one coordinate system κ on Ω, it necessarily must hold on any other coordinate system on Ω. So, if λ is a *field of covectors along* γ (that is, λ is a section of the pullback $\gamma^*(T^*M)$ or, equivalently, λ is a mapping defined on $[a,b]$ such that, for each t, $\lambda(t)$ belongs to $T^*_{\gamma(t)}M$ of *M* at $\gamma(t)$), then the property that λ is a solution of the adjoint equation associated to $\mathbf{f}$ along γ is well defined, if γ is contained in an open set Ω and $\mathbf{f} = (f_1, \ldots, f_m)$ is a basis of sections on Ω. Any field of covectors along γ that has this property will be called an $\mathbf{f}$-*adjoint vector along* γ. We use $\mathrm{Adj}_{\mathbf{f}}(\gamma)$ to denote the set of all $\mathbf{f}$-adjoint vectors along γ. It is clear that $\mathrm{Adj}_{\mathbf{f}}(\gamma)$ is an n-dimensional linear space —where $n = \dim M$—and that, for each $t \in [a, b]$, the map $\lambda \to \lambda(t)$ establishes an isomorphism between $\mathrm{Adj}_{\mathbf{f}}(\gamma)$ and $T^*_{\gamma(t)}M$. A field λ of covectors along γ such that $\lambda(t) \neq 0$ for some $t \in [a, b]$ will be called *nontrivial*. If $\lambda \in \mathrm{Adj}_{\mathbf{f}}(\gamma)$ is nontrivial, then $\lambda(t) \neq 0$ for all $t \in [a, b]$.

We call a $\lambda \in \mathrm{Adj}_{\mathbf{f}}(\gamma)$ $\mathbf{f}$-*H-minimizing* if in addition the control functions u_i are such that, for almost every t, the vector $(u_1(t), \ldots, u_m(t))$ minimizes the linear function $\mathbf{u} = (u_1, \ldots, u_m) \to \mathbf{u}\Lambda_{\mathbf{f}}(s)$ (where $\Lambda_{\mathbf{f}}(s)$ is the column vector with components $\langle \lambda(s), f_1(\gamma(s)) \rangle, \ldots, \langle \lambda(s), f_m(\gamma(s)) \rangle$) on the unit ball $\{v : v_1^2 + \ldots + v_m^2 \le 1\}$. Equivalently, λ is $\mathbf{f}$-*H*-minimizing if, for each t:

$$\sum_{i=1}^m \left\langle \lambda(t), f_i(\gamma(t)) \right\rangle^2 > 0 \ \text{implies}$$

$$u_i(t) = \frac{-\langle \lambda(t), f_i(\gamma(t)) \rangle}{\sqrt{\sum_{j=1}^m \left\langle \lambda(t), f_j(\gamma(t)) \right\rangle^2}} \ \text{for } i = 1, \ldots, m\,.$$

(When $\sum_{j=1}^m \left\langle \lambda(t), f_j(\gamma(t)) \right\rangle^2 = 0$, the $u_i(t)$ can be arbitrary, provided only that they satisfy the constraint $\sum_{i=1}^m u_i(t)^2 \le 1$.)

We can now go one step further, and drop the dependence on the basis $\mathbf{f}$, for *orthonormal bases* $\mathbf{f}$ and $\mathbf{f}$-*H*-minimizing adjoint vectors. To do this, it suffices to show that, if $\mathbf{f} = (f_1, \ldots, f_m)$ and $\mathbf{g} = (g_1, \ldots, g_m)$ are two orthonormal bases of sections on Ω, and λ is $\mathbf{g}$-adjoint and $\mathbf{g}$-*H*-minimizing, then it is also $\mathbf{f}$-adjoint and $\mathbf{f}$-*H*-minimizing. Write $g_i = \sum \alpha_{ij} f_j$, where the α_{ij} are smooth functions on Ω. Then the matrix $A(x) = (\alpha_{ij}(x))_{i,j=1,\ldots,m}$ is orthogonal for each x. On any subinterval I of $[a, b]$ such that $\gamma(I)$ is contained in the image of a chart κ, we can write

$$\dot{\gamma}(s) = \sum_i v_i(s) g_i(\gamma(s)) = \sum_j u_j(s) f_j(\gamma(s))$$

where $u_j(s) = \sum_i v_i(s) \alpha_{ij}(\gamma(s))$. Then the adjoint equation relative to $\mathbf{g}$ says

$$\dot{\lambda}(s) = -\sum_{i=1}^m v_i(s) \lambda(s) \frac{\partial g_i}{\partial \kappa}(\gamma(s))\,.$$

But

$$\frac{\partial g_i}{\partial \kappa}(x) = \frac{\partial(\sum_j \alpha_{ij} f_j)}{\partial \kappa}(x) = \sum_j \alpha_{ij}(x)\frac{\partial f_j}{\partial \kappa}(x) + \sum_j \frac{\partial \alpha_{ij}}{\partial \kappa}(x)f_j(x) \, .$$

The term $\sum_j \frac{\partial \alpha_{ij}}{\partial \kappa}(x)f_j(x)$ is a square matrix whose columns are the partial derivatives $\sum_j \frac{\partial \alpha_{ij}}{\partial \kappa_\ell}(x)f_j(x)$. Suppose we evaluate this matrix at $x = \gamma(s)$, then left-multiply by the row vector $\lambda(s)$, multiply by $v_i(s)$, and sum over i. The result is a row vector whose components are the inner products

$$\langle \mathbf{v}(s), B_\ell(\gamma(s))\Lambda_{\mathbf{f}}(s)\rangle,$$

where $B_\ell = \frac{\partial A}{\partial \kappa_\ell}$, $\mathbf{v} = (v_1, \ldots, v_m)^\dagger$ and, for any m-tuple $\mathbf{h} = (h_1, \ldots, h_m)$ of vector fields, we write

$$\Lambda_{\mathbf{h}}(s) \overset{\text{def}}{=} (\lambda(s)h_1(\gamma(s)), \ldots, \lambda(s)h_m(\gamma(s)))^\dagger \, .$$

(Here we are using $\dagger$ to denote matrix transpose.) Since $A(x)$ is orthogonal, the matrices $B_\ell(x)A(x)^\dagger$ are skew-symmetric. Since

$$\lambda(s)g_i(\gamma(s)) = \sum_j \alpha_{ij}(\gamma(s))\lambda(s)f_j(\gamma(s)),$$

the vectors $\Lambda_{\mathbf{f}}(s)$ and $\Lambda_{\mathbf{g}}(s)$ are related by $\Lambda_{\mathbf{g}}(s) = A(\gamma(s))\Lambda_{\mathbf{f}}(s)$, i.e. by $\Lambda_{\mathbf{f}}(s) = A(\gamma(s))^\dagger \Lambda_{\mathbf{g}}(s)$. Then

$$\langle \mathbf{v}(s), B_\ell(\gamma(s))\Lambda_{\mathbf{f}}(s)\rangle = \langle \mathbf{v}(s), B_\ell(\gamma(s))A(\gamma(s))^\dagger \Lambda_{\mathbf{g}}(s)\rangle \, .$$

Since λ is $\mathbf{g}$-H-minimizing, the vector $\Lambda_{\mathbf{g}}(s)$ is of the form $\rho(s)\mathbf{v}(s)$ for some scalar $\rho(s)$. Therefore

$$\langle \mathbf{v}(s), B_\ell(\gamma(s))A(\gamma(s))^\dagger \Lambda_{\mathbf{g}}(s)\rangle = \rho(s)\langle \mathbf{v}(s), B_\ell(\gamma(s))A(\gamma(s))^\dagger \mathbf{v}(s)\rangle \, ,$$

which is equal to zero because $B_\ell(\gamma(s))A(\gamma(s))^\dagger$ is skew-symmetric. So we have shown that

$$\langle \mathbf{v}(s), B_\ell(\gamma(s))\Lambda_{\mathbf{f}}(s)\rangle = 0 \, .$$

Using this, the equation $\frac{\partial g_i}{\partial \kappa}(x) = \sum_j \alpha_{ij}(x)\frac{\partial f_j}{\partial \kappa}(x) + \sum_j \frac{\partial \alpha_{ij}}{\partial \kappa}(x)f_j(x)$ yields, if we evaluate at $x = \gamma(s)$, left-multiply by $\lambda(s)$, multiply by $v_i(s)$, and sum over i:

$$\sum_{i=1}^m v_i(s)\lambda(s)\frac{\partial g_i}{\partial \kappa}(\gamma(s)) = \sum_{ij} v_i(s)\alpha_{ij}(\gamma(s))\lambda(s)\frac{\partial f_j}{\partial \kappa}(\gamma(s)) \, ,$$

that is

$$\sum_{i=1}^{m} v_i(s)\lambda(s)\frac{\partial g_i}{\partial \kappa}(\gamma(s)) = \sum_{j=1}^{m} u_j(s)\lambda(s)\frac{\partial f_j}{\partial \kappa}(\gamma(s)),$$

since $\sum_i v_i(s)\alpha_{ij}(\gamma(s)) = u_j(s)$. Therefore

$$\dot\lambda(s) = -\sum_{j=1}^{m} u_j(s)\lambda(s)\frac{\partial f_j}{\partial \kappa}(\gamma(s)),$$

so λ is **f**-adjoint as well.

Since λ is **g**-H-minimizing, the vector $(v_1(t),\ldots,v_m(t))$ minimizes the linear functional $\mathbf{v} \to \mathbf{v}\Lambda_{\mathbf{g}}(s)$ on the unit ball of $\mathbb{R}^m$. But then, since $A(\gamma(s))$ is orthogonal, and $u_j(s) = \sum_i v_i(s)\alpha_{ij}(\gamma(s))$, the vector $(u_1(t),\ldots,u_m(t))$ minimizes the linear functional $\mathbf{u} \to \mathbf{u}\Lambda_{\mathbf{f}}(s)$ on the unit ball. So λ is **f**-H-minimizing as well.

It is not hard to see that the t-derivative of the quantity $\sum_{i=1}^{m}\left\langle\lambda(t), f_i(\gamma(t))\right\rangle^2$ is equal to

$$2\sum_{i,j=1}^{m} u_j(t)\left\langle\lambda(t), [f_j, f_i](\gamma(t))\right\rangle\left\langle\lambda(t), f_i(\gamma(t))\right\rangle.$$

Since the vector with components $\langle\lambda(t), f_i(\gamma(t))\rangle$ is a scalar multiple of the vector with components $u_i(t)$, and the matrix

$$\left(\left\langle\lambda(t), [f_j, f_i](\gamma(t))\right\rangle\right)_{ij}$$

is skew-symmetric, the derivative is in fact equal to zero, so the quantity $\sum_{i=1}^{m}\left\langle\lambda(t), f_i(\gamma(t))\right\rangle^2$ is constant. The identity $\Lambda_{\mathbf{g}}(s) = A(\gamma(s))\Lambda_{\mathbf{f}}(s)$ then implies that this constant does not depend on the choice of orthonormal basis.

Summarizing, we have shown that

*the condition that a field of covectors along an E-admissible trajectory γ is both **f**-adjoint and **f**-H-minimizing for some orthonormal basis **f** of sections of E is in fact independent of the basis **f**. This condition makes therefore intrinsic sense, and when we want to verify it or use it we can choose the orthonormal basis arbitrarily. Moreover, the number $\sum_{i=1}^{m}\left\langle\lambda(t), f_i(\gamma(t))\right\rangle^2$ is in fact independent of t, and its value does not depend on the basis either.*

We call a field λ of covectors an *adjoint H-minimizing covector* along an E-admissible trajectory $\gamma : [a,b] \to M$ if, whenever $\Omega \subseteq M$ is open, $\mathbf{f} = (f_1, \ldots, f_m)$ is an orthonormal basis of sections of E defined on Ω, and I is a subinterval of $[a,b]$ such that $\gamma(I) \subseteq \Omega$, then the restriction of λ to I is an $\mathbf{f}$-adjoint $\mathbf{f}$-H-minimizing field of covectors. We call λ *normal* (resp. *abnormal*) if the constant $\sum_{i=1}^{m} \left\langle \lambda(t), f_i(\gamma(t)) \right\rangle^2$ is > 0 (resp. $= 0$). An E-admissible trajectory γ for which there exists a nonzero adjoint H-minimizing covector λ will be called an *extremal*. If λ can be chosen to be normal (resp. abnormal), then γ is called a *normal* (resp. *abnormal*) extremal. Since there may exist more than one nonzero adjoint H-minimizing covector for a given trajectory, it is possible for an extremal to be both normal and abnormal. A *strictly abnormal extremal* is an abnormal extremal which is not also a normal extremal.

The Pontrjagin Maximum Principle says, simply, that

> *Every minimizer is an extremal.*

3. Abnormal extremals in dimension 4

We now let (M, E, G) be a regular sub-Riemannian manifold, such that $\dim M = 4$ and $\dim E = 2$. We let Σ be the associated control problem, whose trajectories are the E-admissible curves γ such that $\|\dot{\gamma}(t)\| \leq 1$ for almost all t.

We necessarily have $\mu_E^1 = 2$, $\mu_E^2 = 3$, $\mu_E^3 = 4$. Moreover, if f and g are two smooth sections of E with domain $\Omega \subseteq M$ that are linearly independent at each point of Ω, then the three vectors $f(p)$, $g(p)$, $[f,g](p)$ span the space $E^2(p)$—and are therefore linearly independent—for every $p \in \Omega$, whereas the five vectors $f(p)$, $g(p)$, $[f,g](p)$, $[f, [f,g]](p)$, $[g, [f,g]](p)$ span the tangent space T_pM for every $p \in \Omega$.

We now determine the abnormal extremals for (M, E, G). Call a smooth section $g \in \Gamma(E)$, with domain $\Omega_g \subseteq M$, an *abnormal infinitesimal generator* (AIG) if

1. $\|g(p)\|_G = 1$ for all $p \in \Omega_g$,

and

2. if $f \in \Gamma(E)$ is any smooth section of E with domain Ω_f, then the vectors $f(p)$, $g(p)$, $[f,g](p)$ and $[g, [f,g]](p)$ are linearly dependent for every $p \in \Omega_f \cap \Omega_g$.

It is easy to see that, if Ω is any nonempty open subset of M, and g_1, g_2 are smooth sections of E with domain Ω that satisfy Condition 2, then g_1, g_2 must be linearly dependent at each point of Ω. (Indeed, if there is a $p \in \Omega$ such that $g_1(p)$ and $g_2(p)$ are independent, then the three vectors $g_1(p)$, $g_2(p)$, $[g_1, g_2](p)$ are linearly independent and each of the two 4-tuples

$$(g_1(p), g_2(p), [g_1, g_2](p), [g_2, [g_1, g_2]](p)),$$
$$(g_1(p), g_2(p), [g_1, g_2](p), [g_1, [g_2, g_1]](p)),$$

is linearly dependent, as can be seen by applying Condition 2 to both g_1 and g_2. Since the first three vectors of these 4-tuples are independent, it follows that both $[g_2, [g_1, g_2]](p)$ and $[g_1, [g_2, g_1]](p)$ are linear combinations of $g_1(p)$, $g_2(p)$, and $[g_1, g_2](p)$. But this contradicts the fact that $g_1(p)$, $g_2(p)$, $[g_1, g_2](p)$, $[g_2, [g_1, g_2]](p)$, and $[g_1, [g_2, g_1]](p)$ span T_pM.)

On the other hand, given any point $p \in M$, we can construct an AIG on a neighborhood of p as follows. Start with any two linearly independent sections f, h, defined on a neighborhood Ω of p. Let ω be a smooth nowhere vanishing 1-form such that the annihilator $\{v \in T_qM : \omega(q)(v) = 0\}$ is exactly $E^2(q)$ for each $q \in \Omega$. Let $g = \alpha f + \beta h$ be a linear combination of f and h with $\mathcal{C}^\infty$ coefficients. Then, modulo members of $\Gamma^2(E)$, the brackets $[g, [f, g]]$ and $[g, [h, g]]$ are given by $\alpha\beta[f, [f, h]] + \beta^2[h, [f, h]]$ and $-\alpha^2[f, [f, h]] - \alpha\beta[h, [f, h]]$. So, if we choose $\alpha = \langle \omega, [h, [f, h]] \rangle$, $\beta = -\langle \omega, [f, [f, h]] \rangle$, we see that $\langle \omega, [g, [f, g]] \rangle \equiv \langle \omega, [g, [h, g]] \rangle \equiv 0$, so that $[g, [f, g]]$ and $[g, [h, g]]$ are both in $\Gamma^2(E)$. From this it follows easily that g satisfies Condition 2. Moreover, g can never vanish, because $g(q) = 0$ would imply $\langle \omega, [h, [f, h]] \rangle(q) = \langle \omega, [f, [f, h]] \rangle(q) = 0$, from which it would follow that both $[h, [f, h]](q)$ and $[f, [f, h]](q)$ belong to $E^2(q)$, so $E^3(q) = E^2(q)$, which is a contradiction.

Finally, it is easy to see that, if a vector field $g \in \Gamma(E)$ satisfies Condition 2, and φ is a smooth real-valued function on the domain of g, then φg satisfies 2 as well. Therefore we can modify g by multiplying it by the smooth function $\|g\|_G^{-1}$, and we obtain a vector field that satisfies both conditions 1 and 2, i.e. an AIG.

So we have shown that every point p has a neighborhood Ω such that there is a $g \in \Gamma(E)$ with domain Ω which is an AIG. For a given Ω, g is obviously unique up to sign. Therefore there is a well defined line subbundle L of E, characterized by the fact that the fiber $L(p)$ at each point $p \in M$ is the linear span of $g(p)$, g being any AIG whose domain contains p. The AIG's are then exactly the smooth sections of L that have length 1 at each point. A global AIG, defined on all of M, may or may not exist, depending on whether the bundle L is orientable.

Now, let $\gamma : [a, b] \to M$ be an E-admissible curve parametrized by arc-length. Assume that γ is an abnormal extremal. Let λ be a nonzero field of covectors along γ which is an abnormal adjoint vector for γ. Let $t \in [a, b]$ and choose, near $p = \gamma(t)$, a basis of sections consisting of an AIG g and a smooth section $f \in \Gamma(E)$ that has length 1 and is orthogonal to g at each point. We can then express the curve $s \to \gamma(s)$, for s near t, as a solution of the system $\dot{x}(s) = u(s)f(x(s)) + v(s)g(x(s))$, where u and v are measurable functions such that $u(s)^2 + v(s)^2 = 1$. The adjoint equation for λ then implies that for every smooth vector field X the derivative of the function $s \to \langle \lambda(s), X(\gamma(s)) \rangle$ is the function $s \to u(s)\langle \lambda(s), [f, X](\gamma(s)) \rangle + v(s)\langle \lambda(s), [g, X](\gamma(s)) \rangle$. The abnormality condition says that the functions $s \to \langle \lambda(s), f(\gamma(s)) \rangle$ and $s \to \langle \lambda(s), g(\gamma(s)) \rangle$ vanish identically. Differentiating these two functions we see that $s \to u(s)\langle \lambda(s), [f, g](\gamma(s)) \rangle$ and $s \to v(s)\langle \lambda(s), [f, g](\gamma(s)) \rangle$ vanish identically. Since $u^2 + v^2 \neq 0$, it follows that $s \to \langle \lambda(s), [f, g](\gamma(s)) \rangle \equiv 0$. One more differentiation implies the identity

$$u(s)\langle \lambda(s), [f, [f, g]](\gamma(s)) \rangle + v(s)\langle \lambda(s), [g, [f, g]](\gamma(s)) \rangle \equiv 0.$$

Since $\lambda(s)$ annihilates the vectors f, g and $[f, g]$, evaluated at $\gamma(s)$, and $\lambda(s) \neq 0$, it follows that the annihilator of $\lambda(s)$ is precisely the linear span of these three vectors, i.e. the subspace $E^2(\gamma(s))$. Therefore the vector

$$V(s) = u(s)[f, [f, g]](\gamma(s)) + v(s)[g, [f, g]](\gamma(s))$$

belongs to $E^2(\gamma(s))$. On the other hand, since g is an AIG, it follows that $[g, [f, g]](\gamma(s)) \in E^2(\gamma(s))$. Since, at each point, the vectors $[f, [f, g]]$ and $[g, [f, g]]$ span E^3 modulo E^2, we can conclude that $[f, [f, g]](\gamma(s)) \notin E^2(\gamma(s))$. But then $V(s)$ can only belong to $E^2(\gamma(s))$ if $u(s) = 0$, in which case $v(s) = \pm 1$. Therefore γ is in fact L-admissible, that is, $\dot{\gamma}(t) \in L(\gamma(t))$ for each t.

Conversely, if $\gamma : [a, b] \to M$ is any L-admissible curve which is parametrized by arc length, then we claim that γ is an abnormal extremal. To see this, we must find a nowhere vanishing covector λ along γ which is an abnormal adjoint vector for γ. Pick a nonzero covector $\bar{\lambda} \in T^*_{\gamma(a)}M$ that annihilates $E^2(\gamma(a))$. We claim that there exists an abnormal adjoint vector λ along γ such that $\lambda(a) = \bar{\lambda}$. To see this, we pick a maximal subinterval I of $[a, b]$ such that $a \in I$ and there is an H-minimizing abnormal adjoint vector λ_I for the restriction of γ to I such that $\lambda_I(a) = \bar{\lambda}$. We claim that $I = [a, b]$. If this is not true, let $\tau = \sup I$, so $a \leq \tau \leq b$. Find an interval J that contains τ in its interior relative to $[a, b]$, and is such that $\gamma(J)$ is contained in an open set Ω on which there exist both a coordinate chart and an orthonormal basis $\mathbf{f} = (f, g)$ of sections of E, such that g is an AIG. Pick a $\theta \in I \cap J$. Solve the adjoint equation with initial condition $\lambda(\theta) = \lambda_I(\theta)$. A solution λ_J exists on J because the adjoint equation is linear with respect to λ, and this solution must agree with λ_I on $I \cap J$, by uniqueness. Our curve γ satisfies, on J, an equation $\dot{\gamma}(t) = v(t)g(\gamma(t))$, where $|v(t)| = 1$ a.e., because γ is L-admissible. If we let $f_1 = f$, $f_2 = g$, $f_3 = [f, g]$, $\varphi_i = \langle \lambda_J, f_i \rangle$, then the functions φ_i satisfy, for some functions ψ_i, the system of differential equations $\dot{\varphi}_1 = -v\varphi_3$, $\dot{\varphi}_2 = 0$, $\dot{\varphi}_3 = \sum_{i=1}^{3} \psi_i\varphi_i$, where the third equation follows from the fact that $[g, f_3]$ is a linear combination of f_1, f_2 and f_3, because g is an AIG. Since λ_J is abnormal on $I \cap J$, the functions φ_1 and φ_2 vanish there. Then the equation $\dot{\varphi}_1 = -v\varphi_3$, implies that φ_3 also vanishes on $I \cap J$. By uniqueness, the φ_i vanish on J. So λ_J is abnormal on J. The field λ of covectors that agrees with λ_I on I and with λ_J on J is therefore an abnormal adjoint vector on $I \cup J$. The maximality of I then implies that $J \subseteq I$. Then $I = [a, b]$, and our conclusion follows.

It is clear that an L-admissible curve parametrized by arc-length satisfies, locally, an equation $\dot{\gamma}(t) = v(t)g(\gamma(t))$, where g is an AIG and the measurable function v takes values ± 1.

Let us call a curve *simple* if it has no double points, i.e. if it is a one-to-one map or, equivalently, if it contains no loops. We call γ *locally simple* if there is a $\delta > 0$ such that the restriction of γ to every interval of length $\leq \delta$ is simple.

It is clear that a curve which is not simple cannot be time-optimal, since by removing a loop one gets a shorter curve with the same initial and terminal points. The equation $\dot{\gamma}(t) = v(t)g(\gamma(t))$, implies that γ is contained in an integral curve of g. More precisely, if we fix a t_0 in the domain of γ, and let $s \to \xi(s)$ be the integral curve of g which goes through $\gamma(t_0)$ when $s = 0$, then we have $\gamma(t) = \xi(V(t))$, where $V(t) = \int_{t_0}^{t} v(s)ds$. The function V is absolutely continuous and satisfies $|\dot{V}(s)| = 1$ for almost all s. If V is not one-to-one, then clearly γ is not simple. For γ to be simple, then V has to be one-to-one, and hence monotonic, which implies that $\dot{V}$

is either ≥ 0 a.e. or ≤ 0 a.e., and if we combine this observation with the fact that $|\dot{V}(s)| = 1$ a.e., we see that either $v \equiv 1$ or $v \equiv -1$. From this it follows easily that *the locally simple abnormal extremals parametrized by arc-length are precisely the curves that satisfy, locally, an equation of the form $\dot{\gamma}(t) = g(\gamma(t))$ or $\dot{\gamma}(t) = -g(\gamma(t))$, where g is an AIG.*

So we have proved:

Theorem 1. *If (M, E, G) is a sub-Riemannian manifold such that $\dim M = 4$ and E is two-dimensional and regular, then there exists a line subbundle L of E such that the abnormal extremals parametrized by arc length are exactly the L-admissible curves parametrized by arc length. An abnormal extremal cannot be locally optimal unless it is simple. The locally simple abnormal extremals parametrized by arc length are precisely the curves that satisfy, locally, an equation of the form $\dot{\gamma}(t) = g(\gamma(t))$ or $\dot{\gamma}(t) = -g(\gamma(t))$, where g is an AIG. In particular, through every point there pass exactly two oriented (or one unoriented) locally simple abnormal extremals parametrized by arc-length.*

4. Optimality

We now prove

Theorem 2. *If (M, E, G) is a sub-Riemannian manifold such that $\dim M = 4$ and E is two-dimensional and regular, then every locally simple abnormal extremal is locally uniquely optimal.*

To prove Theorem 2, we pick a locally simple abnormal extremal $\gamma : [a, b] \to M$. We will show that the interval $[a, b]$ can be covered by open intervals I_ν such that the restriction of γ to every closed subinterval of an I_ν is uniquely optimal. Once this is proved, the local optimality of γ follows by letting $\delta > 0$ be a Lebesgue number of the covering $\{I_\nu\}$. If $a \leq t_1 < t_2 \leq b$, and $t_2 - t_1 \leq \delta$, then the interval $[t_1, t_2]$ is contained in one of the I_ν, and therefore the restriction of γ to $[t_1, t_2]$ is uniquely optimal.

To prove the existence of the I_ν, it suffices to pick a $\bar{t} \in [a, b]$ and find a $\delta > 0$ such that the restriction of γ to the interval $[a, b] \cap [\bar{t} - \delta, \bar{t} + \delta]$ is uniquely optimal. Let $p = \gamma(\bar{t})$. Let Ω be an open subset of M that contains p and is such that on Ω there is an orthonormal basis (f, g) of sections of E such that g is an AIG.

Consider the map Φ defined by

$$\Phi(x_1, x_2, x_3, x_4) = p\, e^{x_3[f,[f,g]]} e^{x_4[f,g]} e^{x_2 g} e^{x_1 f} ,$$

where we are using exponential notation for the flow of a vector field, and have the exponentials act on points on the right, so that $t \to x e^{tX}$ is the integral curve of the vector field X that goes through x at time $t = 0$. Since f, g, $[f, g]$ and $[f, [f, g]]$ are independent at p, the map Φ is well defined on a neighborhood of the origin in $\mathbb{R}^4$, and maps diffeomorphically some cube $C^4(\rho) = \{(x_1, x_2, x_3, x_4) \ : \ |x_i| < \rho$ for $i = 1, 2, 3, 4\}$ onto a neighborhood U of p in M. The inverse map Φ^{-1} defines a chart, with respect to which we are going to identify U with $C^4(\rho)$, so that p becomes the point $(0, 0, 0, 0)$. Clearly, f then just equals $\frac{\partial}{\partial x_1}$. Moreover, g is equal to $\frac{\partial}{\partial x_2}$ whenever $x_1 = 0$. So $g = \frac{\partial}{\partial x_2} + x_1 \sum_{i=1}^{4} \psi_i \frac{\partial}{\partial x_i}$, where the ψ_i are smooth functions. Then

$$[f, g] = \sum_{i=1}^{4} \psi_i \frac{\partial}{\partial x_i} + x_1 \sum_{i=1}^{4} \frac{\partial \psi_i}{\partial x_1} \frac{\partial}{\partial x_i} .$$

In particular, $[f, g] = \sum_{i=1}^{4} \psi_i \frac{\partial}{\partial x_i}$ whenever $x_1 = 0$. On the other hand, $[f, g] = \frac{\partial}{\partial x_4}$ whenever $x_1 = x_2 = 0$. So the functions ψ_1, ψ_2, ψ_3 and $\psi_4 - 1$ vanish when $x_1 = x_2 = 0$.

An easy calculation shows that the x_3-component of $[g, [f, g]]$ is equal, for some smooth function η, to the function $\frac{\partial \psi_3}{\partial x_2} - \psi_1 \psi_3 + x_1 \eta$. When $x_1 = 0$ this component is therefore just equal to $\frac{\partial \psi_3}{\partial x_2} - \psi_1 \psi_3$. On the other hand, the third component of $[f, g]$ is equal to ψ_3 when $x_1 = 0$, whereas the third components of f and g vanish when $x_1 = 0$. Since $[g, [f, g]]$ is in the linear span of f, g and $[f, g]$, we conclude that on the set defined by $x_1 = 0$ the function $\frac{\partial \psi_3}{\partial x_2}$ is equal to ψ_3 times a smooth function. Since ψ_3 vanishes when $x_1 = x_2 = 0$, it follows that $\psi_3 = 0$ whenever $x_1 = 0$. So ψ_3 is in fact equal to x_1 times a smooth function η. Then g has the form

$$g = \frac{\partial}{\partial x_2} + x_1 \sum_{i \in \{1,2,4\}} \psi_i \frac{\partial}{\partial x_i} + x_1^2 \eta \frac{\partial}{\partial x_3} .$$

This means that the trajectories of the restriction to U of the control system Σ are given by the equations:

$$\dot{x}_1 = u + v x_1 \psi_1 \,,$$
$$\dot{x}_2 = v(1 + x_1 \psi_2) \,,$$
$$\dot{x}_3 = v x_1^2 \eta \,,$$
$$\dot{x}_4 = v x_1 \psi_4 \,,$$

where u, v are controls that are required to satisfy $u^2 + v^2 \leq 1$. Our locally simple abnormal extremal γ satisfies the above equations with $u(t) \equiv 0$, and either $v(t) \equiv 1$ or $v(t) \equiv -1$.

We are now in a situation nearly identical to that of [5], and an argument similar to the one given there will enable us to prove optimality.

5. An optimality lemma

In this section we prove a general optimality lemma that transforms the main technical idea of [5] into a widely applicable method. We state once and for all the general version of the lemma that will be used in [6], rather than the slightly weaker result that would suffice for the four-dimensional case considered here.

Lemma. *Let $n \geq 3$, and let Ω be an open subset of $\mathbb{R}^n$. Let φ, $\psi_1, \psi_2, \psi_4, \ldots, \psi_n$, $\eta_1, \eta_3, \ldots, \eta_n$ be smooth functions on Ω, and let Σ be the control system*

$$\dot{x}_1 = u\varphi(x) + v x_1 \psi_1(x) \,,$$
$$\dot{x}_2 = (1 + x_1 \psi_2(x))v \,,$$
$$\dot{x}_3 = v x_1 \psi_3(x) \,,$$
$$\dot{x}_4 = v x_1 \psi_4(x) \,,$$
$$\vdots$$
$$\dot{x}_n = v x_1 \psi_n(x) \,,$$

where

$$\psi_3 = x_1 \eta_1 + x_3 \eta_3 + \ldots + x_n \eta_n \,,$$

and the controls u, v are subject to the constraint $u^2 + v^2 \leq 1$. Let $x^* : [a, b] \to \Omega$ be a trajectory of Σ, corresponding to the control functions $u(t) \equiv 0$, $v(t) \equiv 1$, and starting at a point $\bar{x} = x^*(a) = (0, \bar{x}_2, 0, \ldots, 0)$

(so that $x^*(t) = (0, \bar{x}_2 + t - a, 0, \ldots, 0)$ for $a \le t \le b$). Assume that $\eta_1(x^*(t)) \ne 0$ for $a \le t \le b$. Then x^* is locally uniquely time-optimal for Σ.

Proof. Let $K = \{x^*(t) : a \le t \le b\}$, so K is a compact subset of Ω. Let $U \subseteq \Omega$ be open, and such that $K \subseteq U$ and the closure $\bar{U}$ of U is compact and contained in Ω. Then there exists a constant $C_1 > 0$ such that $\|\dot{x}(t)\| \le C_1$ whenever $x(\cdot)$ is a trajectory of Σ which is contained in U. Let

$$\delta = \min\left\{\|x - y\| : x \in K, y \in \mathbb{R}^n - U\right\}.$$

Then $\delta > 0$. Let $\hat{\tau}_1$ be such that $0 < \hat{\tau}_1 < \frac{\delta}{C_1}$. Then, if $x(\cdot) : [t_1, t_2] \to \Omega$ is a trajectory of Σ that goes through a point of K and is such that $t_2 - t_1 \le \hat{\tau}_1$, it follows that $x(\cdot)$ is entirely contained in U. We let

$$C_2 = \sup\left\{|\psi_i(x)| : x \in U, i = 1, \ldots, n\right\},$$

$$C_3 = \sup\left\{\frac{|\eta_i(x)|}{|\eta_1(x^*(t))|} : a \le t \le b, x \in U, i = 1, \ldots, n\right\},$$

$$\rho = \inf\left\{|\eta_1(x^*(t))| : a \le t \le b\right\}.$$

Let $\hat{\tau}_2 > 0$ be such that, whenever $x, y \in U$ and $\|x - y\| \le C_1\hat{\tau}_2$, then it follows that $|\eta_1(x) - \eta_1(y)| \le \frac{\rho}{4}$. Let

$$\hat{\tau}_3 = \left(4(n - 2)C_2C_3\right)^{-1},$$

and pick a $\hat{\tau}_4 > 0$ such that $\hat{\tau}_4 < (3C_1C_2)^{-1}$. Let $\hat{\tau} = \min(\hat{\tau}_1, \hat{\tau}_2, \hat{\tau}_3, \hat{\tau}_4)$.

Now let $a \le t_1 \le t_2 \le b$ be such that $\tau = t_2 - t_1 \le \hat{\tau}$. Let γ be the restriction of x^* to the interval $[t_1, t_2]$. We will show that γ is uniquely optimal. Assume that $\xi : [s_1, s_2] \to \Omega$ is another trajectory of Σ that goes from $\gamma(t_1)$ to $\gamma(t_2)$ in time $\sigma = s_2 - s_1$ and corresponds to control functions $u_\xi, v_\xi : [s_1, s_2] \to \mathbb{R}$ such that $u_\xi(s)^2 + v_\xi(s)^2 \le 1$ for a.e. s. We will show that $\sigma \ge \tau$, with equality holding iff $\xi(s) = \gamma(s - s_1 + t_1)$ for $s_1 \le s \le s_2$.

Let $\xi(s) = (\xi_1(s), \xi_2(s), \xi_3(s), \zeta(s))$, where $\zeta : [s_1, s_2] \to \mathbb{R}^{n-3}$. Let us assume that $\sigma \le \tau$. Since $\tau \le \hat{\tau}$, and ξ goes through a point of K, it follows (since $\hat{\tau} \le \hat{\tau}_1$) that ξ is entirely contained in U, and the bound $\|\dot{\xi}(s)\| \le C_1$ holds for almost all s. In particular, we have $|\dot{\xi}_1(s)| \le C_1$ for almost all s.

Let $h(s) = \int_{s_1}^s v_\xi(r)dr$. Then $|h(s)| \leq s - s_1$ for all s, so $-\sigma \leq h(s_2) \leq \sigma$. Let $\alpha = \sigma - h(s_2)$, $\beta = \sup\{|\xi_1(s)| : s_1 \leq s \leq s_2\}$. We then have

$$\begin{aligned}
\tau &= x_2^*(t_2) - x_2^*(t_1) \\
&= \xi_2(s_2) - \xi_2(s_1) \\
&= \int_{s_1}^{s_2} \Big(1 + \xi_1(s)\psi_2(\xi(s))\Big)v_\xi(s)ds \\
&\leq \int_{s_1}^{s_2} v_\xi(s)ds + \beta\sigma C_2 \\
&= h(s_2) + \beta\sigma C_2 \\
&= \sigma - \alpha + \beta\sigma C_2 \ .
\end{aligned}$$

On the other hand,

$$\int_{s_1}^{s_2} \xi_1(s)^2 ds = \int_{s_1}^{s_2} \xi_1(s)^2(1 - v_\xi(s))ds + \int_{s_1}^{s_2} \xi_1(s)^2 v_\xi(s)ds \ .$$

The first integral of the right-hand side is bounded by $\beta^2\alpha$. The second integral is equal to

$$\int_{s_1}^{s_2} \xi_1(s)^2 v_\xi(s)\frac{\eta_1(\xi(s_1))}{\eta_1(\xi(s_1))}ds \ ,$$

i.e. to

$$\int_{s_1}^{s_2} \xi_1(s)^2 v_\xi(s)\frac{\eta_1(\xi(s_1)) - \eta_1(\xi(s))}{\eta_1(\xi(s_1))}ds + \int_{s_1}^{s_2} \xi_1(s)^2 v_\xi(s)\frac{\eta_1(\xi(s))}{\eta_1(\xi(s_1))}ds \ .$$

But

$$0 = \frac{\xi_3(s_2) - \xi_3(s_1)}{\eta_1(\xi(s_1))} =$$

$$\int_{s_1}^{s_2} \xi_1(s)^2 v_\xi(s)\frac{\eta_1(\xi(s))}{\eta_1(\xi(s_1))}ds + \sum_{i=3}^{n} \int_{s_1}^{s_2} \xi_1(s)\xi_i(s)v_\xi(s)\frac{\eta_i(\xi(s))}{\eta_1(\xi(s_1))}ds \ .$$

So

$$\begin{aligned}
\int_{s_1}^{s_2} \xi_1(s)^2 ds \leq{}& \beta^2\alpha + \int_{s_1}^{s_2} \xi_1(s)^2 v_\xi(s)\frac{\eta_1(\xi(s_1)) - \eta_1(\xi(s))}{\eta_1(\xi(s_1))}ds \\
&- \sum_{i=3}^{n} \int_{s_1}^{s_2} \xi_1(s)\xi_i(s)v_\xi(s)\frac{\eta_i(\xi(s))}{\eta_1(\xi(s_1))}ds \ .
\end{aligned}$$

For $i \geq 3$, the functions ξ_i satisfy $\dot{\xi}_i = v_\xi \xi_1 \psi_i(\xi)$. Therefore, since $\xi_i(s_1) = 0$, we have

$$|\xi_i(s)| \leq C_2 \sigma^{\frac{1}{2}} \left(\int_{s_1}^{s_2} \xi_1(s)^2 ds \right)^{\frac{1}{2}} .$$

Then

$$\left| \int_{s_1}^{s_2} \xi_1(s)\xi_i(s)v_\xi(s)\frac{\eta_i(\xi(s))}{\eta_1(\xi(s_1))} ds \right| \leq C_2 C_3 \sigma \int_{s_1}^{s_2} \xi_1(s)^2 ds .$$

Also,

$$\left| \int_{s_1}^{s_2} \xi_1(s)^2 v_\xi(s)\frac{\eta_1(\xi(s_1)) - \eta_1(\xi(s))}{\eta_1(\xi(s_1))} ds \right| \leq \frac{1}{4} \int_{s_1}^{s_2} \xi_1(s)^2 ds ,$$

since $|\eta_1(\xi(s_1))| \geq \rho$ and $|\eta_1(\xi(s_1)) - \eta_1(\xi(s))| \leq \frac{\rho}{4}$, because $\|\xi(s_1) - \xi(s)\| \leq C_1 \hat{\tau}_2$ (since $s - s_1 \leq \sigma \leq \hat{\tau}_2$ and $\|\dot{\xi}\| \leq C_1$). So we get the bound

$$\int_{s_1}^{s_2} \xi_1(s)^2 ds \leq \beta^2 \alpha + \left(\frac{1}{4} + (n-2)C_2 C_3 \sigma \right) \int_{s_1}^{s_2} \xi_1(s)^2 ds .$$

Since $\sigma \leq \hat{\tau}_3 \leq \left(4(n-2)C_2 C_3 \right)^{-1}$, we have the bound

$$\int_{s_1}^{s_2} \xi_1(s)^2 ds \leq \beta^2 \alpha + \frac{1}{2} \int_{s_1}^{s_2} \xi_1(s)^2 ds ,$$

from which it follows that

$$\int_{s_1}^{s_2} \xi_1(s)^2 ds \leq 2\beta^2 \alpha .$$

Now let $\tilde{s} \in [s_1, s_2]$ be such that $|\xi_1(\tilde{s})| = \beta$. Since $\xi_1(s_1) = x_1^*(t_1) = 0$, $\xi_1(s_2) = x_1^*(t_2) = 0$, and $|\dot{\xi}_1| \leq C_1$, we have $\tilde{s} - s_1 \geq \frac{\beta}{C_1}$ and $s_2 - \tilde{s} \geq \frac{\beta}{C_1}$. Hence the intervals $I_1 = [\tilde{s} - \frac{\beta}{C_1}, \tilde{s}]$ and $I_2 = [\tilde{s}, \tilde{s} + \frac{\beta}{C_1}]$ are entirely contained in $[s_1, s_2]$. On each of these intervals I_j, $|\xi_1(s)|$ is bounded below by the linear function λ_j which is equal to β at $\tilde{s}$ and to zero at the other endpoint. Clearly, the integral of λ_j^2 over I_j is exactly $\frac{\beta^3}{3C_1}$. So

$$\int_{s_1}^{s_2} \xi_1(s)^2 ds \geq \int_{I_1} \xi_1(s)^2 ds + \int_{I_2} \xi_1(s)^2 ds \geq \frac{2\beta^3}{3C_1} .$$

Combining the upper and lower bounds for $\int_{s_1}^{s_2} \xi_1(s)^2 ds$ we get

$$\frac{2\beta^3}{3C_1} \le 2\beta^2\alpha \ .$$

Therefore

$$\beta \le 3C_1\alpha \ .$$

But then

$$\tau \le \sigma - \alpha + \beta\sigma C_2$$
$$\le \sigma - \alpha + 3C_1 C_2 \sigma\alpha$$
$$= \sigma + \left(3C_1 C_2\sigma - 1\right)\alpha \ .$$

Since $\sigma \le \tau \le \hat{\tau}_4 < (3C_1 C_2)^{-1}$, we have $3C_1 C_2\sigma - 1 \le 0$. Therefore $\tau \le \sigma$.

So we have shown that $\sigma \le \tau$ implies $\tau \le \sigma$. Therefore σ cannot be $< \tau$. So γ is optimal, as stated. Moreover, the inequality $\tau \le \sigma$ is in fact strict unless $\alpha = 0$ (since $3C_1 C_2\sigma < 1$). So $\sigma = \tau$ can only happen if $\alpha = 0$. i.e. if $v_\xi = 1$ a.e. But then $u_\xi = 0$ a.e., and $\xi(s) = \gamma(s - s_1 + t_1)$. So γ is uniquely optimal. ∎

6. End of the proof

We now return to the locally simple abnormal extremal $\gamma : [a, b] \to M$ of §4, which goes through $p = (0,0,0,0)$ at time $\bar{t}$, and recall that we are trying to find a δ such that the restriction of γ to the interval $[a, b] \cap [\bar{t} - \delta, \bar{t} + \delta]$ is uniquely optimal. We observe that

$$\eta(0,0,0,0) \ne 0 \ .$$

(Indeed, $\eta(0,0,0,0)$ is the third component of $[f, [f, g]](0,0,0,0)$. Since the linear span of f, g and $[f, g]$ at $(0,0,0,0)$ is precisely the set of vectors whose third component vanishes, and $[f, [f, g]]$ does not belong to that linear span, our statement follows.)

It follows that the equations given at the end of §4 are exactly of the kind occurring in the lemma, and all the hypotheses of the lemma are satisfied, provided that we first restrict γ to a subinterval $[\tilde{a}, \tilde{b}]$ of $[a, b]$ that contains $\bar{t}$ in its interior relative to $[a, b]$ and is small enough that $\eta(\gamma(t)) \ne 0$ for all $t \in [\tilde{a}, \tilde{b}]$. Then we can apply the lemma and conclude that our trajectory is locally optimal for the system Σ restricted to the domain U. The following remark then concludes the proof.

Remark. Local time-optimality is a "local" property in the following sense. Consider a control system Σ on a manifold M, subject only to the condition that the velocity is uniformly bounded on compact sets, that is, that for every compact subset K of M there is a constant C such that $\|\dot{x}(t)\| \leq C$ for a.e. t, for every trajectory $x(\cdot)$ of Σ which is entirely contained in K. (Here $\|\dots\|$ is the norm with respect to some Riemannian metric on M.) Let Ω be an open subset of M. Let Σ_Ω be the restriction of Σ to Ω. Let $\gamma : [a, b] \to \Omega$ be locally optimal for Σ_Ω. Then γ is locally optimal for Σ. To see this, pick a compact subset K of Ω whose interior contains the set $K_0 = \{\gamma(t) : a \leq t \leq b\}$. Because of the bound on the velocities, there is a $\delta_1 > 0$ such that every trajectory $\tilde{\gamma}$ of Σ that goes through a point of K_0 and is defined on an interval of length $\leq \delta_1$ is entirely contained in K. Since γ is locally optimal for Σ_Ω, there is a $\delta_2 > 0$ such that, if $a \leq t_1 \leq t_2 \leq b$, $t_2 - t_1 \leq \delta_2$, and $\tilde{\gamma} : [\tilde{t}_1, \tilde{t}_2] \to \Omega$ is another trajectory of Σ such that $\tilde{\gamma}(\tilde{t}_1) = \gamma(t_1)$ and $\tilde{\gamma}(\tilde{t}_2) = \gamma(t_2)$, then it follows that $\tilde{t}_2 - \tilde{t}_1 \geq t_2 - t_1$. If we take $\delta = \min(\delta_1, \delta_2)$, then it is clear that, if $a \leq t_1 \leq t_2 \leq b$, $t_2 - t_1 \leq \delta$, and $\tilde{\gamma} : [\tilde{t}_1, \tilde{t}_2] \to M$ is another trajectory of Σ such that $\tilde{\gamma}(\tilde{t}_1) = \gamma(t_1)$ and $\tilde{\gamma}(\tilde{t}_2) = \gamma(t_2)$, then $\tilde{t}_2 - \tilde{t}_1 \geq t_2 - t_1$. (Indeed, if $\tilde{t}_2 - \tilde{t}_1 < t_2 - t_1$, then it would follow that $\tilde{t}_2 - \tilde{t}_1 < \delta_1$, and therefore $\tilde{\gamma}$ is in fact contained in Ω. But then $\tilde{t}_2 - \tilde{t}_1 \geq t_2 - t_1$ because $\delta \leq \delta_2$.)

7. Strict abnormality

We must now take care of the possibility that our abnormal extremals might be normal as well, i.e. might not be strictly abnormal. Suppose g is an AIG, f is a smooth section of E of unit length and orthogonal to g, and γ is an integral curve of g or of $-g$, i.e. a locally simple abnormal extremal. Then a normal H-minimizing covector λ along γ must satisfy $\langle \lambda, f \rangle = 0$, $\langle \lambda, g \rangle = -c$, where c is a strictly positive constant. Differentiation yields $\langle \lambda, [f, g] \rangle = 0$, and then again $\langle \lambda, [g, [f, g]] \rangle = 0$. Since g is an AIG, we have $[g, [f, g]] = \mu f + \nu g + \rho [f, g]$ for some smooth functions μ, ν, ρ. But then $\langle \lambda, [g, [f, g]] \rangle = -\nu c$. If $\nu \neq 0$, this is a contradiction.

Let us call a point $p \in M$ *nice* if the vectors $g(p)$, $[f, g](p)$ and $[g, [f, g]](p)$ are linearly independent, where g is an AIG defined near p, and $f \in \Gamma(E)$ is of unit length and orthogonal to g. Then, if an abnormal extremal goes through a nice point and is not strict, we must necessarily have $\nu = 0$ at that point, contradicting the fact that the point is nice. So we have shown:

Theorem 2. *If (M, E, G) is a sub-Riemannian manifold such that* $\dim M = 4$ *and E is two-dimensional and regular, then every locally simple abnormal extremal that goes through a point p where g, $[f, g]$ and $[g, [f, g]]$ are linearly independent is strictly abnormal.*

8. Conclusion

An Easy Recipe for Producing Millions of Strictly Abnormal Minimizers on General Four-Dimensional Sub-Riemannian Manifolds, Including Lie Groups with an Invariant Metric

Just take any pair of vector fields f, g on a four-dimensional manifold M, such that f, g, $[f, g]$ and $[f, [f, g]]$ are linearly independent everywhere but $[g, [f, g]]$ is a linear combination $\mu f + \nu g + \rho[f, g]$ with smooth coefficients. Make sure that the function ν never vanishes. Define a sub-Riemannian structure on M by letting E be the span of f and g, and declaring f, g to be an orthonormal basis. Then the locally simple abnormal extremals parametrized by arc-length are precisely the integral curves of g. And all these locally simple abnormal extremals are strictly abnormal and locally uniquely optimal.

The regular four-dimensional examples of Montgomery [7], Kupka [3], and Liu and Sussmann [5] can all be obtained in this way.

One can also use the above construction to produce examples of locally uniquely optimal abnormal extremals for invariant sub-Riemannian metrics on Lie groups. It suffices to let G be any four-dimensional Lie group whose Lie algebra L has two generators f and g such that

(i) f, g, $[f, g]$ and $[f, [f, g]]$ form a basis of L,

(ii) $[g, [f, g]]$ belongs to the linear span of f, g, and $[f, g]$,

(iii) $[g, [f, g]]$ does not belong to the linear span of f and $[f, g]$.

(For example, one can take $G = SO(3) \times \mathbb{R}$, and let $f = K_1 \oplus 1$, $g = (K_1 + K_2) \oplus 2$, where K_1, K_2, K_3 are generators of the Lie algebra $so(3)$ of $SO(3)$ such that $[K_1, K_2] = K_3$, $[K_2, K_3] = K_1$ and $[K_3, K_1] = K_2$, and we are identifying the Lie algebra of G with the direct sum $so(3) \oplus \mathbb{R}$. It is easily verified that f, g, $[f, g]$ and $[f, [f, g]]$ are linearly independent, and $[g, [f, g]] = 2f - g$, so all our conditions hold.)

Then we can let E be the subbundle of TG spanned by f and g, and define a sub-Riemannian structure by letting f and g be an orthonormal basis of sections. The integral curves of g are then strictly abnormal locally uniquely optimal trajectories.

References

[1] L.D. Berkovitz, *Optimal Control Theory*, Springer-Verlag, New York, 1974.

[2] R.W. Brockett, "Control Theory and Singular Riemannian Geometry," in *New Directions in Applied Mathematics* (P.J. Hilton and G.S. Young, eds.), Springer-Verlag, (1981).

[3] I. Kupka, "Abnormal extremals," preprint, 1992.

[4] E.B. Lee and L. Markus, *Foundations of Optimal Control Theory*, Wiley, New York, 1968.

[5] W.S. Liu and H.J. Sussmann, "Abnormal Sub-Riemannian Minimizers," Institute for Mathematics and its Applications, University of Minnesota, IMA preprint series # 1059, October 1992. To appear in *Differential Equations, Dynamical Systems, and Control Science: A Festschrift in honor of Lawrence Markus*, K. D. Elworthy, W. N. Everitt and E. B. Lee Eds., M. Dekker, Inc., New York.

[6] W.S. Liu and H.J. Sussmann, "Shortest paths for sub-Riemannian metrics on rank-2 distributions," Mem.-Amer.-Math.-Soc.

[7] R. Montgomery, "Geodesics which do not Satisfy the Geodesic Equations," 1991 preprint.

[8] L.S. Pontryagin, V.G. Boltyanskii, R.V. Gamkrelidze and E.F. Mischenko, *The Mathematical Theory of Optimal Processes*, Wiley, New York, 1962.

[9] R. Strichartz, "Sub-Riemannian Geometry," *J. Diff. Geom.* 24, 221-263, (1986).

[10] R. Strichartz, "Corrections to 'Sub-Riemannian Geometry'," *J. Diff. Geom.* 30, no. 2, 595-596, (1989).

Progress in Mathematics, Vol. 144, © 1996 Birkhäuser Verlag Basel/Switzerland

Stabilization of controllable systems

JEAN-MICHEL CORON[*]

0. Introduction

The goal of this paper is to present a partial survey on the local stabilizability of locally controllable systems.

In section **1**, we recall some well-known conditions for local controllability. In section **2**, we indicate cases where local controllability implies local asymptotic stabilizability by means of stationary feedback laws. In section **3**, we present necessary conditions for the local asymptotic stabilizability by means of stationary feedback laws; these obstructions show that, usually, local controllability does not imply local asymptotic stabilizability by means of stationary feedback laws. But, we see in section **4**, that, at least if the state is of dimension larger than 3, most of the sufficient conditions for local controllability imply the existence of time-varying stationary feedback laws which, locally, stabilized the system asymptotically—and even in finite time. Finally, in section **5**, we describe a method directly inspired from section **4**, which can be useful to prove controllability of distributed control systems; we briefly describe an application of this method to the boundary controllability of the 2-D incompressible Euler equations.

1. Local controllability

Let us consider the following control system

$$\dot{x} = f(x, u) \tag{1.1}$$

where $x \in \mathbb{R}^n$ is the state, $u \in \mathbb{R}^m$ is the control, and $f \in \mathcal{C}^\infty(\mathbb{R}^n \times \mathbb{R}^m; \mathbb{R}^n)$ satisfies

$$f(0, 0) = 0. \tag{1.2}$$

There are various possible definitions for local controllability. Our definition is local in time, state, and control. More precisely, we use

---

[*]CNRS and Centre de Mathématiques et Leurs Applications, ENS Cachan
61 Avenue du Président Wilson, 91435 Cachan, France

Definition 1.1. The systems $\dot{x} = f(x, u)$ is locally controllable (at 0), if for any positive real number ε, there exists a positive real number η such that, for any two points x_0 and x_1 in the ball of $\mathbb{R}^n$ of radius η and center 0, there exists a measurable map $u : [0, \varepsilon] \to \mathbb{R}^m$ such that

$$|u(t)| \leqslant \varepsilon, \qquad\qquad \forall t \in [0, \varepsilon], \quad (1.3)$$

$$\left(\dot{x} = f(x, u(t)), x(0) = x_0 \right) \implies \left(x(\varepsilon) = x_1 \right). \qquad (1.4)$$

By (1.4), we mean that the (maximal) solution of the Cauchy problem $\dot{x} = f(x, u(t))$, $x(0) = x_0$ is defined on $[0, \varepsilon]$ and satisfies $x(\varepsilon) = x_1$.

One does not know any (interesting) necessary and sufficient condition for local controllability but there are powerful necessary conditions and powerful sufficient conditions. For a state of art on this question, see the survey by Kawski [Kaw3], the recent papers [BS2] by Bianchini and Stefani, [AG] by Agrachev and Gamkrelidze, and the references therein. Let us just mention a necessary condition and a sufficient condition.

In order to state the necessary condition, let us denote by $\mathrm{Br}\,(\mathcal{F})$ the set of iterated Lie brackets of the vector fields in the family $\mathcal{F}$. Then, one has the following theorem due to Sussmann and Jurdjevic [SJ].

Theorem 1.2. *Assume that f is analytic. Then, if $\dot{x} = f(x, u)$ is locally controllable, one has*

$$\mathrm{Span}\left\{ h(0) \;;\; h \in \mathrm{Br}\left(\left\{ \tfrac{\partial^\alpha f}{\partial u^\alpha}(\cdot, 0) \;;\; \alpha \in \mathbb{N}^m \right\} \right) \right\} = \mathbb{R}^n. \qquad (1.5)$$

Condition (1.5) will be called the Lie algebra rank condition. Let us remark that this necessary condition is also a sufficient condition in two important cases

(i) Systems without drift: $f(x, u) = \sum_{i=1}^m u_i f_i(x)$. Then (1.5) becomes

$$\mathrm{Span}\left\{ h(0) \;;\; h \in \mathrm{Lie}\left(\{ f_1, \ldots, f_m \} \right) \right\} = \mathbb{R}^n \qquad (1.6)$$

where $\mathrm{Lie}\left(\{ f_1, \ldots, f_m \} \right)$ denotes the Lie algebra generated by $f_1, \ldots, f_m$. The fact that (1.5) (or (1.6)) implies the local controllability (even if f is not analytic) in the case of systems without drift is due to Chow [Ch].

(ii) Linear control systems: $\dot{x} = Ax + Bu$. In this case (1.5) becomes the famous Kalman rank condition

$$\mathrm{Span}\left\{ A^i Bu \;;\; u \in \mathbb{R}^m, i \in [0, n-1] \right\} = \mathbb{R}^n \qquad (1.7)$$

which is equivalent to the local controllability of $\dot{x} = Ax + Bu$ (see e.g. [So2; Chap. 3, Th. 3]).

But, in general, (1.5) does not imply the local controllability of $\dot{x} = f(x, u)$. Let us give a very simple example: $n = m = 1$, $f(x, u) = u^2$; then

$$\frac{\partial^2 f}{\partial u^2}(0, 0) = 2 \neq 0 \tag{1.8}$$

and, therefore, (1.5) holds. But if $\dot{x} = u^2$ then $x(0) \leqslant x(\varepsilon)$ for any positive ε such that x is defined on $[0, \varepsilon]$ and therefore $\dot{x} = u^2$ is not locally controllable. This system is not affine, but one easily checks that, similarly, the affine systems, with $n = 2$ and $m = 1$, $\dot{x}_1 = x_2^2$, $\dot{x}_2 = u$ satisfies (1.5) but is not locally controllable.

For a sufficient condition for local controllability, let us just mention a condition due to Sussmann. For simplicity, let us assume—see [Co3] for the general case—that the system is affine, *i.e.*

$$f(x, u) = f_0(x) + \sum_{i=1}^{m} u_i f_i(x) \tag{1.9}$$

for some $f_0, f_1 \ldots, f_m$ in $\mathcal{C}^\infty(\mathbb{R}^n, \mathbb{R}^n)$. For h in $\mathrm{Br}\left(\{f_0, f_1 \ldots, f_m\}\right)$, let $S_i(h)$ be the number of times that f_i appears in h. For example, if

$$h = [[[f_0, f_1], [f_1, f_2]], f_2],$$

we have

$$S_0(h) = 1, \ \ S_1(h) = 2, \ \ S_2(h) = 2.$$

Then the Sussmann condition [Su4; Th.7.3] is

Theorem 1.3. *Assume that the Lie algebra rank condition (1.5) holds. Assume that there exists θ in $[0, 1]$ such that, for any h in $\mathrm{Br}\left(\{f_0, f_1 \ldots, f_m\}\right)$ with $S_0(h)$ odd and $S_i(h)$ even for all i in $[1, m]$,*

$$h(0) \in \mathrm{Span}\left\{g(0) \ ; \ \begin{array}{l} \bullet \ g \in \mathrm{Br}\left(\{f_0, f_1 \ldots, f_m\}\right), \\ \bullet \ \theta S_0(g) + \sum_{i=1}^{m} S_i(g) < \theta S_0(h) + \sum_{i=1}^{m} S_i(h) \end{array}\right\} \tag{1.10}$$

Then $\dot{x} = f_0(x) + \sum_{i=1}^{m} u_i f_i(x)$ is locally controllable.

The Sussmann condition has been improved by Hermes-Kawski in [HK] and by Bianchini-Stefani in [BS1] and [BS2]. Other sufficient conditions for local controllability can be found in [A], [AG], [BS2], [Kaw1], and [Tr].

Let us remark that the Sussmann condition with $\theta = 0$ is (an improvement of) the Hermes condition (see [He3] or [Su3])—and that the Sussmann condition implies that, as mentioned above, in the case of linear systems or of systems without drift, the Lie algebra rank condition (1.5) is a sufficient condition for local controllability. Indeed, in these cases, any h in $\mathrm{Br}\,(\{f_0, f_1 \ldots, f_m\})$ with $S_0(h)$ odd and $S_i(h)$ even for all i in $[1, m]$ vanishes at 0.

Let us end this section by some examples.

Examples 1.4. We take $n = 2$, $m = 1$, and

$$f(x, u) = f_0(x) + u f_1(x)$$

with

$$f_0(x_1, x_2) = (x_1 - x_2^3, 0), \; f_1(x_1, x_2) = (0, 1) \,.$$

Then $[f_1, [f_1, [f_1, f_0]]](= (-6, 0))$ and f_1 evaluated at $(0, 0)$ span all of $\mathbb{R}^2$. Therefore the Lie algebra rank condition (1.5) is satisfied. Moreover, any h in $\mathrm{Br}\,(\{f_0, f_1\})$ with an odd number of f_0 and an even number of f_1 vanishes at 0. Therefore, by Theorem 1.3 or the Hermes condition, $\dot{x}_1 = x_1^1 - x_2^3$, $\dot{x}_2 = u$ is locally controllable.

Example 1.5. We take $n = 3$, $m = 2$, and consider the control system without drift

$$\dot{x}_1 = u_1, \; \dot{x}_2 = u_2, \; \dot{x}_3 = x_1 u_2 - x_2 u_1 \,. \tag{1.11}$$

So, in this case, $f_0(x) = 0$, $f_1(x) = (1, 0, -x_2)$, and $f_2(x) = (0, 1, x_1)$. Then $[f_1, f_2](x) = (0, 0, 2)$ and therefore $f_1(x)$, $f_2(x)$, and $[f_1, f_2](x)$ span all $\mathbb{R}^3$ (for all x in $\mathbb{R}^3$). Hence, system (1.11) satisfies the Lie algebra rank condition (1.5) and since it has no drift it is locally controllable by Chow's theorem—or by Sussmann's condition. In fact, it is also globally controllable again by Chow's theorem.

2. Sufficient conditions for local stabilizability of locally controllable systems by means of stationary feedback laws

We are looking for a feedback law $u : \mathbb{R}^n \to \mathbb{R}^m$ vanishing at 0 such that 0 is (locally) asymptotically stable for the closed-loop system $\dot{x} = f(x, u(x))$. An important question is the regularity required for u. To impose on u to be of class $\mathcal{C}^\infty$ is too strong. For example, the system, with $n = m = 1$, $\dot{x} = x - u^3$ is locally controllable—this can be proved directly or by noticing that, as shown in section **1**, $\dot{x}_1 = x_1 - x_2^3 \in \mathbb{R}^3$, $\dot{x}_2 = u \in \mathbb{R}$ is locally controllable—but cannot be locally asymptotically stabilized by means of a feedback law $u : \mathbb{R} \to \mathbb{R}$ of class $\mathcal{C}^1$; on the other hand, if one takes $u(x) = (2x)^{\frac{1}{3}}$, which is a continuous feedback, then the origin is a (globally) asymptotically stable point of the closed-loop system $\dot{x} = x - ((2x)^{\frac{1}{3}})^3 = -x$. So we will require the feedback law to be only continuous. One could also look for feedback law which might be discontinuous (but satisfy $u(x) \rightsquigarrow 0$ as $x \rightsquigarrow 0$); we will not do it here, but only mention the papers [He1], [Su1], [FM], [AB], [CS], [CR], and the references therein on this subject. Here, we adopt

Definition 2.1. The system $\dot{x} = f(x, u)$ can be locally asymptotically stabilized by means of a continuous stationary feedback law if there exists a continuous map $u : \mathbb{R}^n \to \mathbb{R}^m$, $x \mapsto u(x)$, vanishing at 0 such that the origin of $\mathbb{R}^n$ is a locally asymptotically stable point of the closed-loop system $\dot{x} = f(x, u(x))$, i.e.

(i) there exists $\varepsilon > 0$ such that

$$\left(|x(0)| < \varepsilon \text{ and } \dot{x} = f(x, u(x)) \right) \implies \left(\lim_{t \rightsquigarrow +\infty} x(t) = 0 \right)$$

(ii) for any $\delta > 0$ there exists $\eta > 0$ such that

$$\left(|x(0)| < \eta \text{ and } \dot{x} = f(x, u(x)) \right) \implies \left(|x(t)| < \delta \quad \forall t \geqslant 0 \right).$$

In Definition 2.1, by $\dot{x} = f(x, u(x))$, we denote *any maximal* solution of $\dot{x}(t) = f(x(t), u(x(t)))$; note that with u only continuous one does not have uniqueness of the solution to the Cauchy problem $\dot{x} = f(x, u(x))$, $x(0) = x_0$. If in (i) one can take $\varepsilon = +\infty$, then we will say that $\dot{x} = f(x, u)$ can be globally asymptotically stabilized by means of a continuous stationary feedback law.

Then one has the following theorem

Theorem 2.2. *Assume that $\dot{x} = f(x, u)$ is locally controllable. Then it can be locally asymptotically stabilized by means of a continuous stationary feedback law in the following cases*

(i) *the linearized control system (around $x(t) \equiv 0$, $u(t) \equiv 0$) $\dot{y} = \frac{\partial f}{\partial x}(0, 0)y + \frac{\partial f}{\partial u}(0, 0)v$, where the state is $y \in \mathbb{R}^n$ and the control $v \in \mathbb{R}^m$, is also controllable,*

(ii) $n = 1$ and f is analytic (or continuous and subanalytic),

(iii) $n = 2$, $m = 1$ and $f(x, u) = f_0(x) + u f_1(x)$ where f_0 and f_1 are analytic function from $\mathbb{R}^2$ into $\mathbb{R}^2$.

Case(i) is classical (see e.g. [I; Chap. 4, Prop. 4.1], [NS; Chap. 10, Th. 10.5], [So2; Chap. 4, Th. 12]).

Case(ii) is a recent result due to Kawski [Kaw2] (see also [He4], [DMK] or [CPr]).

Case(iii) follows directly from [CPr; Lemma 1]. Let us briefly sketch a proof of this case. From the local controllability of $\dot{x} = f(x, u)$ we get that

$$(0,0) \in \overline{\{(x, u) \; ; \; x > 0, \; f(x, u) < 0\}} . \tag{2.1}$$

From the selection curve lemma (see e.g. [Hi; Prop. (3.9)]) we get the existence of an analytic map $\gamma = (\gamma_1, \gamma_2) : [-1, 1] \to \mathbb{R} \times \mathbb{R}^m$ such that

$$\gamma(0) = (0,0) , \tag{2.2}$$
$$\gamma_1(s) > 0 , \qquad \forall s \in (0, 1) , \tag{2.3}$$
$$f(\gamma_1(s), \gamma_2(s)) < 0 , \qquad \forall s \in (0, 1) . \tag{2.4}$$

By the analyticity of γ, (2.2), and (2.3) we get the existence of a continuous map $s : [s, +\infty) \to \mathbb{R}$ such that

$$s(0) = 0 \tag{2.5}$$

and, for $x \geqslant 0$ small enough,

$$\gamma_1(s(x)) = x . \tag{2.6}$$

Let $u_+ : [0, +\infty) \to \mathbb{R}^m$ be a continuous map such that, if $x \geqslant 0$ is small enough,

$$u_+(x) = \gamma(s(x)) . \tag{2.7}$$

By (2.2), (2.5) and (2.7), we get

$$u_+(0) = 0 . \tag{2.8}$$

By (2.4), (2.6) and (2.7) we get, if $x > 0$ is small enough,

$$f(x, u_+(x)) < 0 . \tag{2.9}$$

Similarly, one obtains the existence of a continuous map $u_- : (-\infty, 0] \to \mathbb{R}^m$ such that

$$u(0) = 0 \tag{2.10}$$

and, if $x < 0$ is such that $|x|$ is small enough,

$$f(x, u_-(x)) > 0 \qquad\qquad (2.11)$$

Finally let $u : \mathbb{R} \to \mathbb{R}^m$ be defined by

$$u(x) = u_+(x) \ \text{ if } \ x \geqslant 0, \ u(x) = u_-(x), \ \text{ if } \ x < 0; \qquad (2.12)$$

then u is continuous, vanishes at 0, and by (2.9), (2.11), and (2.12), 0 is locally asymptotically stable point of the closed-loop system $\dot{x} = f(x, u)$. Note that the assumption f analytic—or more generally f continuous and subanalytic—cannot be replaced by f is of class C^∞. Indeed, let $m = 1$ and let $f(x, u) = \sin(\frac{1}{u}) \exp\left(\frac{-1}{u^2}\right)$ if $u \neq 0$, $f(x, 0) = 0$; then $\dot{x} = f(x, u)$ is locally controllable but one easily checks that there exists no continuous map $u : \mathbb{R} \to \mathbb{R}$ vanishing at 0 such that, for some $\varepsilon > 0$,

$$f(x, u(x))x < 0, \qquad\qquad \forall x \in (-\varepsilon, \varepsilon) \backslash \{0\} ; \quad (2.13)$$

therefore $\dot{x} = f(x, u)$ cannot be locally asymptotically stabilized by means of continuous feedback law $x \mapsto u(x)$ (vanishing at 0).

Let us end up this section by mentioning that there have been numerous research articles (e.g. [Ba2], [BI], [CSV], [OS], [So2], [Ts1], [Ts2]), surveys (e.g. [So4], [D]), and even a nice book by Bacciotti [Ba1] published in the recent past on stabilization by means of stationary feedback laws of control systems—which are not necessarily controllable. For output stabilization, see e.g. the paper [TP] by Teel and Praly, the book [I] by Isidori, and the references therein.

3. Necessary conditions for local stabilizability by means of stationary feedback laws

The previous section suggests that, at least if f is analytic, local controllability might imply local stabilizability by means of continuous stationary feedback laws. In 1979, Sussmann has shown in [Su1] that the global statement is false in general and in 1983 Brockett has shown that even the local statement is false. More precisely, he has proved in [Br].

Theorem 3.1. *Assume that $\dot{x} = f(x, u)$ can be locally asymptotically stabilized by means of a continuous stationary feedback law. Then f maps any neighborhood of 0 (in $\mathbb{R}^n \times \mathbb{R}^m$) onto a neighborhood of 0 (in $\mathbb{R}^n$).*

And, as noticed by Brockett in [Br], for the system of Example 1.4, which is locally controllable, $f(\mathbb{R}^n \times \mathbb{R}^m) \subseteq \mathbb{R}^n \setminus (\{0\}^{n-1} \times (\mathbb{R} \setminus \{0\}))$; therefore this system does not satisfy Brockett's condition, *i.e.* f does not map any neighborhood of 0 onto a neighborhood of 0; hence, by Theorem 3.1, this system cannot be locally asymptotically stabilized by means of a continuous stationary feedback law.

In [Z] Zabczyk has observed that from a theorem due Krasnosel'kii—[Kr] or [KZ; Th. 52.1], see also [Hal; Rem. 2]—one can deduce another stronger necessary condition—called the index condition later on.

Theorem 3.2. *Assume that $\dot{x} = f(x, u)$ can be locally asymptotically stabilized by means of a continuous stationary feedback law. Then there exists a continuous map $u : \mathbb{R}^n \to \mathbb{R}^m$ vanishing at 0 such that $f(x, u(x)) \neq 0$ if x is small enough and the index of $x \mapsto f(x, u(x))$ at 0 is $(-1)^n$.*

For a definition of the index see, for example, [KZ; p. 9]. Theorem 3.2 is stronger than Theorem 3.1 since if the vector field $x \mapsto X(x)$ has an index which is not zero then X maps any neighborhood of 0 in $\mathbb{R}^n$ onto a neighborhood of 0 in $\mathbb{R}^n$. But an important disadvantage of the degree condition is that this condition, contrary to the Brockett condition, is not necessary for dynamical stabilizability. Let us recall the definition of dynamical stabilizability

Definition 3.3. The system $\dot{x} = f(x, u)$ can be locally asymptotically stabilized by means of a dynamic continuous stationary feedback law if there exists an integer p such that the control system

$$\dot{x} = f(x, u), \ \dot{y} = v, \tag{3.1}$$

where the state is $(x, y) \in \mathbb{R}^n \times \mathbb{R}^p$ and the control $(u, v) \in \mathbb{R}^m \times \mathbb{R}^p$, can be locally asymptotically stabilized by means of a continuous stationary feedback law.

Then it is proved in [CPr; Prop.1] that there are systems which do not satisfy the degree condition which can be locally asymptotically stabilized by means of a dynamic continuous stationary feedback law.

The Brockett condition, which is, as it follows directly from Theorem 3.1, a necessary condition for local asymptotic stabilizability by means of a dynamic continuous stationary feedback law, can be slightly strengthened in the following

Proposition 3.4. *Assume that $\dot{x} = f(x, u)$ can be locally asymptotically stabilized by means of a (dynamic or not) continuous stationary feedback law, then, for any positive and small enough ε,*

$$f_*\Big(\sigma_{n-1}\big(\{(x, u) \; ; \; |x| + |u| < \varepsilon, \; f(x, u) \neq 0\}\big)\Big) = \sigma_{n-1}(\mathbb{R}^n \backslash \{0\})(\mathbb{Z})$$

$$(3.2)$$

where $\sigma_{n-1}(A)$ is the stable homotopy group of order $(n-1)$ of A—for a definition of the stable homology groups, see e.g. [Wh; p. 552].

Let us notice that (3.2) with homology group instead of stable homotopy group is a weaker condition than (3.2). Condition (3.2) with homotopy group instead of stable homotopy group is stronger than (3.2) but it is too strong: it is not a necessary condition for local asymptotic stabilizability by means of a *dynamic* continuous stationary feedback law. The index condition implies (3.2) (even with homotopy group) and (3.2) implies, if f is analytic, that a "dynamic extension" of $\dot{x} = f(x, u)$ satisfies the index condition.

More precisely one has (see [Co5; Sect.2])

Proposition 3.5. *Assume that f is analytic—or only continuous and subanalytic. Assume that (3.2) is satisfied. Then, if $p \geqslant 2n + 1$, control system (3.1) satisfies the index condition.*

There are analytic locally controllable systems, which satisfy the Brockett condition but which do not satisfy (3.2) and therefore cannot be locally asymptotically stabilized by means of—dynamic or not—continuous stationary feedback laws (see [Co1; Sect.3]). One does not know any analytic locally controllable system which satisfies (3.2) and which is not locally asymptotically stabilized by means of a dynamic continuous stationary feedback law. But the natural guess is that such a system exists; a possible candidate (with $n = 3$, $m = 1$) is

$$\dot{x}_1 = x_3^2(x_1 - x_2), \; \dot{x}_2 = x_3^2(x_2 - x_3), \; \dot{x}_3 = u. \tag{3.3}$$

This control system satisfies (3.2), is locally controllable—it satisfies the Hermes condition— but cannot be locally asymptotically stabilized by means of a continuous stationary feedback law (see [Co1; Sect. 4]); we guess that it cannot also be locally asymptotically stabilized by means of a dynamic continuous stationary feedback law.

We end this section by some comments on homogeneous control systems. Let $r = (r_1, \ldots, r_n)$ be a sequence of n positive real numbers, let $s = (s_1, \ldots, s_m)$ be a sequence of m positive real numbers, and let τ be a real number in $(\min\{r_i \; ; \; i \in [1, n]\}, +\infty)$.

We assume that f satisfies the homogeneity condition

$$f^i\left(\varepsilon^{r_1} x_1, \ldots, \varepsilon^{r_n} x_n, \varepsilon^{s_1} u_1, \ldots, \varepsilon^{s_m} u_m\right) = \varepsilon^{\tau + r_i} f^i(x_1, \ldots, x_n, u_1, \ldots, u_m) \tag{3.4}$$

for all $x = (x_1, \ldots, x_n)$ in $\mathbb{R}^n$, all i in $[1, n]$, and all ε in $[0, +\infty)$ and where f^i is the i-th component of f. A stationary feedback law $u : \mathbb{R}^n \to \mathbb{R}^n$, $x = (x_1, \ldots, x_n) \mapsto u(x) = (u_1(x), \ldots, u_m(x))$ is homogeneous if

$$u_j(\varepsilon^{r_1} x_1, \ldots, \varepsilon^{r_n} x_n) = \varepsilon^{s_j} u_j(x_1, \ldots, x_n) \tag{3.5}$$

for all x in $\mathbb{R}^n$, all j in $[1, m]$, and all ε in $[0, +\infty)$. Kawski in [Kaw4] and Dayawansa-Martin in [DM] have given interesting necessary conditions for asymptotically stabilizability by means of homogeneous continuous stationary feedback laws; see also [D] and the references therein for sufficient conditions. Let us give a new simple necessary condition. Let, for λ in $\mathbb{R}$,

$$f_\lambda^i(x, u) = f_i(x, u) - \lambda r_i x_i \tag{3.6}$$

for all i in $[1, n]$. Then one has, with $f_\lambda = (f^1, \ldots, f^n)$

Proposition 3.6. *Assume that $\dot{x} = f(x, u)$ can be locally asymptotically stabilized by means of a homogeneous continuous stationary feedback law then, for all λ in $[0, +\infty)$, f_λ satisfies condition (3.2)—and therefore also the Brockett condition.*

Proof. Let u be a homogeneous continuous stationary feedback law such that 0 is a locally asymptotically stable point of $\dot{x} = X(x) := f(x, u(x))$. By homogeneity—see (3.4) and (3.5)— we have

$$X_i(\varepsilon^{r_1} x_1, \ldots, \varepsilon^{r_n} x_n) = \varepsilon^{\tau + r_i} X_i(x_1, \ldots, x_n) \tag{3.7}$$

for all $(x_1, \ldots, x_n)$ in $\mathbb{R}^n$ and all ε in $[0, +\infty)$.

Then, by a homogeneous version of a converse of Lyapunov's second theorem due to Rosier [Ro1], there exists V in $\mathcal{C}^1(\mathbb{R}^n; \mathbb{R})$ such that

$$V(0) = 0, \; V(x) > 0, \; \forall x \in \mathbb{R}^n \setminus \{0\}, \tag{3.8}$$

$$V(x) \to \infty, \qquad \text{as } |x| \to +\infty, \tag{3.9}$$

$$\nabla V(x) \cdot X(x) < 0, \qquad \forall x \in \mathbb{R}^n \setminus \{0\}, \tag{3.10}$$

and, for some positive real number k,

$$V(\varepsilon^{r_1} x_1, \ldots, \varepsilon^{r_n} x_n) = \varepsilon^k V(x_1, \ldots, x_n) \tag{3.11}$$

for all $(x_1, \ldots, x_n)$ in $\mathbb{R}^n$ and all ε in $[0, +\infty)$. From (3.8) and (3.11) we get

$$\sum_{i=1}^{m} r_i x_i \frac{\partial V}{\partial x_i}(x) = kV(x) \geqslant 0, \qquad \forall x \in \mathbb{R}^n \tag{3.12}$$

From (3.6), (3.10), and (3.12) we get that

$$\nabla V(x) \cdot f_\lambda(x, u(x)) < 0 \,\forall x \in \mathbb{R}^n \setminus \{0\}, \ \forall \lambda \in [0, +\infty) \tag{3.13}$$

and therefore, by (3.8), (3.13), and the Lyapunov second theorem, 0 is a globally asymptotically stable point of $\dot{x} = f_\lambda(x, u(x))$. The conclusion follows from Proposition 3.4. ∎

Let us apply Proposition 3.6 to a linear system $\dot{x} = f(x, u) = Ax + Bu$, where A is a linear map from $\mathbb{R}^n$ into $\mathbb{R}^n$ and B a linear map from $\mathbb{R}^m$ into $\mathbb{R}^n$. So we take $(r_1, \ldots, r_n) = (1, \ldots, 1)$, $(s_1, \ldots, s_m) = (1, \ldots, 1)$, and $\tau = 0$. In this case Proposition 3.6 reads: assume that $\dot{x} = f(x, u) = Ax + Bu$ can be locally asymptotically stabilized by means of a homogeneous continuous stationary feedback law, then

$$\{(A - \lambda)x + Bu \,;\, x \in \mathbb{R}^n, \ u \in \mathbb{R}^m\} = \mathbb{R}^n, \qquad \forall \lambda \in [0, +\infty). \tag{3.14}$$

Let us recall that $\dot{x} = f(x, u) = Ax + Bu$ can be locally (or globally) asymptotically stabilized by means of a—homogeneous or not—continuous stationary feedback law if and only if it satisfies the Hautus condition

$$\{(A - \lambda)x + Bu \,;\, x \in \mathbb{C}^n, \ u \in \mathbb{C}^m\} = \mathbb{C}^n, \ \forall \lambda \in [0, +\infty) \times \mathbb{R} \subseteq \mathbb{C} \tag{3.15}$$

It would be interesting to know if there is an improvement of Proposition 3.6 which, in the case of linear systems, gives (3.15) instead of (3.14).

Let us finally mention that there are homogeneous control systems which cannot be locally asymptotically stabilized by means of a homogeneous continuous stationary feedback law. A first example with f of class C^k is given by Rosier in [Ro2]. For f analytic a first example has been found by Aeyels, Sepulchre and Mareels in [ASM].

4. Stabilization by means of time-varying feedback laws

We have seen in the previous section that many—even analytic—locally controllable systems cannot be locally asymptotically stabilized by means of a continuous stationary feedback law. The goal of this section is to state theorems which show that, in many cases, this does not hold if one replaces "stationary" by means of "time-varying periodic" *i.e.* the feedback law is now a function of x and t, instead of x only, which is periodic in time.

The interest of time-varying feedback law compared to stationary feedback law has been pointed out by Sontag and Sussmann in [SS] for $n = 1$, by Wang in [Wa] for decentralized linear systems, and by Samson in [Sa] for system (1.11). Let us also mention the work by Kapitsa in 1951 described in particular in [LL; Chap. 5, §30].

Similarly to Definition 2.1, we adopt

Definition 4.1. Let T be in $[0, +\infty)$. The system $\dot{x} = f(x, u)$ can be locally (resp. globally) asymptotically stabilized by means of a continuous time-varying T-periodic feedback law if there exists a continuous map $u : \mathbb{R}^n \times \mathbb{R} \to \mathbb{R}^m$, $(x, t) \mapsto u(x, t)$, vanishing on $\{0\} \times \mathbb{R}$, T-periodic with respect to t, such that the origin of $\mathbb{R}^n$ is a locally (resp. globally) asymptotically stable point of $\dot{x} = f(x, u(x, t))$.

The first result of this section is that, roughly speaking, for systems without drift, controllability implies asymptotic stabilization by means of smooth time-varying periodic feedback laws. More precisely, one has the following theorem, proved in [Co2],

Theorem 4.2. *Assume that* $f(x, u) = \sum_{i=1}^{m} u_i f_i(x)$ *satisfies*

$$\{h(x) \; ; \; h \in \mathrm{Lie}\left(\{f_1, \ldots, f_m\}\right)\} = \mathbb{R}^n , \qquad \forall x \in \mathbb{R}^n \setminus \{0\} . \quad (4.1)$$

Then, for all positive T, $\dot{x} = f(x, u)$ *can be globally asymptotically stabilized by means of a time-varying* T-*periodic feedback law of class* $\mathcal{C}^\infty$.

Note that it follows easily from [Co2] that a local version of Theorem 4.2 holds, *i.e.* if (4.1) holds only for $|x|$ small enough but positive then, for any $T > 0$, $\cdot x = \sum_{i=1}^m u_i f_i(x)$ can be locally stabilized by means of a time-varying T-periodic feedback law of class C^∞. We sketch below a proof of Theorem 4.2.

In order to state the next result which allows a drift term we need the following definition

Definition 4.3. The origin (of $\mathbb{R}^n$) is locally continuously reachable in small time for $\dot x = f(x, u)$ if, for any positive real number τ, there exists u in $C^0\left(\mathbb{R}^n; L^1\left((0, \tau); \mathbb{R}^m\right)\right)$ and a positive real number ε such that

$$\sup\left\{|u(a)(t)| \; ; \; t \in (0, \tau)\right\} \rightsquigarrow 0 \quad \text{as} \quad a \rightsquigarrow 0, \tag{4.2}$$

$$\left(\dot x = f\big(x, u(x(0))(t)\big) \quad \text{and} \quad |x(0)| < \varepsilon\right) \Longrightarrow \left(x(\tau) = 0\right). \tag{4.3}$$

Let us mention that modifying slightly arguments given by Hermes in [He2] and by Kawski in [Kaw3] one can prove (see [Co3]) that "many" sufficient conditions for local controllability imply that the origin is locally continuously reachable in small time. This is in particular the case for the Sussmann condition (Theorem 1.3). In fact, we conjecture a positive answer to the question

Question 4.4. Assume that f is analytic and that $\dot x = f(x, u)$ is locally controllable. Is the origin always locally continuously reachable in small time ?

Note that, as shown by Sussmann in [Su1], the global version of Question 4.4 has a negative answer. Our next result is

Theorem 4.5. *Assume that $n \geqslant 4$, the origin is locally continuously reachable in small time, and the Lie algebra rank condition (1.5) holds. Then, for any positive T, $\dot x = f(x, u)$ can be locally asymptotically stabilized by means of a continuous time-varying T-periodic feedback law u. Moreover, one can require local stabilization in time T i.e. that for some positive real number ε and for any real number s*

$$\left(\dot x = f(x, u(x, t)), \; |x(s)| < \varepsilon\right) \Longrightarrow \left(x(t) = 0 \quad \forall t \geqslant s + T\right). \tag{4.4}$$

For a proof and some results if $n \leqslant 3$, see [Co4]. Let us just give the main idea of the proof of Theorem 4.2. It consists of constructing first a time-varying T-periodic feedback law $\bar{u}$ of class $\mathcal{C}^\infty$ vanishing on $\{0\}\times\mathbb{R}$ such that

$$\left(\dot{x} = \sum_{i=1}^{m} \bar{u}_i(x,t)f_i(x)\right) \Longrightarrow \left(x(0) = x(T)\right) \tag{4.5}$$

and, for any x_0 in $\mathbb{R}^n\setminus\{0\}$, the linearized control system around the trajectory $x(x_0,t)$ defined by $\frac{\partial x}{\partial t} = \sum_{i=1}^{m} \bar{u}_i(x,t)f_i(x)$, $x(x_0,0) = x_0$ is controllable on $[0,T]$. Using (4.5), and this controllability of the linearized systems it is intuitively clear that one can perturb slightly $\bar{u}$ in a suitable way in order to get a time-varying T-periodic feedback law u of class $\mathcal{C}^\infty$ vanishing on $\{0\}\times\mathbb{R}$ such that

$$\left(\dot{x} = \sum_{i=1}^{m} u_i(x,t)f_i(x),\ x(0) \neq 0\right) \Longrightarrow \left(|x(T)| < |x(0)|\right), \tag{4.6}$$

which implies that for $\dot{x} = \sum_{i=1}^{m} u_i(x,t)f_i(x)$, 0 is a globally asymptotically stable point. In order to construct $\bar{u}$ one can proceed in the following way. One looks for $\bar{u}$ of the form

$$\bar{u}(x,t) = a(x)b(t), \qquad \forall(x,t) \in \mathbb{R}^n\times\mathbb{R} \tag{4.7}$$

with $a \in \mathcal{C}^\infty(\mathbb{R}^n;\mathbb{R}) \cap \mathcal{C}^\infty(\mathbb{R}^n\setminus\{0\};(0,+\infty))$, $a(0) = 0$, $b \in \mathcal{C}^\infty(\mathbb{R};\mathbb{R}^m)$, and

$$b(t) = b(t+T) = -b(T-t), \qquad \forall t \in \mathbb{R} \tag{4.8}$$

Note that

$$\bar{u}(x,T-t) = -\bar{u}(x,t), \qquad \forall(x,t) \in \mathbb{R}^n\times\mathbb{R} \tag{4.9}$$

which implies (4.5). Let us recall—see [SM] or [So3; Prop. 3.5.15]—that the time-varying linear control system $\dot{y} = A(t)y + B(t)v$, where the state is y in $\mathbb{R}^n$ and the control v in $\mathbb{R}^m$, is controllable in $[0,T]$ if one has for some $\bar{t}$ in $[0,T]$

$$\mathrm{Span}\left\{\left(\tfrac{d}{dt} - A(t)\right)^i B(t)\big|_{t=\bar{t}} v\ ;\ i \geqslant 0,\ v \in \mathbb{R}^m\right\} = \mathbb{R}^n. \tag{4.10}$$

We choose $\bar{t} = \frac{3T}{4}$—for example—. It is proved in [Co2] that, for generic b in $\mathcal{C}^\infty(\mathbb{R};\mathbb{R}^m)$ satisfying (4.8), then, for a small enough in $\mathcal{C}^0(\mathbb{R}^n\setminus\{0\};(0,+\infty))$ for the $\mathcal{C}^0$ Whitney topology, the linearized control systems around the trajectories $x(x_0,t)$ with $x_0 \neq 0$ satisfy (4.10) and therefore are controllable on $[0,T]$, which ends the proof. ∎

Let us end this section by a few comments

Remark 4.6

(a) In [P] Pomet has introduced an interesting method which allows, in particular, to compute u from $\bar{u}$—see also [CPo].

(b) A generic b in $\mathcal{C}^\infty(\mathbb{R};\mathbb{R}^m)$ may be hard to find! But let us assume that for some integer l iterated Lie brackets of $f_1,\ldots,f_m$ of length at most l span $\mathbb{R}^n$ at each point in $\mathbb{R}^n\setminus\{0\}$ then one can proceed in the following way. First choose a function $c : [\frac{T}{2},T] \to \mathbb{R}$ of class $\mathcal{C}^\infty$ vanishing in a neighborhood of $\{\frac{T}{2},T\}$ and equal to 1 in a neighborhood of $\frac{3T}{4}$. Then, for a polynomial $p : \mathbb{R} \to \mathbb{R}^m$ of degree at most q, let b defined on $[\frac{T}{2},T]$ by
$$b(t) = c(t)p(t)$$
and extend b to all of $\mathbb{R}$ by requiring b to be T-periodic and to be odd with respect to $\frac{T}{2}$ (see (4.8)). Then, it is proved in [Co2] that there is an algebraic subset Σ of the set of polynomial of degree at most $l^2 m^l$ of codimension one such that, for a small enough in $\mathcal{C}^0(\mathbb{R}^n\setminus\{0\};(0,+\infty))$, the linearized control systems around the trajectories $x(x_0,t)$ with $x_0 \neq 0$ satisfy (4.10). Note also that Σ is independent of $f_1,\ldots,f_m$—provided that iterated Lie brackets of $f_1,\ldots,f_m$ of length at most l span $\mathbb{R}^n$ at each point in $\mathbb{R}^n\setminus\{0\}$.

(c) In [Gr; 2.3.8. (E)] Gromov have proved that generic undetermined linear (partial) differential equations are algebraically solvable. Note that (4.10) for all τ in $(0,T)$ is equivalent to the algebraic solvability of $\dot{y} - A(t)y - B(t)v = r(t)$ in $(0,T)$ where $r(t)$ is given and (y,v) are the unknown (see [INS], [Gr; 2.3.8 (B)], and [Co2]). Of course, our linearized control systems are not generic but it turns out that the genericity of b is enough to conclude.

(d) In [So1] Sontag has shown that if $\dot{x} = f(x,u)$ satisfies the Lie algebra rank condition (1.5) then, even if $f(x,u) \neq \sum_{i=1}^m u_i f_i(x)$, there exists a control law $t \mapsto u(t)$ such that the linearized control system around the trajectory $\dot{x} = f(x,u(t))$, $x(0) = 0$ is controllable on $[0,T]$ (if T is small enough in order to avoid blow-up). This in fact still holds for generic u, see [Co3]. Moreover, Sontag in [So5] has shown, in particular, that such a genericity result can also be deduced from a result due to Sussmann on observability [Su2] if f is analytic; he has also deduced from [Su2] that the strong accessibility rank condition on $\mathbb{R}^n$

implies that, still if f is analytic, for generic $t \mapsto u(t)$ the linearized control system around *any* trajectory of $\dot{x} = f(x, u(t))$ is controllable; a similar result also holds with f only of class$\mathcal{C}^\infty$, see [Co4] for a proof and related results.

(e) One can find interesting explicit examples of time-varying stabilizing feedback laws in [Sa], [Po], [MWS] and [SCW].

5. Return method and controllability

In this last section, we present a simple strategy, that we will call the *return method*, directly inspired from the proof of Theorem 4.2, to prove controllability in some cases. We give an example to the controllability of a distributed control system: the 2-D incompressible Euler equations.

The return method is the following one. Let T be positive real number. Assume that one can find $\bar{u} : [0, T] \to \mathbb{R}^m$ such that, with $\bar{x} : [0, T] \to \mathbb{R}^n$ defined by $\dot{\bar{x}} = f(\bar{x}, \bar{u}(t))$, $\bar{x}(0 = 0)$, then $\bar{x}(T) = 0$ and the linearized control system around $(x(t), u(t))$ is controllable on $[0, T]$. Then, clearly, $\dot{x} = f(x, u)$ is, locally in state, controllable in time T; *i.e.* there exists a positive real number ε such that for any x_0 and x_1 in $\mathbb{R}^n$ of norm less than ε, there exists an open loop control $u : [0, T] \to \mathbb{R}^m$ such that

$$\Big(\dot{x} = f(x, u(t)), \ x(0) = x_0\Big) \Longrightarrow \Big(x(T) = x_1\Big). \qquad (5.1)$$

Note that in the proof of Theorem 4.2 that we have sketched in the previous section one does not use Chow's theorem; we do not also use the theorem due to Sussmann-Jurdjevic [SJ] and Krener [K] which tells that the strong Lie algebra rank condition implies that the set of point which can be reached from 0 has a non empty interior. But the proof of Theorem 4.2, which relies on the return method, allows to recover Chow's theorem: for system without drift local controllability implies easily global controllability. Of course, this is not very interesting since there are very simple proofs of Chow's theorem available; see e.g. [Ga; Th. II, 23] or [NS; Prop. 3.15].

For distributed systems x is in an infinite dimensional space and Lie brackets do not seem to be, at least for the moment being, very successful. In this case, the return method can be useful since it allows to reduce in some cases the controllability of a nonlinear problem to the problem of the controllability of a (time-varying) linear control system and such a problem have been very much studied; see e.g. [L1], [L2]—the HUM method—, [Ru2], the references therein and in particular [BLR]. Let us remark that in the infinite dimensional case the controllability of the linearized control system around $(\bar{x}(t), \bar{u}(t))$ does not, in general, imply the local state controllability; but using the usual implicit function theorem or the Nash-Moser method if there is some "loss of derivatives" (see e.g. [Ham]) one can show that this implication still holds in some interesting cases.

As an example, let us briefly explain how the return method can be used to prove boundary controllability for the 2-D incompressible Euler equations on a connected and simply connected bounded domain in $\mathbb{R}^2$. Let Ω be a connected bounded open subset of $\mathbb{R}^2$ of class C^∞. Let Γ_0 be non empty compact submanifold of dimension 1 of $\Gamma := \partial\Omega$; Γ_0 may have a boundary. Our controllability problem is the following one. Let T be a positive real number, let y_0 and y_1 be two maps in $C^\infty(\overline{\Omega}; \mathbb{R}^2)$ such that

$$\operatorname{div} y_0 = \operatorname{div} y_1 = 0, \qquad \text{in } \Omega, \qquad (5.1)$$

$$y_0 \cdot n = y_1 \cdot n, \qquad \text{on } \Gamma \backslash \Gamma_0, \qquad (5.2)$$

where n is the unit normal vector to Γ directed towards the exterior of Ω. Does there exists y in $C^\infty(\overline{\Omega} \times [0, T]; \mathbb{R}^2)$ and p in $C^\infty(\overline{\Omega} \times [0, T]; \mathbb{R})$ such that

$$y_t + (y \cdot \nabla)y + \nabla p = 0, \qquad \text{in } \overline{\Omega} \times [0, T], \qquad (5.3)$$

$$y(\cdot, 0) = y_0(\cdot), \qquad (5.4)$$

$$y(\cdot, T) = y_1(\cdot), \qquad (5.5)$$

$$y(x, t) \cdot n(x) = 0, \quad \forall (x, t) \in (\Gamma \backslash \Gamma_0) \times [0, T], \qquad (5.6)$$

$$\operatorname{div} y(\cdot, t) = 0, \qquad \forall t \in [0, T]. \qquad (5.7)$$

Note that the control, which does not appear explicitly in our formulation, is $y(x, t) \cdot n(x)$ on $\Gamma_0 \times [0, T]$ as well as the vorticity $\omega(x, t)$ on $\Gamma_0 \times [0, T]$ when $y(x, t) \cdot n(x) < 0$; see [Kaz] for example.

If for any T, y_0, and y_1 as above such functions y and p exist, we will say that the 2-D incompressible Euler equation is controllable for (Ω, Γ_0). Our next result gives a partial positive answer in dimension 2 to a conjecture raised by J.-L. Lions in [L3; Sect. 7] (see also [L4; Sect. 4]).

Theorem 5.1. *If Ω is simply connected, the 2-D incompressible Euler equation is controllable for (Ω, Γ_0).*

We briefly sketch the main ingredients of the proof; see [Co7] for more details. First one takes notices that, by straightforward arguments relying on change of scale $\big(y_\lambda(x,t) = \lambda y(x, \lambda t), p_\lambda(x,t) = \lambda^2 p(x, \lambda t)\big)$, and reverse of time $\big(y_-(x,t) = -y(x, T-t),\, p_-(x,t) = p(x, T-t)\big)$, we may assume that $T = 1$, $y_1 = 0$ and y_0 is small (for example in the $\mathcal{C}^2$ norm). There is in fact a slight problem of regularity with respect to time to reduce the general case to $T = 1$, $y_1 = 0$, and y_0 small; we take care of this problem in [Co7]. For simplicity, we will now assume that there exists two disjoint 1-dimensional connected submanifolds of Γ_0 with boundary Γ_0^+ and Γ_0^- such that

$$\{x \in \Gamma_0 \; ; \; y_0(x) \cdot n(x) \neq 0\} \subseteq \big(\Gamma_0^+ \cup \Gamma_0^-\big). \tag{5.8}$$

For the general case, see [Co7]; let us also notice that (5.8) is, of course, satisfied if, for example, $y_0(x) \cdot n(x) = 0$ for all x in Γ_0. We now use the return method. Let $q \in \mathcal{C}^\infty(\overline{\Omega}; \mathbb{R})$ be such that

$$\Delta\theta = 0, \qquad\qquad \text{in } \Omega, \tag{5.9}$$

$$\frac{\partial\theta}{\partial n} = 0, \qquad\qquad \text{on } \Gamma \backslash (\Gamma_0^+ \cup \Gamma_0^-), \tag{5.10}$$

$$\frac{\partial\theta}{\partial n} < 0, \qquad\qquad \text{on } \Gamma_0^+ \backslash \partial\Gamma_0^+, \tag{5.11}$$

$$\frac{\partial\theta}{\partial n} > 0, \qquad\qquad \text{on } \Gamma_0^- \backslash \partial\Gamma_0^-. \tag{5.12}$$

Let α in $\mathcal{C}^\infty([0,1]; [0,+\infty))$ be such that

$$\alpha(0) = \alpha(1) = 0, \tag{5.13}$$

$$\alpha > 0, \qquad\qquad \text{on } (0,1). \tag{5.14}$$

Let M be a positive real number and let $\bar{y} \in \mathcal{C}^\infty(\overline{\Omega} \times [0,1]; \mathbb{R}^2)$ and $\bar{p} \in \mathcal{C}^\infty(\overline{\Omega} \times [0,1]; \mathbb{R})$ be defined by

$$\bar{y}(x,t) = M\alpha(t)\nabla\theta(x), \qquad\qquad \forall(x,t) \in \overline{\Omega} \times [0,1], \tag{5.15}$$

$$\bar{p}(x,t) = -M\dot{\alpha}(t)\theta(x) - (M^2/2)\alpha(t)^2|\nabla\theta(x)|^2, \forall(x,t) \in \overline{\Omega} \times [0,1]. \tag{5.16}$$

Note that we have

$$\bar{y}_t + (\bar{y} \cdot \nabla)\bar{y} + \nabla\bar{p} = 0, \qquad\qquad \text{in } \overline{\Omega} \times [0,T], \tag{5.17}$$

$$\bar{y}(x,0) = 0, \qquad\qquad \forall x \in \overline{\Omega}, \tag{5.18}$$

$$\bar{y}(x,1) = 0, \qquad\qquad \forall x \in \overline{\Omega}, \quad (5.19)$$

$$\bar{y}(x,t)\cdot n(x) = 0, \quad \forall (x,t) \in (\Gamma\backslash\Gamma_0)\times[0,T], \quad (5.20)$$

$$\operatorname{div} \bar{y}(\cdot,t) = 0, \qquad\qquad \forall t \in [0,T]. \quad (5.21)$$

Then, one checks that the linearized control system around $\bar{y}$ is controllable on $[0,T]$ if

$$\nabla\theta(x) \neq 0, \qquad\qquad \forall x \in \overline{\Omega}, \quad (5.22)$$

and M is large enough. Let us emphasize that, if $M = 0$ (*i.e.* $\bar{y} = 0$) or if (5.22) does not hold, the linearized control system around $\bar{y}$ is not controllable on $[0,T]$. One can prove that (5.22) holds by using arguments relying on Morse theory, the topology of Ω, Γ_0^+, and Γ_0^-, and the behaviour of a harmonic function near a critical point. As mentioned above, the fact that the linearized control system is controllable is not enough to conclude that our system is locally controllable "along $\bar{y}$" but, using an extension method similar to the one introduce by Russell in [Ru1], one can prove directly that this local controllability "along $\bar{y}$" holds, which ends the proof. In fact in [Co7] we use a slightly different construction for θ in order to simplify the proof. By local controllability along $\bar{y}$, we mean that if y_0 and y_1 are small enough, there exists y and p close to $\bar{y}$ and $\bar{p}$ and satisfying (5.3), (5.4), (5.5), (5.6) with $\Gamma_0^+\cup\Gamma_0^-$ instead of Γ_0, and (5.7). ∎

Remark 5.2

(a) In [L3] and [L4] J.-L. Lions makes the conjecture that turbulence helps for controllability. Arguments in favor of this conjecture are given in [L3]. The return method provides also a (vague) argument for this conjecture; indeed when there is turbulence there are a lot of periodic trajectories.

(b) If Ω is not simply connected, the conclusion of Theorem 5.1 does not hold in general. Indeed a necessary condition to have controllability of the 2-D incompressible Euler equation for (Ω,Γ_0) is that any connected component of $\delta\Omega$ meets Γ_0: if y_0 and y_1 are such that

$$\int_\Gamma y_0\cdot\delta r \neq \int_\Gamma y_1\cdot\delta r \qquad\qquad (5.23)$$

for a connected component Γ^* of $\partial\Omega$ which does not meet Γ_0, then, by Kelvin's law, there exists no (y,p) satisfying (5.3) to (5.7). The same argument shows that one has never controllability with a distributed control on any closed set included in Ω.

(c) The main problem to extend Theorem 5.1 to the dimension 3—with Ω contractible instead of simply connected—is that the Morse theory does no longer lead to (5.22). The reason is that, roughly speaking, there is no topological obstruction to the existence of two critical points for θ of Morse index 1 and 2 respectively, and, moreover, such indices are possible for a harmonic function in this dimension.

Acknowledgments. The author thanks J. Mawhin for interesting references concerning Theorem 3.2, D. Aeyels and R. Sepulchre for fruitful discussions concerning Proposition 3.6, P.-L. Lions and J.-L. Lions for bringing his attention to the controllability of the equations of incompressible fluids, and J.-C. Saut for useful references concerning these equations.

References

[A] A. A. Agrachev, Newton diagram and tangent cones to attainable sets, Colloque international sur l'Analyse des Systèmes contrôlés, Lyon, June 1990, Birkhäuser.

[AB] A. Andreini and A. Bacciotti, Stabilization by piecewise constant feedback, Proc. of the first European Control Conference, July 2-5 (1991), Vol. 1, p. 474-479.

[AG] A. A. Agrachev and R. V. Gamkrelidze, *Local controllability and semigroups of diffeomorphisms*, Acta Appl. Math., 32 (1993), no. 1, 1-57.

[ASM] D. Aeyels, R. Sepulchre and I. Mareels, On the stabilization of planar homogeneous systems, Preprint, Université Catholique de Louvain, 1993.

[Ba1] A. Bacciotti, Local stabilization of nonlinear control systems, Series on advances in Mathematics for applied sciences, Vol. 8, World scientific, Singapore - New Jersey - London - Hong Kong, 1992.

[Ba2] A. Bacciotti, Linear feedback: the local and potentially global stabilization of cascade systems, Nolcos'92, M. Fliess ed., Bordeaux, June 92, p. 21-25.

[BI] C. I. Byrnes and A. Isidori, Asymptotic stabilization of minimum phase nonlinear systems, *IEEE Trans. Aut. Control*, 36 (1991) p. 1122-1137.

[BLR] C. Bardos, G. Lebeau and J. Rauch, Sharp sufficient conditions for the observation, control and stabilization of waves from the boundary, *SIAM J. Control Optim.*, 30 (1992) p. 1024-1065.

[Br] R. W. Brockett, Asymptotic stability and feedback stabilization, in: *Differential geometric Control Theory*, R. W. Brockett, R. S. Millman and H. J. Sussmann eds, Birkhäuser, Basel - Boston, 1983.

[BS1] R. M. Bianchini and G. Stefani, Sufficient conditions of local controllability, Proc. of 25th Conference on Decision and Control, Athens, Greece, Dec. 1986, p. 967-970.

[BS2] R. M. Bianchini and G. Stefani, Controllability along a trajectory: a variational approach, *SIAM J. Control Optim.*, **31** (1993), no. 4, 900–927.

[Ch] W. L. Chow, Über systeme von linearen partiellen differentialgleichungen ester ordnung, *Math. Ann.*, 117 (1940-41) p. 98-105.

[Co1] J.-M. Coron, A necessary condition for feedback stabilization, *Systems and Control Letters*, 14 (1990) p. 227-232.

[Co2] J.-M. Coron, Global asymptotic stabilization for controllable systems without drift, *Math. Control Signals Systems*, 5 (1992) p. 295-312.

[Co3] J.-M. Coron, Links between local controllability and local continuous stabilization, Preprint Eth-Zürich and Université Paris-Sud, October 1991, and Nolcos'92, M. Fliess ed., Bordeaux, June 92, p. 477-482.

[Co4] J.-M. Coron, Linearized control systems and applications to smooth stabilization, *SIAM J. Control Optim.*, 32 (1994) p. 358-386.

[Co5] J.-M. Coron, Relations entre commandabilité et stabilisations non linéaires, Collège de France Seminar, H. Brezis and J.-L. Lions eds., Research Notes in Math. Pitman, Boston, To appear.

[Co6] J.-M. Coron, On the stabilization in finite time of locally controllable systems by means of continuous time-varying feedback laws, *SIAM J. Control Optim.* 33 (1995) p. 804-833.

[Co7] J.-M. Coron, Contrôlabilité exacte frontière de l'équation d'Euler des fluides parfaits incompressibles bidimensionnels, *C. R. Acad. Sci. Paris Ser.I Math.*, **317** (1993), no. 3, 271–276.

[CPo] J.-M. Coron and J.-B. Pomet, A remark on the design of time-varying stabilizing feedback laws for controllable systems without drift, Nolcos'92, M. Fliess ed., Bordeaux, June 1992, p. 413-417.

[CPr] J.-M. Coron and L. Praly, Adding an integrator for the stabilization problem, *Systems and Control Letters*, 17 (1991) p. 89-104.

[CR] J.-M. Coron and L. Rosier, A relation between continuous time-varying and discontinuous feedback stabilization, *J. Math. Syst., Estimation and Control* (1993) To appear.

[CS] C. Canudas de Wit and O. J. Sordalen, Examples of piecewise smooth stabilization of driftless NL Systems with less inputs than states, Nolcos'92, Bordeaux, June 1990, M. Fliess ed., p. 26-30.

[CSV] R. Chabour, G. Sallet, and J. C. Vivalda, Stabilization of nonlinear two dimensional systems, a bilinear approach, Nolcos'92, Bordeaux, June 1990, M. Fliess ed., p. 230-235.

[D] W. P. Dayawansa, Recent advances in the stabilization problem for low dimensional systems, Nolcos'92, M. Fliess ed., Bordeaux, June 1992, p. 1-8.

[DM] W. P. Dayawansa and C. F. Martin, Asymptotic stabilization of law dimensional systems, in *Nonlinear synthesis, Progress in Systems and Control Theory*, Vol. 9, Birkhäuser, Boston, 1991.

[DMK] W. P. Dayawansa, C. Martin, and G. Knowles, Asymptotic stabilization of a class of smooth two dimensional systems, *SIAM J. Control Optim.*, 28 (1990) p. 1321-1349.

[FM] M. Fliess and F. Messager, Vers une stabilisation non linéaire discontinue, Proc. 9th Conf. Analysis and Optimization of Systems, Lecture Notes Control Inform., Springer, Berlin, 1990.

[Ga] J. P. Gauthier, *Structure des systèmes non-linéaires*, Editions du CNRS, Paris, 1984.

[Gr] M. Gromov, *Partial differential relations*, Ergebnisse der Mathematik und ihrer Grenzgebiete, Folge 3, Bd. 9, Springer-Verlag, Berlin-Heidelberg, 1986.

[Hal] A. Halanay, Solutions périodiques des systèmes non linéaires à petit paramètre, *Atti della Accademia Nazionale dei Lincei*, 22 (1957), p. 30-32.

[Ham] R. S. Hamilton, The inverse function theorem of Nash and Moser, *Bull. Am. Math. Soc.*, 7 (1982) p. 65-222.

[He1] H. Hermes, Discontinuous vector fields and feedback control, in *Differential equations and dynamical systems*, J. K. Hale and J. P. La Salle eds., Academic Press, New York and London, 1967.

[He2] H. Hermes, On the synthesis of a stabilizing feedback control via Lie algebraic methods, *SIAM J. Control Optim.*, 18 (1980) p. 352-361.

[He3] H. Hermes, Control systems which generate decomposable Lie algebras, *J. Diff. Equations*, 44 (1982) p. 166-187.

[He4] H. Hermes, Homogeneous coordinates and continuous asymptotically stabilizing feedback controls, in *Differential Equations, Stability and Control*, S. Elaydi ed., *Lecture Notes in Pure and Appl. Math.*, 127, Marcel Dekker, Inc., (1990) p. 249-260.

[Hi] H. Hironaka, Subanalytic sets, in Number theory algebraic geometry and commutative algebra, in honor of Y. Akizuki, Kinokuniya Publications, Tokyo (1973) p. 453-493.

[HK] H. Hermes and M. Kawski, Local controllability of a single-input affine system, Nonlinear Anal. and Appl., Lecture Notes Pure and Appl. Math., 109 (1987) p. 235-248.

[I] A. Isidori, *Nonlinear Control Systems*, Springer-Verlag, Berlin Heidelberg, 1989.

[INS] A. Ilchmann, I. Nuernberger, and W. Schmale, Time-varying polynomial matrix systems, *International J. Control*, 40 (1984) p. 329-362.

[Kaw1] M. Kawski, Control variations with an increasing number of switchings, *Bull. Amer. Math. Soc.*, 18 (1988) p. 149-152.

[Kaw2] M. Kawski, Stabilization of nonlinear systems in the plane, *Systems and Control Letters*, 12 (1989) p. 169-175.

[Kaw3] M. Kawski, Higher-order small-time local controllability, in *Nonlinear Controllability and Optimal Control*, H. J. Sussmann ed., Monographs and Textbooks in Pure and Applied Mathematics, 113, Marcel Dekker, Inc., New York, 1990, p. 431-467.

[Kaw4] M. Kawski, Homogeneous stabilizing feedback laws, *Control Theory and Advanced Technology*, 6 (1990) p. 497-516.

[Kaz] A. V. Kazhikov, Note on the formulation of the problem of flow through a bounded region using equations of perfect fluid, *PMM U.S.S.R.*, 44 (1981) p. 672-674.

[Kr] M. A. Krasnosel'Skii, *The operator of translation along the trajectories of differential equations*, Trans. Math. Monographs, 19, Providence, Amer. Math. Soc., 1968.

[KZ] M. A. Krasnosel'Skii and P. P. Zabreiko, *Geometrical Methods of nonlinear Analysis*, Springer-Verlag, Berlin Heidelberg, 1984.

[L1] J.-L. Lions, Exact controllability, stabilization and perturbation for distributed systems, *SIAM Rev.*, 30 (1988), p. 1-68.

[L2] J.-L. Lions, *Contrôlabilité exacte*, Masson, Paris, 1988.

[L3] J.-L. Lions, Are there connections between turbulence and controllability?, 9th INRIA International Conference, Antibes, June 12-15, 1990.

[L4] J.-L. Lions, Remarks on turbulence theory, *Turbulence and Coherent Structures*, O. Métais and M. Lesieur eds., Kluwer Academic Publishers, 1991, p. 1-13.

[LL] L. Landau and E. Lifchitz, *Physique théorique*, Mécanique, Ed. Mir, Moscou, 1988.

[MWS] R.M. Murray, G. Walsh, and S.S. Sastry, Stabilization and tracking of nonholonomic control systems using time-varying state feedback, Nolcos 92, M. Fliess ed., Bordeaux, June 92, p. 182-187.

[NS] H. Nijmeier and A. J. Van Der Schaft, *Nonlinear dynamical control systems*, Springer-Verlag, New York, 1990.

[OS] R. Outbib and G. Sallet, Stabilizability of the angular velocity of a rigid body, *Systems and Control Letters*, 18 (1992) p. 93-98.

[P] J.-B. Pomet, Explicit design of time-varying stabilizing control laws for a class of controllable systems without drift, *Systems and Control Letters*, 18 (1992) p. 147-158.

[Ro1] L. Rosier, Homogeneous Lyapunov function for homogeneous vector field, *Systems and Control Letters*, 19 (1992) p. 467-473.

[Ro2] L. Rosier, Homogeneous stabilizable system which cannot be stabilizable by means of homogeneous feedback laws, Manuscript, December 1992.

[Ru1] D. L. Russell, Exact boundary value controllability theorems for wave and heat processes in star-complemented regions, in *Differential Games and Control Theory*, Roxin, Liu, and Sternberg eds., Marcel Dekker, New York, 1974, p. 291-319.

[Ru2] D. L. Russell, Controllability and stabilization theory for linear partial differential equations, Recent progress and open questions, *SIAM Rev.*, 20 (1978) p. 639-679.

[Sa] C. Samson, Velocity and torque feedback control of a nonholonomic art, Int. Workshop in Adaptative and Nonlinear Control: Issues in Robotics, Grenoble, 1990, p. 125- 151, *Lecture Notes in Control and Information Sciences*, Vol. 162, Springer-Verlag, Berlin - New York.

[SCW] R. Sepulchre, G. Campion, and V. Wertz, Some remarks about periodic feedback stabilization, Nolcos'92, M. Fliess ed., Bordeaux, June 92, p. 418-423.

[SJ] H. J. Sussmann and V. Jurdjevic, Controllability of nonlinear systems, *J. Diff. Equations*, 12 (1972) p. 95-116.

[SM] L. M. Silverman and H. E. Meadows, Controllability and observability in time variable linear systems, *SIAM J. Control Optim.*, 5 (1967) p. 64-73.

[So1] E. D. Sontag, Finite dimensional open-loop control generators for nonlinear systems, *Int. J. Control*, 47 (1988) p. 537-556.

[So2] E. D. Sontag, A "universal" construction of Artstein's theorem on nonlinear stabilization, *Systems and Control Letters*, 13 (1989) p. 117-123.

[So3] E. D. Sontag, *Mathematical Control Theory, Deterministic finite dimensional Systems*, Texts in Applied Mathematics 6, Springer-Verlag, New York - Berlin - Heidelberg - London - Paris - Tokyo - Hong Kong, 1990.

[So4] E. D. Sontag, Feedback stabilization of nonlinear systems, in *Robust Control of linear Systems and nonlinear Control*, M. A. Kaashoek, J. H. Van Schuppen and A. C. M. Ran eds., Birkhäuser, Cambridge, MA, 1990.

[So5] E. D. Sontag, Universal nonsingular controls, *Systems and Control Letters*, 19 (1992) p. 221-224.

[SS] E. D. Sontag and H. J. Sussmann, Remarks on continuous feedback, IEEE CDC, Albuquerque, 2 (1980) p. 916-921.

[Su1] H. J. Sussmann, Subanalytic sets and feedback control, *J. Diff. Equations*, 31 (1979) p. 31-52.

[Su2] H. J. Sussmann, Single-input observability of continuous-time systems, *Math. Systems Theory*, 12 (1979) p. 371-393.

[Su3] H. J. Sussmann, Lie brackets and local controllability: a sufficient condition for scalar input systems, *SIAM J. Control Optim.*, 21 (1983) p. 686-713.

[Su4] H. J. Sussmann, A general theorem on local controllability, *SIAM J. Control Optim.*, 25 (1987) p. 158-194.

[TP] A. Teel and L. Praly, Global stabilizability and observability implies semi-global stabilization by output feedbacks, Preprint, Fontainebleau, December 1992.

[Tr] A. I. Tret'Yak, On odd-order necessary conditions for optimality in a time-optimal control problem for systems linear in the control, *Math. U.S.S.R. Sbornik*, 70 (1991) p. 47- 63.

[Ts1] J. Tsinias, Sufficient Lyapunov-like conditions for stabilization, *Math. Control Signals Systems*, 2 (1989) p. 343-357.

[Ts2] J. Tsinias, Stabilization of nonlinear systems by using integrators, Nolcos'92, M. Fliess ed., Bordeaux, June 92, p. 645-648.

[Wa] S. H. Wang, Stabilization of decentralized control systems via time-varying controllers, *IEEE Trans. Aut. Control*, AC 27 (1982) p. 741-744.

[Wh] G. W. Whitehead, *Elements of homotopy theory*, Graduate text in Mathematics, 61, Springer-Verlag, New York Inc., 1978.

[Z] J. Zabczyk, Some comments on stabilizability, *Applied Mathematics and Optimization*, 19 (1989) p. 1-9.

Index

Subscripts refer to authors' names: [B] for Bellaïche, [C] for Coron, [G] for Gromov, [M] for Montgomery, and [S] for Sussmann.

Example: "Chow's theorem, 2_B, 3, 15, 87_G, 95, 113, 380_C", means that the entry appears on pages 2, 3 and 15 of Bellaïche's article; 87, 95 and 113 of Gromov's article; and on page 380 of Coron's article.

order of a
    differential operator (at a point), 44_B
    function (at a point), 35_B
    vector field (at a point), 44_B
overregularity, 258_G

Pansu
    convergence theorem, 94_G
    isoperimetric inequality, 106_G
    rigidity theorem, 112_G
parabolic structure, 282_G
partially horizontal submanifolds, 256_G
periodic feedback, 376_C
Pfaffian system, 234_G
pinching, 140_G
    problems, 142_G
polarization, 85_G
    see also distribution, subbundle
polarizations defined by 1-forms, 96_G
polynomial growth, 94_G
Pontrjagin
    maximum principle, 330_M, 346_S, 351
privileged coordinates, 35_B, 70
pseudo-norm, 43_B

quasi-conformal mappings, 192_G

radial gauge, 313_G
regular
    curve, 327_M
        see also normal path
    distribution, 332_M, 345_S
    extremal, 329_M
    geodesic, 327_M
    minimizer, 327_M
    point, 30_B, 51
    subbundle, 345_S
    see also equiregular polarization
return method, 12_B, 380_C
Rumin
    complex, 201_G
    operator, 202_G
Rumin-de Rham theorem, 202_G

second Lipschitz approx. theorem, 215_G
self-similar nilpotent groups, 135_G
self-similarities, 90_G

semicontinuity of the energy, 190_G
singular
    curve, 327_M
        see also abnormal path
    geodesic, 327_M
        see also abnormal extremal
    minimizer, 327_M
    point, 30_B
Sobolev inequalities, 170_G
stable homology groups, 373_C
strictly
    abnormal, 362_S
        extremal, 351_S
    singular minimizer, 329_M
sub-Riemannian
    distance, 7_B, 325_M
    manifold, 4_B, 345_S
    metric, 6_B, 7
        associated to a system of vector
            fields, 5_B
    structure, 325_M, 345_S
subbundle
    nonholonomic, 345_S
    see also polarization, distribution
sum of squares, 8_B, 152_G
Sussmann condition, 367_C
Sussmann-Stefan theorem, 13_B
systole, 154_G

tangency, 122_G
tangent
    cones, 93_G
    Lie algebra, 48_B
    space
        as a group, 73_B
        in the Gromov sense, 54_B, 71, 73
        of a sub-Riemannian manifold,
            49_B, 71, 73
taut map, 174_G
theorem
    Alexandroff, 265_G
    approximation, 144_G
    ball-box, 71_B, 98_G, 117, 128
    Chow, 2_B, 3, 15, 87_G, 95, 113, 380_C
    connectivity, 86_G
    disk extension, 218_G
    extension, 264_G

Progress in Mathematics

Edited by:

H. Bass
Columbia University
New York
10027
U.S.A.

J. Oesterlé
Dépt. de Mathématiques
Université de Paris VI
4, Place Jussieu
75230 Paris Cedex 05, France

A. Weinstein
Dept. of Mathematics
University of CaliforniaNY
Berkeley, CA 94720
U.S.A.

Progress in Mathematics is a series of books intended for professional mathematicians and scientists, encompassing all areas of pure mathematics. This distinguished series, which began in 1979, includes authored monographs, and edited collections of papers on important research developments as well as expositions of particular subject areas.

We encourage preparation of manuscripts in such form of TeX for delivery in camera-ready copy which leads to rapid publication, or in electronic form for interfacing with laser printers or typesetters.

Proposals should be sent directly to the editors or to: Birkhäuser Boston, 675 Massachusetts Avenue, Cambridge, MA 02139, U.S.A.

L. Kadison, University of Copenhagen, Denmark /
M.T. Kromann, University of Pennsylvania, USA

Projective Geometry and Modern Algebra

1996. 224 pages. Hardcover
ISBN 3-7643-3900-4

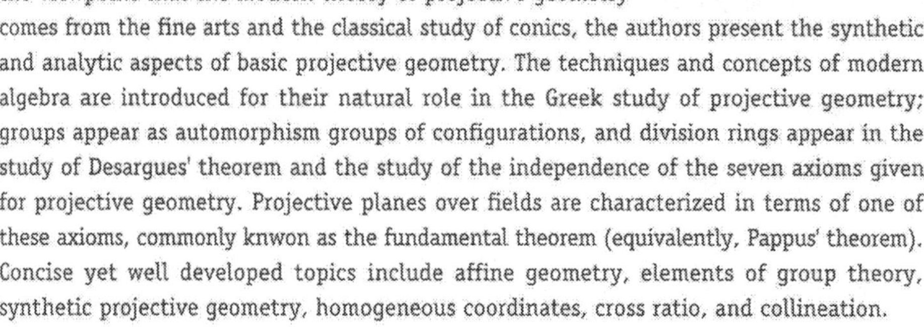

Starting with an engaging historical foreword, which develops
the viewpoint that the modern theory of projective geometry
comes from the fine arts and the classical study of conics, the authors present the synthetic
and analytic aspects of basic projective geometry. The techniques and concepts of modern
algebra are introduced for their natural role in the Greek study of projective geometry;
groups appear as automorphism groups of configurations, and division rings appear in the
study of Desargues' theorem and the study of the independence of the seven axioms given
for projective geometry. Projective planes over fields are characterized in terms of one of
these axioms, commonly knwon as the fundamental theorem (equivalently, Pappus' theorem).
Concise yet well developed topics include affine geometry, elements of group theory,
synthetic projective geometry, homogeneous coordinates, cross ratio, and collineation.

This text is ideally suited to an undergraduate or elementary graduate course intended as
an introduction to modern algebra in the framework of attractive and useful geometric
applications. It has five appendices conics, algebraic curves and Bezout's theorem, elliptic
geometry, ternary rings, and lattices of subspaces that provide the reader with topics for
independent study and the instructor with programs for guided projects. Ample exercises,
figures, solutions, and a detailed index round out this self-contained volume.

*Please order through your
bookseller or write to:*
Birkhäuser Verlag AG
P.O. Box 133
CH-4010 Basel / Switzerland
FAX: ++41 / 61 / 205 07 92
e-mail: farnik@birkhauser.ch
http://www.birkhauser.ch

*For orders originating in the USA
or Canada:*
Birkhäuser
333 Meadowlands Parkway
USA-Secaucus, NJ 07094-2491
FAX: ++1 / 800 / 777 4643
e-mail: orders@birkhauser.com

Birkhäuser 

Birkhäuser Verlag AG
Basel · Boston · Berlin

GPSR Compliance
The European Union's (EU) General Product Safety Regulation (GPSR) is a set
of rules that requires consumer products to be safe and our obligations to
ensure this.

If you have any concerns about our products, you can contact us on

ProductSafety@springernature.com

In case Publisher is established outside the EU, the EU authorized
representative is:

Springer Nature Customer Service Center GmbH
Europaplatz 3
69115 Heidelberg, Germany